This book is to be returned on or before
the last date stamped below.

AIR POLLUTION CONTROL IN TRANSPORT ENGINES

The Institution of Mechanical Engineers

AIR POLLUTION CONTROL IN TRANSPORT ENGINES

A Symposium arranged by the
Automobile Division and the Combustion Engines Group
of the Institution of Mechanical Engineers
9th–11th November 1971

1 BIRDCAGE WALK · WESTMINSTER · LONDON SW1H 9JJ

© The Institution of Mechanical Engineers 1972
ISBN 0 85298 032 9
Library of Congress Catalog Card Number 72-184586

CONTENTS

		PAGE
Introduction		vii
Paper C121	Meeting future emission standards with leaded fuels, by W. E. Adams, H. J. Gibson, B.S., M.S., D. A. Hirschler and J. S. Wintringham	1
Paper C122	Influence of water injection on nitric oxide formation in petrol engines, by K. Zeilinger, Dipl.Ing.	7
Paper C123	Vehicle particulate emissions, by K. Campbell and P. L. Dartnell, C.Eng., M.I.Mech.E.	14
Paper C124	Experiments in the control of diesel emissions, by P. M. Torpey, M. J. Whitehead and M. Wright, B.Sc.	21
Paper C125	Metabolism and dietetics of 'moteur diesel compensé B' and its result on air pollution, by M. G. Brille, A.M., S.I.A., and Y. E. Baguelin, S.I.A.	34
Paper C126	Controlling exhaust emissions from a diesel engine by LPG dual fuelling, by D. Lyon, B.Sc., Ph.D., A. H. Howland and W. L. Lom	42
Paper C127	First results obtained with the carbon monoxide monitoring network installed in Paris, by P. Chovin, Ph.D.	57
Paper C128	The contribution of additives to the elimination of air pollution, by J. W. Dable, C.Eng., M.I.Mech.E., and T. J. Sheahan	62
Paper C129	The influence of fuel composition on automotive exhaust emissions, by B. J. Kraus and G. P. Richard	77
Paper C130	Manifold reaction with exhaust recirculation for control of automotive emissions, by D. R. Fussell, G. P. Richard and D. Wade	89
Paper C131	Exhaust emission studies using a single-cylinder engine, by A. E. Dodd, B.Sc., C.Eng., M.I.Mech.E.	101
Paper C132	The influence of gasoline bulk fuel properties on exhaust emissions under idle and dynamic running conditions, by D. R. Blackmore, M.A., Ph.D.	110
Paper C133	The effect of reduction of gasoline lead content on road anti-knock performance, by J. L. Addicott, C.Eng., M.I.Mech.E., and D. Barker	120
Paper C134	Swirl—its measurement and effect on combustion in a diesel engine, by S. Ohigashi, Y. Hamamoto and S. Tanabe	129
Paper C135	Second generation carburettors for emission control: first stage, by G. L. Lawrence, C.Eng., F.I.Mech.E.	137
Paper C136	Stratification and air pollution, by J. E. Witzky	147
Paper C137	Diesel engine exhaust emissions and effect of additives, by M. S. El Nesr, B.Sc.(Eng.), M.Ph., S. Satcunanathan, B.Sc.(Eng.), Ph.D., C.Eng., M.I.Mech.E., and B. J. Zaczek, Ph.D., C.Eng., M.I.Mech.E.	156

CONTENTS

PAGE

Paper C138   Spatial and temporal history of nitrogen oxides in the spark-ignition combustion chamber, by L. Muzio, Ph.D., H. Newhall, Ph.D., and E. Starkman, B.S., M.S., C.Eng., F.I.Mech.E. 163

Paper C139   Accomplishments of the I.I.E.C. programme for control of automotive emissions, by W. J. Koehl, Ph.D., D. P. Osterhout, B.E., M.S., and S. E. Voltz, B.A., M.A., Ph.D. . . . 172

Paper C140   The mechanisms of soot release from combustion of hydrocarbon fuels with particular reference to the diesel engine, by D. Broome, M.A., C.Eng., M.I.Mech.E., and I. M. Khan, B.Eng. . . 185

Paper C141   Recent automotive air pollution control legislation in the U.S. and a new approach to achieve control: alternative engine systems, by J. J. Brogan, B.M.E., B.M.B.A. 198

Paper C142   Prediction of soot and nitric oxide concentrations in diesel engine exhaust, by I. M. Khan, B.Eng., G. Greeves, B.Sc.(Eng.), Ph.D. (*Graduate*), and D. M. Probert, B.Tech. 205

Paper C143   The sampling and measurement of exhaust emissions from motor vehicles, by J. P. Soltau, C.Eng., F.I.Mech.E., and R. J. Larbey   218

Paper C144   Catalytic reduction of atmospheric pollution from the exhaust of petrol engines, by C. D. Haynes, B.Sc., C.Eng., M.I.Mech.E., and J. H. Weaving, Ph.D., C.Eng., F.I.Mech.E. . . . 232

Paper C145   Petrol injection control for low exhaust emissions, by J. P. Soltau, B.Eng., C.Eng., F.I.Mech.E., and K. B. Senior, B.Sc. . . 241

Paper C146   Exhaust emission control: further experience in the application of exhaust thermal reactors to small engines, by J. W. Wisdom . 253

Paper C147   The design, development and application of exhaust emissions control devices, by A. S. de Forge-Dedman, C.Eng., M.I.Mech.E., and J. A. Howard, C.Eng., M.I.Mech.E. . . . 264

Paper C148   The influence of valve overlap on oxides of nitrogen exhaust emissions, by C. Henault 273

Paper C149   Kinetics of nitric oxide formation in internal-combustion engines, by W. E. Bernhardt, Dr.Ing. 279

Paper C150   Evaporative losses from automobiles: fuel and fuel-system influences, by R. W. Hurn, M.S., and B. H. Eccleston . . . 286

Paper C151   Factors affecting emissions of smoke and gaseous pollutants from direct injection diesel engines, by I. M. Khan, B.Eng., and C. H. T. Wang, B.Sc.(Eng.), C.Eng., M.I.Mech.E. . . . 293

Paper C152   Exhaust emission control system for the rotary engine, by T. Muroki   304

Paper C153   Automotive emission research—A review of the C.R.C. programme, by C. E. Moser, Ph.D. 311

Discussion 319

Communications 346

Authors' replies 352

List of delegates 366

Index to authors and participants 370

Subject index 371

Air Pollution Control in Transport Engines

A SYMPOSIUM was held at the Solihull Civic Hall and the Group Research Centre, Joseph Lucas Limited, Shirley, Solihull, from the 9th–11th November 1971. The symposium was sponsored by the Automobile Division and the Combustion Engines Group of the Institution of Mechanical Engineers and 357 delegates registered to attend.

OPENING OF SYMPOSIUM

The symposium was formally opened on the morning of Tuesday, 9th November, by G. A. Hunt, C.B.E., C.Eng., Managing Director of Chrysler United Kingdom Limited. This was followed by an introductory address by F. J. Lawther, M.B., D.Sc., F.R.C.P., Director of the Medical Research Council Air Pollution Unit.

PRESENTATION OF PAPERS

The papers were divided into eight sessions for presentation and discussion.

On Tuesday, 9th November:

Session 1: Environment, legislation and test methods.
Chairman: J. H. Weaving, Ph.D., C.Eng., F.I.Mech.E.
Papers C143, C141 and C127.

Session 2: Fuels and carburation.
Chairman: J. H. Boddy, B.Sc.(Eng.), C.Eng., M.I.Mech.E.
Papers C129, C132, C128, C133, C135 and C145.

Session 3: Emission formation.
Chairman: N. M. Vulliamy, M.A., C.Eng., M.I.Mech.E.
Papers C151, C142, C140 and C137.

On Wednesday, 10th November:

Session 4: Emission research.
Chairman: J. P. Soltau, B.Eng., C.Eng., F.I.Mech.E.
Papers C153 and C134.

Session 5: Emissions—Oxides of nitrogen.
Chairman: F. J. Wallace, D.Sc., Ph.D., C.Eng., F.I.Mech.E.
Papers C149, C138, C148 and C122.

Session 6: Exhaust treatment.
Chairman: A. Fogg, C.B.E., Ph.D., D.Sc., C.Eng., F.I.Mech.E.
Papers C130, C146, C144, C121, C147 and C139.

Session 7: Emission control.

Chairman: B. W. Millington, B.Sc.(Eng.), C.Eng., F.I.Mech.E.

Papers C125, C126 and C124

On Thursday, 11th November:

Session 8: Engines, particulates and evaporative losses.

Chairman: D. Downs, B.Sc.(Eng.), C.Eng., F.I.Mech.E.

Papers C152, C131, C136, C150 and C123

A summing-up by J. H. Weaving followed and the symposium was brought to a close.

PLANNING PANEL

The members of the Planning Panel responsible for organizing the symposium were: J. H. Weaving, Ph.D., C.Eng., F.I.Mech.E., T. R. Atkinson, J. H. Boddy, B.Sc.(Eng.), C.Eng., M.I.Mech.E., R. A. C. Fosberry, M.A., C.Eng., M.I.Mech.E., B. W. Millington, B.Sc.(Eng.), C.Eng., F.I.Mech.E., J. P. Soltau, B.Eng., C.Eng., F.I.Mech.E., M. Vulliamy, M.A., C.Eng., M.I.Mech.E., and F. J. Wallace, D.Sc., Ph.D., C.Eng., F.I.Mech.E.

C121/71 MEETING FUTURE EMISSION STANDARDS WITH LEADED FUELS

W. E. ADAMS* H. J. GIBSON* D. A. HIRSCHLER* J. S. WINTRINGHAM*

One route to meeting 1975 and 1976 exhaust emission standards involves catalysts and exhaust gas recycle. No catalyst has yet been found that is adequately active and durable, even with lead-free gasoline. Unleaded fuels have many disadvantages: their use causes abnormal octane requirement increases due to deposits, problems of exhaust valve seat and guide wear, depletion of natural resources, and the strong possibility of increasing atmospheric pollution. It is preferable to have an ultra-lean high-compression engine with exhaust gas cycle and a particle trap. This approach gives good driveability, normal performance, minimum fuel economy penalties, and the lowest cost penalty. Many of our principles are applicable to engines of all sizes, using all systems of exhaust control.

INTRODUCTION

THE EMISSION of unburnt hydrocarbons (HC), carbon monoxide (CO), and oxides of nitrogen (NO_x) in the exhaust gases from Otto-cycle gasoline engines has become the most important factor in determining whether or not this type of power plant can survive. In the U.S.A. a segment of the automobile industry has indicated that catalytic treatment of exhaust gases along with exhaust gas recycle will be necessary to meet 1975 and 1976 Federal standards for exhaust emissions. The industry has asked that lead anti-knock compounds be removed from motor gasolines, because lead fouls many catalysts and may foul recycle systems. However, in spite of extensive efforts no catalyst has yet been found that has adequate activity and durability, even with unleaded gasolines.

There are many disadvantages in using unleaded gasolines (1)†. These include:

(1) The octane number requirement increase due to combustion chamber deposits is considerably larger.

(2) A problem of exhaust valve seat and guide wear exists. This has been observed both in privately owned vehicles and in test cars.

(3) If the octane number level of the fuel is allowed to drop as far as the lead removal permits, low-compression engines must be provided. These engines use more fuel, depleting our natural resources. Furthermore, since the mass of exhaust gas is increased, there may be no reduction in the mass emissions—there may even be an increase.

(4) If the octane level of the fuel is retained, more extensive and much more expensive refinery processing is required. This not only requires more crude oil, but results in a more highly aromatic gasoline. The exhaust emission of phenols and polynuclear aromatics is directly proportional to fuel aromaticity. In addition, when used in cars without exhaust treatment devices, such gasolines exhaust more aromatic hydrocarbons which contribute to increased photochemical reactivity in the atmosphere.

THE ETHYL APPROACH

The Ethyl approach retains the advantages of high-compression engines and leaded fuels. High-compression engines would never have been developed if they had not provided higher efficiency and more power per cubic inch of displacement. Even more important is the fact that high-compression engines today use less air and fuel to do any job. Since exhaust emissions are now measured on a mass basis, the smaller mass of emissions with a higher compression ratio becomes a very important consideration. Furthermore, the higher compression ratio can permit operation at leaner air/fuel ratios and allow the use of higher rates of exhaust gas recycle. Both these advantages of high compression are used in our lean-reactor car programme.

Until a few years ago the automotive engineer's prime aim was to produce the maximum possible horsepower. Along with larger engines, carburettor air passages and inlet manifold branches were enlarged to permit high volumetric efficiencies. This trend often resulted in poor atomization of the fuel, poor mixing of fuel with air, and quite uneven distribution of fuel in a multi-cylinder engine. Operation of such engines at mixture ratios much

---
The MS. of this paper was received at the Institution on 10th May 1971 and accepted for publication on 21st June 1971. 23
* *Ethyl Corporation Research Laboratories, Detroit, Michigan, U.S.A.*
† *References are given in Appendix 121.1.*

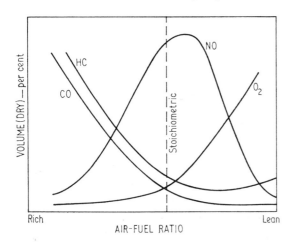

Fig. 121.1. Effect of air/fuel ratio on exhaust emissions

Fig. 121.1 shows the well-known relationship between several exhaust gas constituents and the air/fuel ratio in a single-cylinder engine with good air–fuel mixture preparation before induction. In the multi-cylinder engine with quite different air/fuel ratios in the various cylinders, the exhaust gas can contain large concentrations of CO, HC, and O_2.

Ethyl began work on improved air–fuel mixture preparation and distribution early in the 1960s. Some background work had already been done in our research laboratories in the late 1930s. We also derived a simple way to determine how well air and fuel are mixed, distributed, and burned in a multi-cylinder engine by analysing the exhaust gas from the whole engine for CO_2, CO, and O_2. This work has been reported (2)–(4).

Referring again to Fig. 121.1, it is clear that nitrogen oxides can be minimized by running quite rich or quite lean. Running at rich air/fuel ratios, however, produces substantial quantities of HC and CO. Disposing of these in the exhaust system involves adding air for combustion, with a resulting substantial release of energy. If the disposal is performed in a thermal manifold reactor, high leaner than stoichiometric was impossible, because the cylinder receiving the least fuel misfired. Before this point was reached, engine operation was often ragged, with surging of power.

Fig. 121.2. Experimental 3-venturi carburettor

temperatures and expensive materials are involved. If the disposal is attempted in a catalytic muffler, high temperatures may ruin the catalyst. Additionally, attrition of the catalyst will increase the emission of particulates. Rich air/fuel ratios result in a substantial reduction in fuel economy.

A wiser approach would be to devise a fuel metering system that permits very lean operation. HC, CO, and NO_x are all minimized while fuel economy may be improved. Even with exhaust recycle, fuel economy will be impaired much less than is the case with rich operation and recycle.

This approach was taken by developing a three-venturi carburettor consisting of one small primary venturi and two larger secondary venturis (Fig. 121.2). The secondary venturis are of variable cross-section, provided by controlled pistons. This arrangement provides high-velocity air flow in all three barrels which, coupled with careful attention to the design of the fuel jet discharge geometry, leads to very good fuel atomization and mixing with air. The primary section is extremely simple, with no separate idle system. During idle, the air velocity through the small venturi, coupled with a slightly closed choke, provides adequate fuel flow from the main jet. Multi-cylinder engines equipped with this carburettor system can be run much leaner than has previously been practical and still have good driveability. They can, in fact, run leaner than is necessary to minimize HC emissions and maximize fuel economy. Thus, the 'lean barrier' no longer exists. Even if the rich-mixture approach or the catalytic reactor approach is taken, such improved carburation is highly desirable. Running rich with poor fuel distribution results in excess carbon monoxide (CO) and unburnt hydrocarbons (HC) in some cylinders and excess oxides of nitrogen (NO_x) in others.

Normally, our carburettor is adjusted to run slightly leaner than stoichiometric at idle. Cruise air/fuel ratios are such that 2–3 per cent oxygen is present in the exhaust gases. At full throttle, mixture ratios are slightly richer than stoichiometric. In general, passenger car engines made in the U.S.A. have fairly large displacements. Such engines are not heavily loaded in the current 7-mode, 7-cycle Federal test procedure or in the longer non-repetitive cycle that will be used for 1972-model and later cars. Under nearly all test conditions for emissions, therefore, there is an excess of oxygen in the exhaust gases. If the gases can be kept hot and given time for oxidation of the small amounts of HC and CO present, combustible emissions can be reduced to very low levels.

Reaction in the exhaust system is improved by conserving heat in the exhaust ports by lining them with stainless steel tubes insulated from contact with the cooled walls of the exhaust ports. Further reaction is promoted by substituting larger than normal exhaust manifolds of thin stainless steel (for quick warm-up) for the conventional cast iron manifolds. These manifolds or reactor chambers, shown in Fig. 121.3, are insulated to further conserve heat. Under the relatively moderate temperature

Fig. 121.3. Exhaust port liner

conditions that prevail with the lean engine, 310 stainless steel has proved to be satisfactory.

The very stringent standards for NO_x require recycle of the exhaust gas. Gases are tapped from the exhaust pipe and passed through a heat exchanger (surrounded by circulating engine coolant). The recycle gas effectiveness is greater if the gases are cooled, but the gases must be kept above the dew point to avoid problems due to water. When the engine reaches operating temperature, the gases leave the heat exchanger in the temperature range from 80°C (176°F) to 150°C (302°F). The recycle gas is discharged into the intake system under the carburettor from an annulus that has several orifices to provide good mixing. Flow is controlled by carburettor venturi vacuum, with no flow at idle or full throttle and tailored quantities at intermediate throttle positions. Deposit problems in the system are nearly eliminated by a combination of the use of the heat exchanger, large-diameter tubing, and coating the critical parts with Teflon. Table 121.1 shows the emissions

Table 121.1. Effect of exhaust recycle on emissions

Car: 1969 Pontiac 400-CID V-8; automatic transmission.

Modifications: 3-venturi carburettor; exhaust port liners; exhaust manifold reactors; modified ignition system.

Test conditions: Federal 7-mode, 7-cycle test procedure; mass emissions calculated.

Exhaust emissions	Emissions, g/mile	
	Without recycle	With recycle*
Hydrocarbons	0·53	0·36
Carbon monoxide	8·35	9·13
Nitrogen oxides	3·38	0·87

* Maximum of 30 per cent.

of a lean-reactor car without and with exhaust gas recycle. As the amount of exhaust gas recycle increases, the fuel economy suffers. The relationship of fuel economy to NO_x control with recycle is shown in Fig. 121.4 for three different cars weighing about 4500 lb.

Gasoline motor vehicles were responsible for only 1·8 per cent of all particulate emissions in the U.S.A. in 1968 (*5*). Ethyl has studied particulate emissions from cars since the mid-1950s (*6*). Recent work has shown that lead compounds comprise one-third of the mass of particulate small enough to be airborne (*7*). Particulate emissions from cars run on unleaded fuel are about 60 per cent of those from cars run on leaded fuel, as shown in Table 121.2. Although present levels of lead particulates in the air from vehicle exhaust gas do not pose a threat to health, it may be desirable to have a trapping device in the exhaust system to minimize particulate emissions.

Lead particulates can be agglomerated by cooling the exhaust gases. This is done in our cars by adding fins to as much of the exhaust pipe as is possible in the space available. The gases are cooled to 345°C (650°F) or lower

Table 121.2. Emission of suspended particulates from vehicles

Vehicles: Variety of automobiles, operated in owner-type service throughout their life on commercial leaded or unleaded gasoline; 30 000–100 000 miles of operation.

Fuels: Unleaded commercial premium or leaded commercial gasolines.

Procedure: Total exhaust gas collected in 2500-ft³ black polyethylene bag. Suspended particulate measured by weighing filter sample drawn from bag.

1971 Federal procedure	No. of tests		Particulates, g/mile		Ratio unleaded to leaded
	Leaded	Unleaded	Leaded	Unleaded	
4 cold cycles	16	5	0·512	0·316	0·62
4 hot cycles	17	5	0·240	0·134	0·56
Weighted*	16	5	0·339	0·197	0·58

* Weighted 35 per cent cold and 65 per cent hot.

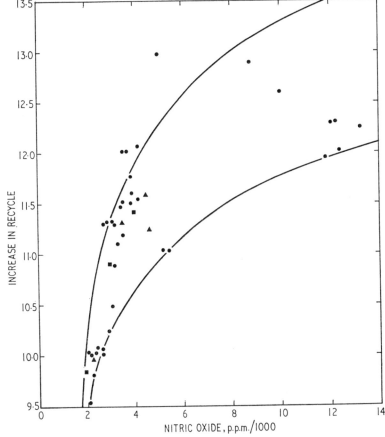

- 383 CID, Ethyl 3-V carburettor, recycle, and modified distributor.
- 400 CID, Ethyl 3-V carburettor, recycle, exhaust reactors, and modified distributor.
- 360 CID, Ethyl 3-V carburettor, recycle, Chrysler exhaust reactors, and modified distributor.

Fig. 121.4. Effect of nitric oxide reduction on fuel economy—nine California cycles (cold start)

Table 121.3. Melting points of lead compounds

Compound	Melting point	
	°C	°F
$PbBr_2$	373	703
$PbClBr$	~400	~750
$PbCl_2$	501	934
PnO	888	1630
$PaSo_4$	1170*	2138

* Started to decompose at 1000°C.

Table 121.4. Effect of test procedure on exhaust emissions

Type of car	Exhaust emissions, g/mile		$\frac{1972}{1971}$
	1971 7-mode, 7-cycle	1972 DDS and CVS	
1970 Pontiac 400-CID V-8			
Hydrocarbons . .	0.25	0.75	3.0
Carbon monoxide . .	5.70	7.15	1.25
Nitrogen oxides . .	0.92	1.5	1.6
1971 Plymouth 360-CID V-8			
Hydrocarbons . .	0.28	0.8–1.0	2.9–3.6
Carbon monoxide . .	7.05	7.0–10.0	1.0–1.4
Nitrogen oxides . .	1.12	1.3–2.0	1.2–1.8

(see Table 121.3 for melting temperatures of some lead compounds), and are then fed into a trap which contains mesh to further promote agglomeration. The gases then pass through a cyclone or an anchored vortex device to precipitate the particulates (Fig. 121.5).

Early in 1970 an experimental car with these modifications (except the particle trap) was very close to meeting the 1975 emission standards as they were then proposed (7-mode, 7-cycle procedure and grammes per mile limits of 0.5 HC, 11.0 CO, and 0.9 NO_2) (8). However, emission test procedures were changed late in 1970 for 1972 and later model cars, and 1975 standards were changed by the Clean Air Act, both becoming more difficult. Table 121.4 shows the emissions from two Ethyl lean-reactor cars as measured by the 1971 7-mode, 7-cycle procedure and by the 1972 procedure, which involves a new dynamometer driving schedule (DDS), and a constant volume sampler (CVS) system. The final column in Table 121.4 shows that the 1972 procedure, compared with the 1971, triples the measured HC and substantially increases CO and NO_2. The new emission standards for 1975 were set at 0.46 g HC/mile, 4.7 g CO/mile, and 3.0 g NO_2/mile. The 1976 standards for HC and CO are the same, but the proposed 1976 standard for NO_2 is 0.4 g/mile.

Table 121.5 shows the exhaust emission levels attained in March 1971 in a Pontiac 400-CID V-8 car. About one-half of the combustible emissions (HC and CO) occur during the first 2 min of the 22-min 50-s test. This is due to engine choking and to the fact that the engine and the emission control systems are cold. Tests were run with LPG as fuel for the first 2 min and no choking. During the first 2 min there was a 60 per cent reduction in HC and an 80 per cent reduction in CO.

In work aimed at reducing exhaust emissions it is important to know how a reduction in total hydrocarbons is reflected in reductions in the more reactive species of hydrocarbon. Reactive hydrocarbons are those which, in the presence of NO_2, ozone, and sunlight, react to form compounds, such as peroxybenzyl nitrate, which are highly irritating to eyes, nose, and throat and which damage vegetation and rubber. Gas chromatograph studies were made of the exhaust from a production car and from our lean-reactor car of the same make and model. Table 121.6 shows the chromatograph data. It can be seen that the weight fraction of the emissions that are non-reactive from the lean-reactor car is more than double that

Fig. 121.5. Experimental dual-vortex particulate trap. 65 per cent reduction in lead

Table 121.5. Emissions performance of Ethyl lean-reactor cars

Exhaust emissions	Emissions, g/mile*			
	Before emission control	Ethyl lean-reactor cars	1975 Federal standards	1976 Federal standards
Unburnt hydrocarbons	16·8	0·6–1·0	0·46	0·46
Percentage reduction	—	94–96·5	97+	97+
Carbon monoxide	125	6·0–10·0	4·7	4·7
Percentage reduction	—	92–95+	97+	97+
Nitrogen oxides	8·6	1·0–1·8	3·0	†
Percentage reduction	—	79–88	65	?

* Measured by 1972 test procedure.
† The Clean Air Act of 1970 requires that 1976 emissions of oxides of nitrogen be 90 per cent less than in 1971-model cars. The latter level is being established.

Table 121.6. Gas chromatograph data

Hydrocarbon reactivity	Umodified car		Lean-reactor car	
	Mole %	Wt %	Mole %	Wt %
Non-reactive				
Methane	16·46	4·48	25·49	9·56
Ethane	1·62	0·83	2·38	1·67
Propane	0·09	0·07	0·18	0·19
Benzene	2·55	3·38	2·92	5·34
Acetylene	10·51	4·65	12·99	14·19
Total	31·23	13·41	43·96	30·95
Quite reactive				
Ethylene	14·37	6·84	21·70	14·23
Propylene	7·42	5·30	8·14	8·01
1-Butene-1,3-butadiene	4·46	4·25	4·40	5·77
2-Methyl-2-butene	0·84	1·00	0·38	0·62
Toluene	8·52	13·32	4·62	9·96
Xylenes	2·67	4·24	1·47	3·23
1,3,5-Trimethylbenzene	0·27	0·55	0·11	0·31
1-Methyl-3-ethylbenzene	0·70	1·43	0·32	0·90
1-Methyl-4-ethylbenzene				
Total	39·25	36·93	41·14	43·03
Intermediate	29·52	49·66	14·90	26·02
Average carbon no.	4·13		3·00	
Hydrocarbons, g/mile	3·46		1·03	
Indices of reactivity				
Maga	0·62		0·53	
General Motors	2·87		2·43	
U.S. Bureau of Mines	1·79		1·06	

for the production car. While there is also a 20 per cent increase in the fraction which is quite reactive, the mass of reactive material is only about one-third that of the unmodified car. Three common indices of exhaust gas reactivity (Maga, General Motors, and U.S. Bureau of Mines) were all reduced in the lean-reactor car.

CONCLUSIONS

The lean-reactor car performs and drives as well as a standard production car but suffers a loss in fuel economy of about 10 per cent in normal owner driving. There is no reason to believe adequate durability cannot be built into the system.

Compared to non-controlled vehicles, the lean-reactor car has eliminated about 95 per cent of the unburnt hydrocarbons, about 90 per cent of the carbon monoxide, and more than 80 per cent of the oxides of nitrogen from exhaust gas. Efforts are continuing to further reduce exhaust emissions. The principles used in Ethyl's lean-reactor car (i.e. good mixture preparation and distribution, conservation of exhaust heat, and careful design of exhaust recirculating systems) are applicable to emission control systems for all engines.

ACKNOWLEDGEMENT

The authors acknowledge the major contributions of Mr Frederick J. Marsee, Program Manager, Ethyl Corporation Research Laboratories, to the development of experimental lean-reactor cars and the work reported in this paper.

APPENDIX 121.1

REFERENCES

(1) FELT, A. E. and KERLEY, R. V. 'Engines and effects of lead-free gasoline', S.A.E. Preprint No. 710367, 1970 (October).
(2) BARTHOLOMEW, E. 'Potentialities of emission reduction by design of induction systems', S.A.E. Preprint No. 660109, 1966 (January).
(3) ELTINGE, L. 'Fuel–air ratio and distribution from exhaust gas composition', S.A.E. Preprint No. 680114, 1968 (January).
(4) ELTINGE, L., MARSEE, F. J. and WARREN, A. J. 'Potentialities of further emissions reduction by engine modifications', S.A.E. Preprint No. 680123, 1968 (January).
(5) *Nationwide inventory of air pollutant emissions—1968*, NAPCA No. AP-73, 1970 (U.S. Dept of Health, Education, and Welfare).
(6) HIRSCHLER, D. A., GILBERT, L. F., LAMB, F. W. and NIEBYLSKI, L. M. 'Particulate lead compounds in automobile exhaust gas', *Ind. Engng Chem.* 1957 **49**, 1131.
(7) TERHAAR, G. 'Composition, size, and control of automobile exhaust particulates', Paper presented at a meeting of the Combustion Institute, Ann Arbor, Michigan, 1971 (30th March).
(8) HIRSCHLER, D. A. and MARSEE, F. J. 'Meeting future automobile emission standards', N.P.R.A. Paper No. AM-70-5, 1970 (April).

C122/71 INFLUENCE OF WATER INJECTION ON NITRIC OXIDE FORMATION IN PETROL ENGINES

K. ZEILINGER*

Reduction of nitric oxides is one of the most difficult tasks in solving emission problems. It is necessary to reduce NO emissions without increasing CO, HC, and b.s.f.c. As water-injection was considered to be a possible solution, an attempt was made to reduce NO emissions and to investigate NO formation more precisely. In these experiments the water was mixed together with an emulsifier and the fuel and was injected with the fuel by means of individual injection nozzles. The results given in the first part of the paper show the influence of different quantities of water on NO emissions at various speeds, loads, and equivalence ratios. The second part of the paper describes the results of analyses of exhaust gas samples taken from one cycle of one cylinder.

INTRODUCTION

THE ADDITION OF WATER to fuel, and the consequent presence of water during combustion, have been investigated to determine the effects of water-injection (a) as a means of reducing the amount of nitric oxide emitted and (b) as a means of investigating more precisely how nitric oxide forms, as this is greatly complicated by the mutual interference of different parameters.

This paper presents the results obtained from measurements on a modern petrol engine with mechanical fuel injection and from calculations of temperature curves, rate of heat release, and the amount of heat released. Water was mixed with the fuel and together they were injected through the individual nozzles into the manifold just ahead of each intake valve.

The results of (a) above were obtained during preliminary tests to part (b). Parts of the experimental work were carried out by Mader (1)† and Fink (2).

SET-UP AND TEST PROCEDURES

Test engine

The test engine was a modern, water-cooled, six-cylinder, four-stroke-cycle petrol engine with a mechanical manifold fuel injection system. The engine, which had the following data, is not standard:

Volume:	$V_H = 141$ in^3 (2306 cm^3)
Bore:	$d = 2.87$ in (72.8 mm)
Stroke:	$h = 3.23$ in (82 mm)
Compression ratio:	$\epsilon = 8.7$

The MS. of this paper was received at the Institution on 18th May 1971 and accepted for publication on 10th June 1971. 33
* Institut für Verbrennungskraftmaschinen und Kraftfahrzeuge der Technischen Universität München, Munich, Germany.
† References are given in Appendix 122.1.

Water injection

In other investigations (3) water was injected separately through an additional nozzle, but in the experiments described in this paper 0.3 vol% emulsifier (Necanil C of BASF) was added to the fuel, which was injected together with the latter through the six individual nozzles of the injection system in front of the inlet valve of each cylinder. In this way an optimal equal distribution not only of the fuel but also of the water could be secured. After having been mixed the emulsion stays constant for approximately 30 min before separation starts again. In order to avoid separation, the emulsion was continually stirred by an agitator.

Exhaust gas analysis

As preliminary tests have shown, it is absolutely useless to measure continuously the exhaust emissions of the whole engine and to correlate the pressure diagrams taken of one cycle of one cylinder only with these exhaust gas data. The variations in the course of combustion from cycle to cycle and from cylinder to cylinder are far too numerous to associate the integral data of the exhaust gas with an arbitrarily selected cycle. The same also holds for the case where the mean values of several cycles are used.

For this reason the pressure diagram of only one cycle has been taken, and an exhaust gas sample of that cycle was selected and analysed. Extraction and analysis of the exhaust gas samples were carried out as detailed below (see Fig. 122.1).

The sample container is a glass cylinder (1) with a movable piston (2). Before extraction (the piston is in the upper final position at that time) the whole space up to the solenoid valve (3) is evacuated by the suction pipe (4).

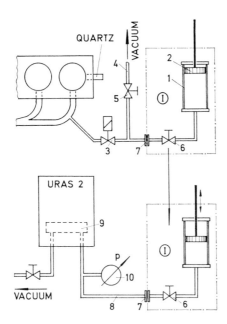

Fig. 122.1. Extraction and analysis of the exhaust gas samples

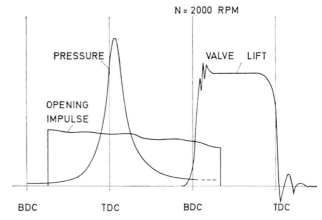

Fig. 122.2. Pressure, opening impulse, and valve lift

The valve (5) is then closed. Extraction time is determined by the duration of the opening impulse of the solenoid valve. During this time the sample container is connected to the exhaust manifold of the first cylinder. Therefore, exhaust gas is sucked in by the previously produced vacuum. The exhaust gas sample is enclosed in the sample container by closing valve (6). Afterwards the whole of unit I is removed from the engine by loosening the coupling (7).

Unit I is then coupled by the same coupling (7) to an infrared absorption spectrometer (URAS 2) as shown in the lower part of Fig. 122.1. All pipes (8) to the left of the valve (6), including the cuvette (9), are then evacuated. After reopening valve (6) the exhaust gas to be analysed flows into the cuvette (9). The pressure during measurement, controlled by a manometer (10), must be equal to the calibration pressure. It is adjusted by moving piston (2). For one analysis a gas volume of about 12·3 in³ (200 cm³) is sufficient. A similar system has already been described in reference (*4*).

Errors in measuring are possible by the fact that the NO-URAS shows sensitivity also for water vapour. This error, however, is the same with all measurements as the tests always take place within the region of water saturation. The sample container furnishes proof since its glass wall is always covered with condensed water—even during investigations with pure fuel. Since only comparative measurements were to be made, elimination of this error was not attempted.

Synchronization

An important requirement for obtaining precise results is the exact synchronization of the pressure diagram and opening impulse of the solenoid valve within one cycle.

The oscillograph is triggered by an inductive pick-up fixed at the crankshaft. The cylinder pressures measured by means of a Kistler quartz pick-up were not recorded continuously. An electronic triggering unit ensures that, after pushing a button, only one pressure diagram of a single cycle is recorded. Simultaneously an opening impulse is sent to the solenoid valve. The valve opens shortly after b.d.c. and is kept open during the exhaust stroke of this cycle. By the vacuum in the sample container the exhaust gas is taken from exactly that cycle, whose pressure diagram was previously stored by the oscillograph. Fig. 122.2 shows an oscillogram with pressure, opening impulse, and valve lift versus crank angle.

Computer program

For the calculations a program developed by Dr K. Lange, described in detail in reference (*5*), could be used. Modifications of this program were introduced where the special data of the test engine and the special test conditions made them necessary.

Temperatures, rates of heat release, and heat releases were calculated in steps of 4° of crank angle, starting from basic input data like those of the engine and fuel and from test data (pressure diagram, b.m.e.p., n, b.s.f.c., etc.). This program has been laid out for tests with pure fuel only. However, the following evaluation of the influence of the added water proves that the latter may be neglected for these calculations.

The maximum water addition is 30 per cent by volume referred to the volume of the fuel. At an assumed fuel volume of some 25 mm³ (i.e. 19 mg) used in the test described here, not more than 5·5 mg of water are present during combustion. The following amounts of heat are necessary for this quantity of water to heat it from ambient temperature to boiling temperature, to evaporate it, and finally to overheat it up to the peak temperature of some 1800°C:

20–100°C:	0·44 cal
evaporation:	2·9 cal
overheating to 1800°C:	5·0 cal

These are altogether some 8 cal. Considering, however, the heat released by the 19 mg fuel (that is, 19 mg × 11·3 cal/mg = 215 cal), not more than 4 per cent of the delivered energy is consumed by the water. Considering, further, that the water quantity is not more than 2 per cent of the total charge, it becomes obvious that such caloric properties as the isentropic exponent k and the gas constant R are only little affected. The error resulting from neglecting the influence of water in the program is definitely much smaller than the errors resulting from inaccuracies in analysing exhaust gas, measuring pressure, and converting the data from the analog to the digital form.

REDUCTION OF NITRIC OXIDE BY WATER INJECTION

Test results

Apart from supplementary devices such as catalytic reactors which influence exhaust gas emission in the exhaust system, only a few possibilities exist which reduce NO emissions from a given engine without too complicated design modifications. These are mainly variation of fuel/air ratio, spark timing, exhaust gas recirculation, and water injection. The experiments reported here were made with the system of water addition described above. As only integral data have been available, measurements of the whole exhaust gas, and not only of the exhaust gas from one cylinder, were made. This method is not only faster but for this case also more reasonable.

Fig. 122.3 shows b.m.e.p., b.s.f.c., and the exhaust gas constituents CO, HC, and NO plotted against the volumetric percentage of the added water referred to the fuel volume. In order to eliminate other influence factors, the equivalence ratio $\Phi = 1$ [Φ = (fuel/air ratio of the mixture)/(stoichiometric fuel/air ratio)] and spark timing, 21° b.t.d.c., are kept constant. Parameters are engine speed, n, and throttle position, α.

For a better comparison the scales of the ordinates are not absolute but are percentage values. The values for 0 per cent water (i.e. pure fuel) are always equated to 100 per cent. The absolute values for 0 per cent water are given in Table 122.1 for all points.

It is evident that b.m.e.p. first increases with increasing water addition, reaches maximum at 10–20 per cent water, and decreases then almost symmetrically. At about 28 per cent water the same b.m.e.p.s are available as at 0 per cent water. Brake specific fuel consumption is always inversely proportional to b.m.e.p., since the fuel rate is kept constant. While CO values do not change measurably, HC values are similar to the b.s.f.c. curves, having a minimum at about 20 per cent water and increasing with higher water percentages. Curve 4 at 20° throttle opening and 3000 rev/min is an exception. At higher loads HC emissions increase (curve 4) in contrast to the other curves for lower loads. A more thorough investigation of this observation could not be carried out since the injection pump at 40 per cent water addition delivers only 60 per cent of the normal fuel rate, so that measurements at higher loads could not be made.

As Fig. 122.3 shows, water injection particularly influences NO emissions. Rather independent of throttle opening and engine speed, NO values decrease to 50 per cent at 30 per cent water addition—at 40 per cent water addition they even decrease to 30 per cent of their original values. Without loss in b.m.e.p. and fuel economy, nitric oxides can be reduced to a level of about 55 per cent by adding 28 per cent water. In this case, there is a decrease in HC by about 10–20 per cent at most operating levels. At higher loads, however, they may increase by about the same amount. An influence of the engine speed could not be observed in these investigations.

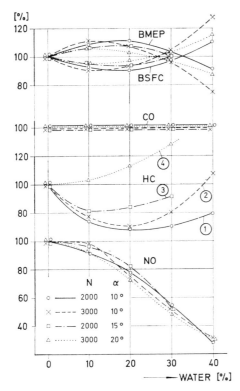

Fig. 122.3. B.m.e.p., b.s.f.c., CO, HC, and NO as functions of water percentage ($\Phi = 1$)

Table 122.1. Absolute values for 0 per cent water for the curves in Fig. 122.3

Curve	B.m.e.p., lbf/in²	B.s.f.c., lb/hp h	CO, %	HC, p.p.m.	NO, p.p.m.
① ○	54·0	0·56	0·2	570	1470
② ×	32·0	0·76	0·2	400	1040
③ □	76·2	0·49	0·2	510	2180
④ △	86·8	0·48	0·2	320	2300

Realization of water injection

At a time when the very strict exhaust gas standards of 1975 are imminent and when there are only very vague ideas, particularly with respect to NO reduction, all possibilities of how to meet these standards should be considered. This also applies to those possibilities that seem to be difficult to realize at a first glance.

Of all the means influencing the composition of exhaust gases before or during combustion, water injection and exhaust gas recirculation both have the advantage of having the greatest influence on NO emissions. Both are able to reduce nitric oxide by 50 per cent or more. Exhaust gas recirculation has the advantage of being realized and adjusted by rather simple means. Its main disadvantage, however, is that it causes a loss in torque at WOT. According to the results of reference (6), about 8–10 vol% of exhaust gas referring to the volume of intake air are to be recirculated to get a NO reduction of 50 per cent. This, however, causes a corresponding decrease in power of 8–10 per cent. Water injection avoids this disadvantage, as the injected water only requires a negligible part of the volume.

The expenditure for water injection could be reduced considerably. The need for emulsification would not exist if water could be introduced to the cylinders, by calibrated holes, separately from the fuel. By this procedure the system could be simplified and made more reliable. The water could be supplied by a piston or diaphragm pump driven by the camshaft. Pressures of only about 15–30 lbf/in^2 are necessary. The water quantity can be varied in relation to the load by using the manifold vacuum. Thus an adjustment in the quantity of water supplied to each operating point is possible. This adjustment can be relatively coarse, as engine performance is almost uninfluenced. In order to avoid difficulties in starting and during warm-up, water circulation is short-circuited and delivery is therefore stopped. The possible danger of a higher corrosion of engine parts could not be confirmed. During the whole period of operation of more than 100 hours, no corrosion was observed.

The main disadvantage is the necessity of a water tank which must have a capacity of about one-quarter of the volume of the fuel tank. Considering the advantages of water injection, however, this should not put it out of consideration.

FORMATION OF NITRIC OXIDES
Introduction

Since the amount of nitric oxides was limited by the U.S.A. exhaust gas standards two years ago, the influences of various engine parameters on NO emissions have been discussed in a large number of publications. It has been shown conclusively that NO is affected by all those parameters which influence temperatures during combustion. A dependency on parameters which do not influence the temperatures could not be verified. Hence, it may be concluded that NO formation depends mainly on the temperature level during combustion. This is confirmed also by chemical considerations (7). More detailed investigations (8) show that nitric oxides, once formed, do not dissociate with decreasing temperatures. Concentrations formed, corresponding to the highest temperature level, remain preserved. The reactions are frozen in. On the basis of these concepts predictions were made of the expected NO concentration by calculating the peak temperatures.

The tests with water injection described here were done with similar intentions. Water injection has the advantage of influencing the NO level within wide limits without simultaneously changing other parameters such as equivalence ratio, spark timing, or i.m.e.p., which might influence NO formations.

Results

Concentrations of NO and CO and pressure diagrams at various equivalence ratios (Φ) and various water quantities were measured for different operating points defined by i.m.e.p. and engine speed n. By measuring CO it was always possible to adjust the same equivalence ratios at the different water rates of 0, 20, and 30 per cent. Furthermore, the throttle position was adjusted in such a way that for 20 and 30 per cent water the i.m.e.ps corresponding to 0 per cent water were also obtained (see Fig. 122.4).

To be able to judge the cyclic variations, five measurements were made for each operating point, but temperature and heat release curves were calculated only for some of them to limit the considerable expense of calculation. The following figures show results of measurements and calculations for some points. They are representative of all other results.

Fig. 122.5 shows the influence of the quantity of water on NO formation at various equivalence ratios. These results, obtained by individual measurements of always one cycle, agree with the results of measurements of the whole exhaust gas described above. Nitric oxide emissions can be reduced to about 50 per cent by adding 30 per cent water nearly independently of the equivalence ratio.

Fig. 122.6 shows NO emissions plotted against the calculated peak temperature, T_{max}. Unfortunately, as

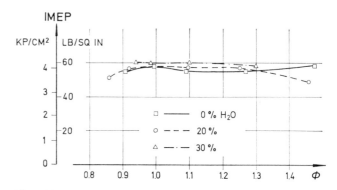

Fig. 122.4. I.m.e.p. as function of the equivalence ratio Φ for different water quantities

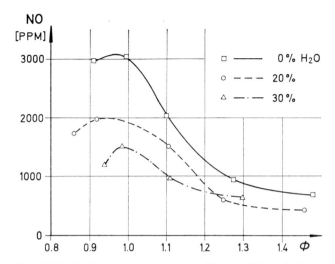

Fig. 122.5. Influence of water quantities on NO emissions at different equivalence ratios Φ

Fig. 122.4 shows, we were unsuccessful in adjusting always exactly the same i.m.e.p. At 30 per cent water the i.m.e.p. is a little higher, resulting in somewhat higher temperatures.

If scatter of the points is neglected, the curves NO $= f(T_{max})$ lie in the hatched, nearly exponential, region according to theory. Considering, however, the individual points, great divergences from the theoretical curve can be observed. For example, points 1 and 2 have nearly the same calculated peak temperature but NO levels differ by 1000 p.p.m. Points 2 and 3 have the same NO level but temperatures differ by more than 100 degC.

The rate of heat release ($dQ/d\alpha$) indicated in Fig. 122.7 allows a much better prediction and thus offers an explanation for the divergences described. In Fig. 122.7, the rates of heat release are plotted for the selected operating points 1, 4, and 5. It is obvious that an early rise in the rate of heat release leads to higher NO levels. The rate of heat release corresponding to point 5 has the latest rise and, consequently, the lowest NO level although point 5 has the highest calculated peak temperature, as shown in Fig. 122.6.

These relations become still clearer, considering the heat release $Q = f(\alpha)$, i.e. the integral of the rate of heat release (see Fig. 122.8). In this diagram it is quite evident that early heat release causes higher NO levels. At the same time the influence of water on the course of combustion is evident. The rate of combustion is reduced by water. This causes a delay of heat release and hence a decrease of the NO level.

Figs 122.9 and 122.10 show the influence of cyclic variations on the rate of heat release and on heat release for the operating point 4. Evidently only individual cycles can be used sensibly for this kind of investigation. Within the cyclic variations of one operating point, NO differences of 100 p.p.m. and more—as other points show—are possible.

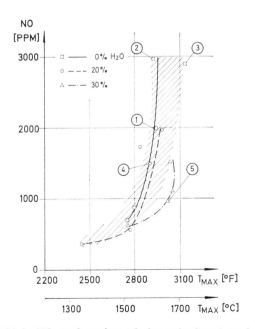

Fig. 122.6. NO as function of the calculated peak temperature T_{max} for different water quantities

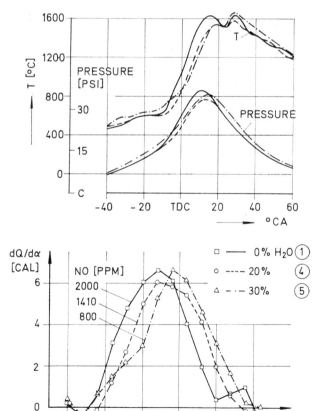

Fig. 122.7. Pressure, temperature, and rate of heat release for different water quantities ($\Phi = 0.91$)

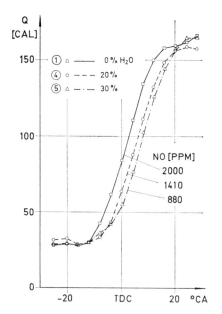

Fig. 122.8. Heat release for the same points as in Fig. 122.7

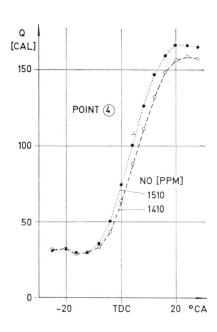

Fig. 122.10. Influence of cyclic variations on heat release for the same point as in Fig. 122.9

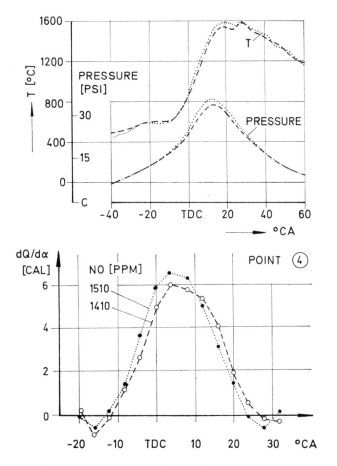

Fig. 122.9. Influence of cyclic variations on pressure, temperature, and rate of heat release ($\Phi = 0.91$)

Here, too, the diagrams of the rate of heat release and of actual heat release prove to be a suitable indication of the NO level. The calculated peak temperature, however, is nearly equal in both cycles and would lead us to expect the same NO level.

CONCLUSIONS

The program for the calculation of temperatures, rates of heat release, and heat releases postulates—as all other programs of this kind published to date—equal pressures and temperatures at each moment in the whole cylinder. Hence, for each crank angle α there is assumed but one value p and one value T. For the pressure, this assumption is correct since pressures propagate with sound velocity. Temperature equalization, however, is relatively slow. There always exists a temperature gradient in the combustion chamber, at least during the period of maximum temperatures and pressures.

The average temperature used in the calculation at any time, therefore, is always lower than the highest temperature occurring at this moment at one or more places in the combustion chamber. Therefore, the calculated peak temperature, T_{max}, is not the *real* peak temperature, T_S, which is responsible for the NO level. This real peak temperature, T_S, can occur at an instant which need not be identical with the instant for which T_{max} is calculated. Moreover, T_S does not exist in the whole combustion chamber but only in a limited region of the gas.

Recent publications (9) (10) show the attempt to calculate a more exact real peak temperature by subdividing the combustion space. As comparisons between theory and measurement in reference (10) show, a better agreement could be obtained. However, an exact calculation of the real peak temperature would require the consideration of a temperature gradient, not just one co-ordinate. It is a three-dimensional problem, being not only influenced by reproducible factors such as combustion chamber design, flame propagation velocity, or heat transfer, but also by statistical factors such as the local distribution and preparation of the mixture—and the latter, especially, can hardly be determined exactly.

Considering, moreover, the very steep rise of NO at higher temperatures (shown in Fig. 122.6), a calculation of NO emissions remains subject to great errors, even if the calculations are made with peak temperatures determined with great effort.

SUMMARY

Investigations described in this paper prove reduction of NO emissions to 55 per cent to be possible by adding about 28 per cent water to the fuel. The values of b.m.e.p., b.s.f.c., CO, and HC do not deteriorate, with the exception of HC at higher loads.

Nitric oxide values taken for an individual cycle correlate to the calculated peak temperature of this cycle within a rather wide scatter. Therefore, a prediction of NO emissions by calculating peak temperatures is only possible within wide limits. The curves of rates of heat release and of actual heat releases allow a better correlation to NO formation.

A calculation of the peak temperatures is extremely complicated, being a three-dimensional problem depending not only on reproducible but also on statistical factors.

Therefore, an exact prediction of NO levels by calculating peak temperatures does not seem to be possible.

APPENDIX 122.1
REFERENCES

(1) MADER, R. and ZEILINGER, K. 'Messung der Stickoxid-Emission eines Ottomotors', *IVKK-Ber.* (No. 293).

(2) FINK, J. and ZEILINGER, K. 'Einfluss der Wassereinspritzung auf die Stickoxid-Emission', *IVKK-Ber.* (No. 319).

(3) NICHOLLS, J. E., EL-MESSINI, J. A. and NEWHALL, H. K. 'Inlet manifold water-injection for control of nitrogen oxides—theory and experiment', S.A.E. Paper 690018, 1969.

(4) MÜLLER, H. 'Untersuchung der zyklischen Variationen im Druckverlauf von Ottomotoren', *XIIth Fisita Conf.*, Barcelona, 1968 **1**, 3

(5) LANGE, K. 'Beitrag zur Untersuchung der Verbrennung im Dieselmotor unter besonderer Berücksichtigung der Wandauftragung des Brennstoffes', Dissertation, Technische Universität München, 1970.

(6) GLASS, RUSSELL, TADE, and HOLLABAUGH. 'Evaluation of exhaust recirculation for control of nitrogen oxides emissions', S.A.E. Paper 700146, 1970.

(7) HOLLEMAN, A. F. and WIBERG, E. *Lehrbuch der anorganischen Chemie* 1960 (Verlag Walter de Gruyter, Berlin).

(8) NEWHALL, H. K. and STARKMAN, E. S. 'Direct spectroscopic determination of nitric oxide in reciprocating engine cylinders', S.A.E. Paper 670122, 1967.

(9) SCHWARZBAUER, G. and GRUDEN, D. 'Brenntemperatur und Wärmefreisetzung im Verbrennungsmotor', *MTZ* 1971, 61.

(10) MUZIO, L. J., STARKMAN, E. S. and CARETTO, L. S. 'The effect of temperature variations in the engine combustion chamber on formation and emission of nitrogen oxides', S.A.E. Paper 710158, 1971.

C123/71 VEHICLE PARTICULATE EMISSIONS

K. CAMPBELL*   P. L. DARTNELL*

Vehicle particulate emissions have been shown to consist of lead, iron, solid carbon, ammonium chloride, nitrate ion, hydrocarbon tars and oil mist. The quantity emitted is affected by exhaust system design, operating conditions, and driving habits. Under certain conditions particulate emissions are greater from unleaded gasoline than from gasoline containing lead. Methods of determining vehicle particulate emissions include prior mixing of the exhaust gas with air and filtration of a small representative sample, admission of all the exhaust gas into a pre-conditioned chamber and withdrawal of samples for examination, and filtration of the total volume of exhaust. A method is being developed to considerably reduce lead emissions. The devices used have no detrimental effect on vehicle performance.

INTRODUCTION

THE EXISTING EMISSION STANDARDS for the gasoline engine, originally proposed in the U.S.A. for 1975 and 1980, include standards for particulates of 0·1 g/mile and 0·03 g/mile respectively. These proposals first focused attention on particulate emissions.

The importance of particulates in the detrimental aspects of overall air pollution is open to debate, but high concentrations can undoubtedly reduce visibility and cause soiling. Some idea of the importance afforded to particulate emissions in the U.S.A. is indicated by these estimated quantities of 'minor' emissions in Los Angeles County (1)†:

Particulates	ton/day
Salt particles from ocean spray	5600
Ground dust from natural sources	3100
Aerosols and vapours from aerosol cans	40
Rubber from vehicle tyres	32
Smoke from cigarettes	23
Jet engine exhaust	11
Footwear use	5·7
Smoke from cigars	4·5
Particulates from refineries	4

Based on the U.S. Department of Health, Education, and Welfare estimates of pollutants, and re-evaluating them in terms of environmental effect (using the Californian determined air quality standards), particulate matter arising from the motor vehicle accounts for only 2 per cent of the total compared with 51 per cent from industrial sources.

Serious proposals have, nevertheless, been put forward by H.E.W. for the control of particulate emissions from the gasoline engine.

Although the standards have been promulgated, they are not, as is the case with other gaseous components of the exhaust, as yet the subject of legislation. For this reason, particulate emissions have not been so extensively studied, and there is a serious lack of information regarding their definition, measurement, and ways of reduction. This paper reviews the information available, including work currently being carried out at Associated Octel.

THE SOURCES AND COMPOSITION OF PARTICULATES IN THE EXHAUSTS OF GASOLINE ENGINES

The source of particulates

Any material passing into the combustion chamber—either with the fuel or the oil—which forms a solid during combustion, would be emitted as particulates. For example, lead, phosphorus, and inorganic additives in the gasoline, and alkaline earth compounds from the lubricating oil. The fuel composition also contributes significant quantities of solid carbon, due to cracking and polymerization arising from incomplete combustion. Particulate material and condensable vapours generate in the combustion chamber and subsequently deposit on the surface of the exhaust system. At some later time, as a result of thermal or mechanical shock, this deposited material flakes off, carrying significant quantities of iron from the surface, or increasing the iron content of the particulate emissions by attrition of the exhaust system. This material ranges in size from flakes several millimetres long to submicron particles. The carbonaceous component varies in consistency from hard solids to heavy liquids such as tars or oil mist. It has been found that unburnt oil passes out of the tail pipe as a very fine aerosol and will pass through several fine filters. For an engine with an acceptable oil

---

The MS. of this paper was received at the Institution on 27th May 1971 and accepted for publication on 5th July 1971. 23
* *The Associated Octel Co. Ltd, Bletchley, Bucks.*
† *References are given in Appendix 123.1.*

consumption, i.e. one pint per thousand miles, the quantity of oil mist emitted is at least 0·02 g/mile. Whether or not this, and the tars from the fuel, are classed as particulates is still to be resolved.

The composition of particulates

The composition of the particulates is related to particle size, engine operating conditions, and fuel composition (2)–(5).

When using leaded gasoline, the larger particulates have a similar composition to exhaust system deposits containing 60–65 per cent lead salts, mainly as the halide, 30–35 per cent iron oxide, and 2–3 per cent carbon (3). The fine particulates differ significantly in composition. A typical analysis of this airborne material is shown in Table 123.1 (5). The finer particles contain less iron and lead and much more carbon.

This difference in composition with changing particle size substantiates the fact that vehicle lead emissions are of two kinds and are from two sources: the larger particles, which are deposits flaked from the exhaust system, and the finer particles, which pass through the system direct from the combustion chamber.

Engine operating conditions mainly affect the particle size which changes the composition as stated previously. Operating conditions also affect the carbon content, e.g. particulates from stop–start type driving are richer in carbon than those from hot, continuous operation. Obviously, oil consumption and composition of the oil (ashless or non-ashless) will also be significant.

Fuel effects on particulate composition are affected by additives or natural components present in the gasoline. The addition of phosphorus or increasing sulphur content causes considerable quantities of the lead to be emitted as a phosphate or sulphate. If no lead is present, the particulates will be mainly carbon.

Some of these factors, besides affecting the composition of the particulates, also affect the quantity, as discussed in the next section.

FACTORS AFFECTING THE QUANTITY OF VEHICLE PARTICULATE EMISSIONS

These are complex and often interrelated. The following factors apply for both leaded and unleaded gasoline, but some of them would be less significant in the absence of lead. These factors include exhaust system design, engine operating conditions, and type of mileage accumulation.

Since condensable components in the exhaust gas adhere to the surface of the exhaust system, the quantity of particulates emitted is affected by the surface area of the system. For a given engine size less particulates are emitted from twin exhaust pipes than from a single pipe. Also, because twin systems tend to run cooler, more condensable material is solidified and retained. New exhaust systems run cooler, due to better heat transfer, and provide a better surface for deposit adhesion, so that particulate emissions are lower with new exhaust systems and tend to increase as

Table 123.1. The composition of the airborne material arising from vehicle particulate emissions

Element or substance identified	%wt
Lead	24·5
Iron	0·9
Chlorine	8·6
Bromine	4·0
Carbon	28·0
Hydrogen	5·8
Nitrate ion	7·3
Ammonia	5·4
Alkaline earths	2·6
TOTAL	87·1

the system ages. Silencer design is also significant; more particulates are retained by a resonant absorbing type of silencer than by the baffle plate design.

Engine operating conditions affect the quantity of particulates emitted. This increases as engine speed and load increase, since not only do the higher exhaust gas temperatures give less opportunity for condensation, but the increased vibrational and thermal shock increase the flaking of deposits, previously laid down, from the exhaust system.

Mileage accumulation condition has a very significant effect on the quantity of particulates emitted. Thus, in stop and start driving the exhaust system never gets hot, and most of the potential solids condense and are retained on the surface. These are emitted at a later stage when the engine is run for longer periods at higher speeds and loads.

One obvious factor which has been expected to influence the quantity of particulate emission is fuel composition.

Since, in general, at least half of the lead entering the engine with the gasoline is emitted at the tail pipe as particulates, it is to be expected that removal of lead from gasoline would radically reduce particulate emissions. This is not necessarily true and depends very much on the vehicle operating conditions (4). This is illustrated in Fig. 123.1, showing the quantity of particulate emissions from leaded and unleaded gasoline during different types of vehicle operation. These ranged from complete choke action on every start, representing winter conditions, to no choke action on every start, representing short trips during the summer. To simulate average vehicle usage a combination of one-third starts with full choke and two-thirds starts with no choke was used. Test lengths ranged from 25 to 200 miles. The results show that when lead is present in the gasoline the total of particulate emissions is not seriously affected by driving conditions, and remains reasonably consistent between 0·3 and 0·4 g/mile. It is of interest that the carbon content of the deposit decreased with increasing choke action. Using the unleaded fuel the weight of particulates increased drastically with increasing use of the choke and, under winter driving conditions, exceeded the quantity of particulate emissions of the leaded fuel.

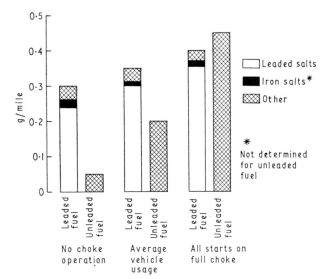

Fig. 123.1. Composition of total particulate emissions

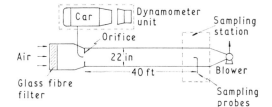

Fig. 123.2. Proportional sampling system for exhaust particulate

This definite lead benefit is due to the lead catalysing the oxidation of the carbonaceous materials, as evidenced by the lowering carbon content of the lead deposits with increasing use of the choke. This catalytic action of lead on carbon is supported by published work on surface ignition and pre-ignition, when it was found that carbonaceous material in the combustion chamber was oxidized by lead in the gasoline.

It is significant that even the lowest quantity of particulate emissions, which is shown in Fig. 123.1 and which was obtained under no choke conditions on unleaded gasoline, exceeds the U.S. proposals for 1980 of 0·03 g/mile.

THE DETERMINATION OF TOTAL PARTICULATE EMISSIONS

As stated previously, particulate emissions range in size from several millimetres to submicron. For this reason the determination of total particulates by filtration of a small sample of the exhaust, which appears to be the simplest method, will always be open to the criticism that it may not be representative of the actual particle composition. In spite of this, because of its simplicity, this method is extremely attractive, particularly with the advent of the 1972 Federal Constant Volume Sampling (CVS) procedure which presents a representative sample of the exhaust for analysis. Such samples would be expected to be deficient in the larger particles.

Habibi has suggested a way to overcome this (4). The apparatus is illustrated in Fig. 123.2. The total exhaust gas stream from the vehicle is introduced into a tunnel and mixed with air, using the constant volume sampling technique. The length of the tunnel was chosen to ensure complete mixing of the air and exhaust streams and to permit gravitational settling of the very large particles (200 to 3000 μm). On completion of the test these can be recovered for examination. A small sample of the air–gas mixture can be withdrawn from the sampling station, just ahead of the blower, for examination of the finer particulate emissions. Habibi has suggested that smaller tunnels, equally effective, can be developed to work in conjunction with the standard CVS rig.

Ter Haar (5) has suggested another method of determining airborne particulates. In this, exhaust gas from the vehicle under test is fed into a large, black polythene chamber with a volume in excess of 2500 ft^3. Black polythene is used to reduce the possibility of photochemical reactions occurring. Before the test, the air in the bag is dehumidified to less than 10 per cent relative humidity to reduce condensation effects. After the admission of the exhaust gas, small samples are withdrawn from the centre of the chamber for examination.

In both of the above-mentioned techniques the small samples are withdrawn through fine filter media such as Millipore AA or Whatman 41. Metallic contents can be determined by classical chemical methods, carbon by burning to carbon dioxide which is then determined, and the total particulates are determined by the weight gain of the filter media.

The determination of particulate emissions by direct filtration of the total volume of the exhaust gas is fraught with many difficulties such as condensation of the water and temperature effects. For metallic constituents of the particulates such as lead and iron, however, it has been found that the design of filter shown in Fig. 123.3 works satisfactorily.

The filter unit is cylindrical in shape, 24 in long and 15 in in diameter. Exhaust gas is fed into the centre of the drum and passes out through two layers of glass fibre filter wrapped around a wire grid support. The glass fibre filter media is the type commonly used for removing dust from air conditioning systems. This filter cylinder is sealed by clips top and bottom and a metal strip along the seams. The whole is then enclosed in an outer case in order to collect the emitting gases, for passage through other filters or into an eductor system.

When used to determine lead or other metallic particulate emissions, all the filter media is first extracted with an organic solvent to remove oil (which can be determined later), macerated into very small pieces, and then boiled in acid. A small sample of the acid extract is then taken for determination of the desired components.

The results indicate that these filters are extremely efficient for removing metallic particulates. For example,

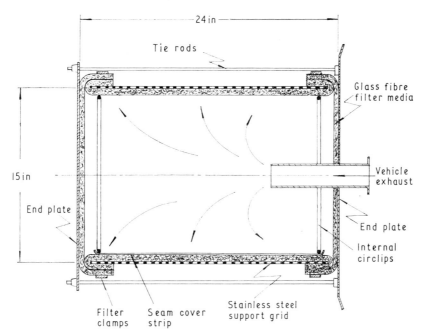

Fig. 123.3. The design of the total filters

if two such filters are used in series, less than 0·2 per cent of the lead emitted with a vehicle's exhaust gas is found in the second filter.

This type of filter, however, has limitations when applied to emissions other than metallic. For instance, when the hydrocarbon wash used for degreasing is examined for tars and oil, as much oil is found on the second filter as on the first, and probably some passes through them both. At the moment, no way has been found of reliably determining the carbon caught on these filters; the major problem is the large mass involved and the uncertainty of representative analysis if small pieces are taken for examination.

One way of measuring total particulates would be to weigh the filter before and after a known volume of exhaust has been passed. Unfortunately, balances which have the required capacity and sensitivity are not readily available on a commercial scale. Precautions would also have to be taken to weigh the filter under standard conditions of dryness and humidity. One further problem is that under the gas temperature conditions to which the filter is exposed, the glass fibre filter media tend to lose the organic binder. This results in a weight loss.

In summary, no easy way has yet been found of determining total particulate emissions other than metallic ones, and it will be interesting to note the method, if any, which is eventually promulgated by the Federal authorities. Equally interesting will be the Federal definition of what constitutes particulates.

WAYS OF REDUCING PARTICULATE EMISSIONS

Although only the U.S. authorities have suggested standards for particulate emissions, and are likely to implement these, vehicle lead emissions have been criticized in other countries such as Sweden and Germany. For obvious reasons, this is the aspect of particulates that has been investigated most intensively by Associated Octel.

It has been shown that it is technically feasible (4) to reduce vehicle lead emissions by 95 per cent, by replacing the standard exhaust system with a sophisticated complex of fluted pipes to cool the exhaust, mesh filled pipes and boxes to cause the lead particles to agglomerate, and a final catchment device in the form of a cyclone. Such a sophisticated system is expensive but may be economically attractive if it lasts the lifetime of the vehicle.

The work at Associated Octel has been based on a somewhat different philosophy. The objective is to develop a simple but effective lead remover that will fit into the standard exhaust system, will replace the standard silencer and have similar noise abatement properties, will be comparable in cost with present silencers, and will require no maintenance and be replaced as part of the normal exhaust system. This would have the benefit of reducing lead emissions in the quickest possible way, since used cars would also be controlled as their exhaust systems were replaced.

Initial development of such devices centred around cyclones or inertia-type devices. Table 123.2 shows that some success was achieved. Lead retentions, ranging from 12·8 to 40·6 per cent, were obtained during tests on a variety of vehicles, but the cyclones tested created considerable back pressure in the exhaust, and in our experience did not give sufficient noise abatement to replace the normal silencer. Because of the success of other techniques, investigation of inertia-type devices has been given lower priority.

One alternative to inertia separators is a form of

Table 123.2

Test	Mileage accumulation procedure	Vehicle	Test mileage	Percentage ingoing lead retained in vehicle			
				Engine	Exhaust system	Device	Total
Vehicles with cyclones	Road test	Ford Cortina 1500	15 162	2·0*	10·7	34·2	46·9
		Ford Cortina 1500	15 000	8·7	8·8	27·0	44·5
		Volkswagen 1600 TLE	15 209	11·2	8·6	22·5	42·3
		Fiat 124	15 131	7·4	15·0	15·4	37·8
		Ford Corsair 2000	15 361	11·0	14·5	12·8	38·3
Cyclone plus agglomerator		Ford Cortina 1500	10 000	9·8	8·9	40·6	59·3
		Volkswagen 1600 TLE	15 000	9·6	3·1	22·7	35·4

* Incomplete analysis data.

filtration, and it is in this area that the greatest overall success has been achieved. Simple cylinders, packed with stainless steel wire wool coated with alumina, greatly reduced lead emissions when used to replace the standard silencer. Fig. 123.4 shows the construction of these cylinders. At this stage of development only cylindrical geometry has been used since this is compatible with many exhaust systems.

Lead emissions are monitored by the total filter using the lead determination procedure detailed earlier. In the actual tests, and to guard against accidental damage to the first filter, two total filters in series were used. These filters, fitted to a car under test, are shown in Fig. 123.5. To date, dynamometer tests have been carried out on two quantity-production European cars.

The first of these was a 3·3-litre vehicle with a six-cylinder engine. The exhaust gases leave the engine through a common pipe which then splits into a twin-exhaust system. The rear silencer only of each exhaust was replaced with an alumina-coated filter. This reduced lead emissions to approximately one-half of those from the standard vehicle over a 12 000-mile test period.

On a smaller vehicle, having a four-cylinder engine of 1200 cm^3 capacity, two tests have been carried out. In the first of these, only the rear silencer was replaced with an alumina-coated filter. The results given in Fig. 123.6 show that this reduced lead emissions to approximately half of those from the standard exhaust system. In the second test, both the front and rear silencers were replaced with alumina-coated filters. The results given in Fig. 123.6 show that lead emissions were reduced to approximately one-fifth of those with the standard exhaust system over a 25 000-mile test period. Ancillary research indicates that with this type of system where the alumina coating is exposed to very hot exhaust gases, chemical catchment of the lead occurs in the first filter. Further tests are proceeding on the road using other vehicles. Future development will investigate such aspects as optimum geometry, fuel composition, and the economics of lead recovery from the alumina coating.

Systems containing a single filter have been tested for noise abatement by a major silencer manufacturer who is co-operating in the development programme, by a major vehicle manufacturer, and by an independent organization. The filter gave good noise abatement equal to that of the exhaust system with the standard silencer. Assur-

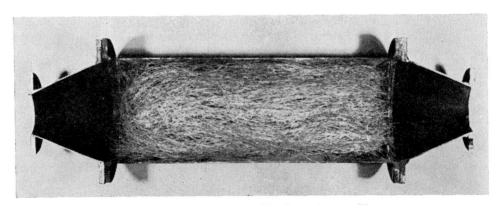

Fig. 123.4. Construction of lead catchment filters

Fig. 123.5. Measuring total particulate emissions on vehicle dynamometer

ance has also been given that incorporation of this type of packing into production silencers would not cause any major manufacturing problems, and should not increase costs significantly. It has also been established that, with such units, there is no measured penalty to vehicle performance as measured on a vehicle dynamometer.

Fig. 123.7 shows the effect on back pressure, fuel consumption, and power output of replacing the rear of two standard silencers by a single alumina-coated filter.

Emission tests show that inclusion of the filter in the exhaust system does not increase the emissions of the other gaseous pollutants, and may even reduce some

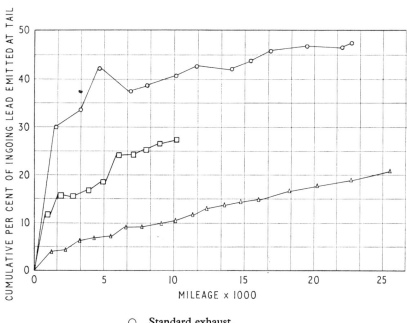

○ Standard exhaust.
□ 1 filter replacing rear silencer.
△ 2 filters replacing silencers.

Fig. 123.6. The effect of alumina-coated filters on vehicle lead emissions from a four-cylinder engine

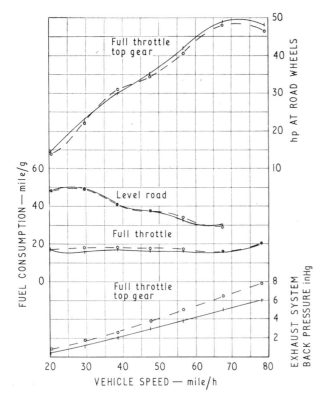

+—+— Vehicle with standard exhaust system with a barometric pressure of 30·25 inHg. Ambient 11°C.

—o—o Vehicle with filter replacing silencer. Barometric pressure 30·00 inHg. Ambient 11°C.

Fig. 123.7. Effect of alumina-coated filters on vehicle performance: 1600 cm³ four-cylinder car

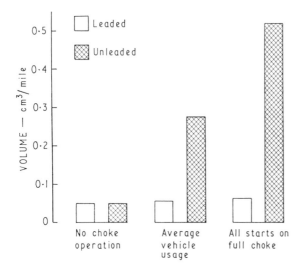

Fig. 123.8. Volume of particulate emissions—leaded and unleaded fuels

minor but odorous compounds, e.g. aldehydes. This aspect, and the long-term durability of the filters, are some of the areas currently being investigated.

As stated previously, the company's prime interest in reducing particulate emissions is in containing the lead, but even with no lead present in the gasoline, curtailing particulate emissions may become a very difficult problem for vehicle manufacturers if, in the future, particulate standards are enforced. For example, there is evidence, as stated earlier in this paper, that the total weight of particulates is similar with both leaded and unleaded gasolines under certain driving conditions. If the difference in density of these two types of deposits is considered, that derived from unleaded gasoline, which is mainly carbon, is much lower in density than that from leaded gasoline. As a consequence, although the weight of particulate emissions is similar for unleaded and leaded gasoline, on a volume basis, a far greater amount results from the use of unleaded fuel. This is indicated in Fig. 123.8. On this evidence, taking lead out of gasoline could increase the particulate emissions of the gasoline engine and increase the general soiling of the atmosphere.

CONCLUSIONS

Vehicle particulate emissions can arise from additives in the fuel and oil, and also include unburnt fuel and oil.

The composition of the particulates changes with size. The quantity emitted is affected by engine operating conditions, mode of mileage accumulation, and fuel composition. Under certain conditions more material is emitted with an unleaded fuel than when lead is present.

No simple way of determining a vehicle's total particulate emissions has yet been found.

A way of drastically reducing vehicle lead emissions has been developed using filters of stainless steel wool coated with alumina. The efficacy, simplicity, and expected low cost of this method offer an attractive alternative to the restriction of the use of lead in gasoline.

APPENDIX 123.1
REFERENCES

(1) MARCHESANI, V. J., TOWERS, T. and WOHLERS, H. C. 'Minor sources of air pollutant emissions', J. Air Pollut. Control Ass. 1970 **20** (No. 1), 19.

(2) MCKEE, H. C. and MCMAHON, W. A. 'Automobile exhaust particulates—source and variation', J. Air Pollut. Control Ass. 1960 **10**, 456.

(3) HIRSCHLER, D. A., GILBERT, L. F., LAMB, F. W. and NIEBYLSKI, L. M. 'Particulate lead compounds in automobile exhaust gas', Ind. Engng Chem. 1957 **49** (No. 7), 1131.

(4) HABIBI, K., JACOBS, E. S., KUNZ, W. G. and PASTELL, D. L. 'Characterization and control of gaseous and particulate exhaust emissions from vehicles', E. I. Du Pont de Nemours & Co. (Inc.), Wilmington, Delaware, 1970. (Paper for presentation to the A.P.C.A. West Coast Section 5th Technical Meeting 8th–9th October, 1970, San Francisco, California.)

(5) TER HAAR, G. L., et al. 'Composition, size and control of automotive exhaust particulates', A.P.C.A. Paper 71-111, 1971 (May).

EXPERIMENTS IN THE CONTROL OF DIESEL EMISSIONS

P. M. TORPEY*   M. J. WHITEHEAD*   M. WRIGHT†

This paper presents some results of a comprehensive investigation into the emissions of gaseous pollutants from diesel engines. Tests on direct-injection and swirl-chamber engines are described, and the characteristics of each compared. The effects of water induction, exhaust recirculation, and change of compression ratio on the swirl-chamber engine were studied.

It is concluded that emissions of oxides of nitrogen are most likely to cause concern: these emissions were found to be very dependent on injection timing. In general, the direct-injection engine emissions were twice as high as those of the indirect-injection engine. Water induction and cold exhaust recirculation were found to be powerful methods of reducing oxides of nitrogen.

INTRODUCTION

PAPERS on diesel emissions are relatively few in number. In Britain two papers by Lunnon (1)‡ and by Speirs and Vulliamy (2) are noteworthy; from Germany, Abthoff's paper (3) deserves considerable attention, while in the United States Valdmanis and Wulfhorst (4) and Marshall and Hurn (5) give a fairly comprehensive introduction to the American scene. Despite the pioneering role of the U.S.A. in setting emission standards for diesel-engined vehicles, surprisingly few papers have been published which give detailed information on the emission levels to be expected from typical automotive diesel engines.

On the basis of these comments, the authors feel that a useful contribution to the literature can be made by presenting a reasonably comprehensive comparison of emissions from typical direct-injection (D.I.) and indirect-injection (I.D.I.) engines. It is hoped that this paper will give sufficient information to permit a crude evaluation of the total mass emissions from diesel-engined vehicles in the U.K. and Europe at the present time. It is further hoped that the information given will allow an estimation of the possible reductions in total mass emissions that could be effected by choice of engine type, choice of operating conditions, use of exhaust recirculation, or water injection.

In the context of this paper, the term 'exhaust emissions' refers to the gaseous pollutants only—viz. oxides of nitrogen (usually termed NO_x), carbon monoxide (CO),

The MS. of this paper was received at the Institution on 17th May 1971 and accepted for publication on 7th July 1971. 22
* *Project Engineer, Ricardo & Co. Engineers (1927) Ltd, Bridge Works, Shoreham by Sea, Sussex.*
† *Development Engineer, Ricardo & Co. Engineers (1927) Ltd, Bridge Works, Shoreham by Sea, Sussex.*
‡ *References are given in Appendix 124.1.*

and unburnt hydrocarbons (unburnt HC). Although smoke is not specifically listed as a pollutant it is understood that no engine condition described herein will cause emission of exhaust smoke in excess of the limits permitted by B.S. AU 141, and when the term 'derating' is used, it shall mean reduction of fuelling such that the smoke limits of B.S. AU 141 are not exceeded.

TEST WORK

The experimental work was carried out, in the main, on the basic Ricardo E16 single-cylinder research engine. This unit has a stroke of 140 mm, and 121-mm bore.

The E16 engine was tested in both D.I. and I.D.I. builds. Table 124.1 gives details of the engine and combustion systems.

SCOPE OF TESTS

On both types of combustion systems the effects of speed, load, and injection timing on emissions of NO, CO, and unburnt HC were examined. In addition to the above, the following conditions were also examined on the I.D.I. engine:

(a) water injection into the inlet manifold,
(b) injection of water–methanol mixture into the inlet manifold,
(c) exhaust recirculation (hot and cold),
(d) variation of compression ratio,
(e) variation of combustion chamber configuration.

ANALYSIS OF EXHAUST GASES

Infrared gas analysers (I.R.G.As) were used for NO and CO. The mode of operation of these analysers has been

Table 124.1. Brief specification of single-cylinder research engines used for emissions tests

Details	Direct injection	Indirect injection
Bore × stroke	120 × 140 mm	120 × 140 mm
Rated speed	2000 rev/min*	2000 rev/min*
Fuel-injection pump	CAV 'B' type	CAV 'B' type
Element diameter	10 mm	8 mm
Nom. injection rate	9·25 mm^3/CR°/litre	3·6 mm^3/CR°/litre
Nozzle	4 holes 0·32 mm ϕ	Pintle
Opening pressure	175 atm.	105 atm.
Compression ratio	15·3:1	16·2:1
Combustion chamber shape	2:1 dia./depth ratio toroidal D.I. chamber dia. 63 mm	Comet V
Fuel	Gas oil (S.G. 0·83)	Gas oil (S.G. 0·83)

* Dynamometer limitation (these engines can be run at 2750 rev/min).

adequately described elsewhere (3) (6). In the NO analyser, the interference due to CO and CO_2 in the sample gas was virtually eliminated by the use of specially prepared filter units. The exhaust gases were dried thoroughly before passing through the I.R.G.A., and all results in this paper are quoted on a 'dry exhaust' basis. (But see later for HC measurement technique.)

The accuracy of the I.R.G.A. depends primarily on the accuracy with which the calibration gases are prepared. In the absence of more sophisticated and expensive apparatus, one of the wet-chemical methods for measurement of oxides of nitrogen can be used to check the calibration gases to an acceptable level of confidence. Reference (3) gives a detailed description of the various methods available for measurement of NO.

The exhaust gas sample is taken from a point about 1 m from the exhaust port.

MEASUREMENT OF UNBURNT HYDROCARBONS

Accurate measurement of total hydrocarbons in a diesel exhaust is extremely difficult. An excellent paper by Springer and Dietzmann (6) explains why I.R.G.As cannot be used on diesel engines, and describes a suitable flame ionization method for HC detection. The hydrocarbon analyser used for the tests described herein was designed and developed by the authors' company, as suitable proprietary equipment was not available at the time: the principle of operation is similar to that of the instrument described in reference (6). A schematic flow diagram of the unit is shown in Fig. 124.1. Basically, the exhaust sample is passed through a hydrogen flame. The change in conductivity, or degree of ionization, of the products of combustion can be taken (after suitable calibration) as a measure of the unburnt hydrocarbon content. It is necessary to reduce spurious responses to other combustibles, or alternatively to apply the relevant correction factors.

The oven and sample line are maintained at 150°C. A sealed carbon-vane pump draws a proportion of exhaust gas from the sample line and passes it through a hydrogen-helium flame. Fine control of gas flows—viz. sample gas, burner fuel mix (H_2–He), and air—is necessary for good results. The response of the instrument to oxygen in the exhaust sample was reduced to negligible proportions by

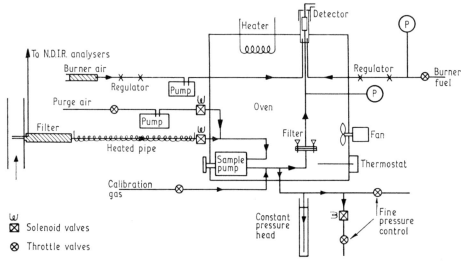

Fig. 124.1. Schematic flow diagram of Ricardo heated flame ionization detector

suitable choice of the hydrogen/helium ratio, and by attention to burner air/fuel ratio, in a manner similar to that described in reference (6).

Marshall and Hurn (7) have published a useful summary of the techniques for diesel emissions measurement. Note that, of necessity, unburnt HC concentrations are measured and expressed on a 'wet' basis.

ENGINE TESTS

Provided that reasonable care is taken with the instrumentation, and particularly with the calibrating gases, then good repeatability of results is fairly easy to achieve for a given set of operating conditions—provided that the engine itself is reliable and repeatable. One of the minor difficulties in emissions measurement (particularly for oxides of nitrogen) is that of defining an optimum injection timing for each engine speed, if both smoke and minimum fuel consumption are to be taken into account. The conventional D.I. engine with air swirl will normally operate with an optimum start of combustion some 5–7 crank degrees before top dead centre (°b.t.d.c.); further slight advance may reduce smoke, particularly if low injection rates are used. The swirl-chamber engine, on the other hand, gives maximum power with combustion starting about t.d.c.; further slight retard will reduce smoke at the expense of increased brake specific fuel consumption (b.s.f.c.) (and reduced power) for a given air/fuel ratio.

Because of the accepted difficulties in measuring smoke density on single-cylinder engines, the optimum timing for these tests was that timing which gave minimum b.s.f.c. In cases where the curve of b.s.f.c. versus timing was fairly flat over a range of 2–3°, the most retarded timing was taken as optimum.

RESULTS

Of immediate interest is a direct comparison between D.I. and I.D.I. combustion systems in the same cylinder sizes. Contour maps of emissions for the two engines operating at optimum conditions for performance are shown in Fig. 124.2.

OXIDES OF NITROGEN EMISSIONS

In general, the D.I. engine gave NO emissions higher than the I.D.I. engine emissions by a factor of two. The I.D.I. engine showed a tendency towards lower NO at low engine speeds, whereas the D.I. had more constant levels over the speed range. Both engines show peaks in the NO curve at fairly lean air/fuel ratios of about 20:1 and the NO levels drop fairly rapidly after the smoke limit has been exceeded.

HYDROCARBONS EMISSIONS

The levels of HC emissions (expressed as p.p.m. C) were higher for the D.I. engine under most operating conditions. A high percentage of the D.I. hydrocarbons is believed to be caused by poorly atomized fuel being blown from the nozzle sac during the non-working stroke. A recent U.S. paper describes how reduction of nozzle sac volume reduces odour and HC emissions (8).

A swirl-chamber engine fitted with a fuel-injection pump that has a load-advance characteristic may have a tendency to misfire at light load and high speed: as a result, hydrocarbon emissions tend to increase under these operating conditions. This condition can be completely cured by slightly advanced light load timings.

As one might perhaps expect, the unburnt HC emissions increase abruptly as the stoichiometric air/fuel ratio is approached. However, this condition is normally not noticeable on the D.I. engine because the permitted smoke level is reached at a higher air/fuel ratio than on the I.D.I.

CARBON MONOXIDE

Due to the large excess of air present during combustion in the diesel engine, the carbon oxidation reaction goes virtually to completion. Significant quantities of CO are not produced until the smoke limit is approached.

For the two engines considered here, CO levels are generally below 0·1 per cent and frequently below 0·01 per cent. The I.D.I. engine shows a tendency for the CO levels to rise in sympathy with the HC when light-load misfire conditions are approached. A very abrupt increase in CO occurs in both engines at full load. The authors consider that, provided engine combustion is good, with a clean exhaust, and provided that engine misfire condition is not approached, then the present levels of CO emissions from diesel engines will meet all foreseeable legislation limits. For this reason, no further discussion of CO emissions will be made. In general, CO rises rapidly as soon as the exhaust gas becomes visible, thus giving a visual indication of its presence.

THE CONTROL AND REDUCTION OF NO AND HC EMISSIONS

The authors do not consider themselves qualified to comment on the world-wide tendency towards progressively more stringent legislation on air pollution from the internal-combustion engine. It is sufficient to note that the State of California has introduced legislative limits which for 1975 and subsequent years will be extremely difficult to meet. As a crude measure of the severity of this legislation, NO emissions from current D.I. engines will have to be reduced by a factor of three in order to meet these limits.

METHODS FOR REDUCING OXIDES OF NITROGEN

In the petrol engine, it is fairly well established that NO is frozen at near-equilibrium levels corresponding to the local maximum flame temperature. From this it would appear that a reduction in flame temperature will reduce the amount of NO formed. In general, this seems to be the case; however, some anomalies have been found (see below on the effect of compression ratio).

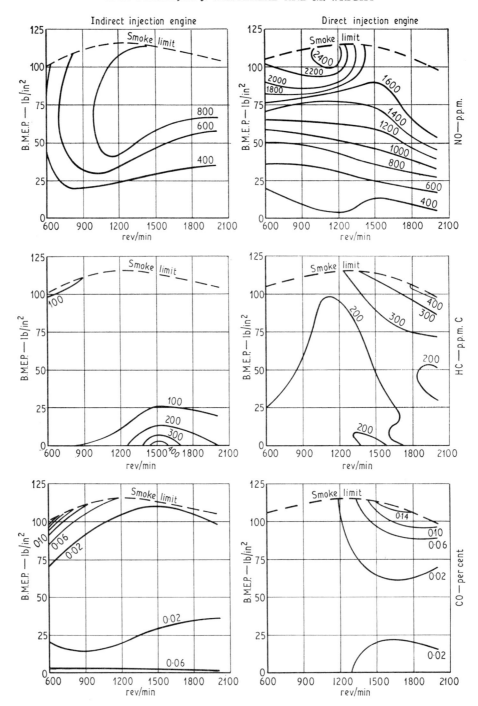

Fig. 124.2. Maps of exhaust emissions

Injection retard

Retarding the injection timing is by now a standard for NO reduction. Fig. 124.3 shows typical sets of results for successive amounts of retard. It can be seen that, although the D.I. engine had higher NO_x levels at optimum injection timing, similar amounts of injection retard bring the peak levels much closer to the I.D.I.

The D.I. engine showed significant reductions in NO levels with a decrease of 55 per cent at 8° retard, at the peak torque speed of 1200 rev/min. However, this reduction in NO levels was also associated with a lowering of rated output by up to 24 per cent at 8° retard (i.e. in order to maintain exhaust condition to B.S. AU 141). At low engine speeds, and particularly at light loads, the engine

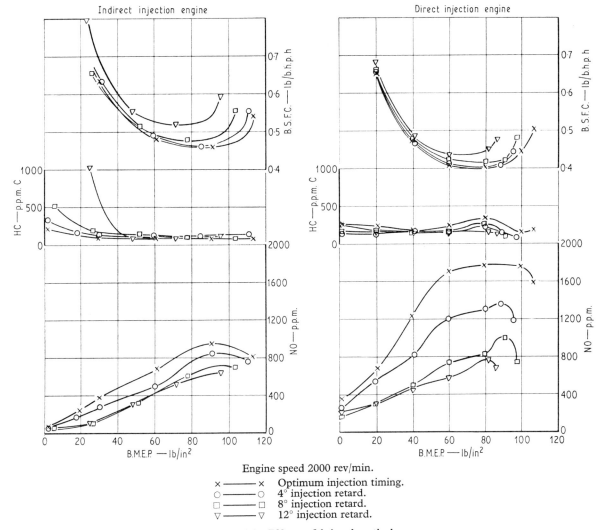

Fig. 124.3. Effect of injection timing

running became so unsteady that 12° retard could not be sustained; but at higher speeds the drop in performance was not so marked. Some form of light load advance would overcome this problem. Neither the HC levels nor the CO levels showed any significant change as a result of retarding the timing.

The I.D.I. engine showed smaller reductions in NO_x levels due to retarded timing, but overall levels were still lower than those of the D.I. The I.D.I. also showed a smaller reduction in rated output.

For both engines it would seem that 8° retard would be the maximum tolerable, since further retard only brings more performance drop without much improvement in emissions. A further limiting factor for the I.D.I. engine was a tendency to induce light-load misfire at very retarded timings, causing high HC levels.

Exhaust gas recirculation

One of the methods of reducing peak diesel cycle temperatures is to introduce a relatively inert substance into the inlet air charge; this will absorb some of the heat energy released during combustion and effect a reduction in temperature. This will, in turn, lead to lower NO_x concentrations.

Cold exhaust recirculation

The exhaust and inlet systems of the I.D.I. engine were interconnected via a valve and an intercooler. By increasing the exhaust back pressure, up to 20 per cent by volume of exhaust at ambient temperature was recirculated into the inlet manifold.

The results (Fig. 124.4) show that successive increases in recirculation gave reductions in peak NO levels of up to 60 per cent at 20 per cent recirculation; this was, however, associated with a reduction in smoke-limited output of about 15 per cent. For practical purposes it would seem that 15 per cent recirculation is all that could be tolerated, giving a 50 per cent drop in NO levels and a 9 per cent drop in peak rating on this particular engine.

A combination of cold exhaust recycling and a 4° retard

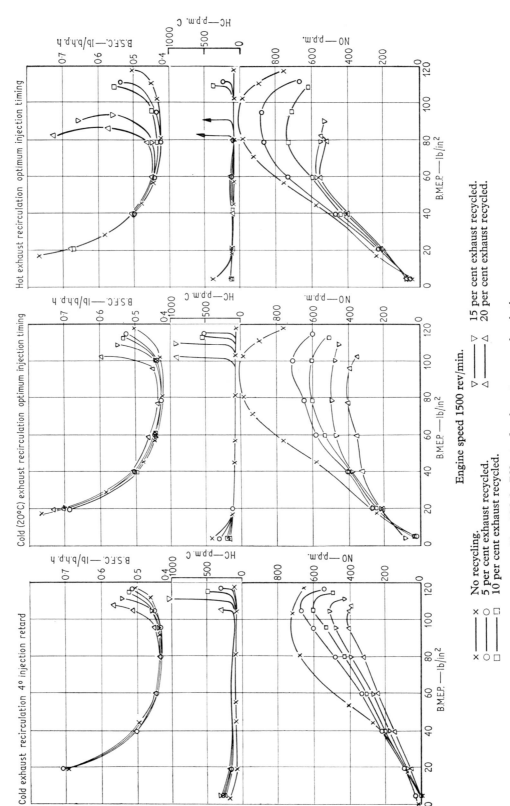

Fig. 124.4. Effect of exhaust gas recirculation

in static injection timing gave proportionately smaller drops in NO levels but did not seem to have such an adverse effect on power and fuel consumption. This may signify a change in optimum injection conditions when employing exhaust recycling techniques.

In order to avoid power loss it should be possible to employ maximum recycling at low loads but limit the quantities recycled at high load. This would be particularly applicable to an I.D.I. system where full-load NO levels are below peak values. (Peak NO levels occur at about 80 per cent load.)

Hot exhaust recirculation

Exhaust was passed by the shortest possible path from exhaust pipe to inlet manifold on the I.D.I. engine, employing exhaust back pressure to obtain up to 20 per cent by volume recirculation.

The results (Fig. 124.4) show that this had less effect on NO levels than cold recycling. It would appear that 10 per cent recycle would be the maximum practically usable, giving a 25 per cent reduction in NO levels but a drop of 12 per cent in peak rating.

It would seem, therefore, that only cooled exhaust recirculation would present worth-while reduction in NO levels.

In a practical installation it would be desirable to use an air-to-exhaust cooler (rather than a heat exchanger using engine coolant), so as to reduce the recycle temperature as near to ambient conditions as possible.

Inlet manifold water injection

A disadvantage with exhaust gas recirculation is that it effectively reduces the charge efficiency of the engine and hence its maximum power output. For this reason, a liquid-phase injection that would vaporize during the compression and combustion processes causing charge dilution would be more suitable. Apart from availability and low costs, the use of water has the advantages of (*a*) a low boiling point, (*b*) a high latent heat of vaporization, and (*c*) a relatively low vapour pressure at inlet temperatures.

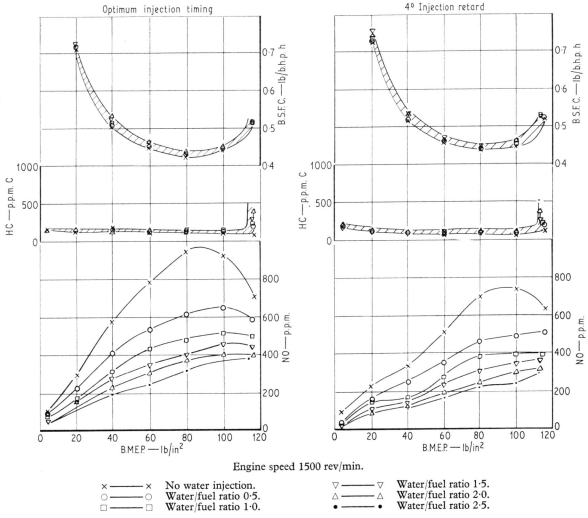

Engine speed 1500 rev/min.

× ———— × No water injection.
○ ———— ○ Water/fuel ratio 0·5.
□ ———— □ Water/fuel ratio 1·0.
▽ ———— ▽ Water/fuel ratio 1·5.
△ ———— △ Water/fuel ratio 2·0.
• ———— • Water/fuel ratio 2·5.

Fig. 124.5. Effect of inlet manifold water injection

Mains water was fed via a pressure-control valve into the inlet manifold of the I.D.I. engine through a calibrated jet. Jet sizes and water pressures were chosen to give water/fuel volume ratios of between 0·5 and 3·0.

The results (Fig. 124.5) show that water was very effective in reducing NO_x levels; a reduction of 60 per cent was obtained with a water/fuel ratio of 2·0:1. Further reductions using this quantity of water, together with a retarded timing, were obtained giving peak NO_x levels of only 300 p.p.m. (on a 'dry' basis).

The high latent heat of vaporization of water should tend to significant cooling of the inlet air charge and hence an increase in volumetric efficiency and power output; however, no increases in power were noted. In fact, the rated power dropped slightly (3 b.m.e.p.) up to a water/fuel ratio of 2·0 and then more quickly until the engine became unsteady at ratios of more than 2·0:1. Water injection has virtually no effect on the HC and CO emissions levels.

Freezing problems could well be encountered using water alone, and a practical solution might have to include some form of antifreeze.

A mixture of 70 per cent water and 30 per cent methanol was tested in a similar manner, but the results showed that severe difficulties were encountered due to the increasing HC levels; these were particularly high at low engine loads.

Although methanol is an excellent fuel in its own right, it would appear that the very weak mixture of methanol and air is too lean to burn completely during the combustion process. It is reported that similarly high HC levels are found with manifold induction of weak mixtures of a gaseous fuel (e.g. on dual-fuel engines) or on engines with fumigation of the intake air.

After having run for approximately 50 h using water injection the engine oil was found to contain 10 per cent water. Although there were no visible signs of excess wear or deposit on any engine parts, the long-term effects of water injection must be investigated before it could be used as a practical method of reducing NO_x emissions.

Combustion chamber redesign and effect of compression ratio

Various swirl chamber configurations and changes in compression ratio have been investigated on the I.D.I. engine and the important results are shown in Figs 124.6 and 124.7. (See Table 124.2 for a list of the engine modifications tested and a summary of the results.)

Although the exhaust emissions from the various configurations at optimum settings were roughly similar, it can be seen that each had a different response to injection retard. In the case of the change in compression ratio (Fig. 124.6 and curve (a) of Fig. 124.7) it will be seen that reducing the compression ratio from 16:1 to 13·5:1 gave little return in terms of reduced NO. The low-ratio engine was difficult to start, and suffered from severe low-load misfire with small amounts of retard. The high-ratio (19·8) engine had short delay periods and a smooth pressure diagram resulting in easy starting and low subjective noise levels.

Further increasing the compression ratio to 23·6:1 had surprisingly little effect on NO levels: however, b.s.f.c. values increased due to increased mechanical losses. At 8° retard, the 23·6:1 engine performed better than the 13·5:1 engine and gave similar NO levels without the tendency to misfire.

In general, the results seem to indicate that there is little to be gained from combustion chamber redesign when emissions levels are compared on an 'equal performance' basis. However, there is every indication that lowering the compression ratio will not give a net improvement, because this configuration is less affected by retarding the timing. Also, noise emissions are becoming more important and should be given consideration.

DISCUSSION

Little need be said about the preliminary tests carried out in order to establish the emission levels of the I.D.I. and D.I. combustion systems in standard builds. These results have been confirmed on other engines, and it is generally

Table 124.2. Modifications to combustion chamber geometry

Specification	'Whirlpool' type chamber	Comet V modification No. 1	Comet V modification No. 2	Comet V modification No. 3	Comet V modification No. 4
Compression ratio	17·4:1	16·2:1	13·5:1	19·8:1	23·6:1
Ratio $\frac{\text{Volume in cyl. head}}{\text{Volume over piston}}$	80/20	25/75	46/54	61/39	73/27
Shape of chamber in head	'Whirlpool'	Externally fitted Comet V	Comet V	Comet V	Comet V
Type of piston	Flat topped	(a) Flat topped (b) Std. Comet V trenches	Deeper Comet V trenches	Flat topped	Flat topped
Piston/head clearance	0·040 in	(a) 0·137 in (b) 0·068 in	0·040 in	0·100 in	0·050 in
13 mode calculations Optimum timing 8° static injection retard	— —	— —	5·75 g/hp h, NO_2 4·72 g/hp h, NO_2	5·76 g/hp h, NO_2 4·78 g/hp h, NO_2	6·89 g/hp h, NO_2 5·32 g/hp h, NO_2

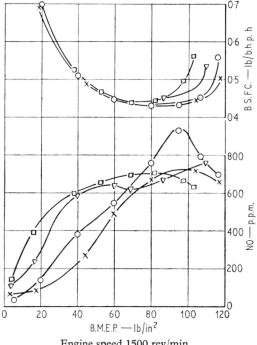

Fig. 124.6. Effect of combustion chamber geometry

Engine speed 1500 rev/min.
× ———— × Standard Comet V, 4° injection retard.
○ ———— ○ 'Whirlpool'-type chamber.
□ ———— □ Modification No. 1(a).
△ ———— △ Modification No. 1(b).

true to say that the D.I. emissions are twice as high as the I.D.I. In general, also, the results obtained by retarding injection timing are considered typical.

It is difficult to find a satisfactory explanation for the widely differing characteristics of the D.I. and I.D.I. engines in respect of NO emissions. It was observed that advancing the injection timing on the I.D.I. to give the same start of combustion as the D.I. (viz. 5–7° b.t.d.c.) increased NO emissions to levels approaching the D.I. One theory that has been put forward is based on the observation that NO levels peak at a lean air/fuel ratio of about 20:1 (about 30 per cent excess air). With richer mixtures, the NO concentration decreases. Now, in the I.D.I. engine all the fuel is injected into the pre-chamber, which on the normal Ricardo Comet chamber will comprise about 50 per cent of the total clearance volume. Consequently, the mixture formed will be over-rich in the pre-chamber. This theory holds that the over-rich mixture inhibits NO formation, and the subsequent complete oxidation of this rich mixture in the main chamber occurs at a temperature which is too low for the formation of significant quantities of NO.

This explanation is considered unsatisfactory. In the first instance, it has been observed experimentally that changing the geometry of the swirl chamber, to permit a high percentage of the total clearance volume (c. 80 per cent) to be contained in the pre-chamber, does not affect the emission characteristics. In this case, the mixture formed in the pre-chamber is quite lean, especially up to 70–80 per cent load. Nevertheless, the NO emissions are considerably below those of a D.I.

It also follows that with a conventional swirl-chamber engine, the pre-chamber runs lean up to approximately 50 per cent load. Nevertheless, it is observed that the I.D.I. engine gives significantly lower NO emissions than the D.I. over the entire load range. In fact, a comparison of the results from the D.I. and I.D.I. engines shows that the D.I. engine at 8° retard gives NO emission levels similar to those of the I.D.I.; at optimum timing under those conditions, the start of combustion for the D.I. is about 2° a.t.d.c. at full load, and the cylinder pressure diagrams are not dissimilar. It seems reasonable, therefore, to conclude that the major advantage of the I.D.I. engine is its ability to run with retarded timings relative to the D.I. and yet have sufficiently high degrees of turbulence in the combustion chamber to give rapid combustion with low smoke levels.

Water injection is the most effective method of reducing NO levels. The results in this paper agree broadly with those of Valdmanis and Wulfhorst (4). Undoubtedly, water injection will pose many development problems. However, these might be minimized by adopting low-pressure injection direct into the cylinder towards the end of the compression stroke. The great attraction of water injection is that really significant reductions in NO levels can be achieved without sacrificing power output—this cannot be said for exhaust recirculation.

The results of the compression ratio tests are extremely interesting. They seem to suggest that a pressure factor must be taken into consideration when considering the mechanisms of NO formation in high-temperature air. It is possible that pressure may inhibit the initiating reaction of oxygen dissociation. It is encouraging to note that the higher compression ratios also lead to reduced noise levels and easy starting. On this particular engine, although friction levels increased with increasing compression ratio, the brake thermal efficiency and net output improved slightly. It is acknowledged that, in general, increasing compression ratio gives a slight reduction in output, because the theoretical improvement in efficiency is outweighed by increased mechanical losses. The effect is more pronounced on the D.I. than on the I.D.I. engine.

The authors do not consider that exhaust gas recirculation is a very satisfactory method of controlling NO emissions; development problems are likely to be as severe as those for water injection, and ultimately a loss in performance must be accepted. The problem of cooling the recycled exhaust to a sufficiently low temperature is likely to be the major stumbling block. Further, in the light of the current U.S. legislation, it is unlikely that 1975 limits could be met by a D.I. engine with exhaust recycle with an acceptable amount of derating.

Throughout this discussion the emphasis has been on NO emissions. Hydrocarbon emissions are relatively insignificant for the diesel engine. It seems that, from the

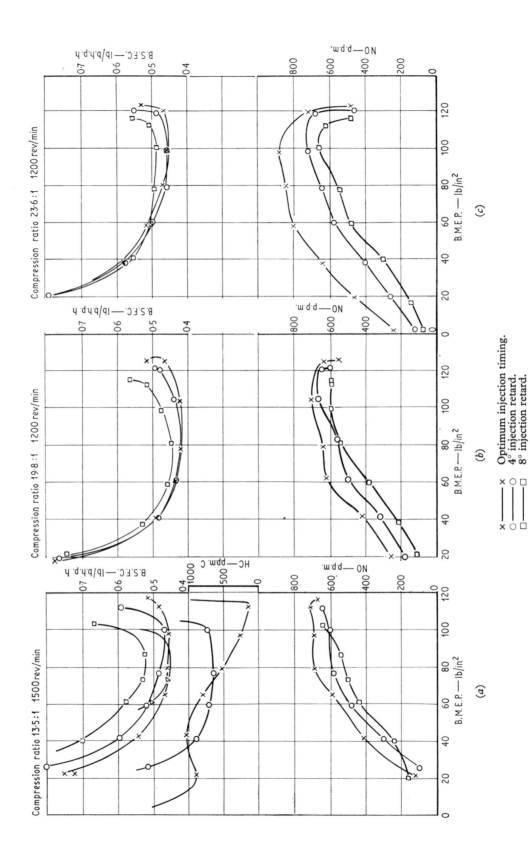

Fig. 124.7. Effect of engine compression ratio

Table 124.3. Estimation of total vehicular emissions

Area	Transport category	Assumed engine capacity, litres	Vehicles in each category, millions	Percentage in each class	Assumed lb.s.f.c., lb/hp h	Assumed brake specific emissions, g/hp h			Gross emissions, thousand tons			Diesel emissions as percentage of total		
						NO_2	CO	HC	NO_2	CO	HC	NO_2	CO	HC
U.K.	Diesel commercial	7.5	1.5	11.5	0.4	9.0	4.0	1	230	102	25	49%	4%	35%
	Gasoline commercial	1.5	0.5		0.7	5.0	30.0	1	235	2 340	47			
	Automotive		11.0		0.7									
Europe	Diesel commercial	12.0	4.0	11.5	0.4	9.0	4.0	1	1 210	535	133	57%	6%	43%
	Gasoline commercial	1.5	1.0		0.7	5.0	30.0	1	895	8 900	177			
	Automotive		4.2		0.7									
U.S.A.	Diesel commercial	14.0	1.0	1.0	0.4	15.0*	4.0	1.5	1 450	385	145	11%	1%	6%
	Gasoline commercial	6.0	16.0		0.55†	11.0	26.0‡	2	11 900	28 000	2200			
	Automotive		83.0											

* The higher NO_2 level of diesel engines in the U.S.A. is due to the higher percentage of turbo-charged, non-aftercooled relative to Europe.
† This specific fuel consumption is a mean figure based on the assumption that the annual U.S. gasoline consumption is divided equally between commercial vehicles and passenger cars (see text).
‡ These results have been taken from new engines. It is probable that the gasoline CO levels quoted are considerably below those of older, poorly maintained engines.

results of these and other tests, I.D.I. emissions of unburnt HC are unlikely to be a problem, even with the current severe U.S. legislation. For the D.I., reduction of nozzle sac volume, elimination of dribble, etc., should reduce HC levels to those of the I.D.I.

To complete this discussion, it would be instructive to use the available data to estimate the total mass emissions of oxides of nitrogen (expressed as NO_2), hydrocarbons, and carbon monoxide from diesel-engined vehicles in Britain, Europe, and the U.S.A. at the present time. These figures are necessarily crude, and are based on the results of tests carried out by Ricardos and others on production engines.

The gross emission data are based on published figures for 1968; the information has been compiled from a variety of sources, and where discrepancies have arisen a mean value has been taken. Calculations have been made for three geographical areas, viz. the U.K., Europe (excluding the U.K.), and the U.S.A.

For each area, and for each transport category (private and commercial), a typical engine size and type have been chosen consistent with the published data available for total vehicle registrations. For these vehicles, typical brake specific emission levels were chosen, and an average load factor assumed. For heavy commercial vehicles it was assumed that the engine worked at 60 per cent of rated load at 60 per cent of rated speed throughout: the corresponding figures for automobiles were 25 per cent load at 50 per cent rated speed. (The choice of these operating conditions may be criticized, but it is suggested that appreciable error in these assumptions will not significantly affect the conclusions—primarily because brake specific emission levels do not vary significantly with load.) Typical brake specific fuel consumptions corresponding to these operating conditions were then chosen, and a relationship established between fuel consumption and emissions. The gross emissions can then be readily calculated from published figures for sales of diesel fuels and gasolines in each area.

Some error may have arisen from the fact that it was not possible to differentiate between gasoline-engined heavy vehicles and gasoline-engined automobiles. In Europe, the percentage of heavy vehicles powered by gasoline engines is very small, but in the U.S.A. the reverse is true. In the absence of more accurate data, it was assumed that half of the total gasoline consumption in the U.S.A. was used by heavy commercial vehicles.

The results of these calculations are shown in Table 124.3. It will be noted that in Europe and the U.K., diesel vehicles form 11·5 per cent of the total vehicle population, yet contribute about 50 per cent of the total vehicular emissions of NO_2, and about 35 per cent of the total HC emissions. As one might expect, emissions of CO from diesel vehicles are minimal.

In order to up-date these estimates to 1971, an annual growth rate of 5 per cent has been assumed.

It must be emphasized that these are necessarily crude calculations, and that there is a real need for a thorough examination of the magnitude of the vehicle pollution problem before any controlling legislation is introduced. It is perhaps worth mentioning that, since the D.I. engine virtually reigns supreme in the heavy vehicle field, gross emission levels could be halved by adopting combustion chambers such as the pre-chamber or swirl chamber. Against this reduction in emissions must be offset the higher fuel consumption (by some 8–10 per cent) of pre-chamber-type engines.

CONCLUSIONS

The authors feel that, at the time of writing, sufficient data are not available to draw very firm conclusions about the potential of the various types of diesel engines for low emissions. However, from the tests described herein, and from other work reported in the literature, the following conclusions may be drawn:

(1) The I.D.I. engine and the D.I. engine differ mainly in respect of NO emissions: this difference can be attributed in the main to the I.D.I.'s ability to run more retarded than the D.I. for the same exhaust smoke condition. NO emissions for the D.I. are approximately twice as high as those for the I.D.I. engine.

(2) Emissions of oxides of nitrogen are extremely sensitive to changes in timing; considerable improvements in NO emissions from current production engines could be effected by closer control of timing over the load and speed range.

(3) Hydrocarbon emissions for I.D.I. engines in their current form are quite low, and should need relatively little attention. For the D.I. engine, HC emissions are more variable, ranging between 200 p.p.m. C and 1500 p.p.m. C. It is reported that elimination of nozzle sac volume will reduce D.I. HC levels to those of the I.D.I.

(4) If existing smoke limits are adhered to, then CO emissions from diesel engines are negligible (certainly in comparison with gasoline engines).

(5) Increasing the compression ratio of the I.D.I. engine resulted in lower emission levels and lower noise, particularly at retarded timings.

(6) Reducing compression ratio increased noise, gave difficult starting, and increased HC emissions at light load.

(7) Induction of water into the intake system reduced NO levels by half with little effect on performance.

(8) Recycling of cold exhaust gases reduced NO by half with 10 per cent power penalty. Hot recycle reduced NO by 25 per cent, also with 10 per cent power loss.

(9) It would appear that, in Europe, the diesel engine contributes between 50 per cent and 60 per cent of the total mass emissions of NO_2 from vehicles.

ACKNOWLEDGEMENTS

The authors wish to thank the Directors of Ricardo & Co. for permission to publish this work, and also their many colleagues who participated in the experimental work described herein.

APPENDIX 124.1
REFERENCES

(1) LUNNON, C. 'Some aspects of diesel exhaust emissions, especially in confined spaces', *Symp. on Critical Factors in the Application of Diesel Engines, Proc. Instn mech. Engrs* 1969–70 **184** (Pt 3P), 106.

(2) SPIERS, J. and VULLIAMY, N. H. F. 'Diesel engine smoke and pollutants', D.E.U.A./Instn Mech. Engrs joint meeting, February 1971.

(3) ABTHOFF, J. 'Measurement and variation of the emissions of oxides of nitrogen from internal-combustion engines', Research Report No. 2–216/1 (Combustion Engine Research Association, Frankfurt).

(4) VALDMANIS, E. and WULFHORST, D. E. 'The effects of emulsified fuels and water induction on diesel combustion', S.A.E. Paper 700736, 1970.

(5) MARSHALL, W. F. and HURN, R. W. 'Factors influencing diesel emissions', S.A.E. Paper 680528, 1968.

(6) SPRINGER, K. J. and DIETZMANN, H. E. 'Diesel exhaust hydrocarbon measurement—a flame ionization method', S.A.E. Paper 700106, 1970.

(7) MARSHALL, W. F. and HURN, R. W. 'Techniques for diesel emissions measurement', S.A.E. Paper 680418, 1968.

(8) FORD, S. and MERRION, H. 'Reducing hydrocarbons and odour in diesel exhaust by fuel injector design', S.A.E. Paper 700734, 1970.

BIBLIOGRAPHY FOR DATA ON VEHICLE POPULATION, FUEL SALES, ETC.

(9) AUTOMOTIVE INDUSTRIES *Engineering specification and statistical issue* 1970 (March).

(10) ASSOCIATED OCTEL *World-wide survey of motor gasoline quality* 1968.

(11) BOARD OF TRANSPORT 'Internal combustion engines', *Business Monitor*, Production Series P. 13, 1968.

C125/71 METABOLISM AND DIETETICS OF 'MOTEUR DIESEL COMPENSÉ B' AND ITS RESULT ON AIR POLLUTION

M. G. BRILLE*   Y. E. BAGUELIN†

This paper deals with an arrangement-exercise on a diesel engine. It was carried out to ease the problems of internal-combustion engines, especially those of pollution and the burning of a wide range of fuels. The temperature and pressure compensations that were carried out contributed to the achievement of these aims and to a significant increase in performance. These developments enable pollution levels to be controlled more closely and will permit future demands to be met more readily. It is considered that the results attained in the exercise justify a true installation study and permit hopes for production.

INTRODUCTION

BUYERS OF road traction continually demand more powerful and reliable engines with minimum weights and volumes; at the same time, legislators stipulate air pollution control regulations of increasing severity.

When facing the difficult compromise involved in this situation, there is a tendency to object to a whole branch of power units and to reconsider the viability of either the internal-combustion engine or the piston engine, or even the thermal engine in general.

However, to attain the objectives it seems prudent to retain for a transitory period the improvements gained on recent engines, especially on the diesel, and to try to control their pollution levels more closely.

From this viewpoint the diesel engine is already considered better than the spark-ignition engine because it emits much less carbon monoxide, which is certainly the main poison. The black or grey smoke, which is a discomfort more than a real poison, has already been reduced to acceptable values. Indeed, these are used as criteria to set the power limits.

The problems arising from two types of emission are still to be solved. The first type concerns the unburnt or not entirely burnt gases, hydrocarbons, and aldehydes produced at low load and speed, when the engine is cold. The second concerns the oxides of nitrogen resulting, at full power, when the engine is hot, from the endothermic combination of excess air components at high temperature. In a given engine, both effects necessarily combine and any step taken to fight the one (for instance, increasing the compression ratio) increases the other.

This dilemma is made more acute with turbo-charged engines, and it occurs in the same way with engines in which the chamber walls have a significant effect on combustion. It applies precisely to the four-stroke, M-chamber engine, and it seemed of particular interest to choose this chamber for the study, taking into account its well-known qualities of silence and efficiency. As the aim was also to reach the best performance, it was turbo-charged, which contributed to the investigations being made under the most difficult conditions.

The steps taken to resolve this dilemma were to introduce (*a*) changeable units to counteract engine variations, and (*b*) a series of automatic compensations in order to lessen the differences between the upper and lower engine operating conditions. The engine design itself was changed as little as possible so that these steps could be applied to any existing four-stroke engine type.

Known methods to change the compression ratio mechanically (by relative displacement of the piston crown or of a combustion-chamber wall) were not versatile enough to meet the objectives as major engine re-design would be required. Furthermore, using these approaches, the outline of the chamber changes with compression ratio and, of course, the supercharge at medium speed is no better. The compensations have therefore been achieved by means exterior to the basic engine, and the new engine unit has been called 'moteur compensé'.

THE 'MOTEUR COMPENSÉ'

The Saviem, type 597, six-cylinder 100/112 engine with a cylinder capacity of 5·27 litre has been chosen as the basic

---
The MS. of this paper was received at the Institution on 5th May 1971 and accepted for publication on 12th July 1971. 33
* *Director, Saviem, 8 quai Gallieni, 92, Suresnes, France.*
† *Advanced Design Engineer, Saviem, 8 quai Gallieni, 92, Suresnes, France.*

engine. It has been highly turbo-charged; consequently, its compression ratio has been fixed at a low value (15:1) to limit the pressures and temperatures at high speed.

First compensation
This concerns only temperature. Installing a heat-exchanger on the air circuit between the turbo-charger and the engine is common practice; this was done first, with the engine circulation water fixed at 80°C. The air supplied by the turbo-charger at full power, at 130°C approximately, was thus significantly cooled (to about 85°C), whereas the air supplied at reduced loads at a temperature near to ambient was significantly heated (to about 75°C). But the compensation thus realized was shown to be insufficient; therefore it was increased by means of the following arrangements, shown in Fig. 125.1.

The air–water heat-exchanger was connected to a secondary water circuit, as was the current oil heat-exchanger (2), so that the oil temperature could also be better controlled. This secondary water circuit (3) is

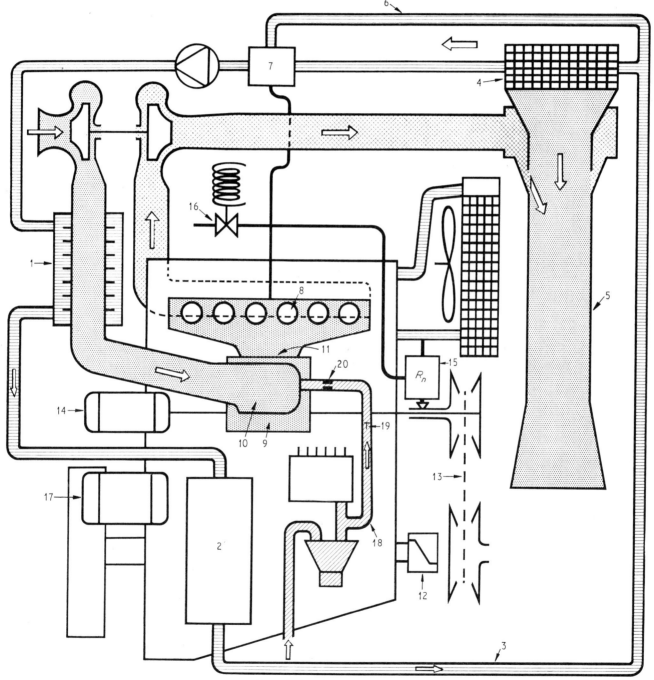

Fig. 125.1. General layout

cooled by a small auxiliary radiator installed laterally. The air which circulates through it is drawn into a form of extractor (5) by the exhaust gases coming from the turbine, which gives the special advantage of proportioning the cooling to the engine power. At very low power, the radiator action is eliminated by a by-pass (6) controlled by a pressure valve (7) dependent upon the manifold pressure. This second circuit temperature is controlled on the vehicle by the valve (7) to make sure that it remains relatively constant for a given size of radiator (4) (60°C) and only attains 90° for low loads and speeds, as may be seen at the top of Fig. 125.2. Under these conditions, at full power, the air after passing through the heat-exchanger is maintained at 65–70°; and at idling, with the short-circuited radiator, it remains at more than 80°C. In every case, the oil is maintained at 105°C approximately.

The inlet air temperature levels are therefore reversed in relation to what normally happens. The same applies in the cylinders at the end of compression. However, as the combustion in the chambers releases much less heat at reduced loads than at full power, this temperature difference at the air inlet must be increased if approximately equal final temperatures are to be realized.

To achieve this, it is possible to reduce the temperature of the secondary water circuit; it is a question of the size of the radiator (4), which is very small to provide 60° water temperature. While still remaining quite small, it can provide a 50°C temperature, which will reduce the inlet air temperature to 60°C approximately. On the other hand, at idling, the air temperature cannot exceed 80°C, for the circuit water must be limited to 90°C. It should be noted that this limit is higher than that of the engine main cooling circuit, which passes through the cylinder head and other hot locations creating vapour bubbles. For the same reason, it is easier to select the secondary circuit rate which has no vapour bubbles to separate. For the heating of air in the heat-exchanger (1), the only heat source is the oil heat-exchanger (2), which uses the heat resulting from mechanical losses of the engine, and piston cooling; at idling, this heat source could not raise the inlet air above 80°C. It was therefore necessary to use other means.

It was quite clear that, at starting, no heat was available from the oil heat-exchanger (2). The above arrangements, which apply to every running speed of the engine, including idling, do not apply to starting, which will be the subject of a special section.

Second compensation

This concerns pressure. In Fig. 125.2 (lower part) the curve (*a*) shows the pressure ratio of the turbo-charger against engine speed at full load. It is well known that such a curve does not give the engine a satisfactory torque characteristic.

Many studies, both old and recent, have had as an objective a better torque curve, even to the extent of constant power, by installing and using in various ways a volumetric blower, generally of the Roots type, either substituted for the turbo-charger or added to it.

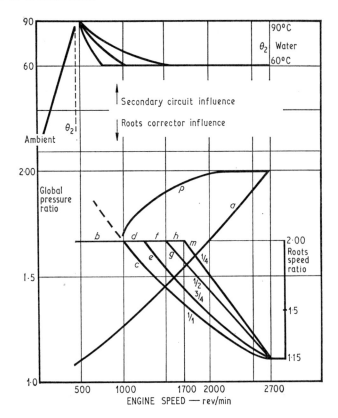

Fig. 125.2. Secondary circuit influence. Roots corrector influence

Constant power did not constitute the aim of this work, which was to re-establish a torque curve at full load, similar to that of a naturally aspirated engine, and to maintain a high supercharge at low load and speed; and as regards the temperature, to complete the action already described.

This pressure compensation has been obtained by using a blower of the Roots type [(9), Fig. 125.1], set as usual between the heat-exchanger (1) and the manifold (8) with the air inlet (10) and the air exhaust (11); its drive shaft is connected to the crankshaft on one side by means of a freewheel (12) and a variable pulley drive (13), and on the other side to the power generator (14). This generator can also be operated as an electric motor.

The speed ratio of the variable pulley drive (13) is determined by the regulator (15), influenced on the one hand by the supercharging pressure and on the other hand by the pressure of the water pump of the main cooling circuit, which serves as a reference for engine speed.

The theoretical result of this arrangement on the Roots speed ratio may be seen in Fig. 125.2 (ordinates on the right). For the 1·15 ratio there is no supercharging, the 2·00 ratio gives the maximum supercharging of the Roots for a given engine speed. At full load (1:1) the ratio variation follows the *c* curve; at three-quarters load it follows the *d*, *e* curves; at one-quarter load, the *b*, *d*, *f*, *h*, *m*, curves.

At full load, the above variation of the speed ratio provides a variation of the final pressure ratio represented by the p curve, which gives a torque curve much more suitable for traction than the a curve.

It is pointed out that for a maximum speed of 2700 rev/min the 1·15 speed ratio is maintained for the whole load range, the Roots is neutralized, and the inlet temperature maintained at around 60°C.

At reduced loads, the speed ratio variation results in pressure increases, implying significant air temperature rises, which are added to the rise in temperature from the secondary circuit, so that the temperature can reach 130°C in the manifold. However, at idling, the maximum speed ratio gives only a small over-pressure because of the internal leakage in the compressor. As this leakage is recycled, the rise in temperature through the blower reaches 30–40 degC, which produces a 110–120°C temperature in the manifold.

The manifold and the Roots, together with its air pipe as far as the heat-exchanger (1), are insulated to avoid losses. This means that at full power an increased cooling through the manifold walls, as usually occurs, is not to be expected.

Starting

It has already been noted that the temperature compensation provides no help at starting, as there is no heat source.

The pressure compensation is in no better position. Whilst at idling the Roots speed of 1000 rev/min (500 × 2) provides some pressure that will start the rise in temperature, this does not occur at starting with a cold engine, at a speed of 200 rev/min (100 × 2), when the relative leakage is very great and any rise in temperature too slow.

The methods tested to reduce this leakage by dimensional changes modify the reliability of the blower in normal operating conditions. It is much more useful to separate the blower speed from the engine speed and to drive the blower by means of the generator (14) temporarily transformed into an electric motor at a speed exceeding 1000 rev/min, which the free wheel (12) permits. The electrical valve (16) can also be opened to provide a higher blower speed. The starting operation is theoretically as follows.

As soon as the starter button has been operated, the electrical valve (16) opens, the generator (14) starts running, and after two or three seconds the starter (17) is engaged and operates the diesel engine; after two or three seconds, the electrical valve (16) shuts.

The air admitted to the engine is then warm enough for ignition to occur, despite the moderate compression ratio.

A pre-mixture

In addition to advantages of a lower pollution due to timing, owners of internal-combustion engines rightly ask for the possible use of various fuels. This possibility is in fact very interesting. Originally, the multi-fuel quality was only appreciated by the army for logistics reasons. Now, two other reasons are added.

The first one again concerns pollution by using very clean fuels, or fuels containing less carbon atoms in their formula. The second one relates to the security of supplies and the possibility for developed countries to alleviate pressures from oil-producing countries by increasing the yield from crude oil as a result of widening the distillation cuts and the possible use of synthetic fuels of various origins.

This multi-fuel quality is also demanded from 'moteur compensé'. Most of the fuels considered, other than diesel oil, have a low cetane number, and whereas they all burn at full load, they do not ignite or else burn incompletely at reduced loads.

Then, the systematic heating and over-pressure, at reduced loads, carried out in the 'moteur compensé' give precisely those characteristics required to burn these fuels. Operating with petrol, which served as a test, was revealed to be satisfactory. However, at very low speeds, some irregularities have shown the bad influence of low turbulence on the ignition delay. A kind of compensation has been obtained by introducing a pre-mixture into the engine at low speeds.

How to reduce the ignition delay by employing an already partially oxidized fuel in the chamber is well known. For instance, in the Vigom process, a small quantity of fuel is injected half a cycle before ignition. In the 'fumigation' system, a weak fuel mixture is injected by a carburettor at inlet. As this 'fumigation' is valueless—and even prejudicial at times—at full power, and the question here is only to compensate the lack of turbulence at low speed, the required mixture has been achieved as follows, and as shown in Fig. 125.1. Some fuel (18) is taken from between the fuel pump and the injection pump, i.e. at a relatively constant pressure of 7 lb/in^2 approximately. After passing through the non-return valve (19) and the calibrated jet (20), the fuel flows through the inlet (10) of the Roots (9), i.e. between the two blowers.

If the air pressure in it was constant and less than 7 lb/in^2, the fuel rate in the by-pass (18) would be constant; thus, the rate per cycle and the mixture richness would be inversely proportional to the engine speed. But as the air pressure increases with the supercharging level, the decrease of the flow rate per cycle is even faster than the proportionality. This rate is reduced to zero when the air and fuel pressures are the same, which occurs at medium speeds; at higher speeds, the valve (19) prevents air injection in the fuel.

The Roots blower provides a good air–fuel mixture; besides, any unmixed fuel that does not immediately reach the cylinders is automatically reintroduced, which contributes to increase its oxidization and its efficiency to reduce the delay.

At idle speed, the pre-mixture richness can be near to stoichiometric and the cylinder injection can be suppressed, the running engine then operating as an Otto cycle with self-ignition. This offers advantages regarding silence. But the exhaust mixture must be examined before generalizing this idling use.

TESTS

The first bench tests began in February 1969; then a second engine was assembled on a 38-ton g.v.w. semi-trailer in November 1970. In the meantime, a third unit was entrusted to Messrs Ricardo in July 1970 in order to verify the earlier results and to examine the exhaust emissions more closely. Thereby, Ricardo was able to carry out useful investigations to improve and develop this process.

Engine No. 1

In its original outline, the prototype engine was cooled by one water circuit, with a 200 litre/min flow rate whose adjustable operating temperature foreshadowed the possibilities of the secondary circuit, as this one could be defined on a vehicle only. The variable pulley drive provided variations ranging from 1·1:1 to 1·6:1.

The level of 200 b.h.p. at 2700 rev/min, 55 m kg at 1800 rev/min, was immediately reached with good smoke and temperature conditions, compared to the basic engine, which developed 125 b.h.p. at 2900 rev/min, when naturally aspirated, with a compression ratio of 17·5. The maximum cycle pressure reached was 120–125 kg/cm², for a ratio less than 10, when operated with petrol as well as gas oil.

Drawing a part of the fuel into the Roots inlet was examined relative to its influence on the combustion

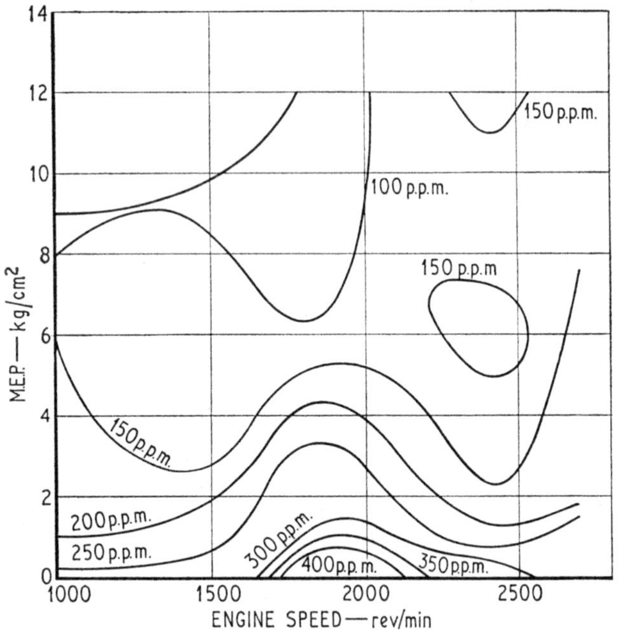

Engine water, 80°.
Secondary water, 60°.

Fig. 125.3. HC emissions map

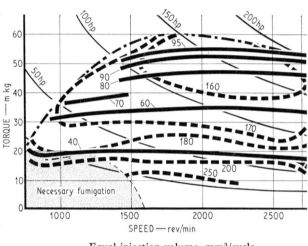

——— Equal injection volume, mm³/cycle.
- - - - Equal consumption, g/hp h.

Fig. 125.5. Petrol performances

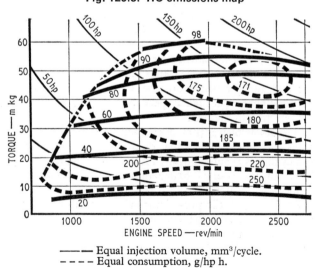

——— Equal injection volume, mm³/cycle.
- - - - Equal consumption, g/hp h.

Fig. 125.4. Diesel oil performances

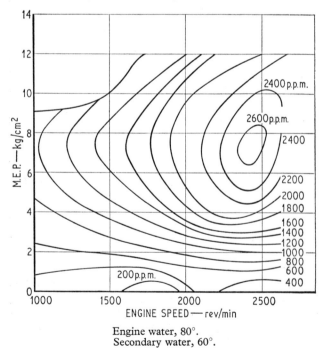

Engine water, 80°.
Secondary water, 60°.

Fig. 125.6. NO$_x$ emissions map

Fig. 125.7. Photo of truck—inlet side

Fig. 125.8. Photo of truck—secondary circuit side

diagram and the consequences of running roughness and noise. A steady idling was obtained without any injection after some empirical selection of timing and fuel recycling. At this stage of the tests, the points to be improved were the starting ability, the smoke at low speed, and the oil temperature, which stabilized in the sump at approximately 140°C, this being too high. The rectifications dealt with the compensation blower tightness and its variable drive. The drive ratio was increased up to 1·8, then to 2.

The hydraulic jack, using the engine oil pressure, was not only regulated by the pressure of supercharging as initially provided, but also the engine speed factor was added, as defined in Fig. 125.2.

With another method, ratios exceeding 2 were attained, and a satisfactory result was obtained in smoke reduction at low speed.

A power-split variable pulley drive providing a ratio varying from 1:1 to 2·32:1 is now being developed and has been partially tested.

Engine No. 2

This is the engine assembled on the above-mentioned vehicle. Operation of the secondary circuit with controlled temperature began with adjusting the section of exhaust gas admission into the air extractor, to determine the possible level of the temperature in the circuit. The adjustment selected enabled an air volume three times the gas to be injected, and the values given at the beginning of the paper were obtained, 60°C water and 105°C oil, whilst at the same time reducing the second radiator surface. The thermostatic regulation gave results near to the curves of Fig. 125.2. By using the air pressure signal of the turbo-charger, the road test showed the usefulness of this device, the effect of which, however, needs to be improved by reducing the thermal inertia of the secondary water circuit. At overrun, the air temperature gains 15°C in some 200 mm of heat exchanger length, which is further increased by the Roots blower. Thus, at minimum injection on the overrun, the engine keeps a thermal level sufficient to receive full injection again.

On the other hand, in transitory operation, the hysteresis affecting the variation of the Roots blower speed ratio is favourable. When accelerating, the effect balances the turbo-charger inertia by providing more air than is necessary; when decelerating, the hysteresis is of no consequence and may even have a favourable effect on the mechanical behaviour.

Last winter, with ambient temperatures around 0°C, the vehicle tests were especially devoted to the solution of the problems of starting. To compensate the Roots tightness, various artificial means were examined. Oil injection of some cubic centimetres into its inlet was made at cranking. The isolation of a three-cylinder row by means of a butterfly valve on each inlet provided the doubling of the stabilizer volumetric compression on the three fed cylinders. The improvement was positive but insufficient.

The proposal, already described, in which the Roots blower will be driven by the engine generator, is now being effected. The insulation of the pipe, the Roots blower, and the manifold is indispensable to this starting. If the results confirm expectations, the compression ratio will be reduced again to 13:1.

The engine endurance is considered satisfactory, after reference to the several hundreds of hours on the test bench and the 4000 km covered by the vehicle. Compared to the same engine with a compression ratio of 17·5 turbocharged by one turbo-charger which develops 155 b.h.p. at 2900 rev/min with a P_2/P_1 ratio of 1·9, the behaviour of the 'moteur compensé' appears especially satisfactory, because of the lowered maximum pressures resulting from the reduced compression ratio, of the levelling of the admission temperature, and of the oil temperature being reduced to 105°C.

Engine No. 3

The tests carried out on the Ricardo benches showed that the M combustion process could be improved so far as the emission ratio is concerned, especially in the distribution of the concentrations when the engine operates. In the initial build the 1973 standards (16 g/hp h, $CH+NO_2$), are met without modifying the general performance and without involving other means.

It should be noted (Fig. 125.3) that the unburnt hydrocarbon emissions are particularly low. The slight increase at light load is attributed to the use of secondary water uniformly kept at 60°C. The shift to 85°C in this area of operation should result in values similar to the rest of the diagrams (Figs 125.4 and 125.5).

The diagram on Fig. 125.6 shows that the NO_x level is quite regular: the mean levels of oxides of nitrogen have been brought back to the usual level of a direct-injection chamber, with an important shift of the maximums at high loads towards the higher road speeds.

These results are satisfactory for the immediate future, as the diminution of unburnt gases, etc., corresponds to an improvement of the combustion, and the diminution of oxides of nitrogen, which on the contrary tends to worsen the efficiency, has been maintained within the required limits, with an overall improvement in the engine performance.

CONCLUSIONS

With the proposed transformations, the engine breathing and temperatures, in a word its metabolism, are controlled in a better way; they appeal to a dietitian, using the same fuel in different ways, as well as various fuels. They offer to the diesel engine—which is scarcely modified, except for the reduction of the compression ratio—improvements in performance, reliability, and pollution.

The means employed are not laboratory conceptions, for they allowed such an engine to be assembled on a commercial vehicle (Figs 125.7 and 125.8). Some developments are still to be made, but prospects for industrial use seem feasible.

In fact, arrangements made to improve only upon pollution would not be economical. It is thus quite satisfactory, if not necessary, that the same means should simultaneously bring an increase of performance contributing to amortize them.

APPENDIX 125.1

BIBLIOGRAPHY

BRILLE, M. and BAGUELIN, Y. 'Moteur diesel compensé', S.I.A. Report dated 5th November 1969 (March 1970).

CHAFFIOTTE, P. and HERRMANN, R. 'La chambre de turbulence à volume variable', S.I.A. Report dated 5th December 1967 (March 1968).

HENAULT, C. and LAGARDE, F. 'Le réallumage', S.I.A. Report dated 4th November 1968 (April 1969).

MANSFIELD, W. P. and DAVIES, S. J. 'Soixante chevaux au litre de cylindrée diesel—Réalisation industrielle grâce à un piston à rapport de compression variable', S.I.A. Report dated 7th February 1967 (July 1967).

TIMONEY, S. G. 'A new concept in traction power plants', *Proc. Auto. Div. Instn mech. Engrs* 1965–66 **180** (Pt 2A, No. 3), 63.

TIMONEY, S. G. 'A compact long-life diesel engine', *Symp. on Transport Engines of Exceptionally High Specific Output, Proc. Instn mech. Engrs* 1968–69 **183** (Pt 3B), 52.

TRYHORN, D. W. 'Transport engines of high specific output', *Gas Oil Pwr* 1968 (December), 308.

TRYHORN, D. W. 'New turbo-charging systems for two-stroke cycle diesel engines', S.A.E. preprint No. 700075, 1970 (January 12th–16th).

CONTROLLING EXHAUST EMISSIONS FROM A DIESEL ENGINE BY LPG DUAL FUELLING

D. LYON*   A. H. HOWLAND*   W. L. LOM*

This paper looks briefly at various proposed techniques to control smoke and gaseous emissions from a diesel engine, and examines in detail one particular method which shows promise—the conversion of auto-diesels to dual fuel operation using liquid petroleum gas as a secondary fuel. Experimental data are presented comparing dual fuel with normal operation on a variety of engines. Direct injection diesels seem to respond excellently to dual fuelling, and this method can be used to control smoke and/or emissions of oxides of nitrogen.

INTRODUCTION

AIR POLLUTION is perhaps the biggest single problem facing the automotive industry in the U.S.A. and is fast becoming of concern in Europe. One of the advantages of the diesel engine is its relatively minor contribution to air pollution. Both unburnt hydrocarbons and carbon monoxide, currently the focus of legislative attack in the U.S.A. and Europe, are so low that until recently little attention has been paid to these emissions from the diesel. However, the California State Air Resources Board has now agreed to a test cycle (1)† and laid down emission standards for new vehicles to become effective in 1973 and further improved in 1975. These standards are designed to bring the diesel engine in line with the proposed standards for gasoline engines.

In addition to gaseous emissions, diesel engines have their own special problem of exhaust smoke which does much to tarnish their good name. Smoke is offensive and unpleasant in nature and the reduction in visibility it presents to overtaking vehicles is a road safety problem. It must be admitted, however, that much of the public concern over diesel smoke stems from the ease with which it can be recognized compared with gaseous emissions.

Smoke legislation and techniques for reducing smoke

Most countries have some form of legislation to deal with smoke and the indications are that this will tighten considerably in the future. In the U.K., for example, the Ministry of Transport will shortly lay regulations to make the British Standard (B.S. AU 141) obligatory for all diesel engines installed in new vehicles.

The MS. of this paper was received at the Institution on 20th May 1971 and accepted for publication on 7th July 1971. 22
* Esso Research Centre, Abingdon, Berkshire.
† References are given in Appendix 126.1.

The findings of a recent survey (2) on a large number of suggested anti-smoke devices gave a pessimistic outlook to the whole question of effective smoke control since few of the devices tested showed any promise.

Engine derating is an obvious means of smoke control but at the penalty of reduced power, which is in direct opposition to the manufacturer's desire to increase the power to weight ratio. Anti-smoke additives do reduce smoke (3) to some extent, but are costly. At the present time, the use of additives is prohibited in the British Standards (B.S. AU 141) test.

Turbo-charging, although primarily used at present to increase the power output of an engine, may be used in the future as a means of reducing smoke. The problems of turbo-charging are reliability and the difficulty of tuning to give good performance over the entire load–speed range. It appears also that turbo-charging may result in increased quantities of nitric oxide in the exhaust (4)–(6).

It is abundantly clear that at the present time no single approach is ideal and much work is needed in all areas to improve performance. The way is still open for a cheap solution.

Legislation on gaseous emissions and techniques to reduce them

The California legislation limits for the diesel engine are quoted in Table 126.1, which also shows how these limits compare with those suggested for the spark ignition engine using rough conversion factors (g/mile to g/b.h.p. h) derived by Bascom and Hass (1).

Obviously, although no attempt has been made to separate nitric oxide and hydrocarbon emissions from the diesel engine the object is to bring diesel emissions into line with those from the gasoline engine.

Considerable effort and a fair degree of progress has been made on the spark ignition side, but there still

Table 126.1. Comparison of predicted diesel and spark ignition emission limits

Mode of operation	Brake specific emissions, g/b.h.p. h			
	CO	HC	NO_2	$HC+NO_2$
1973–74 diesel	40	—	—	16
1975 diesel	25	—	—	5
1972 spark ignition	50	4·8	8·7	13·5
1975 spark ignition	25	1·1	2·0	3·1
1980 spark ignition	10	0·5	0·9	1·4

remains much work to do to meet the levels predicted for 1980. As will be shown later, legislation limits on carbon monoxide emission from the diesel engine appear to pose little problem, whereas nitric oxide emissions must be drastically reduced to meet 1975 levels.

Several methods have been suggested to reduce nitric oxide emissions, including air throttling, exhaust gas recycle, water injection, and injection retarding (**4**)–(**6**). Apart from water injection, which appears to adversely affect hydrocarbon emissions, all the others have a detrimental effect on smoke. Furthermore present-day catalysts for nitric oxide destruction are only really effective in reducing atmospheres and are not suitable for the oxidizing atmosphere of a diesel exhaust.

There is no doubt that diesel exhaust emissions, like those from the gasoline engine, will become more and more important in the future. In the long run, the emissions of oxides of nitrogen may pose a greater threat to the survival of the diesel engine than exhaust smoke. Certainly, it is now no longer sufficient to assess the potential of any pollution control device—whether it be to reduce nitric oxide or smoke—without at the same time determining its effect on the other pollutants.

LPG dual fuelling

In 1964, the Vienna transport authorities initiated trials on the use of liquid petroleum gas (LPG) as a secondary fuel in their diesel buses (**7**). To operate on LPG, the diesel injection pump is partially derated and the gas is introduced into the air intake in proportions necessary to restore the original power. These trials have been so successful that it is planned to convert the entire Vienna transport diesels to LPG dual fuelling.

Several advantages, including significant reductions in smoke, are claimed for dual fuelling compared with operation on pure diesel fuel. Little information has been published, however, on the effect of dual fuel operation on gaseous emissions, and no comprehensive study appears to have been carried out to evaluate this technique thoroughly.

SCOPE OF EXPERIMENTAL STUDIES AT ESSO RESEARCH CENTRE

While dual fuel (LPG) operation appears to offer one route for overcoming the smoke emission problems of diesel engines, a number of important questions remain unanswered. These are:

(1) How significant are the reductions involved?
(2) How does this technique affect gaseous pollutants, e.g. oxides of nitrogen (NO_x) and hydrocarbons?
(3) Is the LPG composition important?
(4) How does this technique affect engine performance?

In 1969 a comprehensive experimental project was started at Esso Research Centre to obtain the answers to these questions. Three lines of attack have been used in this project. First, some preliminary studies were carried out on a single-cylinder Petter AV-1 laboratory engine (direct and indirect injection). Second, to obtain practical experience of dual fuel operation under actual road operating conditions a Morris 3·8-litre diesel van was converted into a 'travelling laboratory' equipped to monitor smoke emission continuously and the important engine variables (fuel consumption, exhaust temperature, and engine speed). This van offered a means of evaluating some of the commercial equipment currently available for converting diesels to dual fuel and has been invaluable in pinpointing practical problems that are likely to occur.

The third line of attack consisted of a detailed test bed study of a Saurer 165 b.h.p. dual fuel diesel engine obtained from Steyr–Daimler–Puch, Austria. This engine is identical with those used in the Vienna municipal transport trials referred to earlier in this paper.

Before discussing the experimental work, there are a number of important points that must be emphasized which can affect dual fuel operation and relate to engine efficiency, exhaust emissions, and knock.

SOME GENERAL FACTORS INFLUENCING DUAL FUEL OPERATION

Thermal efficiency

Under normal single fuel operation, diesel fuel is injected into the compressed air in the combustion chamber as a fine liquid droplet spray. After a short ignition delay the droplets ignite and a heterogeneous flame develops around the droplets. The heterogeneity of the fuel–air mixture has a significant effect on the combustion process in the diesel engine and is the direct cause of smoke (**8**). Fuel concentrations in the combustion chamber range from 100 per cent to zero in regions of pure air remote from the droplets. Under low load conditions with a small diesel injection spray the regions containing no fuel will be greater than under heavy load where most of the chamber will be filled with burning droplets.

In dual fuel operation, gas is normally injected into the air intake to form an air–gas mixture below its limits of inflammability. This means that the air–gas mixture can only burn efficiently in regions close to the burning oil droplets as gas mixture below the limits of inflammability cannot propagate a flame. Under low load conditions, where relatively few burning diesel fuel droplets are available, there can therefore be regions where the added gaseous fuel will not burn and will pass unchanged to the exhaust.

This effect, apart from increasing exhaust emissions, also reduces engine efficiency, as has been demonstrated in studies on dual fuelling using natural gas (**9**). These studies show that with a small diesel charge little of the natural gas burns completely but passes unchanged into the exhaust. At a liquid fuel/air mass ratio of about 0·033 (50 per cent of stoichiometric), however, nearly all the natural gas fed into the air intake burned completely, regardless of its concentration. As most diesels operate at a maximum of 80 per cent stoichiometric this means that to avoid inefficient combustion when the added gas is below the limits of inflammability the liquid fuel/air ratio should be greater than 60 per cent of its maximum value.

This exemplifies an important consideration which will be referred to later in this paper when the engine studies are discussed. Operation of a diesel engine under dual fuel conditions much below 60 per cent of maximum load is likely to be unsatisfactory owing to a loss in thermal efficiency due to poor fuel utilization, and an increase in the quantities of unburnt hydrocarbons, aldehydes, and carbon monoxide in the exhaust.

Knock

A problem which can be of considerable importance in the diesel is that of uncontrolled combustion, i.e. knock. Engines are said to knock when the rate of pressure rise following ignition increases abruptly to very high values—so high, in fact, that continued operation can result in engine damage. Manufacturers therefore rate their engines so that this phenomenon is not encountered over the normal operating range.

When an additional gaseous fuel is introduced to the diesel, care has obviously got to be taken to avoid knock. The type of knock encountered, however, depends on the concentration of gas in the air intake and whether it is above or below the limits of inflammability.

As indicated earlier, below the limits of inflammability large quantities of liquid charge are necessary for efficient gaseous combustion. In this case the burning gas simply augments the normally rapid pressure rise due to diesel fuel combustion. If the rate of pressure rise from a diesel fuel charge is close to the knock limit of the engine, then the contribution from the burning gas may well carry it over the limit. In this type of knock (referred to here as 'diesel type' knock) the magnitude of the diesel charge as well as the gas concentration is important.

Table 126.2 gives details of standard limits of inflammability and the same limits determined under engine conditions for a number of gaseous fuels.

At gas concentrations above the limits of inflammability a very small quantity of liquid diesel fuel charge may be sufficient to trigger the auto-ignition of the entire premixed charge. When knock is encountered it will depend more on the octane quality of the gaseous fuel than on the size of the liquid fuel charge. Under these circumstances the chemical nature of the gaseous fuel will be important and Table 126.3 gives details of the highest useful compression ratios that can be used for a number of fuels (**10**).

Commercial diesel engines normally operate at compression ratios equal to or greater than 16:1. Obviously, therefore, only methane can be used in these engines at gas concentrations above its limit of inflammability.

Table 126.2. Comparison of limits of inflammability

Gas	Lower limit of inflammability by weight in air, per cent	
	Standard limits	Engine limits
Methane	3·0	2·1
Propane	3·3	2·8
Butane	3·7	1·6
Propylene	3·2	—
Benzene vapour	3·7	2·2
Iso-octane vapour	4·3	2·4

EXPERIMENTAL STUDIES ON THE PETTER AV-1 AND THE MORRIS DIESEL VAN

Petter AV-1 studies

Facilities to monitor and control engine variables—e.g. engine speed, water coolant temperatures, fuel–air flows, oil pressures, etc.—were installed on a test bed single-cylinder Petter AV-1 engine. Smoke density measurements were obtained from a Hartridge–BP smokemeter. To ensure reliability of the smoke data, facilities were incorporated to switch rapidly from one gaseous fuel to another without engine shutdown. Gaseous fuels used were metered via a rotameter into the air intake. Combustion pressure measurements were carried out via a piezoelectric transducer in the combustion chamber wall—the pressure signals being fed to an oscilloscope and photographed using a Polaroid camera.

The Petter AV-1 can be operated with either of two different combustion chamber designs—indirect injection (precombustion chamber) or direct injection. Experimental measurements were carried out on both systems.

Dual fuelling in indirect injection Petter AV-1

An immediate increase in engine noise was noted when any of the three main LPG components (propane, butane,

Table 126.3. Highest useful compression ratios of some hydrocarbons

Fuel	Highest useful compression ratio
Methane	17·6
Propane	12·1
Butane	10·4
Iso-octane	10·9

propylene) were added to the air intake of this engine. Pronounced knocking was observed with each gas at concentrations greater than 10–20 per cent of the total fuel depending on load. Under heavy load the condition was most severe.

Apart from knocking, little effect on exhaust smoke or exhaust gas temperature was observed under part-load conditions. However, at high load a reduction in exhaust smoke and exhaust temperature was observed. Butane seemed to be the most satisfactory for reducing smoke, followed by propylene, with propane least. A slight improvement in thermal efficiency was observed with all the gases under study, but this became less pronounced under severe knocking.

Although the gas concentrations in these experiments were less than 1 per cent of the air intake and well below their limits of inflammability, severe knocking was obtained. This effect would therefore seem to be due to a 'diesel type' knock as opposed to a 'gasoline' type. Attempts to eliminate knocking, or at least to extend the knock-free range, met with little success. Retarding injection timing brought only marginal improvements, whereas improving cetane quality by additives showed none whatsoever.

These studies imply that dual fuelling in this engine is not satisfactory because of knock, and raises a question mark over the use of this technique in precombustion or high swirl engines.

Dual fuelling in direct injection Petter AV-1

In the direct injection version of the Petter AV-1, gas concentrations in excess of 50 per cent of the total fuel could be added before any detectable change in engine sound occurred. However, performance results confirmed that dual fuelling is only really thermally efficient under heavy load where efficiency advantages are shown over normal single fuel operation. Fig. 126.1 shows that at the maximum rated power output only methane failed to show any improvement in efficiency.

Under part-load operation a larger heat requirement was needed with dual fuel firing, which became more severe at high gas concentrations in the rough running region. Butane appeared to be utilized more efficiently under part load than propane or propylene, with methane least.

The effect of dual fuel on exhaust temperature appears to mirror engine thermal efficiency. Generally, exhaust gas temperatures under part load remain unaffected during normal engine performance and increase in the rough running region. Only at the maximum rated power were exhaust temperatures reduced, and again methane differed from the LPG components.

Direct injection Petter AV-1 under rough running conditions

Combustion pressure studies were carried out in the direct injection version of the Petter with gas concentrations in the region of 50–70 per cent of the total fuel, i.e. in the

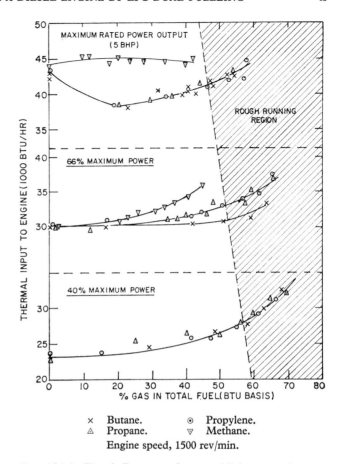

× Butane.   ⊙ Propylene.
△ Propane. ▽ Methane.
Engine speed, 1500 rev/min.

Fig. 126.1. The influence of gas addition on thermal efficiency in Petter AV-1 (direct injection) engine

rough running region. Fig. 126.2 shows combustion pressure profiles obtained with propane, butane, and propylene dual fuel firing compared with the normal diesel fuel curve. The gases all show an increase in ignition delay period. Propane shows the greatest increase, followed by propylene, with butane least. Butane is unusual in that it gives considerably higher maximum pressures while the other gases show maximum pressures lower than that of the original diesel.

Table 126.4 gives the ignition delay period and maximum pressure records for several gas concentrations in the rough running region. Notice that the butane concentration in the intake air exceeds its inflammability limit, whereas propane has not quite reached the limit. The engine limits for propylene are not yet known but they are not expected to differ greatly from propane.

The high maximum pressures recorded with butane result from the rapid combustion of the air–gas mixture immediately the diesel charge ignites. The observed knocking in this case can probably be attributed to a 'gasoline engine' type knock.

The rough running with propane and propylene is not as easy to explain. The engine become difficult to maintain at constant speed and surging is observed. The pressure

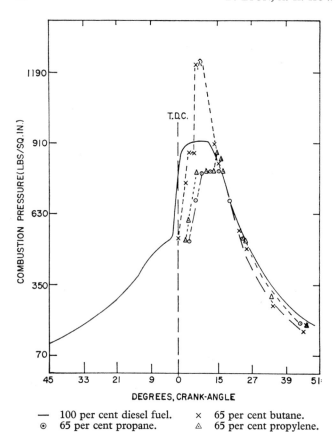

— 100 per cent diesel fuel. × 65 per cent butane.
⊙ 65 per cent propane. △ 65 per cent propylene.
Engine conditions: speed 1500 rev/min; b.h.p. 5·0.

Fig. 126.2. The influence of gas addition on combustion chamber pressure profile of Petter AV-1 (direct injection)

profiles exhibit large variations even between adjacent engine firings, which may be due to the extreme lengthening of the ignition delay observed with these gases.

Table 126.5 shows the influence of diesel injection timing on the onset of rough running with dual fuelling using propane.

In a normal diesel engine, advancing injection timing gives higher maximum pressures while retarding gives

Table 126.4. Combustion pressure profile studies

Gaseous fuel	Percentage gaseous fuel in air, wt/wt	Engine limit of inflammability, wt/wt	Ignition delay, degrees crank angle	Maximum pressure, lbf/in²
None .	0	—	19·6	950
Butane .	1·9	1·6	21·5	1170
	2·4	1·6	22·0	1210
Propane .	1·8	2·8	24·0	950
	2·2	2·8	25·5	745
	2·5	2·8	28·4	600–710
Propylene .	1·8	—	22·0	950
	2·3	—	25·0	745

Air/fuel ratio = 28 : 1.   Injection timing = 22° b.t.d.c.

Table 126.5. Effect of injection timing on rough running with propane gas

Percentage gas in total fuel	Injection timing, degrees b.t.d.c.	Ignition, degrees a.t.d.c.	Ignition delay, degrees crank angle	Maximum pressure, lbf/in²
0	22*	−2·4	19·6	950
71	22	+6·4	28·4	600–710
0	26	−4·7	21·3	975
64	26	+2·4	28·4	815
0	18	+2·4	20·4	780
75	18	+14	32	450–500

* Normal injection timing.

lower. This is also the case with propane dual fuelling when the onset of rough running occurs at slightly lower concentrations with advanced timing and higher with retardation. In the case of retarded injection the engine noise fell away dramatically with gas injection, and it became as quiet as a gasoline engine. This is not surprising since ignition was occurring considerably after top-dead-centre and the engine was essentially operating at a much lower compression ratio. In this mode of operation, however, the efficiency was less than the base diesel case.

Studies on the Morris diesel van

The performance of dual fuelling has been assessed under actual road conditions in the Morris diesel 'travelling laboratory' mentioned earlier in this paper. Although this vehicle is maintained in good condition it is extremely smoky under maximum load, registering 90 Hartridge smoke units over the entire speed range. It therefore represents a stiff test of the potential of dual fuelling to improve this magnitude of smoke emission.

The van is primarily used to evaluate proprietary conversion kits, but studies have also been carried out on gas addition on engine performance under a variety of load and speed conditions. To do this, hill-climbing studies were carried out at several different engine speeds. This allowed the identification of optimum gas concentrations to give smoke reduction without encountering knock or rough running.

The results confirm the findings on the direct injection version of the Petter engine, i.e. that gas composition has little effect on dual fuel performance. Typical results for propane injection are shown in Fig. 126.3, which gives smoke and speed measurements over the hill climb for various percentages of gas in the total fuel. The effect of propane dual fuelling on thermal efficiency also confirmed the Petter results. At some engine speeds, for example, a reduction in the amount of total fuel required was noted (∼10 per cent).

One series of experiments carried out on the Morris van, but not tried on the Petter, compared the anti-smoke

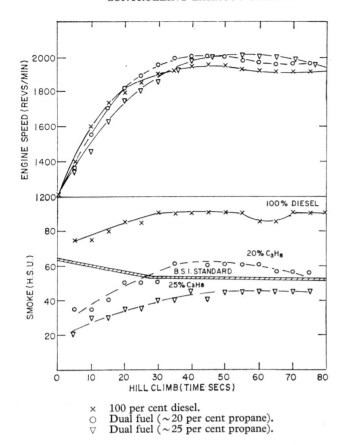

× 100 per cent diesel.
○ Dual fuel (~20 per cent propane).
▽ Dual fuel (~25 per cent propane).

Fig. 126.3. Hill climbing test illustrates how dual fuelling reduces smoke

performance of fuel LPG with alkali earth additive treatment of the diesel fuel (see Table 126.6).

These experiments showed a substantial improvement in smoke reduction of dual fuel LPG over additive treatment. One very interesting implication of these results, however, is that additive treatment can apparently be combined with the dual fuel approach to produce smoke reductions substantially better than either technique used alone. Further investigation of this possibility could be well worth pursuing to determine if it is of general applicability.

Table 126.6. Effect of additive treatment on exhaust smoke from Morris diesel van

Mode of operation	Average exhaust smoke, Hsu	Percentage smoke reduction
Diesel fuel only	76	—
Additive treatment		
0·2 wt% $BaCO_3$ in diesel fuel .	63	} 14
0·2 wt% $CaCO_3$ in diesel fuel .	68	
Butane dual fuel (25%) . .	46	40
Butane dual fuel +0·2 wt% $BaCO_3$ in diesel fuel	28	63

DESCRIPTION AND TEST BED DETAILS OF SAURER DIESEL

General description

This engine is a four-stroke direct injection diesel which can operate on pure diesel fuel or, with the flick of a switch, in a dual fuelling mode with LPG as the secondary fuel. The switch has a twin function: first, it operates a mechanism whereby the diesel fuel maximum supply is downrated by 25 per cent, and second, it allows LPG to be introduced into the air supply in proportions necessary to restore the original power. As designed it is not possible to operate the engine in the dual fuel mode without first derating the diesel injection, and therefore the gas is used to effect a reduction in smoke emission and not as a power booster.

A few of the important technical details of this engine are listed in Table 126.7.

Binary fuelling system

A schematic diagram of the LPG dual fuelling system is shown in Fig. 126.4.

The LPG is stored in a separate fuel tank designed to withstand the high pressures involved. The tank is normally filled to about 85 per cent of its capacity to allow for expansion due to fluctuations in ambient temperature.

The liquid gas is fed to a vaporizer–regulator via a solenoid valve (3) which remains closed when operating on pure diesel fuel or when the engine is not running, but which is open when operating on the twin fuels.

The vaporizer–regulator (1) is a two-stage regulator which converts the high and variable pressurized liquid gas to a constant low pressure (~5 lbf/in² (gauge)) and thence supplies the gas into the air manifold in proportions dependent on vacuum depression. The vaporizer unit is heated by the water coolant system.

A check valve (10), which is operated by the control rod of the diesel fuel injection pump, is incorporated as a safety feature and closes off the gas supply when the engine is not operating. Liquid petroleum gas is finally tuned into the air intake by a gas quantity regulator (7) which is controlled by the accelerator pedal.

Test bed installation

Facilities to monitor and control engine variables such as engine speed, water coolant temperatures, fuel flows, oil

Table 126.7. Saurer 2FU-DG, technical details

Maximum b.h.p.	165
Number of cylinders	6
Bore	120 mm
Stroke	140 mm
Cylinder capacity	9500 cm³
Compression ratio	17:1
Arrangement of cylinders	in-line horizontal
Injection pump	Bosch
Injector nozzle type	Bosch spindle
Injection timing	22° b.t.d.c.
Adjusting range of injection adjuster	12° at the crankshaft

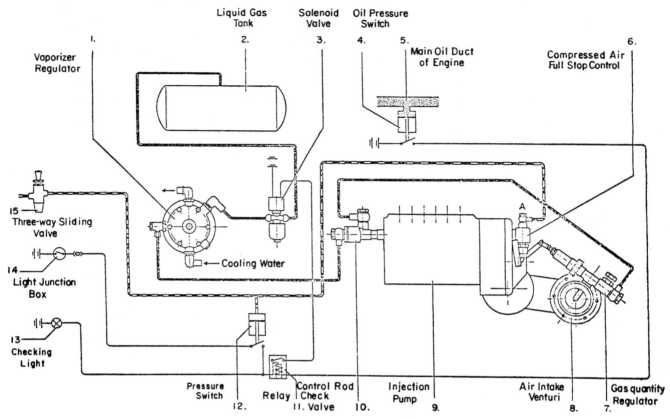

Fig. 126.4. Schematic diagram of Saurer dual fuelling system

pressures, etc., were installed on the test bed. A photograph of the test bed installation, which shows the LPG supply system and also the exhaust gas analytical equipment, is shown in Fig. 126.5. An orifice plate flowmeter was used to measure the air flow through the engine and the power generated was measured on a Froude dynamometer.

Exhaust hydrocarbons were measured using a specially heated sample probe which feeds the exhaust gas to a heated flame ionization detector unit using exhaust back pressure to push the sample gas through the unit. Initial experiments confirmed that the hydrocarbon signal was insensitive to O_2, CO_2, and SO_2 in the proportions normally found in the diesel exhaust. The unit was found to give stable maximum signals provided the entire unit and probe were kept at a temperature between 195 and 300°C.

Non-dispersive infrared equipment was used to measure carbon monoxide and nitric oxide concentrations. Smoke density measurements were obtained continuously from a Hartridge–BP smokemeter.

Low sulphur (0·4 wt%) gas oil was used as the primary fuel throughout the experiment. Commercial grade butane and propane were separately tested as the secondary fuel. Both grades were 95 per cent pure, although the butane contained a mixture of iso-butane and n-butane in the ratio of 2:3.

Both primary and secondary fuel consumptions were measured volumetrically. The LPG was measured in a high-pressure burette. To eliminate boiling from the liquid gas surface, the system was pressurized slightly above ambient pressure using nitrogen gas. Liquid gas density measurements were obtained using a pressure vessel and hydrometer.

ENGINE PERFORMANCE UNDER MAXIMUM POWER

Power and exhaust smoke

The maximum power performance curves are shown in Fig. 126.6 for operation both on pure diesel and in the dual fuel mode. The lines represent the results supplied by the engine manufacturer while the experimental points are the results obtained at the Esso Research Centre. Notice the close agreement obtained in both studies and in particular the close matching of the power curves over the entire speed range for both operational modes.

The agreement on smoke emission is surprisingly good in view of the fact that Bosch units had to be converted into the equivalent in Hartridge smoke units before a comparison could be made. The smoke produced by this engine while operating on pure diesel fails to meet the levels laid down by B.S. AU 141 at all speeds greater than 1400 rev/min, whereas in the dual fuel mode a clear pass at all speeds is obtained.

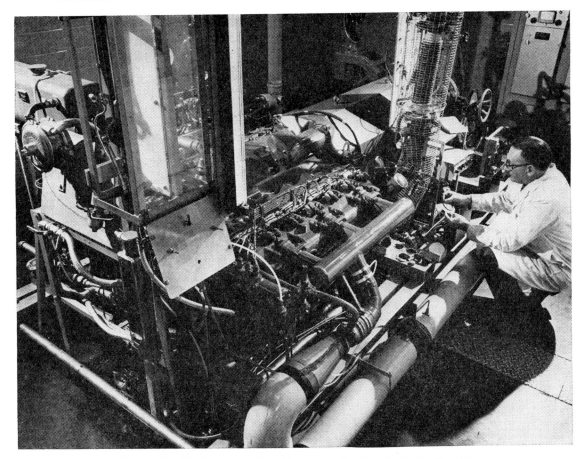

Fig. 126.5. Photograph of the Saurer engine test bed installation

Fuel efficiency and exhaust temperature

The b.s.f.c. and exhaust temperatures at maximum power compared for pure diesel, butane dual fuel, and propane dual fuel are given in Fig. 126.7. In general, dual fuelling results in a greater fuel economy, with butane showing the greatest savings. This is in line with the results obtained in the Petter AV-1 engine. Also consistent with the previous results are the lower exhaust temperatures recorded.

EXHAUST EMISSIONS ON PURE DIESEL FUEL

Carbon monoxide

Fig. 126.8 gives a contour map of carbon monoxide concentrations found in the exhaust of the Saurer engine operating on pure diesel fuel as a function of brake mean effective pressure (b.m.e.p.) and engine speed. The emissions levels, ranging from 200 to 1500 p.p.m., are typical levels obtained from a well maintained engine. The maximum emission levels are obtained at approximately maximum torque and typically drop to a minimum of 50 per cent b.m.e.p. before slightly increasing again under extremely light loads.

Nitric oxide

Fig. 126.9 shows the results obtained from nitric oxide emissions, which again are typical of the levels obtained for a direct injection engine. In general, nitric oxide concentrations in the exhaust increase more or less linearly with b.m.e.p., although at low speeds a maximum in emission is obtained at approximately 75 per cent maximum b.m.e.p. This pattern of a decrease in nitric oxide emission at high fuel/air ratios has been observed by other workers (**4**)–(**6**).

Hydrocarbons

The contour lines for hydrocarbon emissions are given in Fig. 126.10 and again the maximum levels are similar to those reported by previous workers. In this engine maximum hydrocarbon emissions appear at 75 per cent of maximum speed.

Smoke

The contour profile of exhaust smoke in Fig. 126.11 clearly shows the rapidity with which smoke emission increases under heavy load. To bring the smoke emission from this engine within the B.S.I. limits a derating of approximately 6 per cent is required, and to bring the smoke level in line with that produced in the dual fuel mode a derating of at least 15 per cent is required.

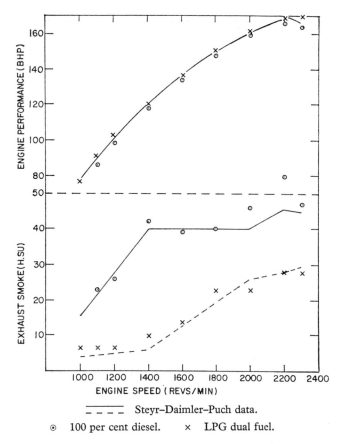

Fig. 126.6. Saurer diesel engine maximum performance curves for power and smoke

– – – – Steyr–Daimler–Puch data.
⊙ 100 per cent diesel.   × LPG dual fuel.

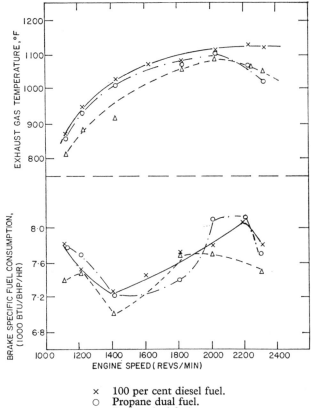

× 100 per cent diesel fuel.
○ Propane dual fuel.
△ Butane dual fuel.

Fig. 126.7. Fuel efficiency and exhaust temperature at maximum power in Saurer engine

ENGINE OPERATION IN DUAL FUEL MODE

Tests have been carried out using both commercial butane and commercial propane as the secondary fuel in the Saurer engine without encountering any problems throughout the entire load–speed range. This is readily understood since the maximum percentage of gaseous fuel in the air charge is 1·3 per cent, well below the inflammability limits of both gases.

In the dual fuel mode, gas was not admitted to the engine at idle, but above this gas was added throughout the entire load–speed range. Since smoke is really only a problem under heavy load it could be argued that gas addition under part load does not serve any useful purpose. Indeed, the results from the Petter studies indicated poor utilization of the gaseous fuel in this mode and no decrease in smoke emission.

By adjustment of the gas quantity regulator it is possible to tune gas into the engine only under heavy load. During this investigation both methods of gas addition were studied. The means of introducing gas throughout the entire load is described as *coarse control*, while tuning the gas into the engine only under heavy load is described as *fine control*. To illustrate the difference between fine and coarse control, Fig. 126.12 shows a plot of fuel consumption as a function of b.m.e.p. for a given engine speed. At maximum b.m.e.p. the gas constitutes 25 per cent of the total fuel in both cases. However, in the fine control method the gas concentration falls to zero just above 60 per cent b.m.e.p. and below this the engine operates entirely on diesel fuel. Operation under coarse control allows only a slight decrease in gas supply with decreasing b.m.e.p., and indeed the proportion of gas to diesel fuel rises to about 40 per cent at 25 per cent b.m.e.p. before falling off at lighter loads. Notice that, consistent with the results obtained in the Petter engine, the fuel consumption under part load is less efficient with coarse control than operation with pure diesel fuel. Dual fuelling using the fine control method makes it possible to obtain the fuel efficiency improvements under heavy load while avoiding the poor efficiency under light load.

Smoke

The smoke emission profiles in Fig. 126.13 show that, in terms of smoke emission, little is gained by operation in the coarse control mode. Indeed, the smoke emission at light loads is somewhat worse than on pure diesel fuel.

Nitric oxide

The concentration of nitric oxide in the exhaust was independent of the nature of fuelling.

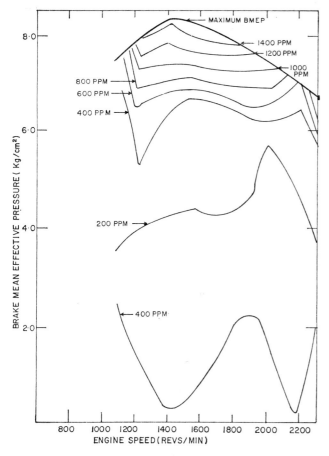

PPM = parts per million by volume.

Fig. 126.8. Saurer engine carbon monoxide emissions on pure diesel fuel

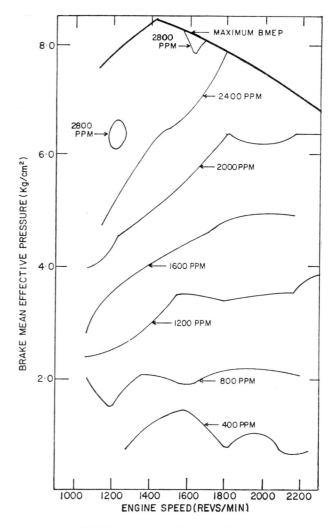

PPM = parts per million by volume.

Fig. 126.9. Saurer engine, nitric oxide emissions on pure diesel fuel

Hydrocarbons

Dual fuelling under light load conditions resulted in large amounts of unburnt hydrocarbons in the exhaust, explaining the fall-off in fuel efficiency. This is best illustrated in Fig. 126.14 which shows exhaust hydrocarbons as a function of b.m.e.p. for a given engine speed. Notice that provided the gas is admitted to the engine only under heavy load, as it is under fine control, the exhaust hydrocarbons are comparable to pure diesel operation. Similar to the results found in the Petter direct injection engine, butane, the more reactive of the two gases, appeared to be better utilized under light load conditions and gave less unburnt hydrocarbons in the exhaust.

Carbon monoxide

The trends reported above for unburnt hydrocarbons were similarly reflected in carbon monoxide emissions. However, the increase in carbon monoxide was less dramatic under light load, indicating perhaps that little oxidation of the gaseous fuel takes place remote from the diesel fuel droplets.

CALIFORNIA TEST CYCLE STUDIES

The previous discussion on gaseous emissions can best be summed up by reporting the results in the form of the California test cycle (Table 126.8).

The gaseous emission levels produced by LPG dual fuelling in the fine control match those produced by operation on pure diesel fuel with the additional benefit of reduced smoke. However, the benefit of reduced smoke from the coarse control mode is offset by the increase in both carbon monoxide and hydrocarbon emissions. Comparing the results for propane and butane, once again the greater reactivity of butane to oxidation and combustion compared with propane is reflected in the lower hydrocarbon but higher carbon monoxide emissions obtained both in the fine and coarse control operation modes.

In terms of legislation limits, Table 126.8 shows that this engine operation on pure diesel fuel fails to meet the

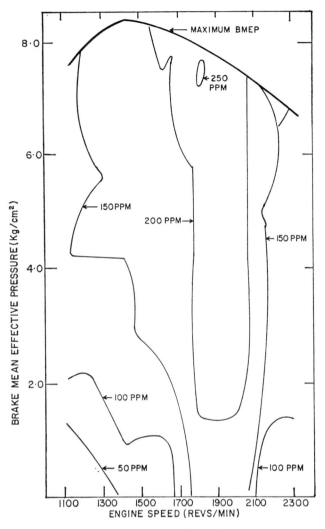

PPM = parts per million carbon.

Fig. 126.10. Saurer engine hydrocarbon emissions on pure diesel fuel

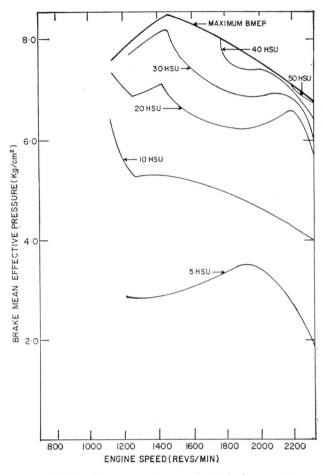

Fig. 126.11. Saurer engine smoke emission on pure diesel fuel

B.S.I. limits on smoke and also the 1973 California standards on total oxides of nitrogen and hydrocarbon emissions. Operation on dual fuel gives a clear pass on smoke but a similar fail on gaseous emissions. However, it will be shown later in this study that it is possible, by a slight retard of diesel injection timing, to sacrifice a small amount of smoke reduction from dual fuelling to obtain a pass on both standards.

Comparison of 'California' brake specific fuel consumption

One obvious question that has still to be answered is: 'How much gas is needed to run a dual fuel engine?' This is particularly relevant to the fine control where gas is only used under heavy load. Obviously, an exact answer will depend on the operating cycle of the vehicles in actual service, and no one cycle can be considered to accurately

Table 126.8. Comparison of gaseous emissions involving California test cycle

Mode of operation	California test emission results, g/b.h.p. h				Maximum smoke
	CO	NO_2	HC	NO_2+HC	
100% diesel	5·2	16·8	0·6	17·4	57
Propane dual fuel (fine)	5·7	16·1	0·9	17·0	28
Propane dual fuel (coarse)	16·0	15·5	6·7	22·2	28
Butane dual fuel (fine)	5·9	16·4	0·7	17·1	26
Butane dual fuel (coarse)	17·4	16·3	5·7	22·0	28

CONTROLLING EXHAUST EMISSIONS FROM A DIESEL ENGINE BY LPG DUAL FUELLING

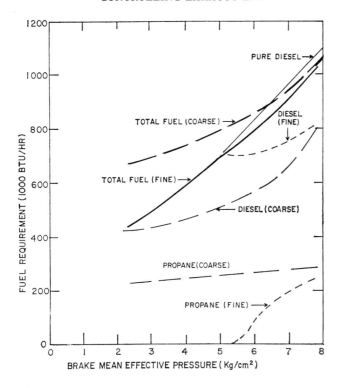

Engine speed, 1800 rev/min.

Fig. 126.12. Fuel requirements for propane dual fuelling in both fine and coarse control compared with pure diesel fuel

Table 126.9. 'California' brake specific fuel consumption on Saurer engine

Mode of operation	'California' b.s.f.c., 1000 Btu/b.h.p. h	Percentage gas of total fuel consumed
100 per cent diesel	9·4	0
Propane dual fuel (fine)	9·3	14
Propane dual fuel (coarse)	9·3	35·5
Butane dual fuel (fine)	9·2	12
Butane dual fuel (coarse)	9·4	38

reflect conditions on the road. However, the California cycle, incorporating as it does 11 distinct operation modes, is an attempt to reproduce on test bed the emissions levels expected from a diesel on the road.

In order to obtain some estimate of how much gas is needed, fuel consumptions have been computed using a similar equation and weighting factors as those used in the California cycle for gaseous emissions, i.e.

$$\text{B.S.F.C.} = \frac{\sum F_n \times WF_n}{\sum BHP_n \times WF_n}$$

where B.S.F.C., the brake specific fuel consumption, is in Btu/b.h.p. h; F_n, the fuel consumption for each mode, is in Btu/h; and where BHP_n is the brake horsepower in each mode and WF_n the weighting factor for each mode.

Table 126.9 shows 'California' b.s.f.c. values for the Saurer engine operating on pure diesel fuel and in the various dual fuel modes, together with the percentage gas used in each case. This shows that the gas consumption varies from 12 to 14 per cent in the fine control to 35 to 38 per cent in the coarse control.

The results expressed in this way also put into perspective the fuel efficiencies obtained from dual fuelling in fine and coarse control. In the coarse control, the efficiency improvements obtained under heavy load are more or less balanced by the losses under light load. The improvement in efficiency using fine control is really only significant with butane dual fuelling. Obviously, depending on how strict a control on gaseous emissions is required, it is possible to vary the gas consumption anywhere between 12 and 38 per cent.

RETARDING INJECTION

Retarding static injection timing

Perhaps the most acceptable method of reducing nitric

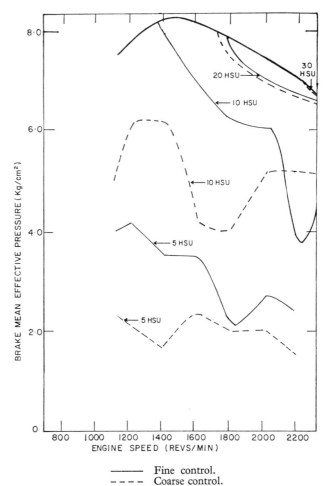

——— Fine control.
- - - - Coarse control.

Fig. 126.13. Comparison of smoke emission with fine and coarse control dual fuelling

oxide emissions open to the industry at the present time is retarding of the diesel injection timing. Unfortunately, as mentioned earlier, this seems to incur the penalty of increased smoke emission together with a loss in power output and fuel efficiency. In order to confirm this observation and also to see how dual fuelling would react to retarding diesel injection, a series of experiments were carried out using the California cycle.

The Saurer engine has a 12° speed advance mechanism which was not altered during the tests. The only variable that was changed was the static injection timing, and the values presented below for injection timing refer to the static timing. No attempt was made to correct for power losses or to increase the maximum fuel consumption whether it be gas or diesel fuel. Comparison tests were carried out on pure diesel fuel and propane and butane dual fuelling in the fine control.

Power and fuel efficiency

The effect of retarding static injection timing on power and thermal efficiency for the three modes of operation is shown in Fig. 126.15. The maximum power on pure diesel fuel fell off almost immediately with injection retarding

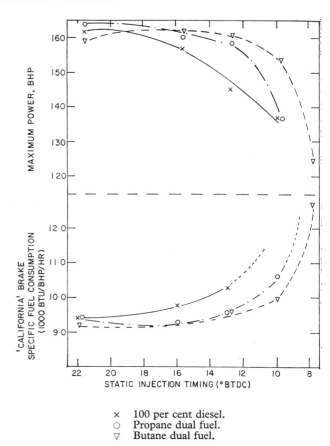

× 100 per cent diesel.
○ Propane dual fuel.
▽ Butane dual fuel.

Fig. 126.15. Effect of retarding static injection timing on maximum power and fuel efficiency

and the engine became more or less inoperable on this fuel below 13° b.t.d.c. static timing. The effect on dual fuel operation was surprisingly different both in terms of power output and fuel efficiency. No loss in power output was obtained using either butane or propane with as much as 9° retarding from the manufacturer's optimum value. With further retarding, propane became less efficient than butane, which was still operable albeit extremely inefficiently at 8° b.t.d.c.

The immediate question is: 'Why the difference?' Perhaps the simplest explanation is that the time required to introduce the fuel into the engine with dual fuelling is less, and hence a shorter combustion time is required. Essentially dual fuelling shortens the diesel injection period by 25 per cent by replacing the last 25 per cent of the diesel charge with gas which is added at the beginning.

As injection timing is retarded, less and less time is available for completion of combustion, and hence the duration of the fuel injection period becomes more and more critical. Any method of reducing this time period should show advantages. Consistent with an improvement due to a shortening of the combustion time are the better results obtained with butane, the more combustible of the two gaseous fuels.

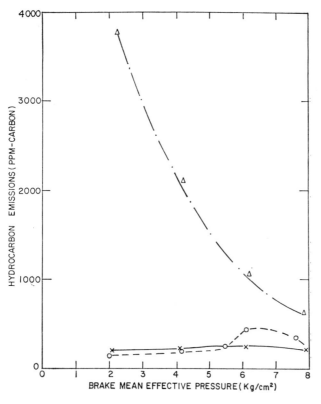

× 100 per cent diesel fuel.
○ Propane dual fuel (fine control).
△ Propane dual fuel (coarse control).
Engine speed, 1800 rev/min.

Fig. 126.14. Dual fuelling results in high hydrocarbon emissions if not restricted to heavy load

Exhaust emissions

The effect of retarding injection timing on gaseous emissions and smoke is shown in Fig. 126.16. Confirming previous observations, smoke increases with injection retarding. Carbon monoxide emissions are initially little affected by retarding injection but eventually are progressively worsened. Emissions of oxides of nitrogen decrease almost linearly with injection timing, whereas hydrocarbons are only slightly increased in the initial stages. This results in a decrease in the sum of these two pollutants as indicated in Fig. 126.16. At 8° b.t.d.c., however, the hydrocarbon emissions rise dramatically, swamping the reduction in oxides of nitrogen.

Also indicated in Fig. 126.16 are the California emission limits for 1973 and 1975 relating to nitrogen oxides and hydrocarbons. Only slight injection retarding in conjunction with dual fuelling is required to bring the gaseous emission levels within California 1973 standards and still maintain the smoke level below B.S.I. limits. Alternatively, more severe injection retarding in conjunction with butane dual fuelling can bring the gaseous emission levels almost in line with California 1975 standards without increasing the smoke level above that of operation on pure diesel fuel.

Engine noise

An additional benefit obtained from retarding injection timing is a significant reduction in engine noise (**11**). The combustion noise level with 8–10° retarded timing was greatly reduced compared with normal timing. No measurement of the noise levels was made but the reduction was sufficient to allow the noise from the cooling fan and air intake to be clearly audible, whereas previously they had been swamped by the combustion noise.

GENERAL CONCLUSIONS

During the past few years a variety of techniques have been proposed to overcome the smoke emission problems of diesel engines. One of the most promising of these is dual fuel operation, in which part of the charge is introduced into the air intake of the engine in the form of LPG.

This study has examined this technique in some detail, and experimental data are given in this paper comparing dual fuel with normal operation on a variety of engines (direct and indirect injection versions of a single-cylinder Petter AV-1, Saurer 165 hp dual fuel diesel, and Morris 3·8-litre diesel).

The main conclusions of this work may be summarized as follows:

(*a*) Smoke emission of diesel engines can be reduced by this technique below the statutory limits proposed for future legislation.

(*b*) Direct injection engines seem to respond excellently to dual fuel, but in indirect injection systems any beneficial effects may be swamped by rough running and knock.

(*c*) Addition of LPG much below 60 per cent of full load is unsatisfactory because of inefficient combustion of the gaseous charge. Optimum performance both from the point of view of reducing emissions and obtaining maximum thermal efficiency occurs when the LPG is introduced only under heavy load conditions.

(*d*) By using dual fuel combined with injection retarding, substantial reductions in gaseous emissions and smoke can be obtained—the California 1973 emission standards were easily passed and the 1975 standards were closely approximated by this technique on the Saurer engine. An additional benefit is a reduction in engine noise.

(*e*) The combined system of anti-smoke additive treated diesel fuel and dual fuel LPG shows advantages in smoke reduction over either technique used alone.

(*f*) Satisfactory performance is obtained with all the main components found in commercial LPG (propane, propylene, butanes). However, there appear to be technical advantages for LPGs containing high proportions of butanes in terms of optimum emission control and engine performance.

In summary, these conclusions suggest that conversion of diesel engines to this type of dual fuel operation must

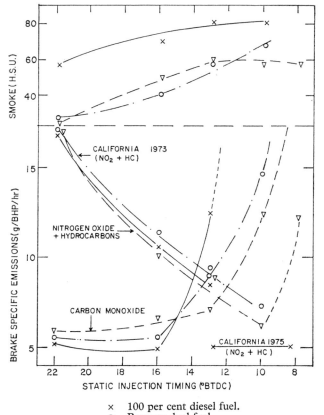

× 100 per cent diesel fuel.
○ Propane dual fuel.
▽ Butane dual fuel.

Fig. 126.16. Effect of retarding static injection timing on exhaust emissions

now be regarded as a contender for overcoming the emission problems of these engines.

ACKNOWLEDGEMENTS

Acknowledgements are due to Mr E. J. Aldridge and Mr J. P. C. Weston for experimental work on the Saurer engine, Mr J. W. Redfern, Mr R. J. Turner, and Mr B. Leech for experimental work on the Morris van and Petter AV-1 engine, and Mr J. R. Gandhi for computer programming of the results. The authors would also like to thank the management of Esso Research Centre for permission to publish this paper.

APPENDIX 126.1
REFERENCES

(1) BASCOM, R. C. and HASS, G. C. S.A.E. Paper 700671, 1970.
(2) SPRINGER, K. J. and HARE, C. T. A.S.M.E. Paper 69-WA/APC-3, 1969.
(3) PEGG, R. E. and RAMSDEN, A. W. *Proc. Int. Clean Air Congress*, London, 1966 (Pt I), Session VI, Paper VI/1.
(4) MCCONNELL, G. 'Oxides of nitrogen in diesel exhaust gas: their formation and control', *Proc. Instn mech. Engrs* 1963–64 **178** (Pt 1), 1001.
(5) NEWHALL, H. K. S.A.E. Paper 670495, 1967.
(6) VALDMANIS, E. and WULFHORST, D. E. S.A.E. Paper 700736, 1970.
(7) *Engineer*, 1969 **227**, 41.
(8) SCOTT, W. M. S.A.E. Paper SP-345 (690002), 1968.
(9) ELLIOTT, M. A. and DAVIS, R. F. *Ind. Engng Chem.* 1951 **43**, 2854.
(10) MOORE, N. P. W. and MITCHELL, R. W. S. *Proc. Instn mech. Engrs–Am. Soc. Mech. Engrs Jt Conf. on Combustion*, 1955, 300.
(11) WATERS, P. E., LALOR, N. and PRIEDE, T. 'The diesel engine as a source of commercial vehicle noise', *Conf. Critical Factors in the Application of Diesel Engines, Proc. Instn mech. Engrs* 1969–70 **184** (Pt 3P), 63.

C127/71 FIRST RESULTS OBTAINED WITH THE CARBON MONOXIDE MONITORING NETWORK INSTALLED IN PARIS

P. CHOVIN*

A network of gauging points for the automatic measurement of the carbon monoxide content of the atmosphere has been installed during 1970 in Paris. Eleven units—each giving the hourly content of pollutants—have been installed at crossroads chosen from among the worst in the capital. The use of several units is by means of a monitor, and the first results will be discussed. In particular, they will be compared with those obtained by an older method, in use for more than 10 years, which provides a useful and immediate sample of the atmosphere at several crossroads in the capital.

INTRODUCTION

FOR MORE THAN 10 YEARS the Central Laboratory of the Paris Police Préfecture has been involved in measuring the air pollution caused by exhaust gases from motor vehicles, taking carbon monoxide as a test substance for this type of contamination. The problems related to this monitoring action are varied and numerous, and only those concerning air sampling and data processing will be discussed in this paper.

AIR SAMPLING METHODS

Air sampling may be accomplished instantaneously or continuously. We have used the former method for 11 years, taking, each month, four samples at each of 317 sites selected in the city as being representative of points with high pollution: street crossings and fairly narrow streets where traffic is often congested.

The 48 items of information obtained yearly in each site—that is, the 15 200 items collected from all the sites in a given year—allow us to draw a pollution map and to compare maps annually. A better comparison is observed if we apply statistical calculation to the total data amassed. In a graph on log-normal paper, the curve of the cumulative frequency distribution versus carbon monoxide concentrations is a straight line with a good approximation and allows us to apply different tests for comparison. As a rule, the difference between two successive annual levels for the 50 percentile has been significant over the last four years and appears to be due to a decrease in pollution rather than pure chance.

This method is also used to survey the concentration of carbon monoxide in traffic tunnels with and without mechanical ventilation. It has been found that, in unventilated tunnels, the annual mean carbon monoxide concentrations are proportional to the lengths of the tunnels—a result which imposes a limit on the tunnel lengths. This limit is approximately one-quarter to one-third of a mile. Beyond this the level of air pollution continues to increase, and the time spent in the tunnel also increases—two circumstances which may be injurious to health.

This method has advantages and disadvantages. It is possible to map the pollution areas. Since only one measuring device is necessary to analyse the numerous samples taken, the cost of equipment is low, but, on the other hand, the method requires the constant services of a car, a driver, and two technicians. In addition, as the samples are taken in a few seconds there can be a great variation between the concentration of carbon monoxide in four samples taken simultaneously at the same site, and the result obtained for a given sample at a given site would be different if the car had arrived at that place a few minutes earlier or later. This is another cause of variation of results, which may be illustrated in calculating the standard deviation of the total data.

To overcome this difficulty I suggested to an O.E.C.D. Commission that the average CO concentration should be measured directly, with an averaging time of 1 hour. This suggestion has been adopted.

To do this it is necessary to integrate the variation of CO concentration, considered as a function of time. This may be performed by one of two methods:

Method (a): Sampling air in plastic bags, with a constant flow rate, during the selected averaging time and later analysing the content of the bags (i.e. 'pneumatic integration' (Fig. 127.1), see Appendix 127.1).

The MS. of this paper was received at the Institution on 7th May 1971 and accepted for publication on 19th July 1971. 33
* *Central Laboratory of the Paris Police Préfecture, France.*

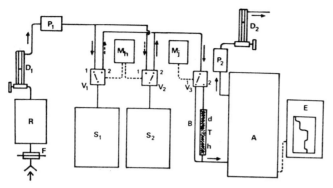

A Infrared analyser.
B By-pass.
D_1 and D_2 Flow meters.
E Recording instrument.
F Filter.
M_h Hourly record switch.
M_j Daily record switch.
P_1 and P_2 Pumps.
R Recording instrument.
S_1 and S_2 Plastic bags.
T Tube enclosure comprising d, drier and h, freezer.
V_1, V_2, V_3 Electromagnetic valves.

Fig. 127.1. Schematic diagram of the apparatus for measuring concentration of carbon monoxide

Method (b): Connecting the output of the analyser to an electronic integrator which delivers a print-out.

At present, six instruments of method (a) are working in six selected sites of Paris and give excellent results. Five instruments of method (b), with electronic integration, have also been installed, but only for a few months, which is not sufficient time to enable us to compare the respective reliability of both types of analyser. Fig. 127.2 shows the distribution of the 11 analysers in Paris.

Of course, method (b) also has advantages and disadvantages. It necessitates the use of a rather expensive apparatus in each site to be monitored. But, conversely, each device need only be visited once a week instead of every day, and by one driver instead of three men, or may not need to be visited at all if the data are transmitted to the laboratory by a special line. However, above all, the collected data offer much greater possibilities of analysis than those resulting from numerous instantaneous sample analyses, and enable valuable information to be obtained

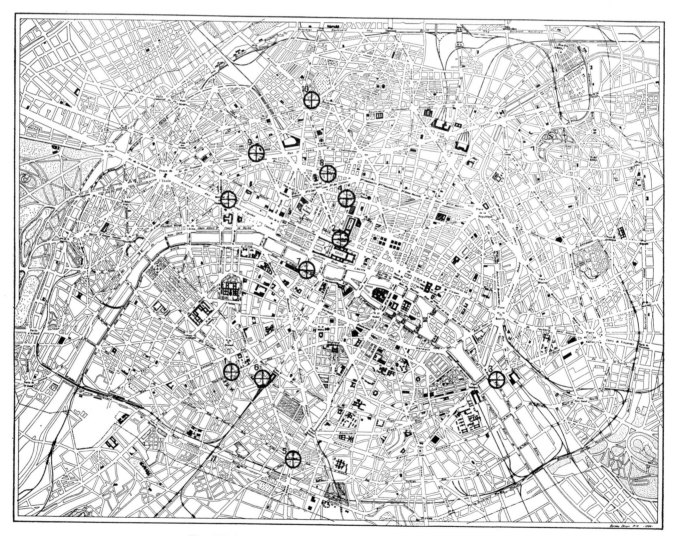

Fig. 127.2. Locations in Paris of 11 automatic analysers

where ambient air quality standards are analysed, or more stringent measures to be taken when such standards, if they exist, are exceeded.

To use this method it is necessary to discuss the conditions leading to comparable results. It is well known that carbon monoxide, as a gas of relatively low molecular weight, diffuses very easily from its emission point. In this respect, the situation is quite different from that arising for SO_2 for which the result of an analysis is, in a small area, largely independent of the sampling site. I would like here to remember the work done by Dr Georgii, of Frankfurt-am-Main, who showed that CO concentration in a street is a sharply decreasing exponential function of the sampling point height above ground level. As a consequence, it is certain (and confirmed by our own experience) that the results of CO measurements brought about in two different locations of the same city, and even more so of two different cities, are not comparable without standardization of the sampling point location.

In this respect, I made the following proposal some years ago. The sampling point must be:

(*a*) near the crossing of two large streets with heavy traffic, but at about 50–60 ft from a corner where traffic lights are in use;

(*b*) in an area where it is prohibited to park, to minimize any screen effect between traffic flow and sampling point;

(*c*) at about 3–4 ft from the kerb of the pavement; and

(*d*) at a height of 5–6 ft above ground level.

We shall only be able to compare our data when this type of standardization is in evidence. I am the first to recognize that it is not always easy to fulfil the above requirements, and having installed 11 devices in Paris I am only too aware of the difficulties encountered!

DATA PROCESSING

It is possible, with this network, to obtain about 100 000 different data annually, each of them being one average hourly CO concentration. It is necessary to submit these data to a mathematical treatment in order to have a good statistical insight into the situation of the pollution by exhaust gases throughout the city.

The calculation program allows us to obtain the following results for each individual analyser, or for those of the whole network:

(1) *For every month:* the successive average, maximum, and minimum hourly concentrations, from midnight to midnight, either for all the days of the month (Figs 127.3 and 127.4 show what happened in one sampling site, crossing of Bd Pasteur and rue de Vaugirard, during June and August 1970, respectively) or for all Mondays, Tuesdays, etc.

(2) *For the whole year:* the same results as in the previous section, the corresponding graphs showing the average, maximum, and minimum hourly concentrations, for all the days of the year or for all Mondays, Tuesdays, etc., of the same period. (Figs 127.5 and 127.6 show the pollution profile corresponding to the data collected during all the Wednesdays and Sundays of the period beginning 15th March 1970 and ending 31st December 1970.)

(3) The average carbon monoxide concentrations during six selected daily periods of eight consecutive

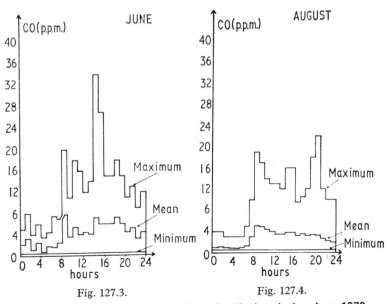

Fig. 127.3. Hourly concentration of pollution during June 1970

Fig. 127.4. Hourly concentration of pollution during August 1970

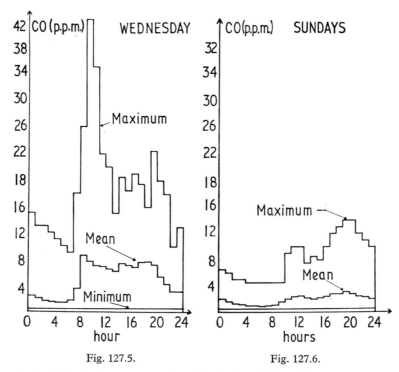

Fig. 127.5. Fig. 127.6.

Fig. 127.5. Hourly concentration of pollution during Wednesdays from 15th March to 31st December 1970

Fig. 127.6. Hourly concentration of pollution during Sundays from 15th March to 31st December 1970

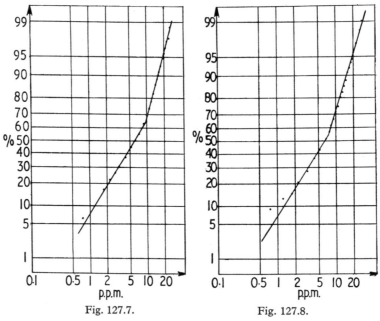

Fig. 127.7. Fig. 127.8.

Fig. 127.7. Cumulative frequency distributions of the hourly concentrations obtained from analyser No. 1 during March 1971

Fig. 127.8. Cumulative frequency distributions obtained from five analysers (Nos 1, 5, 7, 9, and 11) during March 1971

hours are also calculated. These periods correspond to the maximum traffic activity in the streets, as follows:

4 a.m. to noon	noon to 8 p.m.
5 a.m. to 1 p.m.	1 p.m. to 9 p.m.
6 a.m. to 2 p.m.	2 p.m. to 10 p.m.

(4) The computer also gives the cumulative frequency distributions of the hourly concentrations. Fig. 127.7 shows the curve obtained with log-normal paper for one single analyser during March 1971. It may be seen, in this monitoring site, that 50 per cent of the data are lower than 6 p.p.m. and 96 per cent lower than 20 p.p.m. In other words, the pollution level of 6 p.p.m. has been exceeded 50 per cent of the time and the level of 20 p.p.m., 4 per cent of the time. Fig. 127.8 indicates the same curve for five of the 11 analysers installed, and corresponds to the data collected during the same month, March 1971. It may be seen that the result is about the same as previously indicated for one analyser only; the level of 6 p.p.m. is exceeded 50 per cent of the time, and the level of 20 p.p.m., 5 per cent of the time.

Obviously the relationships are not single straight lines, but are formed from two distinct straight lines with different slopes. Accordingly, the Galton–McAllister law does not seem to be accurately followed. This may be explained by the fact that many of the concentrations values are near zero since they were collected during the hours of late night and early morning. It is known that the zero or approaching zero values are not compatible with a log-normal distribution. However, if one eliminates these very low values, the remaining data still give two distinct straight lines. Therefore, we may think that the data belong to two distinct populations. Other hypotheses are not to be rejected. The program gives the β_1 and β_2 Pearson coefficients, but it has not been possible until now to use these coefficients to determine the nature of the distribution laws.

Analogous curves are obtained for the different 8-hourly periods mentioned above.

CONCLUSIONS

Method (b) gives much more accurate results than method (a). Despite the price of the equipment, we prefer this system because the data processing gives more valuable information. As an example, the cumulative frequency distribution curves allow the comparison with any given ambient air quality standard in order to verify that the standard is fulfilled or exceeded, and in this latter case to know during which fraction of the time the standard is exceeded.

The results obtained with this method may help to define standards, and the work was done with this objective in view.

APPENDIX 127.1

PNEUMATIC INTEGRATION

The pneumatic method is now under experiment with a new type of apparatus (see Fig. 127.1). Two plastic bags having a 10–12 litre capacity are connected to a NDIR analyser through three-way electromagnetic valves actuated by an electric clock. With the aid of pumps, one bag is filled up during 1 hour, while the other is being emptied through the analyser. During the following hour the functions of the two bags are reversed, so that the record is a curve with successive steps, the height of each step above the base line being a measure of the average hourly CO concentration during the preceding hour.

To allow for the possibility that the base line may drift, a special device has been added to the apparatus. By means of a three-way valve, the air to be monitored passes once a day—for a period of about 20 min, preferably during the night—over a bed of 'Hopcalite', a well-known catalyst which oxidizes carbon monoxide into carbon dioxide at room temperature. With this method, the air coming to the analyser is free of carbon monoxide. The real base line is the straight line drawn between two consecutive 'zeros' at 24-hour intervals.

In addition, it has been judged necessary to add a freezer to the device, lowering the air temperature to $+2°C$, in order to maintain the water vapour concentration at a constant level.

C128/71 THE CONTRIBUTION OF ADDITIVES TO THE ELIMINATION OF AIR POLLUTION

J. W. DABLE*   T. J. SHEAHAN†

Both fuel and lubricant additives contribute to the elimination of air pollution. This paper deals with the contribution of one specific type of additive, the multi-functional additive for motor gasoline. For the last 15 years, additives have been used in motor gasoline as a means of combating the formation of carburettor deposits which interfere with the proper functioning of the carburettor and result in the necessity of adjustment to avoid rough idling and engine stalling. More recently, carburettor cleanliness and the right mixture of air and fuel have been shown to be important with regard to the production of exhaust pollutants. This has resulted in the design of experiments to investigate the effect of carburettor deposits on exhaust emissions. Data from the experiments described in this paper show that carburettor deposits markedly increase exhaust emissions. However, even in a late model emission-control-equipped automobile the build-up of carburettor deposits can be satisfactorily retarded through the use of polymeric detergent–dispersant gasoline additives. Such additives appear to be compatible with all elements of the vehicle emission control system.

INTRODUCTION

THE SOLUTION to the problem of air pollution from gasoline-powered motor vehicles lies in emission control devices, power unit design changes, and proper care of emission controls. Gasoline composition must continue to be compatible with all modifications. In addition, the gasoline must be formulated to take into consideration the high proportion of 'old' cars with varying degrees of emission control that will remain in service for some time.

Emission controls must continue to function efficiently while the vehicle is in service. Durability and efficiency of the system are determined by proper care. The critical parts of the control system must be kept free of dirt and other debris. This is achieved by using the proper additive-treated gasolines and following prescribed maintenance practices.

EMISSIONS IN THE FIELD

In a recent survey of vehicle emissions in the field (1)‡ the Swedish Ministry of Communications discovered that, out of a sample of 92 cars, the carbon monoxide emission at idle ranged between 0·1 and 11·0 per cent, with an average of 5·8 per cent. Careful carburettor adjustment to 'minimum emission' brought the average figure down to 3·4 per

The MS. of this paper was received at the Institution on 14th June 1971 and accepted for publication on 20th July 1971. 22
* *Technical Service Engineer, Lubrizol International Laboratories, 'The Knowle', Hazelwood, Derby.*
† *Technical Service Engineer, The Lubrizol Corporation, 29400 Lakeland Boulevard, Wickliffe, Ohio, U.S.A.*
‡ *References are given in Appendix 128.8.*

cent. Similar work by the Motor Industry Research Association in the U.K. indicated that the idle carbon monoxide emission ranged between 0·2 and 11·6 per cent with an average of 6·7 per cent. Again, carburettor adjustment brought the average emission down to 4·6 per cent (2).

An earlier survey by the General Motors Corporation of 478 privately owned vehicles from the Los Angeles, California, area also showed significant exhaust emission reductions with carburettor adjustments (3). In pre-1966 vehicles without exhaust emission controls carbon monoxide emissions ranged from 5·9 per cent for cars built by one U.S. manufacturer to 7·4 per cent for another. With carburettor idle speed and mixture adjustments the average carbon monoxide emissions at idle for vehicles built by these two manufacturers were reduced from 5·9 and 7·4 per cent to 4·2 and 4·8 per cent, respectively.

Pre-1966 vehicles built outside the U.S. and comprising less than 8 per cent of the 478-car sample had carbon monoxide emissions at idle speed averaging 6·7 per cent as received and 4·2 per cent after carburettor adjustment.

Exhaust emission control-equipped automobiles in the group of 478 cars also responded to carburettor adjustments. For the five subdivisions (cars made by four U.S. manufacturers and foreign cars), average carbon monoxide emissions at idle ranged from 2·6 to 3·8 per cent. Reductions in carbon monoxide emissions at idle through carburettor adjustment ranged from an average of 0·9 to an average of 2·3 per cent.

The observations by the Swedish Ministry of Communications, the Motor Industry Research Association, and the General Motors Corporation were not accompanied by visual carburettor ratings. Therefore, the extent of deposit build-up in the carburettors is not known.

In the process of locating fleets for emission test purposes in the U.S., employee, police, and taxi vehicles were checked for visual carburettor deposits. As illustrated in Fig. 128.1, two taxi fleets were found to have heavy carburettor throttle body deposits. Both fleets were operating on gasoline containing none to trace amounts of detergent additive. On recording idle speed emissions for some of the cars from one of these taxi fleets, carbon monoxide was found to range from 3·8 to 7·0 per cent with an average for the six cars of 5·4 per cent. Hydrocarbon emissions at idle as p.p.m. carbon (flame ionization detector) ranged from 7200 to 19 000 with an average for the six vehicles of 12 300.

MECHANISM OF THE PROBLEM

Obviously not all emission increases in service are related to the carburettor. The effects of ignition and positive crankcase ventilation system condition on automotive exhaust emissions are well documented (4) (5). It has also been established that exhaust valve condition and the ability for the valve to seat properly influence emission levels (6) (7).

Optimum air–fuel mixture preparation is the single most important function to achieve thorough combustion

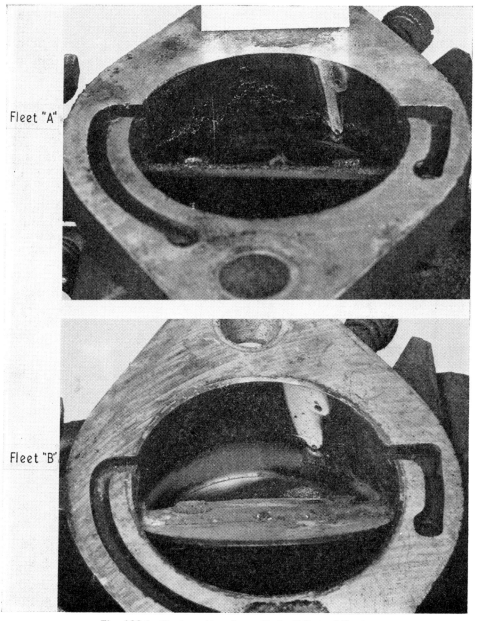

Fig. 128.1. Carburettor deposits in U.S. taxi fleets

in an engine (8). Deposits which build up around the closed throttle plate of the carburettor can upset the air–fuel mixture preparation by restricting air flow to the engine. The effect of this is a reduction in idle speed, a richer mixture, and higher hydrocarbon and carbon monoxide exhaust emissions. Deposits in this area may be compensated for by mechanical adjustment of the throttle plate to restore the original air flow and idle speed, but when this is done the critical relationship between the edge of the throttle plate and the progression jet is changed. Thus, the exhaust emissions may still be high, and adjustment of the mixture volume screw may be necessary. Eventually, of course, no further adjustment of the volume screw is possible and the engine will be idling through the progression jet, with no way of reducing the idle emissions, short of cleaning the carburettor. This problem becomes more acute with some carburettors of recent manufacture, where the mixture adjustments are very limited.

An alternative sequence of events which has been reported to occur on certain fixed-jet carburettors is the progressive restriction of the idle system air bush or jet, which increases the depression existing in the idle speed mixture duct, causing more fuel to be drawn through the idle jet. Again, in practice this is counteracted by turning in the mixture volume screw.

MAINTAINING CARBURETTOR CLEANLINESS—ROAD TEST DATA

Carburettor deposits which built up under accelerated conditions in 1700 miles of urban service caused an average increase of 175 per cent in carbon monoxide emissions at idle in two popular four-cylinder, 1300-cm^3 cars operating in the U.K. Carbon monoxide emissions for these cars are shown in Table 128.1. During this running period there was no carburettor mixture adjustment. Two of the same make of car operating over the same test course but with detergent–dispersant additive in the fuel showed essentially no increase in emissions during the same test period. The complete data from this test appear in Fig. 128.2. At 1700 miles the dirty carburettors from car No. 1

Table 128.1. Carbon monoxide emissions at idle in two cars operating in the U.K.

Car	Initial idle CO emission with clean carburettors, %	Idle CO emission after 1700 miles of urban service, %
1	4·2	9·8
2	3·5	11·3

and car No. 2 (untreated gasoline) were replaced with new carburettors and the test was continued. Again, carbon monoxide emissions at idle speed increased with the untreated gasoline. The carburettors on car No. 3 and car No. 4, which were operating on treated fuel, continued to remain clean and exhibit low emission levels.

The erratic nature of the emission curve from car No. 2 during the second phase of the test may be explained by the fact that a subsequent engine inspection revealed that the crankshaft thrust washers had fallen out, resulting in excessive wear and some piston distress.

Carburettor deposits at each stage of the test are shown in Fig. 128.3. Before the carburettors were finally removed, all cars were run through the Economic Commission for Europe (E.C.E.) Type I and Type II emission tests. Results from these tests are shown in Table 128.2, which

Table 128.2. European road test—E.C.E. test results at end of running period

Vehicle	Fuel	Type I test*		Type II test†
		CO, g	HC, g	CO, %
1	Untreated	422	38·0	11·8+
2	Untreated	413	42·1	7·3
3	Treated	228	8·4	1·6
4	Treated	249	9·8	2·7
Proposed emission limits for test vehicle	Approval vehicle .	134	9·5	4·5
	Production vehicle .	161	12·2	4·5

* Test over four E.C.E. cycles.
† Idle emission measurement at the end of Type I test.

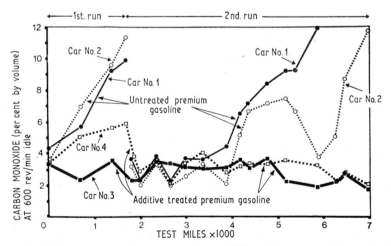

Fig. 128.2. European road test—variation of idle carbon monoxide emission

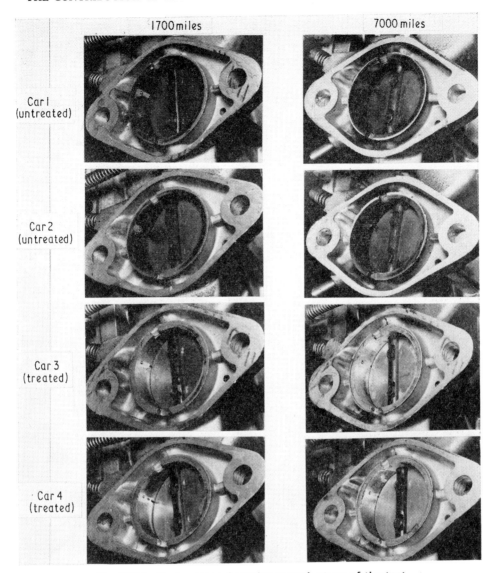

Fig. 128.3. Carburettor deposits at each stage of the test

provides comparisons between the cars at the end of the running period in terms of cycle carbon monoxide and hydrocarbon emissions.

An important aspect of this test concerned the carburettor adjustments which were necessary to maintain the correct idle speed. During the test, after measuring the idle emissions, the throttle plates were adjusted to give an idle speed of precisely 600 rev/min. Then the emissions were again measured. All idle emission data given in Fig. 128.2 were obtained after adjustment. Fig. 128.4 indicates the necessary adjustments during the test.

MAINTAINING CARBURETTOR CLEANLINESS—DYNAMOMETER AND LABORATORY ENGINE TESTS

Prior to the road test work, extensive laboratory dynamometer and engine test studies were conducted. Some of these studies are summarized below. Details of the engine operating conditions, etc., appear in the Appendices. The test procedures all have their own short-comings and their relationship to field performance is still not fully understood. Although test development is still continuing, these tests have proved valuable in establishing the relationship between properly treated gasolines and low exhaust emissions.

Twin carburettor cleanliness tests

A 1500-cm^3, four-cylinder engine was equipped with twin S.U. constant-vacuum carburettors, arranged so that each carburettor independently fed two engine cylinders. Crankcase blow-by (and in some cases a proportion of exhaust gas) was used as the deposit-forming medium. The test was run in two stages. In the first stage, the additive was fed only to the front carburettor. The carburettors were then removed, inspected, and cleaned,

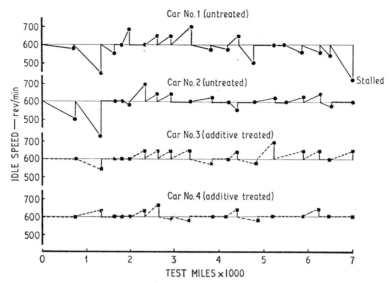

Fig. 128.4. European road test—idle speed variation

Table 128.3. Results from twin carburettor tests using a European engine

Test procedure	Merit rating (10 = clean)	Run 1		Run 2		Average	
		Untreated	Treated	Untreated	Treated	Untreated	Treated
S.U. constant vacuum carburettors	Throttle plate	3·1	10·0	3·2	10·0	3·2	10·0
	Throttle bore*	6·8	10·0	7·9	9·6	7·4	9·8
Weber fixed-jet carburettors	Throttle bore*	3·4	8·6	3·8	9·8	3·6	9·2

* Area downstream of closed throttle plate.

and replaced for the second stage, where the additive was fed only to the rear carburettor.

The final result was expressed as the average of the two runs. A colour factor 'rating' system was used with a black surface being assigned a merit rating of zero, and a completely clean surface a rating of 10. A similar test was also run using twin Weber fixed-jet carburettors. Table 128.3 indicates typical results from these tests.

Another twin carburettor deposit test was carried out using an American V-8 engine equipped with two single-choke carburettors and a special 'Y' manifold (9). Again, a fixed duration test was run and the carburettor deposits were inspected at the end of the test. In this case, however, each carburettor was fitted with a thin aluminium sleeve which formed the throttle bore and which could be removed and weighed. Typical results from tests carried out on this procedure are shown in Table 128.4.

Whilst test procedures employing engines with two carburettors are valuable in the development of improved additives, they have limitations which may restrict their usefulness. It is not possible to evaluate two treatments side-by-side and use a deterioration of engine operation and the production of exhaust pollutants as criteria for carburettor detergency performance. Furthermore, when comparing two additives simultaneously, the use of blow-

Table 128.4. Results from twin carburettor tests using an American engine

Carburettor throttle bore deposits, g		
Untreated fuel	Treated fuel	% Reduction of treated over untreated fuel
20·4	8·2	60

by gases to build deposits may be questioned. Gasoline additives are known to remain in the engine blow-by. Thus, both carburettors may receive the benefit of a mixture of detergents through the deposit-forming medium.

Single carburettor cleanliness tests

Tests have been developed in both European and U.S. engines to evaluate the ability of polymeric detergent–dispersant additives to resist the build-up of carburettor deposits, which in turn cause high exhaust emissions at idle speed. A 767-cm^3, four-cylinder engine fitted with a fixed-jet carburettor was used to develop a cyclic test designed to simulate severe road conditions. The build-up of carburettor deposits was monitored by measuring carbon monoxide emissions at the idle mode. As deposits built up and the idle speed dropped, the throttle plate was gradually opened to maintain the idle speed of 800 rev/min. Typical results comparing the performance of treated and untreated fuels are shown in Fig. 128.5.

A rather specialized test was also developed using the 767-cm³ engine described above. In this case the interest was in the idle system air bush, rather than the more usual throttle plate area. In order to eliminate the effect on

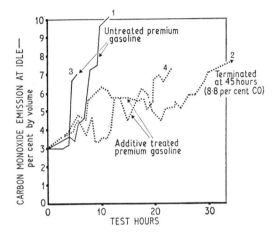

Fig. 128.5. Exhaust emission test—767-cm³ European engine

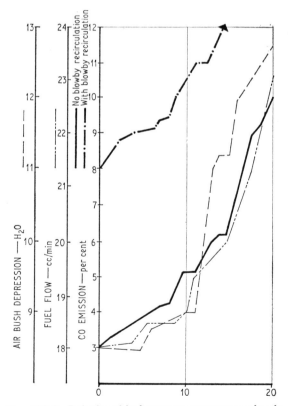

Fig. 128.6. Relationship between parameters in air bush blocking test

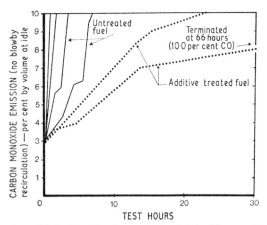

Fig. 128.7. Results from air bush blocking test

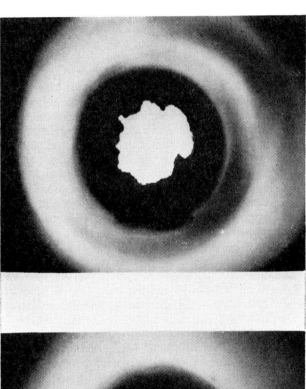

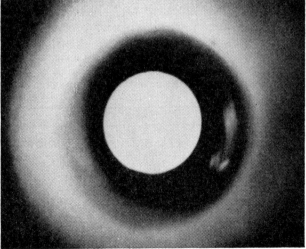

Fig. 128.8. Partially blocked air bush compared with new air bush

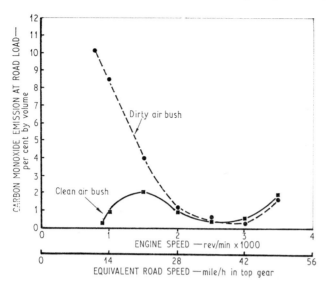

Fig. 128.9. Effect of idle system air bush blocking on road load CO emissions

emissions of any deposits around the throttle plate during idle, so that the effect of air bush restriction could be measured accurately, a hole was cut in the throttle plate and the engine idle with the throttle completely closed. The shape and size of this hole were established after considerable road running with an engine of this type in order to maintain the correct relationship between intake manifold depression and idle system depression. Depression across the air bush, fuel flow through the idle jet, and subsequent idle carbon monoxide emission were monitored throughout each test. These parameters were found to correlate quite well, as illustrated in Fig. 128.6, and this correlation was useful for the control of accurate and repeatable tests. Typical results comparing untreated fuel with the same fuel with additive treatment are shown in Fig. 128.7. The level of air bush deposits corresponding to an idle emission of 10 per cent CO is illustrated in Fig. 128.8.

The effect of air bush deposits on emission levels under road conditions has been investigated. A road load curve was run on the test-bed engine using a perfectly standard clean carburettor and figures of CO emission were obtained. This curve is shown in Fig. 128.9. Superimposed is a curve obtained by substituting a typical 'dirty' air bush from the end of a normal test where the idle CO emission was 10 per cent. These curves indicate the serious effect of air bush deposits on CO emission levels under low speed, city driving conditions.

REMOVING PREVIOUSLY FORMED DEPOSITS

What is termed 'clean-up' testing presents more difficulties than evaluations designed to measure the effectiveness of an additive in preventing or retarding the formation of carburettor deposits. Tests must be designed with greater emphasis on eliminating operating variables which may mask carburettor deposit emission effects. With laboratory dynamometer and engine tests there must be some confidence that the deposits preformed under controlled conditions are representative of deposits that may be found in the field.

ROAD TEST DATA

A road test was carried out in the U.K. where the first phase involved a survey of employee-owned cars based at the Lubrizol Laboratories. Not all the employee cars were inspected, since the survey was on a voluntary basis, but 27 owners offered their cars for consideration. The first inspection was a quick examination of carburettor deposits without removing the carburettors from the engines. From this, 14 cars (52 per cent) were found to have sufficiently heavy deposits to be worthy of emission checking. When the idle emissions of the 14 cars were measured, four cars (15 per cent of the total) were in excess of 10 per cent carbon monoxide, which was the limit of measurement of the instrument. Two cars were rejected for other reasons, and finally eight cars were used for the 'clean-up' test, which involved 'normal' driving using a high treatment of detergent–dispersant gasoline additive. After 50 gal of fuel throughput, idle emissions were again checked. Results from this test are shown in Table 128.5. Overall, the average idle carbon monoxide emission was reduced by 30 per cent, from 7·3 to 5·1 per cent.

Similar work carried out in the U.S. includes a test

Table 128.5. 'Old car' fleet—emission reduction using 50 gal of premium gasoline treated with detergent–dispersant additive

Car	Car type	Carburettor type	Idle CO levels, %		Test miles	% reduction in idle CO level
			Initial ('as found')	Final		
1	Armstrong Siddeley	Stromberg fixed jet	6·6	6·0	457	10
2	Triumph 2000	2 × Stromberg CD 150	9·2	7·5	1409	18
3	Simca 1500	Weber twin choke 28/36 DCB	4·4	1·8	1592	59
4	Cortina 1300	Fomoco 6LC	7·4	4·5	648	39
5	Zodiac Mk III	Zenith 42 W1A	6·3	Test aborted—carburettor adjusted during test		
6	Cortina 1200	Solex B30, PSE1–2	7·4	5·7	1775	23
7	Morris 1000	SU HS 2	8·9	6·1	2230	32
8	Austin A40	Zenith 26 VME	8·5	4·3	1462	50

where three vehicles with 'dirty' carburettors were operated at 'turnpike' speeds for 5000 miles. Two vehicles operated on a gasoline treated with additive, and the third used untreated fuel. Both idle and seven-mode cycle emissions were measured during the test. No significant benefits were shown by the data developed under the seven-mode cycle, but a clean-up trend was shown in terms of idle emissions, and visual deposit levels. The idle emission curves are shown in Fig. 128.10, and the deposit levels are illustrated in Fig. 128.11.

In another exercise, American cars fitted with V-8, 5030-cm^3 engines were run on a road simulator facility which is installed in the Lubrizol Laboratories in Cleveland, Ohio. The road simulator consists of eight chassis dynamometers, or 'lanes', each of which can be programmed by means of a magnetic tape controller, to simulate any desired road service. For the test in question, two of the lanes were programmed identically, using a varied city street and high speed route. The test consisted of running the cars using untreated fuel until emission levels had built up due to deposits, and then switching to the same base fuel treated with detergent–dispersant additive to observe the 'clean-up' effect. The resultant emission curves using seven-mode hot cycle figures are shown in Fig. 128.12.

Fig. 128.13 shows the relationship between carbon monoxide emissions at the idle mode and carbon monoxide under the seven-mode hot cycle. Data are also included on air/fuel ratio. It is clearly evident that carburettor deposits have the most pronounced effect on carbon monoxide emissions at idle speed.

DYNAMOMETER AND LABORATORY ENGINE TESTS

The laboratory engine tests described previously, using both European and U.S. engines, have been used to investigate the 'clean-up' performance of additives. The

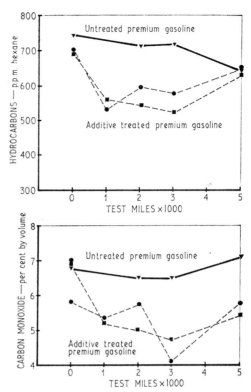

Fig. 128.10. U.S. 'clean-up' road test—emission variation

767-cm^3, four-cylinder engine test was run using an untreated fuel until emission levels had increased sufficiently. Then, without changing the running procedure, the fuel was switched to the same base fuel but treated with detergent–dispersant additive. A typical result is shown in Fig. 128.14 which also shows the carburettor deposits at the end of the 'build-up' and the 'clean-up' stages. In this particular case, after 2 hours of deposit build-up, with high blow-by recirculation to accelerate deposit formation,

Table 128.6. Exhaust emission clean-up test—American 5030-cm^3 engine

Fuel used for clean-up phase	Initial condition			Deposit build-up				Clean-up phase			
	HC as p.p.m. hexane	CO, %v.	Idle speed, rev/min	Test hours	HC as p.p.m. hexane	CO, %v.	Idle speed, rev/min	Test hours	HC as p.p.m. hexane	CO, %v.	Idle speed, rev/min
Commercial premium gasoline—untreated	664	2·8	696	209	1610	7·6	553	25	1548	8·1	620
								47	1332	8·3	590
								72	2400	8·6	530
Same fuel + low treatment of detergent–dispersant	522	2·6	700	120	1900	10·1	519	19	830	6·4	633
								48	774	6·5	644
								92	828	6·5	660
Same fuel + high treatment of detergent–dispersant additive	657	2·7	698	153	1512	8·9	517	21	867	5·6	602
									48	696	4·6	620

Notes: (1) Each test started with new spark plugs in the engine.
(2) Hydrocarbons (HC) were measured by flame ionization detector (FID). The values shown above represent the approximate non-dispersive infra-red (NDIR) equivalent as parts per million (p.p.m.) hexane.
(3) Carbon monoxide (CO) was measured by Orsat.
(4) All emission levels are at idle condition.

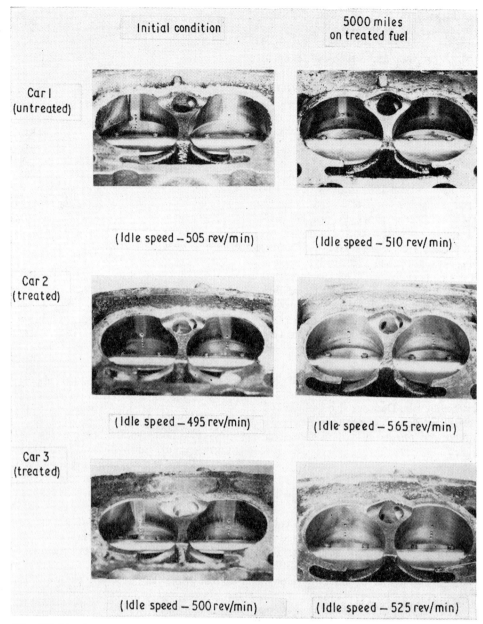

Fig. 128.11. Carburettor deposits and idle speed variation in U.S. 'clean-up' road test

the blow-by recirculation line was disconnected and the test run an additional 7 hours on untreated fuel to consolidate deposits and high emission levels. The fuel was then switched to an additive treatment for the 'clean-up' test.

Fuel flow data, which also appear in Fig. 128.15, illustrate the effect of carburettor deposits on fuel economy at idle speed. Future road and dynamometer test programmes will be designed to investigate more carefully the fuel economy aspect. Deposit removal tests have also been carried out using the American engine described previously. The 5030-cm^3, V-8 engine, equipped with an engine modification-type exhaust emission control system, was operated for about 100 hours (approximately 3000 miles) under cyclic conditions using a plugged PCV system and an unstable fuel to build up carburettor deposits. Without changing the engine oil, the test gasoline was substituted for the deposit build-up fuel, the PCV system was unplugged, and the test continued under the same operating conditions. Carburettor idle speed was allowed to drift downward during the deposit build-up cycle and was allowed to increase during the clean-up cycle. No carburettor mixture adjustments were made throughout the duration of the test. Table 128.6 shows results comparing untreated premium gasoline with the same fuel treated with two levels of additive.

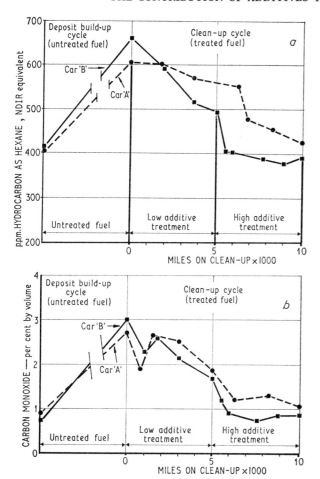

Fig. 128.12. Seven-mode cycle emission from road simulator test

COMPATIBILITY OF DETERGENT–DISPERSANT ADDITIVES WITH EMISSION CONTROL SYSTEMS

As previously stated, new gasolines must be compatible with emission control systems. As new systems are developed they present a real challenge to the petroleum industry to produce gasolines which ensure the efficiency of the control and do not cause harmful side effects.

Steps are already being taken in the U.S. to remove lead alkyl compounds from motor gasolines in preparation for the use of new emission controls. These controls presumably will be exhaust recirculation systems and catalytic reactors. Elements of the current emission control hardware and their interaction with detergent–dispersant gasoline additives are discussed below.

PCV valves

Polymeric detergent–dispersant gasoline additives are known to accumulate in the engine crankcase to supplement the dispersancy of the engine oil. Other investigators have noted beneficial effects for these additives with regard to engine sludge deposit control and PCV valve operation (10). The use of high-quality fuels and lubricants together with improved PCV valve design appear to have reduced the incidence of PCV valve problems.

Evaporative emission control system

Tests have been carried out in the U.S. to determine the effect of gasolines and gasoline additives on the durability of the activated charcoal canister-type control system, which stores vapours when the engine is stationary and then vents them to the induction system when the engine is running. Under certain field conditions it may be possible for liquid fuel to find its way past the liquid–vapour separator and into the canister itself. This possible condition presents two areas of concern: namely, durability of the material of construction and the effect of the liquid fuel on the efficiency and life of the charcoal.

The compatibility of the canister material (Du Pont heat-stabilized Nylon 66) with gasoline and gasoline additives was determined by storing small samples (about 1 g) of the material in the test fuels for 8 weeks at 110–140°F. A change in weight or pliability of the sample was noted at both 4 and 8 weeks' storage. Two base fuels, one leaded and one unleaded, and three different additives were tested all at 100-fold overdose. Even with gross

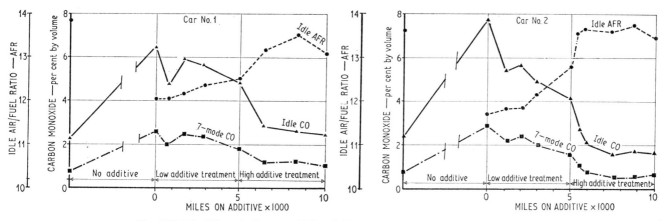

Fig. 128.13. CO emissions and idle air/fuel ratio from road simulator test

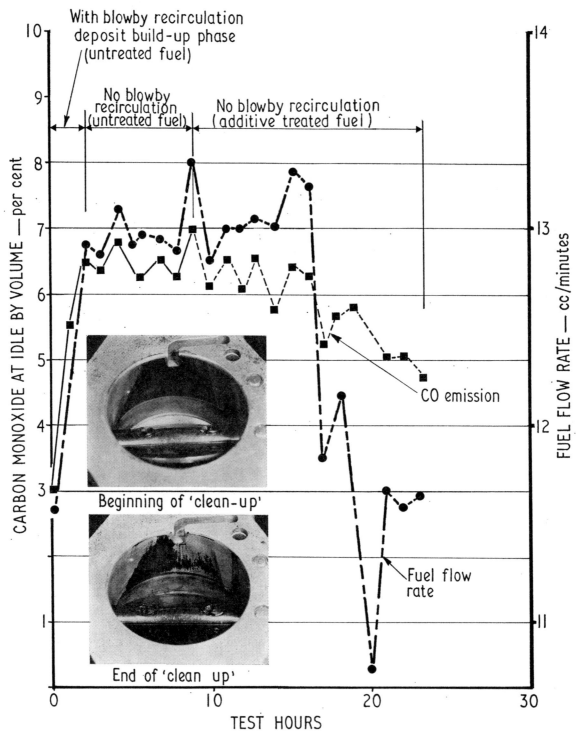

Fig. 128.14. Carburettor 'clean-up' test—767-cm³ European engine

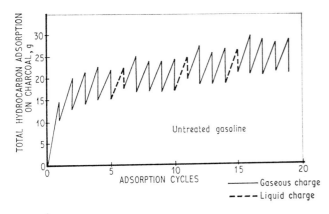

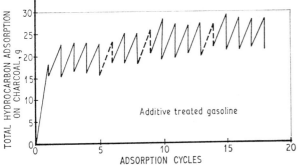

Fig. 128.15. Charcoal canister hydrocarbon adsorption–desorption cycles

Table 128.7. Average hydrocarbon adsorbed–desorbed data—G.M. evaporative emission control system

Additive code	Cycles averaged*	Average hydrocarbon adsorbed	Average hydrocarbon desorbed
None	3–5	7·9	6·9
	6–10	6·9	6·5
	11–14	7·1	6·7
	15–19	7·4	7·1
		7·3†	6·8†
B	3–4	6·8	6·7
	5–9	6·9	6·6
	10–14	7·1	6·8
	15–19	7·0	6·8
		7·0†	6·7†
D	3–4	7·1	6·3
	5–9	7·1	6·6
	10–14	7·2	7·2
		7·1†	6·7†
E	3–5	6·9	6·7
	6–8	6·4	5·9
	9–13	7·1	6·7
	14–18	6·9	6·7
		6·8†	6·5†
F	3–4	6·3	6·3
	5–9	6·5	6·2
	10–14	6·7	6·5
	15–19	7·1	6·9
		6·7†	6·6†

* First two cycles omitted to allow equilibrium conditions to be established.
† Average values for entire test.

over-treatments of the additives tested, these materials did not significantly increase the 0·6–1·1 per cent weight loss observed for the 'base fuels' alone. The test pieces were less pliable after the storage period. This effect was observed with the 'base fuel' as well as the treated gasolines. While it is not possible to accurately predict whether this effect may influence the durability of the component in service, it would seem unlikely that the increased rigidity is significant.

The efficiency of the charcoal to adsorb, store, and desorb vapours was investigated in a series of cyclic bench tests. These tests were designed to determine the effect of liquid fuel on the charcoal. Two conventional detergents and two polymeric detergent–dispersants were tested, but only one base gasoline was used. The base gasoline was carburettor icing reference fuel, selected because of its relatively high volatility (average Reid vapour pressure 12·4 lbf/in², average mid-point 86°C).

The charcoal was obtained from a typical production canister. It was placed in the test canister and gaseous fuel was added via a top nozzle. The addition of gaseous fuel was continued until a flame ionization detector showed the presence of hydrocarbons at a side nozzle (exit). At this point, it was assumed that the charcoal was saturated.

After determining the amount of hydrocarbon adsorbed, the can was purged via the side nozzle with about 1700 cm³ (10 bed volumes) of air per minute for 20 minutes. The canister was reweighed to determine the amount of hydrocarbon desorbed. The adsorption–desorption cycle was repeated a number of times, each adsorption cycle using fresh fuel. Periodically, liquid fuel was added in place of the gaseous fuel in order to simulate a condition where liquid fuel would be forced into the canister.

Table 128.7 summarizes the average amount of hydrocarbon adsorbed and desorbed for base fuel alone and base fuel plus four multi-functional additives. The data indicate that there is no significant trend in the adsorption–desorption characteristics of the charcoal and that the efficiency of the charcoal is not significantly altered. Graphs of hydrocarbon adsorption and desorption showing how the data presented in Table 128.7 were derived are illustrated in Fig. 128.15. For simplicity only one additive treatment is shown, together with untreated fuel. These graphs indicate an initial rapid gain in hydrocarbon weight in the first few cycles, followed by a more gradual weight increase which appears as a step function after each liquid addition. Because of the nature of the gradual weight increase, it appears that it may be due to a small amount of liquid hydrocarbon building up in the bottom or dead space of the test canister where it could be difficult for the air purge to contact efficiently and remove this fuel.

CONCLUSIONS

A series of road, dynamometer, and laboratory engine tests are reviewed which establish that the build-up of deposits in the carburettor causes a deterioration in engine operation and an increase in the quantity of pollutants formed in the exhaust. The control of carburettor deposits is achieved through the use of chemical additives in the gasoline.

The additives investigated are polymeric detergent-dispersants. They are shown to be superior carburettor detergents on both 'keep-clean' and 'clean-up' cycles. These additives are shown to be compatible with both the PCV system and a charcoal canister system used for evaporative emission control.

Proper care of emission controls is an integral part of the programme to reduce pollutants from motor vehicles. The petroleum industry must continue to respond to the challenge of providing fuels which keep the control system from fouling and are compatible with the hardware itself. As compared to the development and refinement of emission control systems, the use of polymeric detergent-dispersant gasoline additives is a minor aid to reduced air pollution. However, the use of additives in general (both fuel and lubricant) is an integral part of ensuring control system efficiency in vehicle service.

APPENDIX 128.1

TWIN CARBURETTOR TEST IN EUROPEAN ENGINE (S.U. CARBURETTORS)

Engine: 1499 cm^3, four cylinders, in-line.

The test was run in two stages. In stage 1, the additive was fed to the front carburettor only. The carburettors were then removed, rated, photographed, and cleaned, and replaced for stage 2, when the additive was fed to the rear carburettor only. The engine was run continuously during each stage under the conditions shown in Table 128.8.

Table 128.8. Test data

Conditions	Details
Test duration	16 h/stage
Engine speed	2500±50 rev/min
Coolant outlet temperature	90±3°C
Oil gallery temperature	75±5°C
Inlet air temperature	58±2°C
Inlet manifold temperature	80±2°C
Exhaust back pressure	4 inH_2O
Manifold depressions	6 inHg

The carburettors and manifolds were contained in a thermostatically controlled box. Into this box all crankcase blow-by was fed, together with a proportion of exhaust gas.

APPENDIX 128.2

TWIN CARBURETTOR TEST IN EUROPEAN ENGINE (WEBER CARBURETTORS)

Engine: 1499 cm^3, four cylinders, in-line.

The test was run in two phases. In phase 1, the treated fuel was fed to the front carburettor only. The carburettors were then removed, rated, photographed, and cleaned, and replaced for phase 2, when the treated fuel was fed to the rear carburettor only. The engine was run to the cyclic procedure shown in Table 128.9, each stage comprising eight cycles, or 16 hours. The test result was expressed as the average of the two runs.

The carburettors and manifolds were contained in a thermostatically controlled box. Into this box all crankcase blow-by was fed, together with a proportion of exhaust gas.

APPENDIX 128.3

TWIN CARBURETTOR ENGINE TEST (AMERICAN V-8 ENGINE)

Engine: V-8 equipped with two single-barrel carburettors.

The engine was operated under cyclic conditions consisting of 8 min at 650 rev/min idle and 0·5 min at 2000 rev/min. The crankcase blow-by gases were directed into the air horns of the test carburettor to accelerate deposit build-up. Base fuel and treated fuel or two different treatments were evaluated simultaneously. Fuel A was run through carburettor 1 and fuel B through carburettor 2 for 20 hours. Then after weighing and cleaning and replacing the special aluminium sleeves, which fitted into the carburettor throttle bores, the fuels were reversed and the test was repeated. The performance of the fuel was expressed as the average of the two runs in terms of the percentage reduction in deposit weight, compared with the base fuel.

Table 128.9. Engine cyclic procedure

Stage no.	Duration, min	Speed, rev/min	Dynamometer load	Oil temp., °C±5	Coolant outlet temp., °C±5	Fuel times for 12·5 ml flow, sec±1	Manifold depressions, inHg±1	Exhaust back pressure, inH_2O
1	15	2800	Vary as necessary to maintain correct speed	90	85	9	—	—
2	60	1000±10		75	85	6·0	15·0	5·0
3	15	2800		90	85	9	—	—
4	30	Shut-down		—	—	—	—	—

Table 128.10. Test cycle

Stage no.	Blow-by recirculation	Duration, min	Speed, rev/min	Dynamometer load	CO emission, %	Manifold depression, inHg	Coolant outlet temp., °C	Oil temp., °C	Air intake temp., °C
1	Yes	50	800	Set up initially at stage 3. Do not adjust	8	18	70	70	40
2	No	2	4000		—	—	70	70	40
3	No	8	900		3 initially	18	70	70	40

APPENDIX 128.4

EXHAUST EMISSION TEST—EUROPEAN ENGINE

Engine: 767 cm^3, four cylinders, in-line, with single-barrel fixed-jet carburettor.

The test cycle was as shown in Table 128.10. After the initial setting, which was made at stage 3 conditions, the dynamometer load was not adjusted throughout the test. Every hour, during stage 3, the idle speed was returned to exactly 900 rev/min by cracking open the throttle plate a little more. The CO emission level was measured at this point.

APPENDIX 128.5

CARBURETTOR DEPOSIT AND EXHAUST EMISSION TEST (AMERICAN 5030-CM^3, V-8 ENGINE)

This test procedure used an American 5030-cm^3, V-8 engine in normal form. The test was cyclic, 4 min at 600 rev/min idle and 1 min at 2000 rev/min (light load). The exhaust emissions were measured every 24 hours, and if the idle speed dropped to 480 rev/min or less, the carburettor idle throttle adjusting screw was opened and the emission levels were remeasured. The curves of exhaust emission levels against time were plotted to determine the relative importance of the test fuels.

No adjustment to the carburettor was permitted after the initial set-up. The above were initial limits and CO emission, fuel flow rate, and air bush depression were allowed to increase as deposits built up.

APPENDIX 128.6

IDLE SYSTEM AIR BUSH FOULING AND EXHAUST EMISSION TEST

Engine: 767 cm^3, four cylinders, in-line, with non-standard, single-barrel, fixed-jet carburettor.

The carburettor had a non-standard slotted throttle plate, designed to ensure that throttle plate area deposits did not contribute to exhaust emission increase. The carburettor controls were set to give a CO emission of 3 per cent without blow-by recirculation, and then run with blow-by connected until deposits had formed in the idle system air bush. The initial setting conditions were as shown in Table 128.11.

Table 128.11. Initial setting conditions

Conditions	Details
Engine speed	800 ± 10 rev/min
Manifold depression	13·8–14·2 inHg
Fuel flow rate	88–92 s/25 ml
Dynamometer load	Varied to maintain speed
CO emission (with clean air)	$3·0 \pm 0·2$ per cent
CO emission (with blow-by)	$8·0 \pm 0·2$ per cent
Coolant outlet temperature	$70 \pm 3°C$
Oil temperature	$70 \pm 3°C$
Air intake temperature	$40 \pm 3°C$
Crankcase pressure	0·2 inH_2O
Blow-by recirculation	Adjust to achieve initial 8 per cent CO
Depression across air bush	$8·3 \pm 0·2$ inH_2O
Oil pressure	15 ± 3 lbf/in^2
Exhaust back pressure	5 inH_2O maximum

APPENDIX 128.7

CARBURETTOR CLEAN-UP AND EXHAUST EMISSION TEST—AMERICAN ENGINE

Engine: 5030 cm^3, V-8.

The engine was operated for about 100 hours (approximately 3000 miles) under cyclic conditions of 1 min at idle followed by 1 min at 2000 rev/min; 30 b.h.p. to build carburettor deposits. Deposit build-up was accentuated by using Cat Distillate fuel and operating with a plugged, positive crankcase ventilation system. Crankcase oil was an SAE 30 lubricant of Supplement 1 performance level.

Without changing the engine oil, the test gasoline was substituted for the Cat Distillate fuel, PCV system was unplugged, and the test continued 48 hours (approximately five tankfuls of fuel) under the same operating conditions.

Hydrocarbons (HC) and carbon monoxide exhaust emission measurements were made at the start and completion of the deposit build-up cycle and periodically on each of the test gasolines. All measurements were made with the PCV system unplugged. The carburettor was not removed from the engine throughout the duration of the test. Carburettor idle speed was allowed to drift downward during the deposit build-up cycle and was allowed to increase during the clean-up cycle. No carburettor mixture adjustments were made throughout the duration of the test.

APPENDIX 128.8

REFERENCES

(1) 'Exhaust gases from gasoline-powered cars', Investigation by the Swedish Ministry of Communication, Exhaust Group.
(2) HAYNES, C. D. and SOUTHALL, M. 'Exhaust emissions of vehicles tested to the draft European procedure', *Symp. on Motor Vehicle Air Pollution Control, Proc. Instn mech. Engrs* 1968–69 **183** (Pt 3E), 95.
(3) DICKINSON, G. W., ILDVAD, H. M. and BERGIN, R. J. 'Tune-up inspection a continuing emission control', *S.A.E. Int. Automot. Engng Congress, Detroit* 1969 (January).
(4) BRUBACHER, M. L. and GRANT, E. P. 'Do exhaust controls really work?—second report', *S.A.E. West Coast Meeting, Portland, Oregon* 1967 (August).
(5) EBERSOLE, G. D. and HOLMAN, G. E. 'Lubricant-closed PCV system relationships influence exhaust emissions', *S.A.E. Automot. Engng Congress, Detroit* 1968 (January).
(6) GALL, J. M. and OLDS, D. A. *Vehicle maintenance for low emissions—a customer education problem* 1971 (January) (Soc. Automot. Engrs, Detroit).
(7) NIEPOTH, G. W., RANSOM, G. P. and CURRIE, J. H. 'Exhaust emission control for used cars', *S.A.E. Int. Automot. Engng Congress, Detroit,* 1971 (January).
(8) SCHNEIDER, H. R., TOOKER, R. S. and KING, J. B. 'The 1970 General Motors emission control systems', *S.A.E. Automot. Engng Congress, Detroit* 1970 (January).
(9) FUCHS, E. J. 'Laboratory engine method for measurement of detergency effectiveness of gasoline additives on carburettor throttle body deposits', *S.A.E. Summer Meeting* 1961.
(10) ASSEFF, P. A. 'Multifunctional gasoline additives reduce engine deposits', *S.A.E. Meeting* 1966 (August).

THE INFLUENCE OF FUEL COMPOSITION ON AUTOMOTIVE EXHAUST EMISSIONS

B. J. KRAUS* G. P. RICHARD†

This paper discusses a study to investigate the effects of changes in fuel composition on the emissions of carbon monoxide, hydrocarbon, and nitric oxide in vehicle exhausts. When the fuel affects carbon monoxide and hydrocarbon emissions, it is through effects on mixture strength and/or cylinder to cylinder fuel distribution. Fuel viscosity, vapour pressure, and specific gravity have been identified as parameters affecting emissions, but the magnitude and direction of their influence depend on the type of induction system in use. Emissions controlled cars operate under lean conditions either by virtue of engine and fuel system design, or by virtue of air injected into the exhaust, and for the current generation of such vehicles changes in fuel quality within the normal commercial range have only a small effect on emissions. Minor mechanical adjustments produce very much larger effects.

INTRODUCTION

LEGISLATION HAS BEEN DEVELOPED, first in the U.S.A. and subsequently in some parts of Europe and other areas, which is designed to regulate the emission of carbon monoxide, hydrocarbons, and, in the U.S.A., nitrogen oxides, in the exhausts of motor vehicles. This legislation has resulted in a great deal of research into engine modifications which will control exhaust composition with a minimum of disadvantages regarding vehicle costs and performance. Very considerable progress has been made in this work, and it is now clear that suitable engine modifications can result in the elimination of some 90 per cent of the pollutants emitted by spark ignition engines (**1**)‡ (**2**). In view of this success it may be asked whether it is necessary to consider the effect of fuel conposition on exhaust emissions. This topic is, however, of some importance since little is known of the manner in which emission controlled vehicles interact with fuels of different composition, and it is conceivable that future sophisticated emission control techniques could require specially tailored fuels. In addition, there is a wide belief, in an increasingly pollution conscious public, that an ideal fuel can be developed which will reduce automotive emissions to very low levels.

Much of the work so far reported on the effects of fuel composition on exhaust emissions has been carried out in the U.S.A., and the principal objective of such work has been the determination of the manner in which fuels of different types affect exhaust reactivity; that is, the distribution of various hydrocarbon species in the exhaust, each individual hydrocarbon being more or less reactive in the complex process of producing photochemical smog (**3**)–(**6**). Two of these investigations—Dishart and Harris (**3**) and Stone and Eccleston (**4**)—did in fact report results on the emissions of carbon monoxide and nitric oxide, which indicated that changes in fuel composition have no practical influence on these pollutants.

Work has also been carried out by Bailey (**7**) on the influence of fuel composition on carbon monoxide in the exhausts of five European vehicles equipped with emission controls. Three of these vehicles were fitted with variable jet carburettors, common in the U.K. but somewhat rare in other areas. Of the remaining two cars, one had a fuel injection system and one had a fixed jet carburettor of a type common in Europe. All five cars showed statistically significant effects of front end volatility on carbon monoxide emissions at idle, but the various driving modes were not investigated.

Overall, the limited information available up to the present time suggests that for some vehicles at least variations in fuel quality have a definite effect on exhaust composition, but since the sample of vehicles examined was reasonably small, and for the most part involved U.S. designs, it was considered necessary to carry out an investigation on a representative sample of European vehicles, varying in engine size, in the method by which

---
The MS. of this paper was received at the Institution on 24th May 1971 and accepted for publication on 22nd July 1971. 23
* *Esso Research and Engineering Co., Linden, New Jersey. (Temporarily assigned to Esso Standard Italiana, Fiumicino, Rome, Italy.)*
† *Esso Research Centre, Abingdon, Berks.*
‡ *References are given in Appendix 129.2.*

fuel is metered, and in emission control systems. This paper describes the results of this investigation which involved 37 European vehicles and was carried out at Esso Research Laboratories in France, West Germany, Italy, and the U.K.

TEST FUELS USED IN THE PROGRAMME

Twelve test fuels were blended in order to study the effects of Reid vapour pressure (R.v.p.), front and mid-volatility, specific gravity, viscosity, and fuel hydrocarbon composition on exhaust emissions. The objective was to obtain a research octane number of 98 minimum, and to blend fuels whose R.v.p., volume distilled at 70°C, specific gravity, and viscosity were essentially independent of one another. In the event, the properties of the available blend stocks were such that this was not possible, and the fuels were prepared with R.v.p. relatively independent of the volume distilled at 70°C, but with many of the other properties inter-correlated to a greater or lesser degree.

Table 129.1 indicates the range of the variations in test fuel properties, and for comparison the range in quality of typical U.K. and European gasolines is also stated.

In most instances the variation in the experimental fuel properties is close to that found in commercial gasolines in the U.K., a significant exception being the volume distilled at 70° or 100°C. For these two quality points the test fuels cover a wider range than is normally found in the U.K. In other parts of Europe, where climate is markedly different, motor gasolines which have volatilities approaching the extremes of the test fuels can be found, but in each individual locality the volatility extremes are no more widely separated than in the U.K.

The inter-correlations found between the test fuels are shown in Table 129.2.

The table shows that the fuel vapour pressure was relatively free from correlation with most other test fuel properties, although correlations significant at the 90 per cent level were found between R.v.p., and volume distilled at 70°C, fuel viscosity, and the olefins content of the fuel. Other fuel properties were, however, quite extensively inter-correlated, the correlation coefficients in a number of instances being high enough to indicate a level of confidence in excess of 99.9 per cent ($|r| > 0.708$). It was appreciated that these inter-correlations, which are unfortunately difficult to avoid when blending fuels from normal gasoline components, would render the identification of fuel effects due to any one quality parameter very difficult.

EXPERIMENTAL VEHICLES AND PROCEDURE

A complete list of the vehicles used in the investigation is given in Table 129.3, which includes brief details of their carburation and emission control systems. Thirty-seven cars were tested; the engines varied in capacity from 500 to 2800 cm^3. All were equipped with emission controls to meet the U.S. requirements for their year of manufacture. In many cases emission control was achieved essentially by lean operation, improved carburation, and spark timing, but the test sample also included vehicles with exhaust air

Table 129.1. Properties of test fuels and typical U.K. and European gasolines

Fuel property	Test fuels		Typical U.K. gasolines		Typical European gasolines	
	Maximum	Minimum	Maximum	Minimum	Maximum	Minimum
Reid vapour pressure, kg/cm^2	0.86	0.42	0.71	0.53	0.97	0.42
Volume distilled at 70°C, per cent	41.0	16.0	32.0	28.0	40.0	20.0
Volume distilled at 100°C, per cent	64.0	38.0	52.0	48.0	70.0	50.0
FBP, °C	196	184	207	178	220	154
Specific gravity	0.781	0.729	0.761	0.738	0.78	0.70
Kinematic viscosity, cS at 20°C	0.62	0.49	—*	—*	—	—
Olefins, per cent	26.4	2.6	30.0	2.0	36.0	nil
Aromatics, per cent	53.2	26.9	53.0	30.0	50.0	8.5
C/H ratio	7.4	6.1	—*	—*	—	—

* Values not normally quoted for commercial gasolines. The test fuels are, however, believed to cover ranges typical of commercial fuels.

Table 129.2. Matrix of correlation coefficients for the fuel properties

Property		1	2	3	4	5	6	7	8
Reid vapour pressure	1	1.000	—	—	—	—	—	—	—
Volume distilled at 70°C, per cent	2	0.497*	1.000	—	—	—	—	—	—
Volume distilled at 100°C, per cent	3	−0.181	0.543*	1.000	—	—	—	—	—
Specific gravity	4	−0.233	−0.625†	−0.752†	1.000	—	—	—	—
Kinematic viscosity at 20°C	5	−0.516*	−0.828†	−0.628†	0.767†	1.000	—	—	—
Olefins, per cent	6	0.562*	0.654†	0.165	−0.416	−0.654†	1.000	—	—
Aromatics, per cent	7	−0.062	−0.573*	−0.791†	0.942†	0.669†	−0.438	1.000	—
C/H ratio	8	−0.027	−0.149	−0.515*	0.836†	0.406	−0.179	0.816†	1.000

* Denotes a correlation statistically significant at the 90 per cent level, but not at 95 per cent.
† Denotes a correlation statistically significant at the 95 per cent level or above.

Table 129.3. Details of cars tested in programme

Car	Carburettor or fuel system type	Emission control system
1	Variable jet—Stromberg	Clean air package (lean operation)
2	Variable jet—S.U.	Exhaust air injection
3	Variable jet—S.U.	Exhaust air injection
4	Variable jet—S.U.	Duplex manifold
5	Variable jet—S.U.	Duplex manifold
6	Variable jet—Stromberg	Exhaust air injection
7	Variable jet—Stromberg	Exhaust air injection
8	Variable jet—S.U.	Clean air package
9	Variable jet—Stromberg	Clean air package
10	Variable jet—Stromberg	Clean air package
29	Fixed jet—Solex	Clean air package
11a	Fixed jet—Solex	Clean air package
11b	Fixed jet—Solex	Clean air package
11c	Fixed jet—Solex	Clean air package
11d	Fixed jet—Solex	Clean air package
12	Fixed jet—Solex	Clean air package
13	Fixed jet—Solex	Clean air package
14a	Fixed jet	Exhaust air injection
15a	Fixed jet—Solex	Clean air package
15b	Fixed jet—Solex	Clean air package
16	Fixed jet—Solex	Clean air package
17	Fixed jet—Weber	Clean air package
18	Fixed jet—Weber	Clean air package
19a	Fixed jet—Weber	Clean air package
19b	Fixed jet—Weber	Clean air package
19c	Fixed jet—Weber	Clean air package
20	Fixed jet—Weber	Clean air package
21	Fixed jet—Weber	Clean air package
22	Fixed jet—Weber	Clean air package
14b	Fixed jet	Exhaust air injection
23	Fixed jet—Solex	Clean air package
24	Fixed jet—Ford	Exhaust air injection
25	Fuel injection—mechanical	Clean air package
26	Fuel injection—mechanical	Clean air package
27a	Fuel injection—electronic	Clean air package
27b	Fuel injection—electronic	Clean air package
28	Fuel injection—electronic	Exhaust air injection

injection systems. Twenty-two of the test cars used fixed jet, and 10 used variable jet carburettors. Five cars had fuel injection systems, two of which were mechanically controlled and three of which used electronic control.

The cars were tested with the engine hot, using the European emissions test procedure. The use of the complete cycle, starting from a cold engine, was ruled out on the grounds that this would be unacceptably time consuming in a programme using so many vehicles and test fuels. While it was recognized that this choice of procedure could possibly prevent the observation of fuel volatility effects during cold engine operations, it was also known that the reproducibility of the test was likely to be improved by the omission of cold cycles. Overall it was felt that the advantages of using a hot start procedure outweighed the disadvantages.

Duplicate tests were run on each fuel in each car, the fuels being tested in random order. Starting with a warmed-up engine, the vehicle under test was run through three cycles of the European test procedure, the exhaust from the final two cycles being collected in a plastic bag for analysis. It would have been desirable to use a single material for the sample bags, but unfortunately the various laboratories involved in the work were differently equipped in this respect. To minimize effects of varying bag permeability, analyses of the gases were carried out as soon as possible after sampling. Carbon dioxide, carbon monoxide, and hydrocarbons in the exhaust were measured for all of the tests carried out. Nitric oxide emissions were also measured in the majority of the cars, but equipment limitations at the time of the experiments prevented a completely full cover of this pollutant.

Records were also kept of the fuel used in each test, the exhaust volume, and the oxygen content of the exhaust, in order to facilitate the calculations of air/fuel ratio and the mass emissions from each test. Mass emissions could not be obtained for all the vehicles since it was not possible to measure exhaust gas volume in all the tests. Where available, mass emission data show the same correlations as the corresponding concentrations. To permit consideration of all the vehicles here, the results will be discussed in terms of concentrations only.

Analysis of the exhaust gas for CO_2, CO, hydrocarbon, and NO was carried out using non-dispersive infrared analyses (NDIR), while oxygen concentration (where measured) was obtained using paramagnetic or polarographic instruments. The exhaust hydrocarbons for three of the vehicles tested were measured both with the NDIR analyser, and by means of a Beckman Flame Ionization Detector (FID analyser).

Where data on exhaust gas oxygen concentrations were available, air/fuel ratio was calculated from the exhaust gas analysis using the method of Spindt (8). In other cases, an alternative calculation, detailed in Appendix 129.1, was employed.

Emissions data obtained in the tests were corrected for air dilution before submission to statistical analysis, the correction factor being calculated from the following expression:

$$F = 14\cdot 5/(\% \ CO_2 + 0\cdot 5\% \ CO + 10\cdot 8\% \ HC)$$

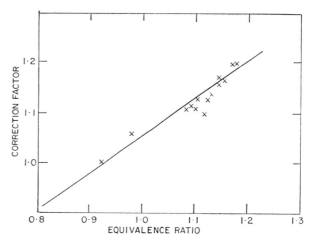

Fig. 129.1. Relation of correction factor to equivalence ratio (no air dilution)

It should be noted that this correction factor is itself a measure of air/fuel ratio—a parameter which has a very considerable effect on emissions from automotive engines. Fig. 129.1 shows a plot of correction factor against equivalence ratio (the ratio of the measured air/fuel ratio to the stoichiometric requirement) in the case of a typical vehicle which does not employ exhaust air injection.

It is apparent from this that the application of the correction factor to exhaust emissions tends to distort the true relation between mixture strength and percentage volume of pollutant in the exhaust. Over a range of 0·8 to 1·2 in equivalence ratio, however, the factor may change from 0·9 to about 1·2, and the magnitude of this change is not considered sufficiently great to introduce a serious error.

EXPERIMENTAL RESULTS AND DISCUSSION

Effects of mixture strength on emissions

It is well known that mixture strength has a profound effect on the emissions of all three of the major exhaust pollutants (**9**), and in the present programme a quantitative examination of its effect was considered important for two reasons. First, small variations in the manner in which a vehicle is driven during the test cycle can affect the mean air/fuel ratio over which it operates, and so produce changes in emissions which may obscure the effects of fuel quality parameters. Second, fuel physical properties can themselves be expected to affect the metering of the fuel to the engine, thus changing the mixture strength. In the examination of the effect of mixture strength on emissions, errors due to changes in the stoichiometric air requirement of the various fuels were avoided by expressing mixture strength in terms of equivalence ratio—that is, the quotient of the observed and the stoichiometric air/fuel ratio.

Carbon monoxide and mixture strength

Carbon monoxide in the exhaust was found in these experiments to be very highly correlated with equivalence ratio, correlation coefficients generally having a value of the order of 0·9. The relation between exhaust carbon monoxide and equivalence ratio is illustrated for a number of cars in Fig. 129.2, from which it can be seen that changes in carbon monoxide are very well explained by variations in equivalence ratio.

It is concluded that within the accuracy of the experiment, no fuel type is better than another, provided that it is operated at the same equivalence ratio, and that, so far as this pollutant is concerned, fuel properties only affect emissions in so far as their physical properties modify the metering of the fuel to the engine.

The results of many of the tests carried out fall essentially along the same curve, but as can be seen from Fig. 129.2, the data from a number of vehicles, particularly in the lean mixture region, were displaced. Such deviations can occur for several reasons other than experimental

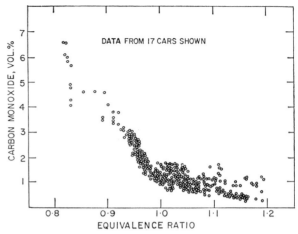

Fig. 129.2. Variation of carbon monoxide with equivalence ratio

error. First, two cars running at the same overall mixture strength over the test cycles can give rise to different CO levels because they may run at different mixture strengths during individual driving modes. Second, different CO levels can arise as a result of differences in the cylinder to cylinder distribution of fuel and air. Finally, in the case of fuel injection cars the engine pumps air only during deceleration, and in such cases the overall equivalence ratio will be greater than the true operational level corresponding to the driving modes in which both fuel and air are pumped.

Examination of the results from individual cars indicated that the test fuels produced in general a variation of about 0·05–0·1 in the equivalence ratio at which a vehicle operates, compared with a test to test variation of the order of 0·01–0·02 for a single fuel. The effect of this change in mixture strength on the emissions of carbon monoxide was very much a function of the mean mixture strength at which the vehicle was operated. For rich operating cars (equivalence ratio around 0·9) the overall variations in test fuel quality resulted in changes of about 3·0 per cent by volume in CO emissions. For lean operating cars (equivalence ratio 0·05–1·2) the variation in CO due to the extremes in test fuel quality was around 0·5 per cent by volume. In terms of change in mass emissions, the effect of fuel quality on the lean cars is only one-sixth of that found for the richest operating cars, and the magnitude of this effect is very considerably less than the vehicle to vehicle variation in the relations between CO emissions and equivalence ratio.

Effect of mixture strength on hydrocarbons and nitric oxide

In many cases a comparison of hydrocarbon and nitric oxide emissions with equivalence ratio for individual vehicles did not show a clear relation between mixture strength and emissions. If, however, the results from several cars are plotted together, as in Fig. 129.2, the fundamental relation of emissions to mixture strength begins to appear.

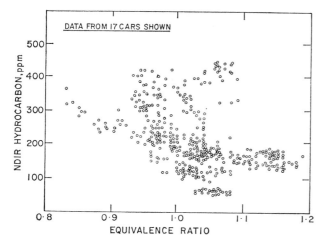

Fig. 129.3. Variation of NDIR hydrocarbon with equivalence ratio

Fig. 129.3 illustrates this in the case of hydrocarbon emissions. Although the results show a very considerable scatter, it can be seen that, as is to be expected, hydrocarbon concentration in the exhaust drops steadily as equivalence ratio is increased.

Systematic work (10) on vehicles running under constant conditions has in the past indicated that an increase in hydrocarbon emissions can be expected at equivalence ratios above about 1·15, due to the onset of misfire. No indication of such an effect can be seen from the cars considered here, but this could be because no vehicle tested operated at much above this limit. As might be expected the change in hydrocarbon per unit change in equivalence ratio appears to be greater for operations on the rich side of stoichiometric than for those on the lean side.

The correlation between nitric oxide and mixture strength obtained for 11 cars in the present series of tests is illustrated in Fig. 129.4.

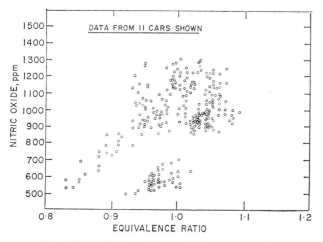

Fig. 129.4. Nitric oxide emissions at varying equivalence ratio

Although the correlation was rather poor the figure shows that nitric oxide emissions increase steadily with mixture strength up to an equivalence ratio of the order of 1·0. At some point beyond this they would be expected to start to decrease, but the present data do not show a clearly defined maximum.

At equal equivalence ratios, exhaust hydrocarbons and nitric oxide vary considerably from car to car, as indicated by the scatter of the points in Figs 129.3 and 129.4. This is by no means surprising, since it is well known that hydrocarbon and nitric oxide emissions are significantly changed by many operating variables in addition to mixture strength. Examples are spark timing, combustion chamber design, valve overlap, and fuel carbon/hydrocarbon ratio (3) (9)–(11).

The effect of fuel properties on emissions

The data presented in the previous section show clearly that operating mixture strength has a most important effect on exhaust emissions, particularly under somewhat rich mixture operation, and it is apparent that changes in fuel quality which affect the metering of the fuel to the engine, or its distribution to the various cylinders, will result in alterations to the concentrations of pollutants emitted by the vehicle. In the case of carbon monoxide the relation between mixture strength and exhaust concentration is so strong that on these, and on general theoretical grounds, it is difficult to see how fuel quality can affect this pollutant by any mechanism other than alteration in mixture strength. However, where hydrocarbons and nitric oxide in the exhaust are concerned it must be recognized that the chemical nature of the fuel could well modify the emissions, quite apart from the effect of fuel physical properties on mixture strength.

To find whether any particular fuel property has a consistent effect on the emissions of the various pollutants over a whole car population, the emissions data obtained in these tests were correlated against the various fuel physical and chemical properties, using the technique of single independent variable linear regression analysis, in order to find the fuel parameters showing correlations with emissions at least at the 95 per cent confidence level. This procedure shows up correlations, even when non-linear, except in cases where maxima or minima exist within the experimental range of the independent variable, and previous examination of the data had shown that in the present case this complication would not arise.

For this series of tests the value of the correlation coefficient at the 95 per cent confidence level was $|r| = 0.389$, and at this level the independent variable under investigation accounts for about 15 per cent of the variation in the dependent variable. Tables 129.4–129.6 show the values of correlation coefficients obtained in this work which were above this minimum, and in addition indicate for each series of tests the range in emissions found as a result of the use of different fuels. These results are discussed in more detail below.

Table 129.4. Regression of exhaust carbon monoxide against fuel properties

Vehicle	Range of CO values, vol.%	R.v.p.	% Distilled at 70°C	Specific gravity	Viscosity	% Olefins	% Aromatics	C/H
Variable jet carburettors								
1	0·4–1·3	0·487	0·644	(0·723)	(0·749)	0·444	(0·669)	(0·540)
2	0·2–0·4	0·783	0·677	—	—	0·793	—	—
3	0·7–1·8	—	0·667	(0·826)	(0·668)	0·401	(0·756)	(0·490)
4	0·6–1·4	0·443	0·738	(0·751)	(0·806)	0·739	(0·719)	(0·410)
5	0·9–1·8	0·444	0·868	(0·780)	(0·914)	0·658	(0·689)	—
6	0·5–0·9	0·743	0·610	(0·610)	(0·707)	0·661	(0·514)	(0·422)
7	0·4–0·8	0·517	0·755	(0·838)	(0·908)	0·627	(0·752)	(0·530)
8	1·3–2·0	—	0·662	(0·824)	(0·804)	—	(0·730)	(0·402)
9	2·7–6·6	0·766	0·730	(0·641)	(0·792)	0·865	(0·556)	(0·422)
10	0·6–1·0	0·487	0·869	(0·822)	(0·824)	0·675	(0·804)	(0·514)
Fixed jet carburettors								
29	1·4–2·2	(0·580)	(0·770)	0·410	0·590	(0·440)	0·400	—
11a	1·9–3·1	(0·598)	(0·415)	—	0·639	(0·460)	—	—
11b	1·5–2·2	—	(0·540)	0·472	0·744	(0·441)	0·400	—
11c	0·8–1·6	(0·520)	(0·540)	0·544	0·659	—	0·426	—
11d	1·2–2·5	(0·470)	(0·710)	0·580	0·710	(0·610)	0·560	—
12	0·9–1·1	(0·419)	(0·408)	0·404	0·521	(0·648)	0·411	—
13	1·4–2·5	—	(0·633)	0·553	0·648	—	0·380	—
14a	2·3–3·2	(0·744)	—	—	0·502	(0·372)	—	—
15a	1·4–2·3	—	(0·509)	—	0·468	(0·427)	—	—
15b	1·0–2·2	0·809	—	—	—	—	—	—
16	1·0–1·6	0·605	—	—	—	—	0·398	—
17	1·7–2·8	0·410	—	—	—	—	—	—
18	1·7–3·3	0·690	—	—	—	—	—	—
19a	1·1–2·2	0·577	—	—	—	—	—	—
19b	1·7–2·2	—	—	—	—	—	—	—
19c	2·0–2·8	—	—	—	—	—	—	—
20	2·5–3·2	—	—	—	—	0·580	—	—
21	1·3–2·6	—	—	—	—	—	—	—
22	0·7–1·2	—	—	—	—	—	—	(0·480)
14b	0·4–0·7	—	—	—	—	—	—	—
23	1·1–1·8	—	—	—	—	(0·456)	—	—
24	1·5–2·3	—	—	—	—	—	—	—
Fuel injection (mechanical)								
25	0·4–1·1	(0·610)	(0·580)	0·840	0·810	(0·620)	0·730	0·690
26	0·3–1·4	—	(0·731)	0·897	0·787	(0·592)	0·908	0·575
Fuel injection (electronic)								
27a	1·3–1·8	(0·430)	—	—	—	—	—	—
27b	0·4–0·7	(0·743)	—	—	—	(0·577)	—	—
28	2·3–3·4	(0·610)	—	—	0·630	—	—	—

★ Figures in parentheses indicate negative correlation coefficients.

Carbon monoxide

Table 129.7 summarizes the results of the work discussed above, and it will be seen that the correlations obtained were very much dependent on the fuelling system of the cars investigated. The table shows that CO correlated with fuel chemical properties as well as with its physical nature, but at least to some extent these relationships arise as a result of the extensive inter-correlations which exist between fuel physical and chemical properties. In the case of three vehicles correlations were found to exist between CO and a single chemical property. It is believed that such correlations arose by chance.

For the large majority of vehicles having variable jet carburettors, all of the fuel parameters under investigation can be correlated with carbon monoxide, and for those cars decreasing volatility, or increasing specific gravity or fuel viscosity, resulted in a decrease in exhaust carbon monoxide. The correlation coefficients obtained for R.v.p. versus CO were in general rather lower in value than those for the other physical properties, but in all cars but two were well above the 95 per cent level of confidence. Since in fact R.v.p. correlates rather poorly with the other fuel properties, it is believed that these correlations with CO represent a genuine fuel quality effect, and it is thought that high R.v.p. leads to increased CO by increasing the rate of evaporation from the carburettor bowl.

Specific gravity, volume distilled at 70°C, and fuel viscosity in the test fuels all show strong inter-correlations, and it is therefore difficult to define the part that these variables play in the formation of CO in the exhaust. The correlation found for viscosity with CO was, however, somewhat better than those for the other variables, and

Table 129.5. Regression of exhaust hydrocarbon against fuel properties

Vehicle	Range of HC values, p.p.m.	Correlation coefficients at or above 95% confidence level*						
		R.v.p.	% Distilled at 70°C	Specific gravity	Viscosity	% Olefins	% Aromatics	C/H
Variable jet carburettors								
1	98–167	—	0.516	(0.870)	(0.625)	—	(0.798)	(0.708)
2	65–159	(0.851)	(0.837)	—	0.651	(0.692)	—	—
3	108–304	—	0.606	(0.871)	(0.768)	0.435	(0.862)	(0.607)
4	90–134	—	—	(0.474)	—	—	(0.439)	—
5	124–231	—	—	—	—	—	—	—
6	155–288	(0.506)	—	—	—	—	—	—
7	95–195	—	—	(0.580)	—	—	(0.550)	(0.415)
8	118–196	—	—	(0.563)	—	—	(0.474)	(0.479)
9	201–374	0.404	0.558	(0.776)	(0.671)	0.625	(0.732)	(0.631)
10	50–65	—	—	(0.571)	—	—	(0.449)	(0.725)
Fixed jet carburettors								
29	403–605	—	—	—	—	—	(0.580)	(0.430)
11a	150–297	—	—	—	—	—	—	(0.553)
11b	189–314	—	—	—	—	—	—	—
11c	110–270	—	—	—	—	—	—	—
11d	213–293	—	—	—	—	—	—	—
12	330–450	—	—	(0.542)	—	—	(0.541)	(0.735)
13	145–245	(0.358)	—	—	0.369	—	—	—
14a	159–420	(0.634)	—	—	—	(0.445)	—	—
15a	127–313	0.399	0.427	(0.704)	(0.450)	—	(0.618)	(0.555)
15b	135–220	—	—	(0.693)	(0.383)	—	(0.703)	(0.525)
16	130–165	—	0.391	(0.735)	(0.405)	—	(0.717)	(0.630)
17	248–444	—	0.440	(0.560)	—	—	(0.740)	(0.700)
18	154–253	—	0.450	(0.870)	(0.580)	—	(0.820)	(0.800)
19a	183–296	0.557	—	—	—	—	—	—
19b	203–292	—	—	(0.820)	(0.610)	—	(0.840)	(0.780)
19c	227–354	—	—	(0.670)	—	—	(0.600)	(0.870)
20	274–411	—	—	(0.750)	(0.540)	—	(0.780)	(0.840)
21	246–436	—	0.500	(0.810)	(0.510)	—	(0.750)	(0.690)
22	130–240	—	—	(0.870)	(0.430)	—	(0.820)	(0.920)
14b	80–145	—	—	—	—	—	—	—
23	280–390	—	—	(0.758)	(0.622)	0.422	(0.769)	(0.644)
24	262–453	—	—	(0.730)	(0.429)	—	(0.665)	(0.578)
Fuel injection (mechanical)								
25	152–254	(0.800)	(0.590)	—	0.530	(0.580)	—	—
26	140–185	(0.581)	(0.593)	—	0.616	(0.929)	0.388	—
Fuel injection (electronic)								
27a	156–235	(0.400)	—	(0.620)	—	—	(0.680)	(0.540)
27b	140–195	—	—	(0.681)	(0.204)	—	(0.581)	(0.794)
28	169–262	—	—	—	—	—	—	—

* Figures in parentheses indicate negative correlation coefficients.

for this reason an independent check was made on the effect of viscosity.

For this purpose, one of the test fuels (no. 2) was treated with a polymeric additive which increased the viscosity but made negligible alterations to the other fuel properties. Several fuels of enhanced viscosity were prepared in this matter and tested on three vehicles, including one of the test cars. In each case, carbon monoxide emissions were shown to be sensitive to fuel viscosity. The results obtained on test car no. 3 are illustrated in Fig. 129.5.

It thus appears that cars using variable jet carburettors are affected principally by fuel viscosity, with vapour pressure playing a subsidiary role in the determination of carbon monoxide exhaust concentrations. Fig. 129.6 illustrates the effects of these two variables in the case of a rich mixture operating car which is particularly sensitive to changes in fuel quality.

Of the 21 fixed jet cars tested, eight showed no correlation with fuel physical properties at the 95 per cent confidence level. The remainder showed correlations involving one to four of these parameters. No single fuel property appeared as a consistent factor in the control of exhaust carbon monoxide, and in most cases the statistical significance of the correlations was lower than in the case of the variable jet cars. Nine cars showed a positive correlation with viscosity, and either a negative or a very low correlation with R.v.p. These vehicles also showed correlations of carbon monoxide with volume distilled at 70°C and specific gravity, and, as in the case of viscosity, the correlations were opposite in sign to those found for variable jet cars. The remaining five fixed jet cars showed

Table 129.6. Regression of exhaust nitric oxide against fuel properties

Vehicle	Range of NO values, p.p.m.	Correlation coefficients at or above 95% confidence level*						
		R.v.p.	% Distilled at 70°C	Specific gravity	Viscosity	% Olefins	% Aromatics	C/H
Variable jet carburettors								
1	1250–1525	—	—	—	—	—	—	0·402
2	1030–1244	—	—	—	—	—	—	—
3	521–904	—	(0·491)	0·807	0·543	—	0·762	0·563
4	1093–1290	—	0·412	(0·613)	(0·561)	—	(0·556)	(0·457)
5	863–1077	—	—	—	(0·390)	—	—	—
6	525–701	—	—	—	—	—	—	—
7	585–745	—	—	—	—	—	—	—
8	765–1153	—	0·515	(0·487)	(0·516)	—	(0·450)	—
9	520–930	(0·735)	(0·638)	0·599	0·687	(0·824)	0·490	0·404
Fixed jet carburettors								
11a	862–1026	—	—	—	—	0·406	—	—
11b	732–956	—	—	—	—	0·460	—	—
11c	861–1087	—	0·416	—	—	—	—	—
15a	1257–1484	—	—	—	—	—	—	—
17	1109–1418	—	—	0·560	0·520	(0·420)	0·570	0·460
18	498–656	(0·410)	—	0·530	—	—	—	0·590
19a	492–708	—	—	—	—	—	—	—
20	783–1271	—	—	—	—	—	—	—
21	850–1126	—	—	—	—	—	—	—
22	1109–1269	—	—	(0·610)	(0·490)	—	(0·550)	—
24	647–1008	—	—	—	—	—	0·395	—
Fuel injection (mechanical)								
25	974–1230	—	—	—	—	—	—	—
Fuel injection (electronic)								
27a	856–985	—	—	—	—	—	—	—

* Figures in parentheses indicate negative correlation coefficients.

a positive correlation of carbon monoxide with R.v.p. and no correlations with other physical parameters.

It is thought that the behaviour of the group of nine cars showing a positive correlation between viscosity and carbon monoxide could be explained by the fact that an increase in fuel viscosity will tend to impair fuel atomizations and thus lead to problems of fuel distribution between cylinders. Since a number of these cars also showed rather less significant correlations between specific gravity and carbon monoxide, it it also possible that the effect is due to the fact that increasing specific gravity tends to increase mass fuel flow, and this leads to increased mixture strength. This latter explanation seems, however, to be unlikely on the grounds that most of the cars showed a negative correlation between R.v.p. and carbon monoxide, and in the series of fuels used in this work a significant negative correlation existed between R.v.p. and viscosity, while little correlation could be found between R.v.p. and specific gravity. In two cars it was possible to calculate the equivalence ratio for all the tests carried out, and in these cases little correlation could be found between equivalence ratio and CO emissions, or between fuel physical properties and equivalence ratio.

Regarding the final group of five cars, it appears most likely that the positive correlation of carbon monoxide with R.v.p. is due to mixture enrichment resulting from

Table 129.7. Correlation of fuel properties with carbon monoxide emissions

Fuel system	No. of cars		No. of cars whose CO emissions correlate with stated property at or above the 95% level*						
	Total	Those showing no correlation at 95% level	R.v.p.	% Distilled at 70°C	Specific gravity	Viscosity	% Olefins	% Aromatics	C/H
Variable jet carburettor	10	nil	8	10	(9)	(9)	9	(9)	(8)
Fixed jet carburettor	22	8	5, (6)	(8)	6	9	1, (8)	7	(1)
Fuel injection (mechanical)	2	nil	(1)	(2)	2	2	(2)	2	2
Fuel injection (electronic)	3	nil	(3)	nil	nil	1	1	nil	nil

* Figures in parentheses indicate negative correlation coefficients.

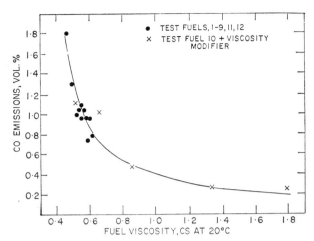

Fig. 129.5. Effect of fuel viscosity on CO emissions, car no. 3

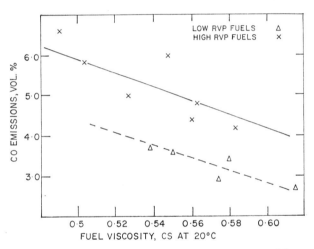

Fig. 129.6. Effects of viscosity and R.v.p. on CO emissions, car no. 9

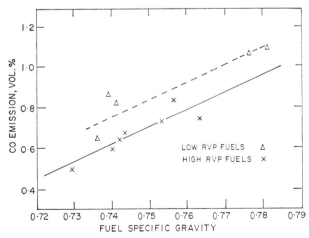

Fig. 129.7. Effects of specific gravity and R.v.p. on CO emissions, car no. 5

the increasing rate of evaporation of the fuel from the carburettor bowl.

Carbon monoxide from the two vehicles fitted with mechanically operated fuel injection systems was significantly correlated with all of the fuel variables, the CO concentration increasing with increasing specific gravity, viscosity, aromatics, and C/H ratio, and with decreasing volatility and olefin content. These cars have a fuel pump which delivers a constant volume of fuel at a given engine speed, and in these cases it is likely that specific gravity, which shows the best correlation with carbon monoxide, is the factor which most influences emissions, due simply to the fact that an increase in specific gravity will result in an increased mass of fuel delivered to the engine. In view of the fact that vapour pressure is not well correlated with specific gravity, the correlation recorded for CO and R.v.p. in one of these cars must be regarded as probably genuine, but no explanation can be offered for the effect. The individual effects of specific gravity and R.v.p. on this car are illustrated in Fig. 129.7, which also serves to emphasize that in spite of the definite relation which this lean operating car shows between exhaust CO and fuel quality, the overall effect of extremes in fuel quality on its emissions is low.

Cars having electronically operated fuel injection are not affected by fuel specific gravity. However, they do respond to R.v.p., and in one case to fuel viscosity. These cars have a constant pressure fuel injection system, controlled by a pressure regulator and a return line to the tank. Since, for these fuels, viscosity increases with specific gravity, an increase in fuel specific gravity is accompanied by a tendency for pressure in the fuel line to rise. The regulator responds by spilling an increased proportion of fuel back to the tank, and the mass charge to the cylinders is more or less unaffected. The negative correlation of R.v.p. with CO could possibly be explained by a similar mechanism.

Hydrocarbon emissions

The results of regression of fuel parameters against hydrocarbon emissions are summarized in Table 129.8. In contrast to the carbon monoxide results, variable jet cars do not show a marked and uniform effect of fuel quality on hydrocarbon emissions. On the other hand, more fuel factors correlate with hydrocarbons in fixed jet cars than was observed for CO. Although, in a number of cases, factors which tend to increase CO in the exhaust also increase hydrocarbons, an outstanding difference between the hydrocarbon and CO data is that in the former case increased fuel aromatic content almost invariably results in reduced hydrocarbon as measured by the NDIR analyser.

Out of 37 cars tested, 22 showed a negative correlation of NDIR hydrocarbon with fuel aromatics, 13 showed no correlation, and in one case, a car with mechanically controlled fuel injection, the correlation was positive. As might be expected, this result was reflected in the correlation of hydrocarbons with specific gravity, viscosity and

Table 129.8. Correlation of fuel properties with exhaust hydrocarbons

Fuel system	No. of cars		No. of cars whose hydrocarbon emissions correlate with stated property at or above the 95% level of confidence*						
	Total	Those showing no correlation at 95% level	R.v.p.	% Distilled at 70°C	Specific gravity	Viscosity	% Olefins	% Aromatics	C/H
Variable jet carburettor	10	1	1, (2)	3, (1)	(7)	1, (3)	2, (1)	(7)	(6)
Fixed jet carburettor	22	4	2, (2)	5	(13)	10, (1)	1, (1)	(14)	(15)
Fuel injection (mechanical)	2	nil	(2)	(2)	nil	1	(2)	1	nil
Fuel injection (electronic)	3	nil	(1)	nil	(2)	(1)	nil	(2)	(2)

* Figures in parentheses indicate negative correlations.

C/H ratio. The variation of NDIR hydrocarbons with fuel aromatics content is illustrated in Fig. 129.8.

In the consideration of this result, it is necessary to take careful account of what the NDIR analyser actually measures. It is able to sense only hydrocarbon and not oxygenated derivatives; further, its sensitivity depends very much on the type of carbon to hydrogen bond. Saturated hydrocarbons are detected with about equal sensitivity, with the exception of methane, which is not seen. It is less sensitive to olefins and is relatively insensitive to aromatics. Thus the correlations noted above may merely be due to the fact that aromatic fuels increase the aromatic content of the exhaust hydrocarbons, rather than to some property of aromatics which reduces exhaust hydrocarbon. Such behaviour has been observed in the past, using gas chromatographic measurement of exhaust hydrocarbon (3). To check this point, hydrocarbon measurements on three of the vehicles were made using both the NDIR analyser and an FID instrument. The latter responds to hydrocarbons by the detection of ions in a hydrogen flame, and is sensitive to all carbon-bearing compounds. Fig. 129.9 shows the mean values of the ratio of the FID to the NDIR results, plotted against the paraffin content of the fuels. This ratio increases sharply as fuel paraffin content is reduced, and it is concluded that in fact the hydrocarbon composition of the fuel determines to a large extent the hydrocarbon species in the exhaust, and that no significant reduction in exhaust hydrocarbon content is obtained by the use of fuels having high aromatic or olefin content.

Nitric oxide

In some 41 per cent of the cars examined for nitric oxide emissions, no correlation significant at the 95 per cent confidence level could be found between nitric oxide emissions and fuel properties. For the remaining vehicles some correlations were demonstrated, but for any given variable the directions of the effect varied from car to car, and no fuel property having a consistent effect on nitric oxide in the exhaust was found. This situation is summarized in Table 129.9.

As was shown in Fig. 129.4 there is a general tendency for nitric oxide emissions to increase to a peak level as mixture strength is decreased from rich operation to about the stoichiometric point, and to some extent this tendency is reflected in the correlations found, particularly in the case of the richer operating cars. However, in addition to mixture strength, fuel chemical composition can be expected on theoretical grounds to affect nitric oxide emissions, since any increase in C/H ratio should increase flame temperature. In fact, such an effect has been reported by Dishart (10) and by Ninomiya and Golovoy

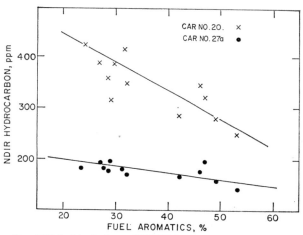

Fig. 129.8. Variation of NDIR hydrocarbon with fuel aromatics

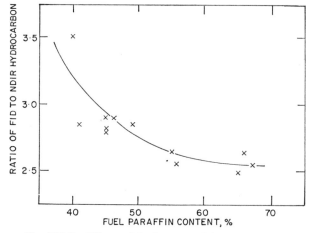

Fig. 129.9. Effect of fuel composition on exhaust hydrocarbon analyses

Table 129.9. Correlation of fuel properties with nitric oxide emissions

Fuel system	No. of cars		No. of cars whose NO emissions correlate with stated property at or above the 95% level of confidence*						
	Total	Those showing no correlation at 95% level	R.v.p.	% Distilled at 70°C	Specific gravity	Viscosity	% Olefins	% Aromatics	C/H
Variable jet carburettor	9	3	(1)	2, (2)	2, (2)	2, (3)	(1)	2, (2)	3, (1)
Fixed jet carburettor	11	4	(1)	1	2, (1)	1, (1)	2, (1)	2, (1)	2
Fuel injection (mechanical)	1	1	nil	nil	nil	nil	nil	nil	nil
Fuel injection (electronic)	1	1	nil	nil	nil	nil	nil	nil	nil

* Figures in parentheses indicate negative correlations.

(11). Thus nitric oxide formation is governed partly by fuel physical properties, the magnitude and direction of which can vary from car to car, and partly by chemical properties which must of themselves be inter-correlated with the physical parameters. In practice, it can be expected that these parameters will sometimes complement and sometimes oppose one another, and in these circumstances it is not surprising that no single fuel parameter can be found which has a consistent and unidirectional effect on nitric oxide emissions.

CONCLUSIONS

(1) This work on a representative sample of European cars fitted with emission controls has shown that, in cars where fuel composition has an influence on emissions of exhaust pollutants, this can to a large extent be accounted for in terms of the effect of fuel physical properties on the mixture strength at which the vehicle operates, or in terms of their effect on cylinder to cylinder mixture distribution. Comparable changes can be produced by minor changes in the mechanical adjustment of the vehicle.

(2) For vehicles fitted with variable jet carburettors, mixture strength, and therefore exhaust carbon monoxide and hydrocarbon, are significantly affected by fuel viscosity, and to a lesser extent by fuel vapour pressure (R.v.p.). Increasing viscosity or decreasing R.v.p. produces leaner mixtures and lower exhaust carbon monoxide.

(3) Many vehicles using fixed jet carburettors appear to be insensitive to changes in fuel physical properties. Where emissions are affected by fuel quality, it appears that viscosity and R.v.p. have the greatest influence on emissions, but the effect of these parameters varies in magnitude and direction from vehicle to vehicle, and it is apparent that these variables can interact with different parts of the fuel system in manners which oppose each other.

(4) Vehicles with constant volume fuel injection systems are sensitive to fuel specific gravity, an increase in which increases the mixture strength and results in increased carbon monoxide and hydrocarbon. On the other hand, those having constant pressure fuel injection systems appear to be consistently sensitive to fuel vapour pressure, an increase in which leads to reduced carbon monoxide emissions.

(5) Nitric oxide formation in a vehicle exhaust can be varied by changes in both the chemical and physical nature of the fuel. The magnitude and direction of the effects on these parameters vary from car to car, and no single fuel parameter has a unique effect on nitric oxide emissions from all cars.

(6) For any type of fuel system practically significant changes in exhaust pollutants per unit change in mixture strength only occur in vehicles operating under overall rich conditions. Current emission controlled vehicles operate under overall lean conditions either by virtue of engine and fuel system design, or by injection of air into the exhaust system, and for such vehicles changes in fuel quality within the range of variations encountered in commercial fuels produce an effect on emissions which is small compared with test reproducibility, and with the effect of small variations in vehicle tune.

(7) For a normal vehicle population the variation in response of individual cars to changes in fuel composition is such that it is not possible to formulate a fuel, using commercial gasoline components, which will on average reduce the level of pollutants in vehicle exhausts.

ACKNOWLEDGEMENTS

The authors gratefully acknowledge the work of some 15 members of the staffs of the Esso Research Laboratories in France, West Germany, Italy, and the U.K. who carried out the measurements reported in this paper, and in particular that of Dr H. M. Ashton who co-ordinated the work in its early stages and formulated the test fuels. Thanks are also due to the many organizations and private individuals who kindly made vehicles available for the study.

APPENDIX 129.1

CALCULATION OF AIR/FUEL RATIO FOR MEASUREMENTS OF CARBON DIOXIDE, CARBON MONOXIDE, AND HYDROCARBON ONLY

Calculated air/fuel ratio

Air/fuel ratio
$$= \frac{2 \cdot 42r[3C/r + M(3/r - 0 \cdot 8) + H(32 \cdot 4/r - 8 \cdot 1) + 100]}{(C + M + 10 \cdot 8H)(r + 1)}$$

where C is the percentage of CO_2 in the exhaust, M the percentage of CO in the exhaust, H the percentage of hydrocarbon in the exhaust (NDIR measurement), and r the carbon/hydrogen ratio of the fuel.

Theoretical air requirement of fuel

Air requirement (weight to weight fuel) $= \dfrac{11 \cdot 5(r+3)}{(r+1)}$.

APPENDIX 129.2

REFERENCES

(1) FUSSELL, D. R. 'Atmospheric pollution from petrol and diesel engined vehicles', *Petrol. Rev.* 1970 **24**, 192.

(2) LANG, R. J. 'A well-mixed thermal reactor system for automotive emission control', S.A.E. Paper 710608, 1971.

(3) DISHART, K. T. and HARRIS, W. C. 'The effect of gasoline hydrocarbon composition on automotive exhaust emissions', *Proc. Div. Refining, Am. Petrol. Inst.* 1968 **48**, 612.

(4) STONE, R. K. and ECCLESTON, B. H. 'Vehicle emissions vs. fuel composition', *Ibid.* 1968 **48**, 705.

(5) STONE, R. K. and ECCLESTON, B. H. 'Vehicle emissions vs. fuel composition. II', *Ibid.* 1969 **49**, 651.

(6) DIMITRIADES, B., ECCLESTON, B. H. and HURN, R. W. 'An evaluation of the fuel factor through direct measurement of photochemical reactivity of emissions', *J. Air Pollution Cont. Assn.* 1970 **20**, 150.

(7) BAILEY, C. L. 'The influence of motor gasoline characteristics upon CO emissions under engine idle conditions', *Symp. Motor Vehicle Air Pollution Control, Proc. Instn mech. Engrs* 1968–69 **183** (Pt 3E), 108.

(8) SPINDT, R. S. 'Air fuel ratios from exhaust gas analysis', S.A.E. Paper 650507, 1965.

(9) HULS, T. A., MYERS, P. S. and UYEHARA, D. A. 'Spark ignition engine operation and design for minimum exhaust emissions', S.A.E. Paper 660405, 1966.

(10) DISHART, K. T. 'Exhaust hydrocarbon composition: its relation to gasoline composition', *Proc. Div. Refining, Amer. Petrol. Inst.* 1970 **50**, 514.

(11) NINOMIYA, J. S. and GOLOVOY, A. 'Effects of air fuel ratio on composition of hydrocarbon exhaust from isooctane, diisobutylene, toluene, and toluene–n-heptane mixture', S.A.E. Paper 690504, 1969.

C130/71

MANIFOLD REACTION WITH EXHAUST RECIRCULATION FOR CONTROL OF AUTOMOTIVE EMISSIONS

D. R. FUSSELL*   G. P. RICHARD*   D. WADE†

This paper summarizes work carried out in Esso Research Laboratories in the U.S.A. and in Europe on the development of advanced systems for the control of automotive emissions. The paper discusses work on thermal reactors in conjunction with exhaust recycle, and it is shown that considerable progress has been made in the development of reliable control systems which use these principles. While much of the early development work was carried out using U.S. vehicles, an account is given of the development of a reactor and recycle system in Europe on an MGB car. It is concluded that the use of this system provides a means to approach the very low levels of emission forecast for the U.S.A. in 1976. However, it has only been possible to show that the hydrocarbon target can be positively met, and where carbon monoxide (and particularly nitric oxide) emissions are concerned, no system has as yet been found which, when modified for production cars, would completely meet the 1976 targets.

INTRODUCTION

THE PRESENT GENERATION of emission controlled cars falls into two main categories. In the first, frequently referred to as 'engine modification systems', low emissions are achieved mainly by means of lean mixture operation, complemented by modifications to spark timing, and by improvements to carburation, fuel distribution between cylinders, and to engine breathing. Many of these modifications are also applied to the second category, but in addition air is injected at the exhaust ports to promote final combustion of carbon monoxide (CO) and hydrocarbon (HC) in the exhaust manifold.

Such measures are quite sufficient to meet the current requirements of U.S. legislation and the proposed European emissions standards, but it is apparent that more effective control of carbon monoxide, hydrocarbon, and in particular nitrogen oxides will be required, at least in the U.S. market, by the year 1975.

This is illustrated in Table 130.1, which compares 1971 to 1976 U.S. requirements with emission results obtained from a Fiat 125 car (1)‡, modified and carefully tuned for lean operation, and fitted with exhaust gas recycle for nitric oxide (NO) control. Also shown are results from an air injected MGB, also carefully tuned, but with no exhaust gas recycle.

Both vehicles easily satisfy current emission standards,

The MS. of this paper was received at the Institution on 7th June 1971 and accepted for publication on 22nd July 1971. 23
* *Esso Research Centre, Abingdon, Berks.*
† *Esso Research and Engineering Co., Linden, New Jersey, U.S.A.*
‡ *References are given in Appendix 130.1.*

and it is to be expected that they would also approach the 1974 requirements, in spite of the fact that a new and more severe test procedure is to be introduced in 1972. Emissions levels from these cars, however, fall very far short of the requirements for 1976, particularly when it is remembered that the standard achieved on a prototype car

Table 130.1. Emissions from current generation pollution controlled vehicles

Item	Emissions, g/mile		
	CO	HC	NO_x as NO_2
U.S. 1971 requirement (2) . .	23	2·2	4·0*
U.S. 1974 requirement (3)† .	39	3·4	3·0
U.S. 1976 requirement (4)† .	4·7	0·46	0·4
Estimated engineering standard to meet 1976 requirements . .	2·1	0·21	0·25
Fiat 125 using engine modification plus exhaust gas recycle (1970 Test procedure) . . .	14·7	1·1	1·1
MGB, using air injection and spark retard for nitric oxide control (1970 Test procedure) . . .	10·1	0·6	2·8

* California only.
† These standards refer to tests carried out using the CVS procedure to be introduced in 1972. This procedure is more severe than the 1970 procedure currently in use, but no exact equivalence can be quoted between the two. The 1974 requirements using the CVS procedure are not intended to be significantly more severe than the 1971 requirements for carbon monoxide and hydrocarbons.

must be much superior to the legal limit if production cars based on the prototype are to meet that limit in a consistent manner.

For the engine modification car, the weighted mean CO concentration by the 1970 Federal procedure is just under 1 per cent, and while it may be possible to obtain some improvement by further refinements to overall engine design, it is considered that the problem of obtaining even distribution of the fuel between cylinders, plus that of providing for all driving conditions the right mixture strength at the right time, makes the probability of meeting the 1976 Federal requirements, using engine modifications alone, extremely low.

Thus, it appears that to meet the severe requirements of the future, motor manufacturers must look to systems which destroy pollutants outside of the combustion chamber, using either thermal or catalytic means. Such systems may also use the principle of exhaust gas recycle for NO control.

The authors' laboratories in the U.S.A. and Europe have therefore conducted a series of practical studies on the application of exhaust gas recycle, plus thermal or catalytic destruction of pollutants in specially designed exhaust systems. It is the purpose of this paper to review the overall results of these studies and, in particular, to discuss their application to the control of emissions from a typical European sized car.

CATALYTIC DESTRUCTION OF POLLUTANTS

In principle, catalytic systems for the disproportionation of NO under reducing conditions, and the subsequent oxidation of residual CO and HC, appear to offer a great deal of promise although they suffer from the disadvantage that many catalysts are easily poisoned, particularly by lead compounds in the gasoline. In the U.S.A., however, low lead gasolines are becoming available and this disadvantage is becoming less important. Many catalysts are available which are capable of oxidizing CO and HC, but those which will reliably disproportionate NO under the varying conditions of vehicle operation are less common. U.O.P. (5), however, have developed a platinum based catalyst which has given promising results, while at Esso Research (6) (7) we have found Monel metal to be most effective for NO reduction. Currently, the system which appears to give most promise in U.S. cars is a dual bed arrangement with 4·5 lb of Monel, followed by 3·5 lb of platinum on alumina. NO is reduced by the Monel, the reducing gases being supplied by a fairly rich engine operation, and residual CO and HC are oxidized over the platinum catalyst with the aid of injected secondary air. Tests on a 307 CID V-8 Chevrolet car showed that emissions could be reduced to a level close to the U.S. 1976 standards, as is indicated in Table 130.2.

Although the results appear promising, a number of problems remain to be solved before the system can be considered commercially viable. First, using a dual bed system, warm-up of the oxidation catalyst is slow and, in

Table 130.2. Dual bed catalyst system performance

Item	Emissions (1970 Federal procedure) (2)		
	CO, per cent	HC, p.p.m.	NO, p.p.m.
Base car	1·3	100	850
Car equipped with catalyst	0·3	37	88
Target level	0·2	20	100

consequence, it is remarkably difficult to reduce CO and HC to the required levels. Second, some NO is always reduced to ammonia over the Monel catalyst, and re-oxidation back to NO over the platinum catalyst tends to keep NO emissions rather high unless very accurate air/fuel ratio control is practised. Finally, physical deterioration of the catalyst after a relatively short mileage (10 000–15 000) means that an uneconomically short catalyst replacement time is needed.

These problems are not considered to be insurmountable, but it is likely that considerable development work will be required for their solution. Moreover, slow warm-up and ammonia formation are problems likely to be general in all catalyst systems. In these circumstances it appears that, for some years at least, non-catalytic control systems are likely to be of major importance for the control of automotive emissions.

DEVELOPMENT OF EXHAUST GAS RECYCLE FOR CONTROL OF NITRIC OXIDE (8) (9)

As part of a NAPCA sponsored programme, exhaust gas recirculation was first installed and evaluated on two U.S. production cars, a 1967 Plymouth Fury equipped for HC and CO control using the engine modification route, and a 1968 Chevrolet Impala, which used air injection for HC and CO control.

The recycle systems used for both vehicles were essentially similar. The recycle gas was taken from the exhaust pipe close to the silencer, and was introduced into the constant depression area of the fixed jet carburettor via a calibrated orifice. This arrangement ensured a roughly constant proportion of recycle to primary air supply. Adjustment of the recycle rate was achieved simply by changing the orifice size. Micro-switches on the accelerator linkage operated an on–off valve in the recycle line, and could be adjusted to switch the recycle on or off during any selected driving mode. For normal automatic operation the system was arranged to give no recycle during idle, or at wide open throttle. Recycle was operative during all other driving modes. This system of operation was selected because early tests showed that the application of recycle during idle leads to rough engine operation, and at the same time is hardly necessary since under these conditions NO emissions are low. During wide open throttle operations the application of recycle caused a significant loss of power. For example, the time for acceleration from 0 to 60 mile/h could be increased by as much as 25 per cent using about 13 per cent recycle. Although NO emissions are high during this mode of operation it was reasoned that little time is normally spent at wide open throttle, particularly during town driving, and that on balance it is preferable to retain peak power rather than to go for minimum NO emissions during full throttle driving.

An important effect of the recycle system in these cars was that its use decreased power for a given throttle setting, so that for any given speed, recycle demanded a higher throttle opening, and a consequent decrease in inlet manifold vacuum. In consequence, it was necessary, when recycle was applied, to make adjustments to the carburettor power jet setting, and to the vacuum advance response of the distributor.

After the necessary adjustments had been made it was possible to show that for both cars the application of about 13 per cent recycle resulted in a reduction of about 60 per cent in NO emissions with only a minimal effect on driveability. The maximum recycle rate that could be used while retaining acceptable driveability was 15–17 per cent.

The effects of changes in ignition timing and of recycle rate on NO emissions were shown to be essentially independent. Recycle at constant ignition timing drastically lowered NO with a small fuel consumption penalty, while spark retard at constant recycle rate lowered NO at a higher fuel penalty. This effect is illustrated in Fig. 130.1 and it can be seen that, by the employment of recycle combined with some spark advance, it should be possible to obtain a sizeable reduction in NO emissions without a fuel penalty.

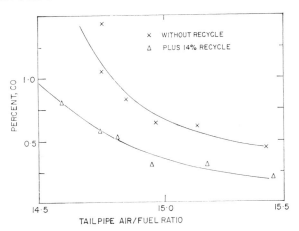

Fig. 130.2. Effect of recycle on CO emissions (Plymouth Fury)

Recycle rate appeared to have little effect on HC emissions, but in general it appeared that the use of recycle tended to reduce CO emissions, as is shown in Fig. 130.2 where, for a series of steady-state runs on the Plymouth vehicle, CO concentration is plotted against the tailpipe air/fuel ratio. This effect is probably due to the fact that the use of recycle can improve cylinder-to-cylinder fuel distribution.

To test the durability of the recycle system, six vehicles were equipped with recycle, using systems similar to those employed for the laboratory cars, and were subjected to some 52 000 miles of mixed city and suburban driving. During this period NO emissions were effectively controlled, and no adverse effects on CO and HC emissions were noted. For both engine modifications and air

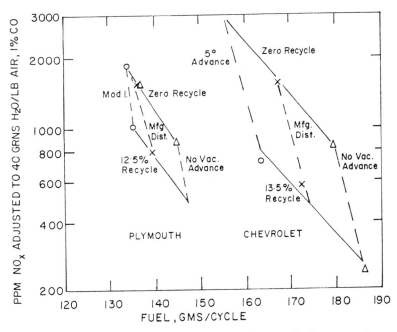

Fig. 130.1. Effect of recycle and ignition timing on fuel consumption

injection cars, engine wear and cleanliness were found to be normal for the driving regime employed.

THE THERMAL REACTOR PLUS EXHAUST GAS RECYCLE FOR TOTAL POLLUTANT CONTROL (10) (11)

In order to achieve really efficient oxidation of CO and HC in the exhaust system, it is necessary to supply an excess of air well mixed with the unburnt material leaving the combustion chamber, to maintain the gas mixture at a temperature such that rapid oxidation can take place, and to provide a reaction volume large enough to ensure that the reactions have sufficient time to go to the desired degree of completion. The most satisfactory approach to this problem is to run the engine somewhat rich, so that a good supply of combustibles in the exhaust is ensured, and to inject air into the system close to the exhaust ports. A number of workers have explored the design conditions which ensure efficient reactions, and the du Pont Organization (12) (13) in particular has shown that thermal reactions are a promising route to CO and HC control.

To explore the potential of this technique of pollutant control, exhaust manifold reactors designed by du Pont were fitted to a 1967 Chevrolet 283 CID V-8 vehicle, originally equipped with simple air injection to meet the 1967 California emissions requirements. In addition, the car was fitted with an exhaust gas recycle system for the simultaneous control of NO emissions.

The system finally adopted is shown schematically in Fig. 130.3. The reactors were supplied with air injected into the exhaust close to the exhaust ports from the original air pump, and a synchronizing valve was included into the system so that air was injected at individual ports as the exhaust valve opened. To achieve optimum CO and HC emissions, particularly during accelerations, it was found that rich operation of the engine was necessary, and this resulted in reactor temperatures in excess of 980°C during high-speed cruise. In order to maintain a maximum level of around 900°C, it was decided that the mixture should be leaned off during this driving mode, and this was achieved by injecting air into the inlet manifold. The supply of air for this purpose was controlled by a diaphragm valve actuated by the exhaust pressure. At road loads in excess of 45–50 mile/h the valve opened and air was injected into inlet manifold. Leaning out of the mixture during accelerations was prevented by placing a restriction in the control line which delayed the opening of the valve for about 20 s.

The recycle system was sized to give 11 per cent recycle, and was controlled by a diaphragm valve actuated from the inlet manifold vacuum. The valve was closed at idle and wide open throttle by means of a bleed system which broke the vacuum. This consisted of a plunger inside a straight passage open at both ends. The middle of the passage was connected to the diaphragm chamber by the bleed line. The plunger was actuated directly by the throttle movement, and the linkage was arranged so that the bleed line was only blocked off during part throttle operation, thus allowing the engine vacuum to act on the diaphragm and open the recycle valve.

Finally, it was found that the use of recycle when the engine was cold slowed down the warm-up of the reactor, so that it became desirable to eliminate recycle during the time that the choke was in use. A further bleed point was therefore installed in the recycle control circuit, which was closed by a small rubber seal when the choke was not in use. A mechanical linkage direct to the choke plate was used to open this seal during choked operation, thus shutting off the recycle.

In addition to the above, several modifications were made to the engine operation. The basic timing was retarded by 2° to increase exhaust temperature and help reduce CO emissions during acceleration. Second, the vacuum advance was inactivated during warm-up. Finally, the off idle jets were slightly enlarged to richen the mixture at the beginning of accelerations.

Driveability of the vehicle was not impaired by the emission control systems, but fuel consumption was increased by 16 per cent. This appeared to be due mainly to the enrichment of the mixture and the spark retardation employed.

Emissions from the modified car are summarized in Table 130.3, which also shows, for comparison, the performance of the base vehicle.

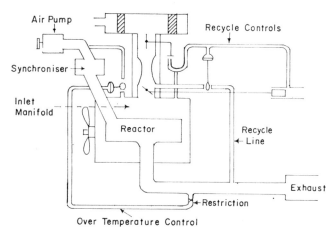

Fig. 130.3. Thermal reactor plus exhaust recycle on a Chevrolet car

Table 130.3. Performance of reactor manifold, plus exhaust recycle on Chevrolet car

Item	Emissions (1970 Federal procedure) (2)		
	CO, per cent	HC, p.p.m.	NO, p.p.m.
Base car, with simple air injection, composite emissions	0·78	325	1380
Modified car, cold cycle emissions	0·53	—	141
Modified car, hot cycle emissions	0·13	< 10	131
Modified car, composite emissions	0·27	10	135

The bulk of the CO emissions were from the cold cycles, and it is to be expected that the Chevrolet car would suffer a considerable increase in overall emissions if tested by the new 1972 Federal procedure which places considerable emphasis on cold start emissions. Thus there is a considerable incentive to find means by which the rate at which the reactor warms up can be increased.

REACTOR PLUS RECYCLE ON A EUROPEAN CAR

The work outlined above has given very promising results in U.S. sized vehicles, and it was therefore considered desirable that the basic concepts of thermal reactors and exhaust recirculation should be evaluated in a vehicle typical of European manufacturers. For this purpose an air injected car was desirable, and the choice fell on the U.S. export version of the MGB. The engine displacement (1798 cm^3) and vehicle weight (1040 kg) were fairly typical of European cars, and although the car is generally regarded as a sports saloon the engine is not particularly highly tuned. In addition, there were obvious advantages in terms of spare part availability, and the opportunity for discussion with the manufacturers in the choice of a car built close to the Abingdon Research Centre.

After a running-in period of around 1000 miles, emissions on the base car were determined, and a number of adjustments were made to obtain optimum emissions consistent with acceptable performance. The emissions from the car at this stage are shown in Table 130.4, and are compared with the 1970 U.S. Federal standards.

The effect of an enlarged exhaust manifold on emissions

During initial work on the MGB it was found that for operations at constant air/fuel ratio the exhaust gas from the siamesed port of cylinders 2 and 3 was 40–60 degC hotter than that from the separate exhaust ports of cylinders 1 and 4. Moreover, emissions of CO and HC from cylinders 2 and 3 were markedly lower than those from numbers 1 and 4, despite the fact that mixture strengths at all cylinders were approximately constant. These observations indicated that a considerable improvement in emissions could be obtained by working at higher temperatures, and this effect plus that of increased residence time of the exhaust gas in the manifold was initially studied by the installation of a larger, insulated, and thin walled exhaust manifold.

The volume of the new manifold was about twice that

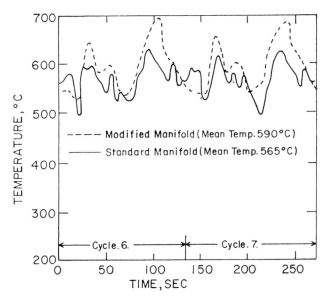

Fig. 130.4. Higher temperatures maintained in modified manifold

of the original cast iron manifold, and because of the thinner walls and insulation it warmed up more quickly and removed less heat from the exhaust gas. Hot cycle temperature data obtained using the standard and modified manifolds are shown in Fig. 130.4.

The temperature in the modified manifold is, on average, 25 degC greater than in the standard equipment. A greater range of temperatures was experienced using the modified manifold, and this is probably due to its lower mass (6·5 lb compared with 15·5 lb for the original manifold). The installation of the modified manifold gave rise to a considerable improvement in CO and HC emissions, as given in Table 130.5.

Effect of ignition timing and mixture strength on emissions

The effect of spark retard on emissions from the MGB was tested by disconnecting the vacuum advance mechanism. At the same time, a series of emissions tests was run at a number of different mixture strengths. The results obtained are summarized in Table 130.6.

CO emissions in this series of tests were rather higher than those reported in Table 130.5 due to slight errors in the resetting of the carburettors after a sequence in which rich mixtures were used. The work showed that simple

Table 130.4. Emissions from air injection MGB

Item	Emissions (1970 Federal procedure) (2)		
	CO	HC	NO$_x$ as NO$_2$
1970 Federal requirement	23	2·2	—
MGB car with simple air injection	16·4	1·3	5·4

Table 130.5. Effect of enlarged exhaust manifold on MGB emissions

Manifold	Emissions (1970 Federal procedure) (2), g/mile		
	CO	HC	NO$_x$ as NO$_2$
Standard manifold	16·4	1·3	5·4
Modified manifold	10·1	1·0	5·7

Table 130.6. Effect of ignition timing and mixture strength on MGB emissions

Item	Mean exhaust manifold temp. (hot cycles)	Emissions (1970 Federal procedure) (**2**), g/mile		
		CO	HC	NO_x as NO_2
Standard ignition and mixture strength	615°C	16·8	0·82	5·4
No vacuum advance	655°C	16·8	0·63	2·8
No vacuum advance, and enriched by 3/4 air/fuel ratio .	—	42·7	1·1	1·3

removal of the vacuum advance reduces nitrogen oxides by about 45 per cent and HC by 20 per cent, while CO is unaffected. Enrichment of the mixture by air/fuel ratios of 3/4 further reduces nitrogen oxides to about 24 per cent of their original value, but at this point exhaust CO and HC are increasing rapidly, and in the MGB such enrichment caused a significant decrease in driveability.

Gas temperature measurements taken in the exhaust manifold showed that the retarded spark increased the gas temperature by about 40 degC, bringing the average hot cycle temperature close to 650°C. It is this effect which is probably responsible for the effectiveness of spark retard in the reduction of HC emissions.

Overall, the results obtained using the vehicle equipped with the modified exhaust manifold suggested that the increased temperature and residence time which could be obtained using a thermal reactor designed specifically for the car should result in very considerably improved CO and HC emissions. In addition, such a reactor would enable richer mixtures to be employed, and this, combined with spark retard, would provide some control of emissions of NO.

Design and construction of a thermal reactor

The reactor is illustrated in Figs 130.5 and 130.6. For this experimental design it was desired to use a material able to withstand very high temperatures in an oxidizing atmosphere and able to resist corrosion from lead compounds in gasoline. An 80–20 nickel chromium alloy (Nimonic 75, m.p. ~1380°C) was selected to satisfy these requirements.

The gauge used was 20 s.w.g. (0·9 mm), the thinnest usable material consistent with adequate mechanical strength. The volume of the reactor core, or central chamber, is approximately 1·6 litres, which is large enough to contain the complete exhaust gas from one combustion chamber. In this design the gases from the port, mixed with extra air from the injectors, pass down a double skinned pipe into the central chamber of the manifold where most of the reaction takes place; from here the gases escape through longitudinal slots and pass into the first concentric chamber where they provide insulation for the next charge entering the reaction zone. The gases finally leave via more longitudinal slots into the outer chamber and thence into the exhaust system. External insulation of the outer skin is provided by a 0·5-in layer of ceramic fibre blanket sheathed in aluminium foil, while two circumferential convolutions, also in the outer skin, provide flexibility to absorb the thermal expansion. Despite efforts to minimize the weight of the reactor, it still weighs over 20 lb—5 lb more than the original cast iron manifold.

Thermocouple points are located near the outlet from the reactor core and just above the flange between the

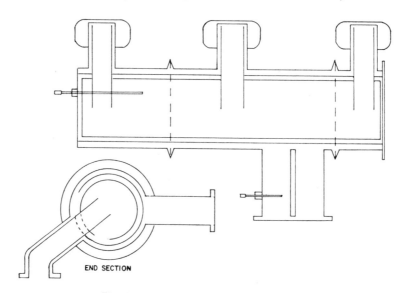

Fig. 130.5. MGB reactor manifold

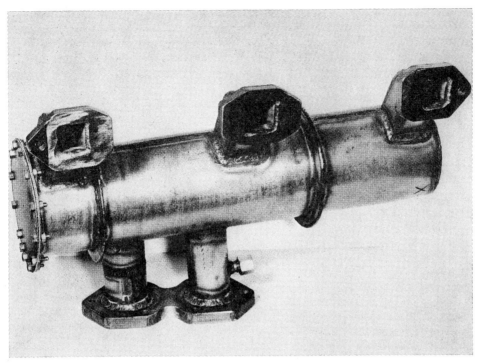

Fig. 130.6. MGB reactor manifold (before insulation)

reactor and the normal exhaust system. In addition to monitoring temperature for development purposes, the output of one of these couples is used during on the road operations to actuate an electronic relay which operates a warning light when the temperature of the reactor exceeds 950°C.

Effect of operating temperature on emissions from reactor

Temperatures recorded throughout the 7-cycle U.S. Federal test are reproduced in Fig. 130.7 for the thermal reactor and the earlier enlarged, insulated manifold. In these and subsequent tests the vacuum advance mechanism was left disconnected to take advantage of the decrease in NO emissions so obtained.

Compared with the earlier manifold, it will be seen that the reactor maintains a condition more conducive to effective oxidation of CO and HCs. Its overall temperature is about 200 degC hotter, and it is less subject to wide fluctuations in temperature. This must be due partly to effective insulation, partly to the exothermic reactions taking place within, and partly to the mass of metal used

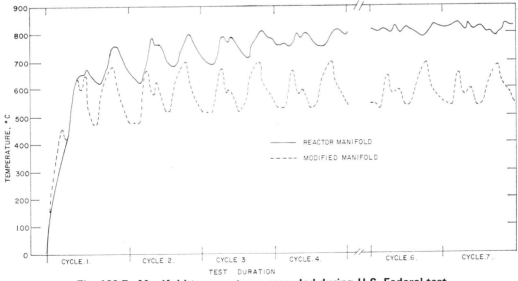

Fig. 130.7. Manifold temperatures recorded during U.S. Federal test

Table 130.7. Reactor temperature and emissions (1970 Federal procedure) (2)

Cycle No.	Mean reactor temp., °C	CO, per cent	HC, p.p.m.	NO, p.p.m.
1	510	1·06	95	753
2	710	0·44	40	672
3	760	0·46	17	700
4	780	0·37	4	650
6	810	0·20	1	705
7	820	0·19	2	748

Table 130.8. Performance of reactor and enlarged manifold at various ignition settings

Static timing, °b.t.d.c.	Vacuum advance	Maximum gas temp., °C	Emissions (1970 Federal procedure) (2), g/mile		
			CO	HC	NO$_x$ as NO$_2$
Enlarged manifold					
17	Yes	615	10·1	0·8	5·4
Reactor					
12·5	No	890	3·2	0·3	2·1
17	No	875	4·0	0·2	2·4
20	No	837	4·5	0·2	3·0
22·5	No	837	4·4	0·2	3·4

in the reactor itself. It does not, however, attain its maximum operating temperature until the fifth driving cycle, whereas the earlier manifold reaches its operating range by the end of the second cycle.

A comparison of reactor temperature and exhaust emissions during the 1970 Federal test is shown in Table 130.7.

It will be seen that both CO and HC emissions are highest during the first cycle; this is to be expected. CO adopts an intermediate level during the remaining three warm-up cycles, but drops sharply in the hot cycles. This trend has been observed consistently over many tests. HCs, however, drop rapidly throughout the cold cycles and are approaching a minimal level by the end of the third cycle. A mode-by-mode analysis reveals that even in the second cycle, two-thirds of the corrected, weighted, HC emissions are accounted for by the 15 mile/h gear change peak and the two decelerations. NO emissions, on the other hand, are maintained at a fairly consistent level throughout the test and are apparently not affected by either engine or reactor temperature.

If we now compare the reactor temperature to the cycle HC and CO emissions it appears that temperatures in excess of about 700°C are necessary for good HC oxidation, whereas low CO emissions require temperatures in excess of about 800°C. The dependence of good CO oxidation on adequate reactor temperature is discussed further in a later section.

Effect of spark timing on reactor performance

The effects of changes in spark timing were explored in a series of tests in which the static timing was varied to cover the range 12·5°–22·5° b.t.d.c. Vacuum advance was disconnected throughout these tests. The results obtained are shown in Table 130.8 which includes, for comparison, results obtained using the enlarged exhaust manifold. The improvement in CO and HC emissions brought about by the higher operating temperature and increased volume of the reactor is clear and impressive.

Apart from a slightly higher HC emission at 12·5° b.t.d.c. due to misfiring during the first cold cycle acceleration caused by the extent of the spark retard, HC emissions from the reactor remained essentially constant as the spark was advanced. NO emissions, as was expected, increased with increasing spark advance.

Contrary to experience with conventional exhaust manifolds, it was found that CO was affected by spark timing and increased with increasing spark advance. This effect is due to the influence of spark timing on reactor temperature, which drops towards the critical temperature for efficient CO oxidation with increasing advance.

Influence of mixture strength on reactor emissions

Fig. 130.8 summarizes a series of tests in which the air/fuel ratio at which the car operates was varied. Air/fuel ratio was calculated from the tailpipe emissions using the method of Spindt (14), and combustion chamber ratio was calculated from the tailpipe ratio by correcting for the known volume of air injected into the reactor. The results shown in the figure are based on bag analysis of the exhaust.

The figure shows that reactor temperature rises steadily as the air/fuel ratio is decreased from 16·0 to 12·5:1. As air/fuel ratio decreases from 16·0 to about 14·0, CO in the exhaust rises. Between 14·0 and 13·0 a marked fall in CO occurs, followed by a gentle increase as the air/fuel ratio falls towards 12·0:1. This behaviour is at first sight surprising but it is thought to be due to changing conditions in the reactor. When the mixture is weak, e.g. around 16·0:1, the exhaust gas leaving the combustion chamber is cooler than at rather richer mixtures. Increasing the mixture strength to around 14:1 not only increases the concentration of CO, but also increases the temperature in the reactor to about 750°C. At this temperature it is believed that the rate of oxidation of CO has increased sufficiently to cause the CO emissions to decrease. As the oxidation of CO proceeds at an increasing rate, so will the temperature in the reactor; this accelerating rate of temperature rise is also apparent in Fig. 130.7. CO emission continues to drop until enrichment beyond 13:1 causes a slight up-turn in CO emission, possibly due to reactor inefficiency.

This effect is not seen for HC emissions. It appears that at reactor temperatures where CO is becoming difficult to oxidize, the conversion of HCs is still efficient. Weakening the mixture beyond about 14:1 results in the onset of

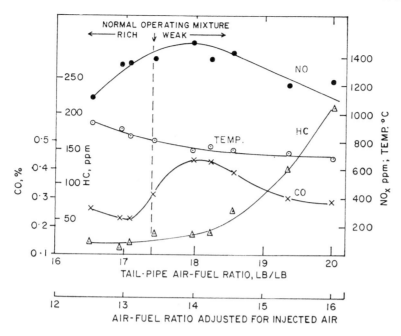

Fig. 130.8. Effect of mixture strength on hot cycle emissions and maximum reactor temperature

engine misfire, with a consequent increase in HC emissions both into and out of the reactor.

As was expected, NO emissions reach a peak at a value close to the stoichiometric air/fuel ratio. It is apparent from the figure that a very rich mixture would be needed to yield a large reduction in NO emissions. Such enrichment would lead to driveability problems, high reactor temperature, and a marked increase in fuel consumption. It was considered that in this car, although a slight enrichment could be made in order to reach a minimum CO emission, it would be necessary to install exhaust recycle to control NO emissions.

Construction of recycle system

Exhaust gas is taken from a point about 3 ft back from the down pipe flange, and at a point where the two separate exhaust pipes join. It is fed back into the intake side of the engine via a turbine type flow meter and control valve mechanism.

Recycle flow control valves, one for each carburettor, are actuated by a vacuum motor which receives its signal from a vacuum tapping in the induction system just upstream of the throttle plate. Throttle positions encountered during the driving cycle cause a depression in the vacuum motor and the valves are then opened. This does not apply, however, to the idle and low-speed cruise conditions when recycle seriously disturbs the smooth operation of the engine, nor at the more advanced throttle positions when power output would be affected. A valve with a calibrated leak in the vacuum sensing line slows the response of the motor such that the car can take up the drive from the engine and be moving forward before the recycle flow commences, yet enables the vacuum motor to close quickly during decelerations. In this way good driveability during accelerations is achieved together with minimum HC emissions during decelerations.

The control valves are conventional 20-mm diameter poppet valves. Exhaust gas approaches from the back of the valve and enters the constant depression (c.d.) area of the twin S.U. carburettors. The valve faces are approximately flush with the wall of the c.d. area and are constantly swept with fuel during operation. It is hoped that this arrangement will help to keep both the faces and seats free of deposits.

In order to accommodate the valve mechanism between the carburettor bridge and the butterfly, the c.d. area has been extended by means of a steel lined Sindanyo spacer. Sindanyo replaces the conventional insulating spacer normally fitted to this car. The throttle butterfly is re-installed nearer to the engine, and this provides sufficient room to install the valve.

Fig. 130.9 shows a schematic diagram of the recycle system and its controls, while in Fig. 130.10 a photograph of the engine space after installation of reactor and recycle systems is shown.

Emissions obtained using reactor plus exhaust recycle

The average cold start emission data from seven U.S. Federal 7-mode cycle tests, together with the ranges of emissions, are quoted in Table 130.9. The range does not represent test repeatability since between each test adjustments were made to improve the overall system operability and design. Nevertheless, it does show that low emission levels are being consistently achieved, while the more recent tests are tending to be at the lower emission levels.

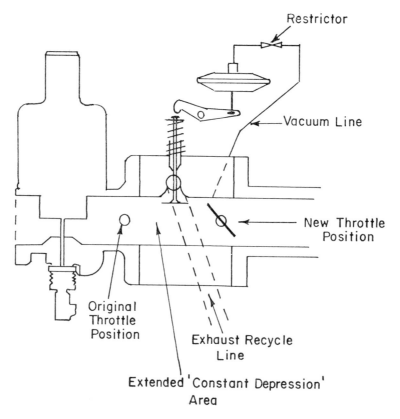

Fig. 130.9. MGB exhaust recycle control system

Fig. 130.10. Exhaust recycle system installed in MGB car

Table 130.9. Reactor and recycle combined give low HC/CO/NO emissions

Item	Emissions (1970 Federal procedure) (2)		
	CO	HC	NO_x as NO_2
Grams per mile (mean)	4·03	0·27	0·84
Range . . .	4·66/3·49	0·44/0·18	0·94/0·67

General performance of car using thermal reactor plus recycle

Fuel consumption tests run on the dynamometer under the conditions of the 1970 Federal test procedure indicate that when the car is set up for optimum reduction of all pollutants, fuel consumption is increased by 10–15 per cent as compared to the base vehicle. About 30–50 per cent of this increase is due to the use of exhaust recycle plus some spark retard, and the remainder is due to the fact that best emissions are obtained using rich carburation. Under steady throttle, low cruise speed conditions, where the vacuum advance would normally operate, the spark retard caused by disconnection of this control increases fuel consumption further, and in this particular driving mode the fuel consumption penalty totals around 20–25 per cent.

Nevertheless, on the road the fuel consumption is still acceptable. Over a distance exceeding 3000 miles of town and country driving, a fuel consumption of around 27 mile/gal was returned. This compares well with published road test data for overall consumptions ranging between 21 and 27 mile/gal for standard production MGB GT cars not fitted with emission controls (**15**) (**16**).

Driveability of the car is good in all driving modes, although some effect of recycle can be felt when warming up under cold conditions, e.g. at ambient temperature around 5°C.

Wide open throttle acceleration time has been increased slightly by the installation of reactor and recycle; for example, the time for acceleration from 15 to 50 mile/h has gone from 11·6 to 12·3 s. In addition, the top speed of the car has been reduced from 106 to around 100 mile/h. Since recycle is turned off at wide open throttle, these changes must be attributed to the changes in carburation employed to improve all round emissions control, and to the influence of the reactor on engine breathing.

This particular car has not accumulated sufficient mileage to permit any comments on the durability of the reactor systems, or on possible problems of inlet manifold deposits due to the use of recycle. Regarding the latter point, however, another European development car fitted with recycle has shown no major problems during a period of more than 15 000 miles of mixed road and dynamometer work, lasting over two years and operating on normal commercial fuels.

No problems of overheating have been observed as a result of the use of retarded ignition timing, and the use of the reactor has not resulted in raised under-bonnet

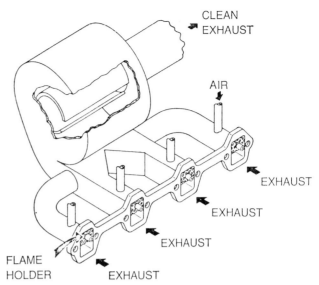

Fig. 130.11. The well-mixed thermal reactor (RAM)

temperatures. In fact, the skin temperature measured on the outside of the reactor is around 250°C, and is lower than that of the standard cast iron exhaust manifold. Since the surface area of the reactor is only about 15 per cent more than that of the standard manifold, the total heat transfer to the engine compartment may even be less than is experienced using a standard manifold.

RESULTS USING A WELL-MIXED THERMAL REACTOR (RAM)

Work carried out on both U.S. and European vehicles has shown clearly that the time taken to achieve optimum operating temperature is a major factor limiting the effectiveness of thermal reactor systems, and in view of this further development has been carried out in which the concept of recirculation of the exhaust gases within the reactor has been exploited. In addition, it has been found that warm-up time is improved by the installation of flame holders at the engine exhaust ports. This new design, known as RAM (Rapid Action Manifold), is shown schematically in Fig. 130.11.

Active work on the reactor is still in progress, but preliminary results obtained on a Chevrolet vehicle are very promising, as is indicated in Table 130.10 (**11**).

These results, obtained using the 1972 procedure, which is very severe on cold start emissions, show that the 1976

Table 130.10. Performance of well-mixed reactor (RAM) plus exhaust recycle

Item	Emissions (1972 Federal procedure) (**3**), g/mile		
	CO	HC	NO_x as NO_2
Test vehicle . . .	3·7	0·08	0·72
1976 engineering target .	2·1	0·21	0·5
1976 U.S. standard (**4**) .	4·7	0·46	0·4

standards can be approached using the RAM reactor, although the engineering targets which are necessary to ensure consistent attainment of these standards have not been achieved. A disadvantage of this system is that rich operation is required to obtain optimum emissions, and on this vehicle fuel consumption was increased by about 20 per cent.

CONCLUSIONS

The work discussed in this paper has shown that while current emissions control methods for CO and HC will satisfy the proposed European standards, and those proposed for the U.S. up to 1974, it will probably be necessary to introduce further specific control methods for NO emissions before 1974. For the years 1975 and beyond, U.S. emissions requirements are very severe, and advanced control methods must be developed to meet these requirements.

Catalytic reduction of NO, followed by oxidation of CO and HC is a promising approach to this problem, but work to date has shown that a considerable amount of development work will be required to overcome catalyst warm-up and durability problems.

The combination of thermal reactor for the destruction of CO and HC with exhaust gas recycle for NO control has been shown to be a practicable technique for U.S. and European vehicles.

By careful attention to reactor design, these techniques can yield emission levels which approach the very severe 1976 U.S. standards, but as yet no system has been found which will consistently meet these standards.

Experience with reactor plus recycle in both U.S. and European vehicles has shown that the use of the techniques is compatible with good vehicle performance, but they do have the disadvantage that fuel consumption is likely to be increased by as much as 20 per cent over all driving modes.

Operational experience, including over 50 000 miles of normal town driving, indicates that exhaust recycle systems will not give rise to major problems when installed on production vehicles. At the present time, however, insufficient experience has been gained to draw conclusions regarding optimum construction materials and the durability of reactor systems.

Further work to optimize the emissions from the RAM reactor on U.S. and European vehicles is planned, together with an investigation of reactor durability using various construction materials. In view of the very considerable fuel economy debit which results from the use of thermal reactors it is also planned to continue active development of catalytic systems for emissions control.

Finally, it must be noted that if the emission levels quoted in this paper are to be reached and good vehicle performance is to be retained, then very careful tuning is required. To maintain a vehicle in this condition will require first-class maintenance, and the general introduction of vehicles having sophisticated emissions controls into any country must be accompanied by measures which ensure adequate engine maintenance if the controls are to remain fully effective.

APPENDIX 130.1
REFERENCES

(1) KRAUS, B. J. and CAMARSA, M. Private communication.
(2) *U.S. Federal Register* 1968 **33** (No. 108, 4th June).
(3) *U.S. Federal Register* 1970 **35** (No. 219, 10th November).
(4) *U.S. Federal Register* 1971 **36** (Nos 39 and 40, 26th and 27th February).
(5) *Air/Water Pollution Report* 1970 (25th May) (Business Publications Inc.).
(6) BERNSTEIN, L. S., et al. S.A.E. Paper 710014, 1971.
(7) BERNSTEIN, L. S., et al. Tech. Session on Engine and Emission Control Studies Section, The Combustion Institute, 1971 (23rd March).
(8) GLASS, W., et al. S.A.E. Paper 700146, 1970.
(9) MUSSER, G. S., et al. S.A.E. Paper 710013, 1971.
(10) GLASS, W., et al. S.A.E. Paper 700147, 1970.
(11) LANG, R. J. S.A.E. Paper 710608, 1971.
(12) CANTWELL, E. N., et al. *S.A.E. Trans.* 1966 **174**.
(13) CANTWELL, E. N., ROSENLUND, I. T., BARTH, W. J. and ROSS, S. W. 'Recent developments in exhaust manifold reactor systems', *Proc. Auto. Div. Instn mech. Engrs* 1969–70 **184** (Pt 2A), 268.
(14) SPINDT, R. S. S.A.E. Paper 650507, 1965.
(15) *Motor road tests* 1966 edition.
(16) *Motor* 1970 (31st October), 17.

# C131/71 EXHAUST EMISSION STUDIES USING A SINGLE-CYLINDER ENGINE

A. E. DODD*

Tests are reported in which different combustion chambers were fitted to a single-cylinder engine. Eleven chambers were investigated including bath-tub, wedge, hemispherical, bowl-in-piston, and annular designs. In addition, a limited study of the effects of ignition timing, knock, compression ratio, and coolant temperature are detailed. Performance data, in terms of brake mean effective pressure and specific fuel consumption, and emission concentrations of hydrocarbons (by non-dispersive infrared equipment and flame ionization analyser), nitric oxide, and carbon monoxide are given. The effects of combustion chamber configuration, knock, and coolant temperature are generally small but significant while ignition timing and compression ratio are of much greater importance.

INTRODUCTION

IN ORDER THAT optimum emission control may be achieved, basic data are required on the effects of internal engine design on exhaust emissions from European-sized engines. Accordingly, an investigation of such effects was initiated using a 325-cm^3, single-cylinder engine, which permitted the elimination of manifold distribution effects. Data obtained on this engine on the influence of the state of the mixture at entry to the cylinder have already been published (1)† while the results of a limited investigation of the reactivity of these emissions have also been reported (2).

Although small in some cases, the effects of engine design may not be negligible since, in areas where the problem of pollution is not severe, an acceptable level of emissions could perhaps be obtained without the necessity for additional, more expensive control features. This paper deals principally with the effects of combustion chamber design; in addition, data are presented on the variation of emissions with coolant temperature, ignition timing, knock, and compression ratio. Further information, more detailed than can be included here, will be found in (3).

SCOPE OF THE INVESTIGATION

Full details of the engine and analysis equipment, together with a description of the rigorous analysis techniques adopted, have already been published (1)–(3). In summary, the rear cylinder of a 1·3-litre (73·7-mm bore, 76-mm stroke) four-cylinder engine was used, and different cylinder heads were clamped on to the cylinder by means of a system of beams and tie rods. Since the original camshaft was retained, valve timing remained constant for all combustion chambers, although valve sizes and lifts varied with the individual cylinder heads. Valve timings measured with the engine cold and with a valve clearance of 0·25 mm (0·010 in) were as follows:

Inlet opens 22° b.t.d.c.
Inlet closes 60° a.b.d.c.
Exhaust opens 54° b.b.d.c.
Exhaust closes 15° a.t.d.c.

Separate cooling of the cylinder block and heads was employed and coolant and oil pumps were remotely driven. A transistorized ignition system provided a repeatable, high-energy spark, thus permitting the maximum amount of weak-mixture running. Except where indicated, ignition timing was adjusted manually to minimum advance for best torque.

Analysis of the exhaust for carbon monoxide (CO), carbon dioxide (CO_2), nitric oxide (NO), and unburned hydrocarbons (HC) was made by non-dispersive infrared (NDIR) equipment, while a flame ionization analyser (FIA) was used to give a measure of total HC emissions.

The programme of work reported here was carried out using a single batch of a reference fuel (Table 131.1) which was stored and fed to the engine from sealed drums to minimize the loss of the light fractions. Three test speeds, 1750, 2250 and 3000 rev/min, were standardized, and measurements at each speed were carried out at full throttle and at a constant brake mean effective pressure (b.m.e.p.) of 4·0 kg/cm^2 (56·6 lb/in^2). The effects of mixture condition have been shown to be small (1), but to eliminate variations that might result from wet mixtures

---

The MS. of this paper was received at the Institution on 24th May 1971 and accepted for publication on 22nd July 1971. 33
* Senior Research Engineer, The Motor Industry Research Association, Lindley, near Nuneaton, Warwickshire.
† References are given in Appendix 131.1.

Table 131.1. Brief inspection data for reference fuel AF 1881

Details	Measures
Specific gravity at 60°F	0·740
Distillation	
Initial boiling point, °C	30
Percentage distilled at 70°C . . .	33
Percentage distilled at 100°C . . .	53
Percentage distilled at 140°C . . .	82
Percentage distilled at 160°C . . .	93
Final boiling point	193
Hydrocarbon analysis	
Aromatics, % volume	32·5
Olefins, % volume	18·0
Saturates, % volume	49·5
Carbon/hydrogen ratio	7·01
Reid vapour pressure at 100°F, lb/in^2 . .	11·0
Octane number, research	100
Octane number, motor	89
Lead content, g/litre	0·8
Phosphorus content, p.p.m.	15
Theoretical air required, lb/lb fuel . .	14·45

in the different inlet ports all tests reported here were carried out using a fully vaporized mixture.

Eleven combustion chambers were tested, seven from production engines, while the remainder were bowl-in-piston and 'bump-on-piston' designs (4) that were thought to be of particular interest. Diagrammatic representations of the combustion chambers are presented in Fig. 131.1, while Table 131.2 gives the physical data for the various chambers. To give an indication of the reliability of the results, two of the chambers, A and B, were subjected to repeat tests. Within the dimensions of the cylinder block, the bowl-in-piston, disc, and annular designs could only be accommodated by using short connecting rods. These were obtained by cutting and welding standard rods, which were afterwards polished and crack-detected. The very much reduced fatigue life of these components was allowed for, each welded rod being discarded after tests on two combustion chambers.

A limited investigation of the effects of ignition timing and knock on emissions was carried out at 1750 rev/min, full throttle, at air/fuel ratios of nominally 14 and 17:1. The tests were carried out on a combustion chamber with stabilized deposits; this ensured that any possibility of beneficial effects due to the removal of deposits by knock could be studied.

A series of tests was also carried out to determine the effects of compression ratio on emissions. To minimize any changes that might arise from changes in quench thickness, etc., the simple disc chamber K was chosen for this work. Three compression ratios, 8·4, 7·5, and 6·5:1,

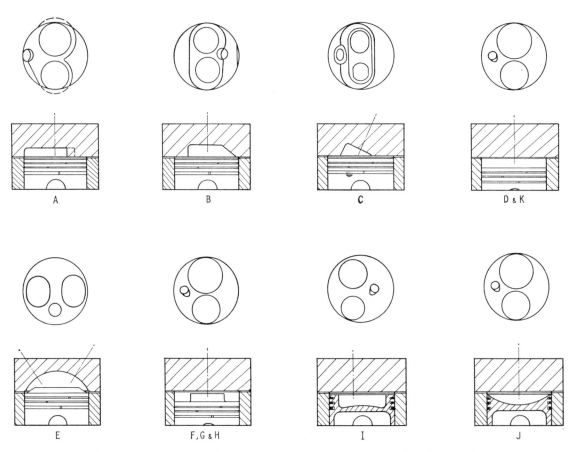

Fig. 131.1. Diagrammatic representations of combustion chambers tested

Table 131.2. Combustion chamber details

Combustion chamber	Chamber type	Compression ratio	Surface area*		Surface/volume ratio*		Squish area		Quench thickness, mm
			Without crown land, cm^2	With crown land, cm^2	Without crown land, cm^{-1}	With crown land, cm^{-1}	cm^2	As percentage of cylinder area	
A	Bath tub	8.5:1	115.3	149.8	2.67	3.39	8.9	20.9	1.0
B	Bath tub	8.5:1	107.8	142.3	2.49	3.23	9.8	23.0	1.5
C	Wedge	8.5:1	110.2	144.7	2.55	3.28	13.7	32.2	3.0
D	Disc	8.5:1	106.8	141.3	2.51	3.21	Disc thickness, 9.9 mm		
E	Hemispherical	8.5:1	122.5	157.0	2.83	3.56	—	—	—
F	Bump-on-piston	8.5:1	120.1	154.6	2.78	3.50	11.1	26.0	5.1
G	Bump-on-piston	8.5:1	119.5	154.0	2.76	3.49	13.7	32.2	5.9
H	Bump-on-piston	8.5:1	121.2	155.7	2.80	3.52	10.0	23.6	3.4
I	Bowl-in-piston	8.5:1	117.4	151.9	2.71	3.44	19.0	44.4	1.0
J	Bowl-in-piston	8.5:1	97.5	132.0	2.25	3.05	5.0	11.8	2.5
K	Disc	8.4:1	109.6	144.1	2.45	3.22	Disc thickness, 11.1 mm		

* Piston at t.d.c.

were investigated and repeat tests were carried out at the two extreme ratios.

The effect of cylinder-head temperatures was investigated by measuring emissions at three head coolant temperatures, 50, 80, and 125°C; at each temperature, the cylinder block was maintained at 80°C. Combustion chamber C was used for these tests, which were made at nominal air/fuel ratios of 13.5 and 17:1.

RESULTS AND DISCUSSION
Repeatability and deposit effects

In earlier reports of this work, the problems of obtaining stability of deposits and details of the repeatability of measurements have been discussed at length. Unfortunately no means of improving repeatability have been found, and the variability in the establishment of deposit stability is still present. For details of repeatability, reference should be made to the earlier reports; for the present work, an indication of repeatability is given by quoting mean values of repeat tests and giving individual repeat measurements where appropriate.

Effects of combustion chambers

In tests to evaluate the different combustion chambers it would have been impracticable, within one engine, to have each chamber operating in exactly the same configuration as in the production engine from which it was taken. This, however, would not have been too desirable, since the comparison of combustion chambers would have been complicated by the indeterminate effects of compression ratio, valve timing, etc.

With the arrangements used for the present tests, valve timing and compression ratio were virtually constant, but the chambers were not all used without geometrical alteration. Variations in 'bumping clearance' and in the surface of the piston crown occurred when compared with the original engines from which the chambers were taken —but these were regarded as the lesser of two evils. A compression ratio of 8.5:1 was chosen as a reasonable compromise, and adjustments were made by varying gasket thickness or grinding metal from the cylinder-head face as appropriate. In addition, any effects on emissions of inlet ports, other than that resulting from port-induced swirl in the cylinder, were minimized by running each combustion chamber configuration on a fully vaporized mixture.

The limited amount of effort which could be expended on this series of tests precluded the investigation of variables such as humidity, and, since NO emissions have been shown to be most affected by humidity, the values of peak NO presented here have been corrected according to the data presented by Robison (5) to a humidity of 60 gr/lb. The corrections, which were less than 5 per cent, did not produce any changes in the order of merit of the combustion chambers.

Specific emission measurements in terms of g/b.h.p. h were also examined, since the different combustion chambers, with their individual valve sizes and lifts, produced different full-throttle power outputs at each engine speed. The results, when expressed in this way, however, produced no significant differences in the order in which the combustion chambers were ranked, and data have thus been presented as concentration levels of pollutants. Obviously, half the emission measurements obtained were effectively specific emissions, since the part-load tests were carried out at a constant b.m.e.p. so that only the full-throttle results would have been modified by power output.

Data on the 11 combustion chambers considered have been given in Fig. 131.1 and Table 131.2, but some further elaboration is desirable. In all cases, the pistons used employed the same ring-pack and crown-land dimensions so that simple surface areas and volumes suffice to characterize the combustion chambers. Also, since it has been possible to maintain a substantially constant compression ratio, the surface area is also representative of that favoured parameter, surface-to-volume ratio. Squish areas have, in some cases, been indeterminate, e.g. the hemispherical chamber where there is no area of constant clearance, while in the disc chamber squish area would most correctly be considered to be zero.

Combustion chambers A, B, C, D, E, I, and K were all designs from current European production engines, while J was tested to investigate the emission performance of a bowl-in-piston design with minimum squish area. Chambers D and K were basically similar and differed mainly in the position and size of the valves and location of the spark plug. F, G, and H were tested to investigate a hypothesis that current bowl-in-piston designs did not utilize gas movement to the best advantage in minimizing emissions. A conventional bowl-in-piston design produces quench areas that are situated in the remote 'corners' of the combustion chamber where gas movement is relatively low. The bump-on-piston design produces an annular combustion chamber in which the remote corners are subjected to a higher degree of turbulence. At the same time, the quench area is to the centre of the chamber, where it would be scoured by the turbulent gases as the piston descends. Unburned mixture in this layer is thus more likely to be entrained and oxidized within the cylinder.

The variations in full-throttle output between combustion chambers are indicated in Fig. 131.2, which compares the b.m.e.p's produced at the three test speeds at air/fuel ratios of 13·5 and 18:1. In addition to variations in output, differences in economy would also be expected, and the full-load specific fuel consumptions achieved by each combustion chamber at the same mixture strengths are presented in Fig. 131.3, while Fig. 131.4 gives the equivalent data for the constant-load operation of the engine. In Figs 131.2, 131.3, and 131.4, data from the repeat tests on chambers A and B have been plotted separately to give some indication of the repeatability of measurements. In addition, the relative performance of the chambers with regard to output and economy may be extracted and studied in conjunction with their emission-producing tendencies, which will now be considered.

Results on each chamber were characteristically similar

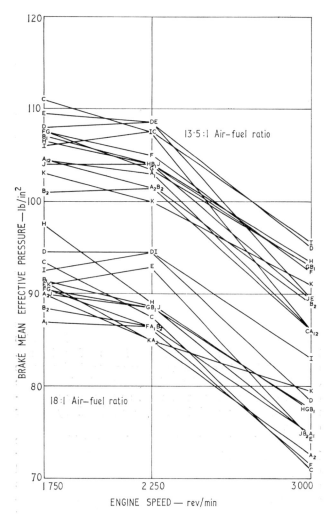

Fig. 131.2. Full-load b.m.e.p., at two mixture strengths and three test speeds, for the 11 combustion chambers of Table 131.2. Subscripts indicate initial and repeat tests

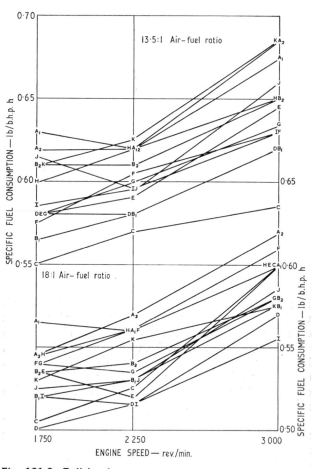

Fig. 131.3. Full-load economy, at two mixture strengths and three test speeds, for the 11 combustion chambers of Table 131.2. Subscripts indicate initial and repeat tests

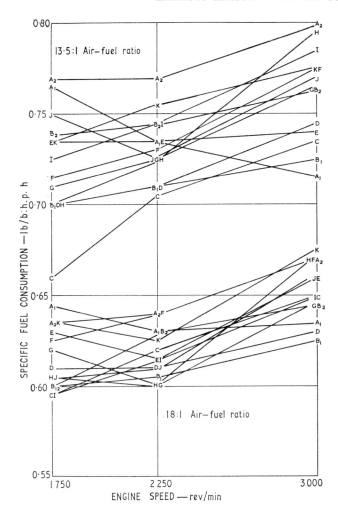

Fig. 131.4. Half-load, constant-b.m.e.p. economy at two mixture strengths and three test speeds, for the 11 combustion chambers of Table 131.2. Subscripts indicate initial and repeat tests

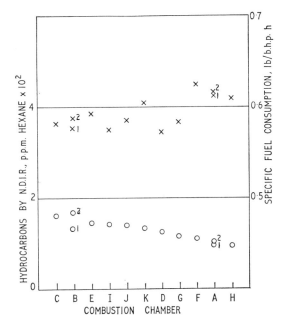

Fig. 131.5. Merit order of combustion chambers based on mean minimum hydrocarbon emissions by NDIR together with corresponding mean specific fuel consumptions

to those presented previously (1), and from such data the minimum HC and maximum NO emissions were extracted. The minimum concentration of HC generally occurred at an air/fuel ratio of about 18:1, and the peak of NO emission at around 16·5:1; at the part-load b.m.e.p. the minimum mass emission of HC invariably coincided with the maximum of NO. Carbon monoxide levels at mixture strengths weaker than stoichiometric were too low to be measured with accuracy on the equipment set to full scale at 12 per cent. An attempt to evaluate the relative emissions of NO over the mixture range produced results that were very similar to those obtained by considering peak values only. The peak values were probably the most reliable values obtained in each test, since a search was made at each speed and load to pinpoint the maximum; thus, only these have been used in the subsequent analysis.

Overall mean values of minimum HC (by NDIR) and peak NO for the 11 chambers have been plotted in order of reducing pollutant concentration, together with the corresponding mean specific fuel consumptions, in Figs 131.5 and 131.6. The repeat tests on chambers A and B are shown separately as an indication of repeatability; clearly this was not ideal, but the differences between repeat tests were appreciably less than the maximum

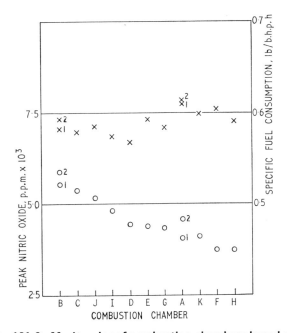

Fig. 131.6. Merit order of combustion chambers based on mean of peak nitric oxide emissions together with corresponding mean specific fuel consumptions

differences between chambers. There is a general tendency for economy to be worse with those chambers that produced lowest emissions, suggesting that the elimination of pollutants within the combustion chamber will not necessarily result in a more efficient engine.

Comparing combustion chambers for individual pollutants, chambers B and C produced the highest NO emissions whilst chambers F and H produced the lowest. The ratings of chambers with regard to HC differed slightly depending on whether NDIR or total HC was taken as the criterion, but basically A and H produced least and B, C, and E emitted most HC. In terms of overall emissions, there appears to be little to choose between chambers H and A at the head of the list. Data on CO emission have not been detailed here since, in theory, changes in chamber design should have no effect on CO level. A brief analysis, however, seemed to indicate that the chambers which were best from HC and NO viewpoints produced slightly higher levels of CO, and vice versa.

One characteristic that might be expected to emerge from this series of tests is the tolerance of each chamber to weak mixtures. The repeatability in establishing the weak limit was not too good, but overall averages of the limits at different speeds and loads were calculated for the chambers tested. In general, the air/fuel ratio at the weak limit increased with load and a reduction in engine speed; but, on average, the differences in weak limit between the 11 chambers were not great. The tolerance of the combustion chambers to weak mixtures could not be related to emission performance.

In designing a combustion chamber for low HC emission, Scheffler (6) has suggested that a low surface area for a given clearance volume has merit, and the present results were thus examined against this parameter. Scheffler's results were obtained in vehicles fitted with various engines having different compression ratios, etc., tested over the (then) California cycle, and the two sets of data are, perhaps, not strictly comparable. However, Fig. 131.7 indicates how the emissions from the 11 chambers relate to surface-to-volume ratio over the relatively narrow range of 3·0 to 3·6 cm^{-1} (7·6 to 9·1 in^{-1}). The scatter of results is similar to Scheffler's, but whereas his results suggest a fairly steep rise in HC emission with increase in ratio, the present data show a trend in the other direction. The relationship with overall average of NO peaks follows the expected pattern with an increase in surface area followed by a decrease in NO. Carbon monoxide emission was apparently not entirely independent of surface-to-volume ratio and overall average concentrations at a mixture strength of 12:1 tended to rise very slightly with increased surface area. No interdependence was noted between the emission level of any pollutant and the other possible design parameters, quench thickness, squish area, or any combination of the two.

From an overall emission viewpoint, the annular chamber H was marginally best with the bath-tub chamber A close behind. The low HC levels of H tend to confirm the

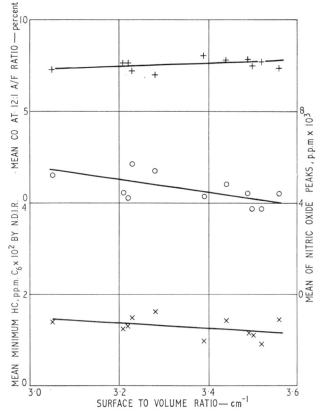

Fig. 131.7. Variation of emissions with surface-to-volume ratio

hypothesis above, which suggests that the reductions are the result of entrainment of the quenched gases due to more consistent turbulence throughout the chamber, while the low NO levels have been shown to be related to increased surface area. No consideration has been given at this stage to octane requirement or to thermal stresses within the piston in such a design.

Effect of ignition timing and knock

The easiest way of producing different intensities of knock was by advancing the ignition, and the investigations of spark timing and knock were therefore combined. In order to obtain the maximum information from the tests, operation under knock was extended so that any possible change in emissions resulting from the removal of deposits by knock could also be identified.

Tests were carried out on what was basically a disc chamber formed by the flat cylinder head of chamber K and a piston with annular waves in the crown. This chamber was produced as part of another investigation, the results of which are not reported here.

At 1750 rev/min, full throttle, the ignition was retarded by some 10–15° from the optima at air/fuel ratios of nominally 14 and 17:1 and advanced in stages until heavy knock was obtained, when the process was reversed. In tests at the richer mixture, small variations in air/fuel ratio

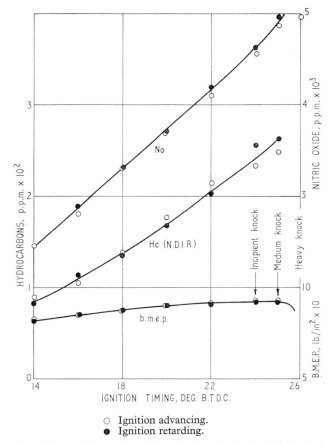

○ Ignition advancing.
● Ignition retarding.

Fig. 131.8. Effect of ignition timing on hydrocarbon and nitric oxide emissions at an air/fuel ratio of 17:1

between. No systematic change in emission levels was noted after this prolonged period of knocking, however, and the inference must be that if changes in deposits were occurring they were either too small or of such a nature as to have no effect on emissions.

Effect of compression ratio

The work to evaluate the effect of compression ratio on emissions was carried out using a disc combustion chamber so that the variations in chamber configuration with compression ratio would be at a minimum. Originally tests at compression ratios of 8·4, 7·5, and 6·5:1 were planned; but because of poor repeatability, and an inability to achieve deposit stability at 7·5:1 within a reasonable time, only tests at 8·4 and 6·5:1 were completed. Two full series of tests were carried out at each compression ratio, and the results in terms of minimum HC and peak NO levels are presented in Figs 131.9 and 131.10 respectively. The values from the two tests at each ratio have been plotted and joined together while the mid-points of the connecting lines at each compression ratio have also been joined. Thus, Figs 131.9 and 131.10 give an indication of the repeatability of results and the trends observed. The

occurred, and these were eliminated by correcting the results to a constant CO concentration of 1·3 per cent. Little difficulty was experienced in maintaining a near-constant mixture strength at 17:1, but the engine refused to run continuously with heavy knock at this weak mixture.

The results of the above tests showed that CO emissions were unaffected by ignition timing or knock, but changes in NO and HC emissions were observed. With incipient and medium knock, the HC emissions were completely normal; but with heavy knock, increased levels and wild fluctuations occurred. In general, NO levels fell slightly when heavy knock occurred. Fig. 131.8 gives the variations of HC and NO emissions with ignition timing at the weak mixture as an example, indicating that a retarded spark appreciably reduced emissions. Similar results were obtained at the rich mixture, but the reduction in NO was much less marked. It is interesting to note that the present data show a much greater emission dependence on ignition timing than had been previously measured (1), but there were differences in combustion chamber, engine speed, and mixture strengths between the two series of tests.

The results presented in Fig. 131.8, as indicated, were obtained as the timing was advanced and retarded with a period of some 20 min running at various knock intensities

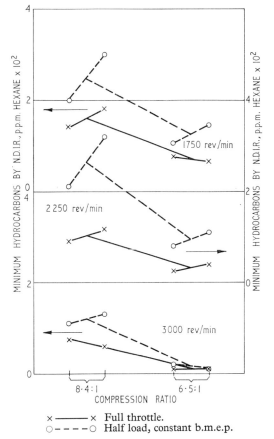

×———× Full throttle.
○-----○ Half load, constant b.m.e.p.

Fig. 131.9. Mean minimum hydrocarbon emissions obtained from tests at the three speeds with compression ratios of 8·4 and 6·5:1. Disc chamber K (Table 131.2)

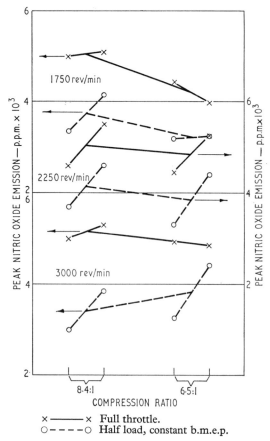

×———× Full throttle.
○-----○ Half load, constant b.m.e.p.

Fig. 131.10. Mean peak nitric oxide emissions obtained from tests at the three speeds with compression ratios of 8·4 and 6·5:1. Disc chamber K (Table 131.2)

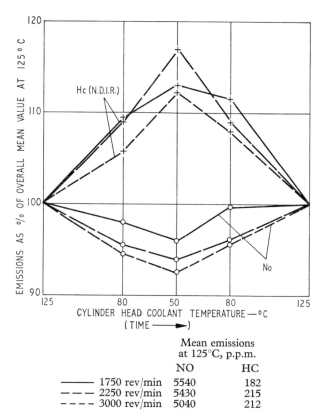

	Mean emissions at 125°C, p.p.m.	
	NO	HC
——— 1750 rev/min	5540	182
– – – 2250 rev/min	5430	215
- - - - 3000 rev/min	5040	212

Fig. 131.11. Effect of cylinder-head coolant temperature on hydrocarbon and nitric oxide emission at an air/fuel ratio of 17:1. Wedge chamber C (Table 131.2)

poor repeatability of the NO peak masked any small change that might have resulted from a reduced compression ratio, although the majority of tests show a downward trend, but a marked decrease in HC emissions was produced, particularly at the highest engine speed, at which minimum levels of the order of 10 p.p.m. were noted.

Effect of cylinder-head temperature

The ability with this single-cylinder engine to run with different coolant circuits for cylinder head and block was used to determine the influence of cylinder-head temperature on the level of pollutants emitted. To avoid the possibility of scuffing, the coolant temperature at outlet from the cylinder block was maintained throughout at 80°C while, with ethylene glycol in the cylinder head circuit, head coolant temperatures were varied between 50 and 125°C. In tests using chamber C at air/fuel ratios of approximately 13·5 and 17:1 at each of the three test speeds, the engine was stabilized with a head coolant outlet temperature of 125°C and emission levels noted. The cylinder head was then cooled to 80°C and a further series of readings obtained before reducing the temperature to 50°C. After stabilizing temperatures and noting the emissions at 50°C, the temperature was allowed to rise again until the 80° and 125°C measurements had been repeated. During each of these temperature cycles, which lasted for some 2 hours, the air/fuel ratio varied slightly but the results were corrected, using earlier emission concentration–air/fuel ratio characteristic data.

The results of these tests were similar at the two mixture strengths, and Fig. 131.11 gives the variations with head temperature of mean minimum HC (by NDIR) and peak NO at 17:1 air/fuel ratio as an illustration. Again, a measure of the repeatability is obtained by comparing the two results at 80°C. Clearly, the trends are as one would predict, with an increase in NO and a decrease in HC emissions following an increase in head temperature. For both mixture strengths it has been possible to derive mean rates of change of pollutants with temperature, and these are presented in Table 131.3. Mixture strength had little effect on the way in which NO and HC by NDIR varied with coolant temperature, but the change in HC at the weak mixture, as indicated by FIA, was double that at the rich mixture.

CONCLUSIONS

By confining tests to a single-cylinder engine, investigation of the effects on emissions of various factors has been possible with minimum interference from other variables.

Table 131.3. Effect of cylinder-head coolant temperature on emissions

Pollutant	Rate of change of emissions, percentage per 10 degC increase	
	At 4·4% CO*	At 17:1 A/F ratio
NO	−0·75	−0·77
HC by NDIR	+1·68	+1·91
HC by FIA	+0·75	+1·47

* Approximately 13·5:1 A/F ratio.

Eleven combustion chambers have been compared and the lowest overall emissions were obtained with an annular chamber formed between a flat head and a bump on the piston. A bath-tub chamber also performed well, but differences between all chambers were only moderate. A trend was noted for chambers that produced lowest emissions to have the highest specific fuel consumptions. With the conventional combustion chambers, an increase in surface-to-volume ratio gave a slight reduction in hydrocarbon and nitric oxide emission.

Ignition timing was found to have a marked effect on emissions with the combustion chamber tested; large reductions in hydrocarbons and nitric oxide resulted from a modestly retarded timing at a weak mixture. A smaller reduction in nitric oxide was obtained at a rich mixture.

Incipient and medium knock produced no changes in emissions, but heavy knock resulted in a slight reduction in nitric oxide and a large increase in hydrocarbons. Operation for some 20 min under knock did not modify the stabilized deposits sufficiently to affect emissions obtained subsequently without knock.

A reduction in compression ratio from 8·4 to 6·5:1 using a disc combustion chamber to minimize chamber shape effects, produced considerable reductions in hydrocarbons; poor repeatability masked any effect on nitric oxide emission.

Cylinder-head coolant temperature was investigated over a range of 50 to 125°C and mean rates of change of pollutant concentration with temperature were determined. As expected, hydrocarbon emissions increased and nitric oxide levels fell as coolant temperature was reduced.

ACKNOWLEDGEMENTS

The author wishes to express his thanks to the Motor Industry Research Association for permission to publish the information contained in this paper and to the Long Term Research Panel of B.T.C. for their practical help and advice. The able assistance of Z. Holubecki in carrying out the tests reported here is also gratefully acknowledged.

APPENDIX 131.1

REFERENCES

(1) DODD, A. E. and WISDOM, J. W. 'Effect of mixture quality on exhaust emissions from single-cylinder engines', *Symp. on Motor Vehicle Air Pollution Control, Proc. Instn mech. Engrs* 1968–69 **183** (Pt 3E), 117.

(2) LYTOLLIS, I. D. 'Atmospheric pollution from petrol engines: effect of mixture quality on the reactivity of hydrocarbons from a single-cylinder engine', M.I.R.A. Report No. 1971/1, 1970 (November).

(3) DODD, A. E. and HOLUBECKI, Z. 'Atmospheric pollution from petrol engines: effect of combustion chamber design and other variables on the emissions from a single-cylinder engine', M.I.R.A. Report No. 1971/7, 1971 (June).

(4) U.K. Provisional Patent No. 38103, 7th August 1970.

(5) ROBISON, J. A. 'Humidity effects on engine nitric oxide emissions at steady-state conditions', S.A.E. Paper No. 700467, 1970.

(6) SCHEFFLER, C. E. 'Combustion surface area, a key to exhaust hydrocarbons', S.A.E. Paper No. 660111, 1966.

C132/71 THE INFLUENCE OF GASOLINE BULK FUEL PROPERTIES ON EXHAUST EMISSIONS UNDER IDLE AND DYNAMIC RUNNING CONDITIONS

D. R. BLACKMORE*

The concentrations of automobile exhaust emissions of carbon monoxide, hydrocarbons, and nitric oxide have been measured as a function of three bulk gasoline properties (volatility, viscosity, and specific gravity) in an experiment with many fuels. Tests were carried out on a number of emission controlled cars, typical of current European manufacture and covering various common types of carburation and emission control systems. Quantitative relationships have been derived statistically between the concentration of each pollutant and each fuel property for each car under both idle and dynamic running conditions. The results agreed in general with those from previous work. When combined with typical variations in properties of current commercial fuels, the results yield a measure of the changes in emissions that can result from bulk fuel changes. Different carburation and emission control systems show somewhat different sensitivities of their emissions to changes in fuel properties.

INTRODUCTION

VARIATIONS IN EXHAUST EMISSIONS with bulk fuel properties are usually considered to be small compared with variations that can be effected by many different modifications of the engine. Nevertheless, such variations with bulk fuel properties are real and need to be quantified both to indicate the extent of any benefits achievable through fuel development, and to be aware of the implications of other changes which may be called for—for example, greater control of fuel metering that may be necessitated by current and future air pollution legislation. Also, in order to help control evaporative loss emissions, it is possible that even lower fuel volatility may be required, e.g. in California.

A number of different studies of aspects of this problem have been undertaken in recent years, mainly in U.S. cars (1)–(4)†. In Europe, one recent co-operative study (5) under the auspices of the British Technical Council (B.T.C.) measured variations of carbon monoxide (CO %vol.) at idle in five different cars and with five different fuels (embodying simultaneous changes in the three bulk properties: volatility, specific gravity, and viscosity). The present work enlarges the scope of that programme by measuring the concentrations of emissions of both carbon monoxide (CO) and hydrocarbon (HC), as well as those of nitric oxide (NO) in most cases. Further, the car test procedure included the measurement over the U.S. Federal (1970) cycle and was not restricted just to the idle mode. The number of fuels was increased to 15 and then to 30 in order to disentangle as far as possible the correlations between the three bulk properties.

EXPERIMENTAL

Cars

Ten emission controlled cars of different sizes were included in the programme; they were chosen to cover as far as practicable the combinations of carburettor type and emission control system commonly found in European cars (Table 132.1). A brief specification of the cars is given in Table 132.2.

Fuels

The test gasolines were selected to give series in which certain bulk properties varied but in which other properties did not. Chemical composition (in terms of aromatic, olefinic, and saturate fractions) could not be kept constant. This is a sound approach if, as is believed to be largely the case, the main variations in emissions have their origins in such physical processes as fuel metering and distribution, but less sound if considerations of chemical combustion are dominant. Accordingly, the test gasolines were formulated around a single full-range platformate of low

---

The MS. of this paper was received at the Institution on 24th May 1971 and accepted for publication on 26th July 1971. 33
* *Shell Research Limited, Thornton Research Centre, P.O. Box 1, Chester CH1 3SH.*
† *References are given in Appendix 132.1.*

Table 132.1. Test cars classified by type of carburettor and emission control system

Type of emission control	Carburettor type			
	Fixed jet	Constant depression (variable jet)		Petrol injection
		Twin	Emission controlled	
Air injection .	Car 9	Cars 6, 10	—	—
Engine modification	Cars 2, 8	Car 7	Car 5	—
'Duplex' manifold .	—	Car 3	Car 4	—
Petrol injection .	—	—	—	Car 1

Reid vapour pressure (R.v.p.), to which were added other blending components to give the desired properties.

In fact, the gasolines, all of which were leaded to give Research octane numbers (R.o.n.) of >99, were blended into two groups. The first group of five gasolines was designed to cover a variation in viscosity from 0·55 to 1·0 cS, whilst R.v.p. and specific gravity were kept constant. (Throughout this paper viscosity refers to kinematic viscosity. Because fuel density is constant in this group of five gasolines, the absolute viscosity varies in exactly the same way as kinematic.) This was achieved for the first two of the 10 cars by adding a thickener (aluminium octoate) to the base platformate and butanizing to 10 lb R.v.p. After certain difficulties had arisen from the inhomogeneity of these gasolines, with consequent gumming of the carburettor on car 2, the remaining eight cars were tested with a set of five gasolines whose viscosity variation over the same range as before was achieved by blending cyclohexane with the base platformate.

The second group of 25 gasolines comprised a 'square' in which R.v.p. varied from 6·5 to 15 lb and the specific gravity from 0·69 to 0·78 at 60/60°F, the viscosities remaining essentially constant. This was achieved by the blending of alkylate and the base platformate and by adding butane to obtain the desired R.v.p. (In the case of car 1 only, this square arrangement was not used. Instead two sets of five fuels were blended to give respectively a variation of R.v.p. from 5 to 15 lb by butanizing the base platformate, and a variation of specific gravity from 0·69 to 0·78 by blending the base platformate with the alkylate. The disadvantage of using these two sets of five fuels is that a total of 40 tests—i.e. four replications with each of the two sets of five fuels—was required for car 1 on the dynamometer: the same information with the same precision could be obtained from 25 tests for the square—i.e. no replication—with each of the other cars.)

The target properties of the blended test fuels and the properties of the principal blending components are given in Table 132.3. The properties of the blended gasolines (as used, for example, by car 4) are given in Table 132.4.

One of the problems met with while these tests were run over a period of several months was the deterioration of the fuels, especially with respect to their R.v.p. In spite of the precaution being taken of storing the fuel cans overnight in a cold room, and of the fuels being used only when cooled to prevent loss of the light ends, the deterioration was such that fuels had to be reblended and their properties (particularly R.v.p.) remeasured at intervals during the test.

Experimental procedure

All the cars selected for test were 1968–70 models, emission controlled to U.S. specification. Each car was set

Table 132.2. Brief specifications, dynamometer settings and U.S. Federal (1970) exhaust emission test results of the cars

Car	Engine size, cm^3	Carburettor	Brake load at 50 mile/h, lb	Inertia load, lb	Mass emissions			Concentration emissions		
					CO, g/mile	HC, g/mile	NO_2, g/mile	CO, %	HC, p.p.m. C_6	NO, p.p.m.
1	1584	Petrol injection	72	2400	17·1	2·51	3·24	1·27	349	1560
2	5735	Fixed jet	130	3600*	31·7	2·41	—	1·33	190	—
3	1986	Twin variable jet—Duplex	73·5	3200	39·4	3·97	3·17	2·27	429	980
4	4235	Variable jet 'emission controlled'—Duplex	50	3200	23·0	2·41	2·46	1·32	261	883
5	1296	Variable jet 'emission controlled'	50	2000	13·7	2·07	2·92	1·29	366	1720
6	1798	Twin variable jet—Air injection	75	2800	12·6	3·20	4·36	0·848	403	1840
7	1978	Twin variable jet	70	3200	12·3	3·81	2·82	0·711	412	1080
8	2286	Fixed jet	155	3600*	25·9	2·25	5·53	1·33	217	1930
9	1599	Fixed jet—Air injection	55	2400	22·5	1·88	1·65	1·67	260	828
10	6230	Twin variable jet—Air injection	110	3600*	21·3	2·00	3·50	0·834	147	905
Legal limits 1970 U.S.					23	2·2	—	1·33†	238†	—
1971 Calif.					23	2·2	4·0	1·33†	238†	1410†

* Indicates that maximum inertia load of the Schenck chassis dynamometer was used.
† For a typical 3000-lb car.

Table 132.3. Target properties of blended test fuels

Viscosity set

Set	R.v.p., lb	Viscosity at 20°C, cS	Specific gravity, 60/60°F
1 Vi	10	0·55	0·78
2 Vi	10	0·60	0·78
3 Vi	10	0·70	0·78
4 Vi	10	0·80	0·78
5 Vi	10	0·95	0·78

Volatility–specific gravity square (Viscosity constant at 0·55 cS at 20°C)

R.v.p., lb	Specific gravity, 60/60°F				
	0·69	0·71	0·73	0·75	0·78
6·5	1	2	3	4	5
8	6	7	8	9	10
10	11	12	13	14	15
12·5	16	17	18	19	20
15	21	22	23	24	25

Volatility set (only for car 1)

Set	R.v.p., lb	Viscosity, cS	Specific gravity, 60/60°F
1 Vo	5	0·60	0·77
2 Vo	7·5	0·60	0·77
3 Vo	10	0·60	0·77
4 Vo	12·5	0·60	0·77
5 Vo	15	0·60	0·77

Specific gravity set (only for car 1)

Set	R.v.p., lb	Viscosity, cS	Specific gravity, 60/60°F
1 SG	10	0·60	0·69
2 SG	10	0·60	0·71
3 SG	10	0·60	0·74
4 SG	10	0·60	0·76
5 SG	10	0·60	0·78

Properties of principal blending components

Component	R.v.p., lb	Viscosity at 20°C, cS	Specific gravity, 60/60°F
Platformate base	4·9	0·635	0·788
Alkylate	6·4	0·631	0·694
Cyclohexane (+butane)	9·9	0·956	0·755
Butane	60·0 (blending R.v.p.)	—	—

Table 132.4. Properties of blended gasolines used for car 4

Viscosity set	R.v.p., lb	Viscosity at 20°C, cS	Specific gravity, 60/60°F
1 Vi	10·8	0·594	0·773
2 Vi	10·8	0·640	0·769
3 Vi	10·7	0·700	0·765
4 Vi	10·6	0·800	0·762
5 Vi	10·6	0·964	0·758

Volatility-specific gravity square	R.v.p., lb	Viscosity at 20°C, cS	Specific gravity, 60/60°F
1	5·55	0·631	0·693
2	(5·57)	0·622	(0·714)
3	(5·58)	0·613	0·739
4	(5·59)	0·620	(0·760)
5	5·60	0·628	0·784
6	7·50	0·616	0·691
7	(7·72)	0·606	(0·713)
8	7·95	0·595	0·734
9	(8·12)	0·598	(0·760)
10	8·40	0·600	0·776
11	10·50	0·593	0·690
12	(10·55)	0·581	0·710
13	10·65	0·581	0·730
14	(10·72)	0·583	0·751
15	10·80	0·584	0·775
16	12·70	0·577	0·684
17	(12·60)	0·570	(0·704)
18	12·50	0·563	0·725
19	(12·40)	0·568	(0·746)
20	12·30	0·572	0·767
21	15·1	0·555	0·678
22	(14·95)	0·548	(0·694)
23	14·85	0·541	0·717
24	(14·7)	0·548	(0·740)
25	14·6	0·554	0·759

Figures in parentheses are interpolated values.

up in the 'as found' condition on a Schenck chassis dynamometer with the inertia load and road load set for 50 mile/h as specified by the Federal procedure (Table 132.2). Before the fuel tests, the exhaust emissions were measured by the full U.S. Federal test (1970) (i.e. cold start and standard Indolene fuel) and the results are also given in Table 132.2. The results show that all the cars gave emissions within, or very close to, the legal requirements.

For the fuel tests, each car was tested over the U.S. Federal cycle in a fully warmed-up state, the warm-up being effected by a 15–30 min cruise at 40 mile/h until ambient air, air cleaner, and sump oil temperatures were constant. For each fuel, emission concentrations of CO_2, CO, HC, and NO were measured continuously on a non-dispersive infrared (NDIR) spectrometer over five cycles and those in the idle mode were picked out from the last four idling periods and tabulated. Also, the emissions from the last two 'hot start' Federal cycles were collected in a 12 ft × 8 ft p.v.c. bag and subsequently measured, this being a much quicker way to integrate the emissions than the specified method of treating the record charts. The procedure is quite satisfactory for a fuel-to-fuel comparison.

Particular care was taken to monitor the operating temperatures of ambient air, air cleaner, and sump oil.

As far as possible, these were kept constant, but small variations were inescapable, particularly as the whole dynamometer room warmed and cooled by a few degrees each day. All three temperatures were measured by thermocouples and recorded on a continuous multi-point recorder throughout each day of testing. Fuel temperatures in the carburettor were not measured; variations in temperature were expected to be small once a car was fully warmed up.

Data reduction

The first aim of the statistical treatment was to obtain the best fitting and most significant regression equations between emission concentration and fuel properties for each car in each driving mode. Subsequent inspection of the results showed that further statistical treatment was necessary to make the results internally consistent.

For those fuel series with a single variable (i.e. those in groups of five), an analysis of variance was carried out in the first instance to identify which variables were significant. For instance, the effects of viscosity (linear and quadratic), of blocks (i.e. the average of all five fuels), of blocks–viscosity interaction and of air cleaner temperature were all explored. The latter was sometimes identified (somewhat surprisingly since temperature changes of usually less than 5 degC were observed) as a relevant parameter after a correlation between emission concentration and test sequence number had been found for car 2 and for car 3. Once the fuel property had been established as a significant variable, a regression was carried out to establish the size of the effect. This regression was usually linear, but in some cases with viscosity a quadratic regression clearly gave a better description.

For the volatility (R.v.p.)–specific gravity fuels square, linear regressions only were used. For each pollutant and driving mode the procedure was to carry out three regressions with a single variable (R.v.p., specific gravity, and viscosity in turn), then three regressions with the variables taken in pairs, and finally a regression with all those variables together. The equation which gave the best fit (least residual standard error (r.s.e.)) with the statistically most significant variables was taken as the 'best' equation (with certain reservations—see later). Finally, another regression involving the variables in this 'best' equation together with the air cleaner temperature as another variable was carried out to see if any improvement was brought about. If so, then this regression was taken as the new and 'best' equation.

Table 132.5. 'Best' regression equations for car 4

Fuels	Emission	Mode	Regression equation				R.s.e.
			Constant term	Linear viscosity term (per cS)	Quadratic viscosity term (per cS^2)	Air cleaner temp. term (per °C)	
Five fuel viscosity series	CO	Idle	+18.1*	−37.5*	+19.6*	0	0.50
		Cycle	+4.84*	−9.15*	+4.57+	0	0.08
	HC	Idle	+6.07**	−1228**	+748**	0	4.4
		Cycle	−192+	+534**	—	+6.56*	26.5
	NO	Idle	+61.4	+195	−179+	−2.88**	6.7
		Cycle	+1510**	−594+	—	0	144

Fuels	Emission	Mode	Constant term	R.v.p. term (per lb)	Specific gravity term (per unit)	Viscosity term (per cS)	Air cleaner temp. term (per °C)	R.s.e.
25 fuel volatility–specific gravity square	CO	Idle	{ +11.8*** { +22.3***	+0.118*** 0	−13.1*** −14.1***	0 −14.6***	0 0	0.62 0.61
		Cycle	{ +4.76*** { +9.28***	+0.052*** 0	−5.34*** −5.80***	0 −6.22***	0 0	0.118 0.127
	HC	Idle	+37.5	0	0	+245**	0	21.6
		Cycle	−124	0	0	+1293***	−11.6+	36.4
	NO	Idle	−187*	0	+188*	0	+5.82**	25.7
		Cycle	−12 200*	−133	0	−16 530*	0	201
25 fuel volatility–specific gravity square, allowing for viscosity effect	CO	Idle	+10.8***	+0.002	−14.0***	—	0	0.604
		Cycle	+4.18***	+0.022**	−5.56***	—	0	0.121
	HC	Idle	+82.6***	−4.48***	0	—	0	24.2
		Cycle	+60.8*	−5.40*	0	—	0	39.4
	NO	Idle	—	—	—	—	—	—
		Cycle	+63.3	−8.8	0	—	0	21.7

RESULTS

Viscosity series

The best regression coefficients for emission concentration against fuel viscosity are listed in Table 132.5 for CO, HC, and NO, car 4 being taken as a typical example. It was sometimes found that a quadratic equation, e.g.

$$CO_{idle} = a + b\,(visc) + c\,(visc)^2$$

gave a better fit than a linear one, e.g.

$$CO_{idle} = d + e\,(visc)$$

No great importance should be attached to the constants in either of the equations. If the air cleaner temperature was significant, the corresponding coefficient is shown too. Measures of the significances are given by the conventional starred code (i.e. + indicates 90 per cent significance, * 95 per cent, ** 99 per cent, and *** 99.9 per cent).

The need for the quadratic equation is particularly evident in the results for CO at idle in car 4 (see Fig. 132.1). The curvature is such that a linear regression is unable to give a significant coefficient. A quadratic equation fits well.

R.v.p.–specific gravity experiments

For the R.v.p.–specific gravity fuels square, only linear regressions were carried out based on the equation

$$CO_{idle} = a + b\,(R.v.p.) + c\,(s.g.) + d\,(visc) \\ + e\,(\text{air cleaner temperature})$$

Since this is a multi-dimensional equation, no two-dimensional representation of the results, analogous to Fig. 132.1, is possible. The coefficients (and significances) of R.v.p., viscosity, specific gravity, and air cleaner temperature for the best regression equations for car 4 are given in Table 132.5.

One surprising feature appeared in the analysis of the fuels square. This was the frequent emergence of viscosity as a significant parameter, even though the square was designed to have a constant viscosity. (In fact, viscosity did vary slightly between 0.54 and 0.63 cS.)

A second, and more disturbing, feature also came to light: there was sufficient negative correlation between the R.v.p. and viscosity (Table 132.4) to cause noticeable effects. This arose because of changes in content of butane which has low viscosity and high volatility. Often an equation with R.v.p. as a variable gave practically as good an r.s.e. fit as one with viscosity, so that a reliable choice was not possible, in principle. Moreover, when the coefficient of the viscosity term in the 'best' equation from the fuels square was compared with the coefficient obtained from a linear regression from the five-fuel viscosity series, it was frequently found that the two values were not consistent. For instance, CO results from the bag method for car 8 give from the fuels square, in addition to a specific gravity term, a viscosity coefficient of +5.41 CO%/cS when only the viscosity term is included (r.s.e. 0.236), and a viscosity coefficient of +15.2 CO%/cS when both R.v.p. and viscosity terms are included (r.s.e. 0.235). When the more reliable direct five-fuel viscosity series is used, the linear coefficient is +0.91 CO%/cS.

In order to resolve the cases in which these and similar inconsistencies occurred, the following procedure was adopted. The viscosity effect (linear, or quadratic as the case may be) is allowed for in the fuels square data (for a

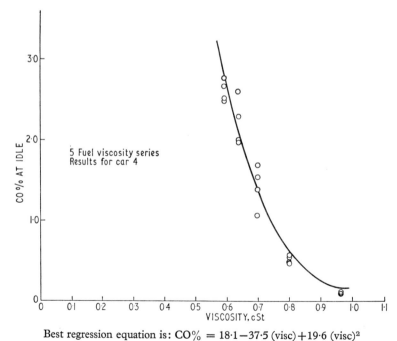

Best regression equation is: $CO\% = 18.1 - 37.5\,(visc) + 19.6\,(visc)^2$

Fig. 132.1. Typical data points for CO%, at idle, plotted against viscosity

given car, pollutant, and mode) by regressing an equation of the type

$$CO - CO' = a + b(\text{R.v.p.}) + c(\text{s.g.})$$

where CO' is given by the best equation—linear or quadratic—from the five-fuel viscosity series. The reliability of this procedure depends on the following assumptions being sufficiently valid:

(a) The viscosity effect from the five-fuel series with varying viscosities is caused by increases in a given component (cyclohexane), and the viscosity effect in the fuel square, which arises from changes in components other than cyclohexane, is assumed to give rise to an equivalent change in emissions. Since it is believed that the effects arise from physical metering effects rather than from chemical effects, this assumption is probably valid.

(b) The five-fuel series is designed to have a constant specific gravity of 0·78, while the specific gravity of the fuels square varies between 0·69 and 0·78. It is assumed that viscosity variations at s.g. 0·78 apply equally well throughout the square, i.e. that there are no strong viscosity–s.g. interactions. Similarly, the same assumption applies to viscosity–R.v.p. interactions; such interactions are likely to be of secondary importance.

(c) The range of viscosity in the five-fuel series is 0·55–0·95 cS, and in the fuels square is considerably smaller (0·54–0·63 cS). It is assumed that the equation derived over the first range should reasonably apply with sufficient accuracy over the second, for which only a small extrapolation is necessary.

The results of this correction are shown in Table 132.5 for car 4, where it can be seen that usually there is little change in the s.g. term, but a larger one (decrease) in the R.v.p. term. Since this is a modification in the 'best' equation (chosen essentially by its low r.s.e.), it is not surprising that in many cases there is a marginal increase of the r.s.e. In only one case (car 3, HC at idle) is there a large increase of the r.s.e., but closer inspection reveals that the five-fuel series peredicts a minimum in the emission of HC at a value of the viscosity very close to the values in the fuels square. In many cases the r.s.e. does indeed improve, and this is probably related to the insertion of a quadratic viscosity correction in what is otherwise a totally linear regression.

DISCUSSION

Effect of current commercial fuel property variations

The regressions listed in Table 132.5 may be reduced to more manageable form by taking the 'best' equations and coefficients (or the viscosity-corrected versions), and calculating the variation in emission when typical figures for the commercial ranges of fuel property are inserted. For R.v.p., a typical range is 5 lb; for specific gravity 0·05; and for viscosity 0·10 cS. (When reference is made to the quadratic terms in Table 132.5, the actual minimum and maximum values of 0·50 and 0·60 cS need to be used.)

The results for all 10 cars and all three emissions are shown in Table 132.6 for the idle mode, and in Table 132.7 for the dynamic U.S. Federal cycle (hot start).

Idle mode

The results for the idle mode confirm, but a little less emphatically, the B.T.C. conclusion for cars at idle: namely, that CO emission increases with R.v.p. for variable jet carburettors, and decreases for fixed jet ones. Both HC and NO emissions are very insensitive to volatility changes. As expected, increase of viscosity gives a decrease of CO, sometimes quite strongly (cars 4 and 5). Increase of viscosity gives little change in NO, and a variable effect with HC. Some cars (cars 3, 6, and 7) give strongly positive HC increases—these cars have, in common, twin variable jet carburettors. Other cars show

Table 132.6. Sensitivity of emissions to fuel property changes (emissions from the idle mode)

Car	Carburettor	Emission control	Volatility (per 5 lb R.v.p.)			Viscosity (0·50–0·60 cS)			Specific gravity (per 0·05)		
			ΔCO, %	ΔHC, p.p.m.	ΔNO, p.p.m.	ΔCO, %	ΔHC, p.p.m.	ΔNO, p.p.m.	ΔCO, %	ΔHC, p.p.m.	ΔNO, p.p.m.
1	Petrol injection	Petrol injection	+0·30	0		−0·31	0		0	0	
2	Fixed jet	Engine modification	−1·51*	−8‡		−1·02	0		0	−13	
3	Twin variable jet	Duplex	−0·72*	−74*		−1·97†	−305†		−0·61*	0	
4	Variable jet 'emission controlled'	Duplex	+0·01	−22*	0	−1·60†	−41†	−1‡	−0·70*	0	+38‡
5	Variable jet 'emission controlled'	Engine modification	−0·02*	−130*	0	−1·42†	−201	−10	−0·46*	0	+14
6	Twin variable jet	Air injection	−0·15*	0‡	−6	−0·94†	(+659)‡	0	−0·25*	−182‡	+22
7	Twin variable jet	Engine modification	+0·10*	−97*	0	−0·77†	+367‡	−11	−0·40*	0	+21
8	Fixed jet	Engine modification	−1·02*	−27*	0	−0·20	−65†	0	+0·51*	+57*	+6
9	Fixed jet	Air injection	−0·07	0	0	−0·08‡	0‡	0	+0·10	−2	+1
10	Twin variable jet	Air injection	+0·01*	−10	0	−0·77 †‡	0	−57†‡ (0)	−0·19*	−21	+13
Average			−0·30	−37	−1	−0·91	−25	−11	−0·20	−16	+17

* Corrected for viscosity effect.
† From quadratic viscosity equation.
‡ Air cleaner temperature significant.
Viscosity results in parentheses are taken from fuels square data.

Table 132.7. Sensitivity of emissions to fuel property changes (bagged emissions for two U.S. Federal (1970) cycles—hot start)

Car	Carburettor	Emission control	Volatility (per 5 lb R.v.p.)			Viscosity (0·50–0·60 cS)			Specific gravity (per 0·05)		
			ΔCO, %	ΔHC, p.p.m.	ΔNO, p.p.m.	ΔCO, %	ΔHC, p.p.m.	ΔNO, p.p.m.	ΔCO, %	ΔHC, p.p.m.	ΔNO, p.p.m.
1	Petrol injection	Petrol injection	+0·14	0		−0·12	0		0	+33	
2	Fixed jet	Engine modification	−0·24*	0		−0·26*	0		0*	−67	
3	Twin variable jet	Duplex	+0·13*	−75*		−0·42	−72†		0*	−93*	
4	Variable jet 'emission controlled'	Duplex	+0·11*	−27*	−44*	−0·41†	+53‡	−59	−0·28*	0*	0*
5	Variable jet 'emission controlled'	Engine modification	+0·17*	−27*	+48	−0·16‡	−79†	0	+0·15*	0*	−56
6	Twin variable jet	Air injection	−0·01*	−105*	+42	−0·11	−6·6†	0	0*	0*	0
7	Twin variable jet	Engine modification	+0·03*	−58*	0	−0·13	+69	−42	−0·06*	−56*	+128
8	Fixed jet	Engine modification	−0·21	0	0	+0·09	−6·6†	0	+0·13	−7	+106
9	Fixed jet	Air injection	−0·15	−22*	+92	0	+37	0	0	−108	+106
10	Twin variable jet	Air injection	+0·29	0	−75	−0·04‡	+59	0‡	0	−74	+127
Average			+0·03	−31	+9	−0·16	+7	−14	−0·01	−37	+59

* Corrected for viscosity effect. † From quadratic viscosity equation. ‡ Air cleaner temperature significant.

negative HC changes (cars 4, 5, and 8)—the first two have variable jet 'emission controlled' carburation in common, plus the curious and unexplained fact that both show a positive effect of viscosity on HC in the fuels square experiment. The effect of increases in specific gravity is to give decreases in CO for all variable jet carburetted cars, and increases for two of the three fixed jet ones (cars 8 and 9). All cars showed small increases in NO, and all (except car 8) decreases or no change in HC.

U.S. Federal cycle (hot start)

The results for the dynamic U.S. Federal cycle closely parallel those from the idle mode. Increase of volatility gives an increase in CO for all variable jet carburettors (except car 7) and a decrease in CO for fixed jet carburettors. (Car 1 (petrol injection) behaves here and elsewhere more like a variable-jet car than a fixed-jet, in disagreement with the B.T.C. findings on one fuel injection car, albeit of different type.) Increase of volatility gives some decreases in HC (strongly for car 6) and some decreases in NO (strongly for cars 4 and 5, both with variable jet 'emission controlled' carburettors). Increase of viscosity gives the expected decrease in CO (except for car 8 when it increases very slightly), an increase in HC (except, notably, for car 5), and little change in NO (decreases were reliably found only for cars 4 and 7). Increase of specific gravity gave little change in CO, decreases in HC (except for car 1), and increases in NO (except for car 5).

Comparison with U.S. legislative limits

The results for typical commercial fuels in cars tested (hot start U.S. Federal cycle) can be compared with U.S. legal limits proposed in California for 1971 for a typical 3000-lb car. These limits are 1·33% CO, 238 p.p.m. HC, and 1410 p.p.m. NO. (The actual U.S. Federal test results for the 10 cars are given in Table 132.3, and are not much different from the actual legal limits.) Against these levels, CO is most affected (up to 25 per cent of the legal limit) in current typical commercial fuels by viscosity and volatility variations, HC is roughly equally influenced (up to about 40 per cent of the legal limit) by all three properties, and NO is most strongly influenced by specific gravity only up to 10 per cent of the legal limit although this limit is due to be reduced to the greatest extent in the next few years.

Characteristics of different carburettor and emission control systems

Of the carburettor/emission control combinations tested in the dynamic mode, the petrol injection system is notably insensitive to all fuel properties. The few cars with twin variable jet carburettors were all insensitive to the influence of specific gravity on CO, and the two variable jet 'emission controlled' carburetted cars were insensitive to the influence of volatility on HC. Few generalizations could be established for the behaviour of the different emission control systems, as opposed to carburation type.

Comparison with other work

The results from the present work are compared with those from other investigations in Table 132.8 in summary form. Also given are the theoretical calculations (based in turn on laminar or fully turbulent flow models) of equivalence ratio, from increases (richening) in which it is inferred that CO and HC will increase and NO will decrease. Such an inference can only be approximately correct as no account is taken of other factors besides carburettor metering, e.g. fuel atomization, distribution, etc. (Indeed, experimentally (Table 132.5) it is found that increases in CO, increases in HC, and decreases in NO do not always occur together.) These calculations show that both variable and fixed jet carburettors respond to viscosity (i.e. neither carburettor type gives fully turbulent flow behaviour). However, these calculations have their

Table 132.8. Summary of fuel effects from several different investigations

Increase of	Emission	This work Idle	This work Cycle	Clarke (1)	API project (2)	Dishart and Harris (3)	Ebersole and McReynolds (4)	B.T.C. project (5)	Theoretical Laminar	Theoretical Turbulent
Viscosity	CO	Dec. (large)	Dec.					Insens.	Dec. (large)	Inc. (small)
	HC	Inc. (twin var. jet) / Dec. (others)	Inc.						Dec. (large)	Inc. (small)
	NO	Dec. (small)	Insens.						Inc. (large)	Dec. (small)
	Δe.r. (per 0.1 cS)								−16%	+1%
Volatility	CO	Inc. (var. jet) / Dec. (fixed jet)	Inc.	Dec.				Inc. (var. jet) / Dec. (fixed jet)	Zero	Zero
	HC	Insens.	Dec.	Dec.	Insens.	Insens.	Dec.		Zero	Zero
	NO	Insens.	Dec.		Dec.	Insens.			Zero	Zero
	Δe.r. (per 5 lb R.v.p.)				Insens.				0%	0%
Specific gravity	CO	Dec. (var. jet) / Inc. (fixed jet)	Insens.					Insens.	Inc.	Zero
	HC	Dec.	Dec.			Insens.			Inc.	Zero
	NO	Inc. (small)	Inc.			Inc.			Dec.	Zero
	Δe.r. (per 0.05)								+3.3%	0%
Mode		Idle	Cycle	Cycle	Cycle	Cycle	Cycle	Idle		
Test apparatus		10 cars		8 cars	16 cars	16 cars	2 cars	5 cars		

limitations in that they do not predict any effects of volatility change. Clearly, a more complex model (e.g. a two-phase flow) is needed.

A recent study by Stone and Eccleston (2), under the auspices of the API, measured variations in exhaust (and evaporative) emissions for typical city driving conditions as a function of fuel volatility and front-end olefin content. Six fuels were tested in 16 cars at controlled ambient temperatures between 20 and 95°F. All the cars had fixed jet carburettors. Emissions of CO, nitrogen oxides (NO_x), and aldehydes were relatively unaffected by fuel modification. A volatility reduction gave rise to an increase in exhaust HC, in agreement with the present work. However, because of a contribution from evaporative losses, it gave an overall decrease in vehicle HC emission, in terms of both amount and reactivity.

Another recent volatility study by Clarke (1) concluded that decrease of front-end volatility (as measured by R.v.p. and as distinct from mid-fill or back-end volatility) also gave an increase in exhaust HC. This experiment also showed, tentatively, an increase in CO, in contrast with the results of the present work.

Fuel properties may affect exhaust emissions, not by virtue of the chemical composition, but by the effect on fuel metering. Thus, Dishart and Harris (3) assert that small increases in NO_x concentrations with large increases in aromatic content may be attributed to this cause, since aromatic content correlates with specific gravity. The present work is in agreement with this. As far as the relationship between increase in specific gravity and exhaust HC is concerned, there is some slight disagreement. Dishart and Harris find exhaust HC to be insensitive, but the present work finds that it decreases. It is generally agreed, however, that CO is relatively insensitive to changes in specific gravity: this has been attributed to the fact that carburettors commonly meter at constant equivalence ratio. Even if increases in specific gravity cause fuel mass flow to increase, this is almost exactly compensated by a concurrent increase in stoichiometric fuel/air ratio of the fuel.

The effect of fuel viscosity, and in particular the fact that increase in temperature decreases fuel viscosity by greater amounts than the normal variations in the viscosities of commercial fuels, has led to some considerable attention being focused on the question of empirical fuel temperature compensators on carburettors. However, there appears to be little published quantitative work on the effect of variations in fuel viscosity on emissions, although it is qualitatively known that viscosity increases will lead to leaner mixtures and decreases in CO and HC emissions.

Relevance to the practical situation

This work was undertaken to see if any changes in fuel properties could contribute to a reduction in emissions. Although the present sample of 10 cars does not represent a true statistical sample of European cars, it may be used as an indication of the practical situation.

What is shown is that possible variations in commercial fuel properties give undramatic but significant variations in emissions. A low-volatility fuel will usually reduce CO at the expense of HC and NO; a high-viscosity fuel will reduce CO at the expense of HC with a concomitant reduction in NO; while a high-specific-gravity fuel will reduce HC with an increase in NO and little change in CO. A fuel with all three of these properties incorporated would give a net reduction in CO and probably in NO and HC as well.

Against this must be set the fact that the composition of the fuel is a compromise against meeting other very necessary performance and cost requirements. Taking into account the size of the effects that can be obtained by varying the fuel properties and also that the effects are not in the same direction for all cars, there is little room to make changes that could lead to a noticeable reduction in emissions.

CONCLUSIONS

(1) The exhaust emissions from 10 emission controlled cars, typical of current European manufacture, have been measured as a function of bulk fuel properties both at idle and for the 'hot start' U.S. Federal cycle.

(2) Quantitative relationships between emission (CO, HC, and NO) and fuel property (R.v.p., viscosity, and specific gravity) have been derived statistically and, from these, the size of the emission changes for typical changes in current commercial fuel properties have been evaluated.

(3) Against the levels of the 1971 Californian limits, variations in the properties of current commercial fuel can give rise to CO changes of 25 per cent of the legal limits (due to R.v.p. or viscosity), to HC changes of about 40 per cent (R.v.p., viscosity, or specific gravity), and to NO changes of about 10 per cent (specific gravity).

(4) Different carburation and emission control systems show somewhat different (both in sign and magnitude) sensitivities in emissions to changes in fuel properties. These differences make the exploitation of these results in terms of producing a 'low emission' gasoline unavoidably dependent on the characteristics of future local car populations.

ACKNOWLEDGEMENTS

I am very grateful to the various divisions of the British Leyland Motor Corporation, Ford Motor Company, and Shell Mex and BP for their loans of emission controlled cars, to D. T. S. Cuthbertson and G. Humphreys for their most able technical assistance, to J. A. Keene and Miss E. M. Jones for the considerable statistical analyses, and to I. C. H. Robinson and R. Lindsay for their helpful advice and encouragement.

APPENDIX 132.1

REFERENCES

(1) CLARKE, P. J. 'The effect of gasoline volatility on emissions and driveability', S.A.E. Paper 710136, 1971.

(2) STONE, R. K. and ECCLESTON, B. H. 'Vehicle emission vs. fuel composition', Part I, *Proc. Am. Petrol. Inst., Div. Refin.* 1968 **48**, 705; Part II, *Proc. Am. Petrol. Inst., Div. Refin.* 1969 **49**, 651.

(3) DISHART, K. T. and HARRIS, W. C. 'The effect of gasoline hydrocarbon composition on automotive exhaust emissions', *Proc. Am. Petrol. Inst., Div. Refin.* 1968 **48**, 612.

(4) EBERSOLE, G. D. and MCREYNOLDS, L. A. 'An evaluation of automobile total hydrocarbon emissions', *S.A.E. Trans.* 1967 **75**, 114.

(5) BAILEY, C. L. 'The influence of motor gasoline characteristics upon carbon monoxide emissions under engine idle conditions', *Proc. Instn mech. Engrs* 1968–69 **183** (Pt 3E), 108.

C133/71

THE EFFECT OF REDUCTION OF GASOLINE LEAD CONTENT ON ROAD ANTI-KNOCK PERFORMANCE

J. L. ADDICOTT*   D. BARKER†

The use of lead alkyls in motor gasoline has become a subject of increased attention throughout the world, and it is possible that their application may be restricted in some countries. Such restrictions may arise for two reasons: the desire to minimize the emission of lead into the atmosphere, and to allow the use of certain types of exhaust reactors. A study has been made of the effect of gasoline containing lead at various concentrations (0–0·7 g Pb/litre) on the anti-knock performance of car engines. The work was conducted in two ways: (1) by octane requirements surveys in which a limited number of fuel series were tested in a wide range of cars; (2) by determining the anti-knock behaviour of a wide range of gasoline compositions in selected cars at high and low engine speeds. The work has shown that, for a given level of research ON, a reduction in gasoline lead content generally resulted in reduced road anti-knock performance. Some additional comment is made on engine performance, thermal efficiency, and effect on exhaust emission levels, consequent upon the use of gasoline of reduced lead content.

INTRODUCTION

LEAD ALKYL anti-knock additives, first discovered by General Motors in 1921, have been widely used in motor gasolines for many years. For example, in the U.K., tetraethyl lead (TEL) was first used commercially in 1933, whilst tetramethyl lead (TML) has been used (in motor gasoline) since 1959. There is now a growing interest in the commercial availability, in certain parts of the world, of gasolines of lower lead content than has hitherto been usual. The reasons for this interest are:

(a) Concern that lead emitted from vehicle exhausts may have a deleterious effect on human and other life forms.

(b) To prevent the rapid reduction in efficiency by lead poisoning of certain types of catalyst systems which, it is reported, might be fitted to vehicles in order that they comply with very stringent exhaust emission regulations in the U.S.A.

Whilst the validity of these reasons may be open to debate, it is nonetheless certain that the use of lead in motor gasolines will be reduced, or may even cease, in certain countries.

Until recently, there was little interest in the anti-knock performance of gasolines containing relatively small (<0·4 g Pb/litre) concentrations of lead alkyls. However, because of the growing possibility that such fuels will be required in the future, British Petroleum has investigated in some detail the anti-knock performance of both unleaded gasolines and gasolines that contain a relatively low concentration of lead alkyls. The results of this work are considered in three sections in this paper, as follows: (1) laboratory octane numbers; (2) octane requirement determinations; (3) fuel ratings.

LABORATORY OCTANE NUMBERS

The effect of changes in lead content on the anti-knock performance, under standard test conditions using a single-cylinder CFR variable compression ratio engine (1)‡, of several typical gasoline components is shown in Fig. 133.1. As can be seen, the increases in octane numbers provided by the lead alkyl anti-knocks, TEL and TML (which are those most commonly used in commercial gasolines), are in all cases useful. For example, the increase in research octane number (1) provided by 0·7 g Pb/litre ranges from 4 ON to 19 ON for the components illustrated.

A finished motor gasoline is manufactured by suitably interblending these and other components, in the proportions dictated by the product quality requirements of a given market and by the processing patterns in a particular refinery. Table 133.1 shows the loss in research octane number that would occur if the lead compounds were

---

The MS. of this paper was received at the Institution on 19th May 1971 and accepted for publication on 26th July 1971. 33
* *Technical Services Branch, British Petroleum Co. Ltd, Britannic House, Moor Lane, London, E.C.7.*
† *Technologist, British Petroleum Research Centre, Chertsey Road, Sunbury-on-Thames, Middlesex.*

‡ *References are given in Appendix 133.1.*

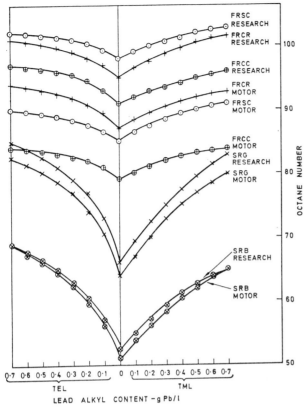

Fig. 133.1. Lead response of typical gasoline components

FRSC Full range steam-cracked gasoline.
FRCC Full range catalytically cracked gasoline.
FRCR Full range catalytic reformate.
SRG   Straight-run gasoline.
SRB   Straight-run benzene.

Table 133.1. Effect of lead anti-knocks on research octane number of typical U.K. gasolines

Grade of gasoline*	Research ON of a typical current formulation	Research ON of same formulation if lead addition omitted
Two star	92	86
Four star	98	92
Five star	100	95

* B.S.I. grade system of specification B.S. 4040.

omitted from three current typical U.K. gasoline formulations. It can be seen that a considerable loss in research octane number would occur if unleaded gasolines were required in the U.K. tomorrow. The installation of the relatively complex equipment that would be required in refineries to produce unleaded gasolines in the same quantity and having the same research octane numbers as today's four and five star gasolines would be a costly and lengthy process.

ROAD TEST PROGRAMMES—GENERAL

It has previously been shown that severe conditions for the occurrence of knock in spark-ignition engines of road vehicles are most likely to occur when accelerating from low engine speeds, and/or when at constant velocity in the mid and upper engine-speed ranges. Briefly, the reasons for these conditions being critical now follow.

Low speed—accelerating

When the throttle is opened rapidly at a low engine speed the inlet manifold vacuum falls and there is a tendency for the higher boiling fractions of the gasoline to condense on the walls of the induction system. Under these conditions the cylinders will receive a charge that is deficient in the high octane number high boiling fractions. The prevention of knock for a given engine design will then depend on the anti-knock quality of the lower boiling fractions of the fuel that enter the combustion chambers.

The difference between the research octane number of the full-range gasoline and the research octane number of a light fraction of the fuel, e.g. the fraction of the gasoline distilling up to 100°C, or the first 50 per cent volume distilled, frequently termed delta octane number, is a commonly used criterion for indicating the susceptibility of a fuel (of given research octane number) to undergo spark knock under these conditions. Because of its low boiling point (110°C) TML is a particularly useful material for producing fuels of low (i.e. good) delta ON (100°C).

Medium and high constant speed

Spark knock is most likely to be induced at high temperatures and pressures within the combustion chamber. These circumstances are found when a fully warmed-up engine is running with the throttle wide open and the loading on the engine maintaining speeds between those giving maximum torque and maximum power. The resistance to knock in multi-cylinder engines at medium and high speeds for fuels at constant research octane number can be assessed by the laboratory motor method octane rating (I), by sensitivity (research octane number minus motor method octane number) and/or olefin concentration (2).

INVESTIGATIONS BY OCTANE REQUIREMENT SURVEYS

Fuels

In the years 1967 and 1968 the octane requirements of over 200 cars were determined with four fuel series. Two series contained a high lead concentration of 0·7 g Pb/litre whilst the other two were made without using lead. A wide contrast in gasoline composition was achieved by preparing non-olefinic fuels in both leaded and unleaded forms (both referred to as Type A) from straight-run and catalytically reformed basestocks and olefinic fuels of Type B in leaded and unleaded forms from straight-run and cracked basestocks.

Table 133.2. Basestocks of the fuel series used for determining the octane requirements of cars in terms of highly leaded and unleaded gasolines

Items	Type A					Type B				
Blending route	Straight-run gasolines and catalytic reformate			C_5/C_6 isomerate, straight-run gasolines and catalytic reformate		Straight-run, catalytically cracked and steam-cracked gasolines and catalytic reformate				
Lead alkyl type	TML and TEL			—		TEL			—	
Lead concentration, g Pb/litre	0.70			Nil		0.70			Nil	
Research octane number	90.4	102.4	104.7	90.4	98.8	91.0	99.4	102.1	88.3	99.6
Sensitivity	6.6	7.8	9.5	7.7	10.1	8.1	13.5	11.5	6.5	13.4
Delta ON (100°C)*	13.8	13.8	13.0	12.4	11.8	3.8	4.2	6.1	7.6	2.1
Olefin content, % volume	0.5	1.0	1.5	1.0	0.5	19.5	34.0	25.5	9.5	23.5

* Delta ON (100°C) is the difference between the research octane number of the fuel and the research octane number of the fraction distilling up to 100°C.

Table 133.3. Basestocks of the fuel series used for determining the octane requirements of cars in terms of gasolines of high and low lead content

Items	Type A_1				Type B_1			
Blending route	Straight-run gasoline and catalytic reformate				Straight-run gasoline catalytically cracked and steam-cracked gasoline			
Lead alkyl type	TML and TEL		TML and TEL	TML	TEL	TML	TEL	TEL
Lead concentration, g Pb/litre	0.70		0.22	0.22	0.47	0.70	0.22	0.22
Research octane number	87.8	101.8	89.2	102.0	91.2	101.8	91.4	100.8
Sensitivity	5.2	8.9	6.4	9.2	9.9	13.8	10.6	14.1
Delta ON (100°C)	11.9	15.0	11.8	15.0	3.8	5.8	4.3	4.4
Olefin content, % volume	0.5	0.5	0.5	0.5	29.5	23.5	25.0	24.0

Data on the fuel basestocks are shown in Table 133.2; the series was prepared by interblending the basestocks to produce a range of fuels spanning the octane number range in increments of 1 octane number. Within each series the lead level was constant, changes in research octane number being obtained by varying the hydrocarbon composition.

In the years 1969 and 1970 further octane requirement determinations were made with more than 180 European cars. However, the interest was changed to examine the differences between car octane requirement levels obtained when using gasolines with comparatively high lead concentrations of 0.47–0.7 g Pb/litre and gasolines of low lead concentrations of 0.22 g Pb/litre. Again, four series of fuels were prepared with widely different hydrocarbon compositions (Table 133.3), and these have been classified as Types A_1 and B_1, being similar in formulation and laboratory anti-knock characteristics to Types A and B respectively of Table 133.2. As before, each fuel series was blended in 1 octane number increments, enabling the car requirements to be determined to the nearest 0.5 octane number.

Octane requirement test procedure

Octane requirements were determined by accelerating the cars according to a procedure agreed upon by the Co-operative Octane Requirement Committee (CORC) (3). Briefly, this consisted of accelerating the cars in top gear with the engine throttle wide open from start speeds of 1000 rev/min, 1500 rev/min, and from 50 mile/h (80 km/h) and determining the highest octane number in each fuel series at which knock was audible. A high proportion of these tests were carried out with the cars on a vehicle dynamometer, and relatively few tests were made on the road or on test tracks. When a vehicle dynamometer was used, precautions were taken to ensure that the car's performance matched that obtained on the road.

It had been established (4) that the octane requirements of cars run at constant speed can be higher than those attained when accelerating. Therefore, when vehicle dynamometers and suitable cars were available, octane requirements were determined also at constant speed. The critical constant speed for knock for each car was determined, using a range of highly olefinic fuels before commencing test work (4).

Car selection

The cars for octane requirement determination were, in nearly all instances, tested in the same year as their manufacture. A wide selection of mainly European cars were tested with a preponderance of British and German models, some French, Italian, and Swedish cars; a few Japanese cars were also tested. No attempt was made to match the car models selected for test in proportion to the

THE EFFECT OF REDUCTION OF GASOLINE LEAD CONTENT ON ROAD ANTI-KNOCK PERFORMANCE

car population. Performance standards were checked and ignition timing set to the manufacturers' recommended value. Generally, each car had accumulated at least 5000 miles before being tested, and in no case had it covered less than 2000 miles.

All the cars in these surveys had accumulated mileage by running on leaded gasoline. However, the relative differences between the anti-knock behaviour of fuels with high, low, and no lead addition are little affected by the formation of leaded deposits in the combustion chambers (see below).

Results of the octane requirement surveys

From Table 133.2 it can be seen that the highest research octane numbers for the two unleaded reference fuel series were only 98·8 and 99·6. To determine the requirements of many cars, fuels of higher octane numbers were desirable. These could have been produced, but not using the available components while maintaining fuel characteristics comparable with the series that contained lead at a concentration of 0·7 g Pb/litre. In practice, existing refinery processing plant is not capable of producing 99/100 research ON unleaded fuels in sufficient quantities to satisfy European demands for premium grade gasolines.

By referring to Table 133.4 the following observations on the results of this work can be made.

(a) Generally under accelerating conditions, more cars preferred the highly leaded fuel series than the unleaded or low-leaded fuels. That is, a majority of cars had lower octane requirements when using the highly leaded gasoline series than they had when using the unleaded or low-leaded series.

(b) The same conclusion holds true for the results of the constant-speed work, but the effect is even more pronounced (83 per cent of the test cars preferred the highly leaded fuels).

(c) Under accelerating conditions, the proportion of the cars showing lower octane requirements on the highly leaded fuel series appears to be independent of the fuel blend route (i.e. olefinic or non-olefinic).

(d) From Fig. 133.2 it can be seen that as the start speed for accelerations increased from 1000 rev/min through to 3000 rev/min (or 50 mile/h) so the superior anti-knock performances of the leaded fuel series of Types A and B became more pronounced.

INVESTIGATION BY FUEL-RATING TESTS
Low-speed accelerating conditions

In order to examine the effect of changes in lead content on gasoline anti-knock performance under low-speed accelerating conditions, a series of test fuels having the characteristics shown in Table 133.5 was rated in 42 test cars using the modified Uniontown technique (5). Details of the engines of the cars are given in Table 133.6, and the results are summarized in Table 133.7 and Fig. 133.3.

It should be noted that all the fuels were nominally of

Table 133.4. Summary of octane requirement results

Gasoline characteristics (before addition of lead)	High lead/unleaded investigation			High lead/low lead investigation		
	Percentage of cars showing preference for:					
	High lead 0·7 g Pb/litre	Unleaded	No preference	High lead 0·47–0·7 g Pb/litre	Low lead 0·22 g Pb/litre	No preference
Types A and A₁ *Advantages:* Low sensitivity, low olefin content with good resistance to knock at high speed. *Disadvantages:* Comparatively poor octane number of the light fractions giving poor resistance to knock when accelerating cars from low engine speeds.	*Accelerating conditions 53 (2·2) ——100% = 216 cars——	22 (1·7)	25	*Accelerating conditions 41 (1·3) ——100% = 185 cars——	24 (1·4)	35
Types B and B₁ *Advantages:* Good octane number distribution through the boiling range with good resistance to knock under low-speed accelerating conditions. *Disadvantages:* High sensitivity, high olefin content giving poor resistance to knock at high speeds.	*Accelerating conditions 51 (2·1) ——100% = 223 cars——	13 (1·2)	36	*Accelerating conditions 44 (1·5) ——100% = 172 cars——	14 (1·5)	42
	†Constant speed conditions 83 (2·4) ——100% = 103 cars——	6 (1·2)	11	†Constant speed conditions 64 (1·5) ——100% = 87 cars——	7 (1·5)	29

* Results obtained by considering the highest octane requirement when cars were accelerated from 1500 rev/min and 50 mile/h in top gear.
† Results obtained by considering the highest octane requirement when cars were run at constant, critical speeds in top gear.
Figures in ( ) are the mean differences between the octane number requirements obtained using the fuels of high lead content and unleaded, or of low lead content, respectively.

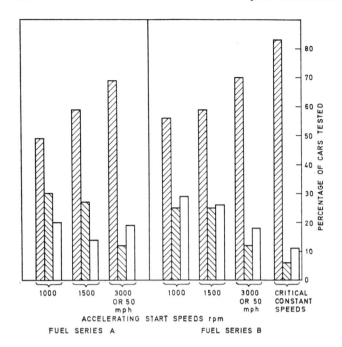

Percentage of cars with octane requirements less when using leaded than when using unleaded gasoline.

Percentage of cars with octane requirements greater when using leaded than when using unleaded gasoline.

Percentage of cars where octane requirements were the same with leaded and unleaded gasolines.

Fig. 133.2. Effects of engine speed on car octane requirements in terms of gasolines with 0·7 g Pb/litre and unleaded

99 research ON. Fuels L1–L4 (leaded) and U1–U4 (unleaded) were essentially of the non-olefinic catalytic reformate/straight-run type, whilst fuels L5–L6 (leaded) and U5–U6 (unleaded) were made using significant quantities of cracked components and are therefore olefinic.

Constant-speed conditions

For the investigation into spark knock occurring under constant-speed, high-load conditions, a series of 66 gasolines—all nominally of 97 research ON and embracing a range of lead contents from 0 to 0·53 g Pb/litre (both TEL

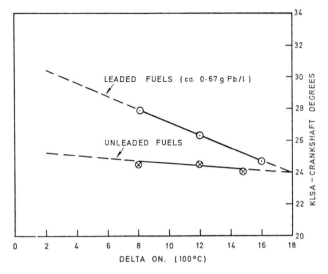

Fig. 133.3. The effect of the presence or absence of lead alkyl anti-knocks on the low-speed anti-knock performance of motor gasolines containing 0 to 1 per cent volume olefins

and TML) and olefin contents from 0 to 25 per cent volume—was used. Brief data on these fuels are given in Table 133.8; whilst data on the nine test cars are shown in Table 133.9; and the results of the constant-speed knock-limited spark advance (KLSA) ratings are summarized in graphical form in Fig. 133.4, which shows the average effect of changes in gasoline lead content on the constant-speed KLSA of the test cars. KLSA is the minimum spark advance required to induce trace knock.

Results of rating programmes

The results of both the high constant-speed and the low-speed rating programmes suggest the following for fuels of constant research octane number.

(*a*) A reduction in lead alkyl concentration results in a reduction in anti-knock performance. It can be seen from Fig. 133.3 that the superior anti-knock performance of leaded fuels appears to be most pronounced in fuels of good octane number distribution (i.e. of low delta ON (100°C)).

(*b*) The effect of changes in lead content is more

Table 133.5. Inspection data on low-speed knock test fuels

Reference No.	L1	L2	L3	L4	L5	L6	U1	U2	U3	U4	U5	U6
TEL content, g Pb/litre	0·40	0·27	0·12	0·47	0·55	0·67	0	0	0	0	0	0
TML content, g Pb/litre	0·27	0·40	0·55	0·20	0·12	0	0	0	0	0	0	0
FIA analysis												
Aromatics, % volume	39·5	42·0	44·5	37·5	35·0	31·0	59·0	56·5	54·0	56·5	57·5	42·5
Olefins, % volume	0·5	0·5	0	8·0	16·5	30·5	1·0	1·0	0·5	5·0	14·0	30·0
Saturates, % volume	60·0	57·5	55·5	54·5	48·5	38·5	40·0	42·5	45·5	38·5	34·5	27·5
CFR ratings												
Research octane number	99·1	99·1	99·1	99·3	99·3	99·1	98·9	99·0	99·2	98·9	98·8	98·7
Sensitivity	7·6	7·6	7·6	9·1	10·2	11·3	10·4	10·2	10·1	10·8	11·4	12·5
Delta ON (100°C)	16·0	12·0	8·1	12·0	8·0	4·3	14·8	12·0	8·8	12·0	8·0	3·7

Table 133.6. Engine specification data for low-speed modified Uniontown rating work

Model code	Capacity, cm³	Bore and stroke, mm	Compression ratio, :1	Carburettor make/model and No.
1	1696	80 × 84·4	11·0	Solex/35 PDSIT-5/1
2	1760	81·5 × 84·4	10·6	Solex/3232 DIDZA/1
3	1573	84 × 71	8·6	Solex/38 PDSI/1
4	1573	84 × 71	9·5	Solex/40 PHH/2
5	1990	89 × 80	8·5	Solex/40 PDSI/1
6	1988	87 × 83·6	9·0	Solex/38 PDSIT/2
7	2197	87 × 92·4	9·0	Stromberg 175 CDS/1
8	2306	82 × 72·8	9·0	Zenith/35-40 JNAT/2
9	2496	82 × 78·8	9·0	Zenith/35-40 JNAT/2
10	2496	82 × 78·8	9·0	Zenith/35-40 JNAT/2
11	2496	82 × 78·8	9·5	Fuel injection
12	843	65 × 63·5	9·3	Weber 30 DICI/1
13	1298	80·98 × 62·99	9·0	Ford/1
14	1599	80·98 × 77·62	9·0	Weber/1
15	2000	93·66 × 72·42	8·9	Weber 32 DIF4/1
16	2994	93·66 × 72·42	8·9	Weber 40 DFAI/1
17	875	68 × 60·4	10·0	Solex 30 PIH5/1
18	1496	81·5 × 71·6	8·4	Zenith 150 CDS/1
19	1724	81·5 × 82·5	9·2	Zenith 150 CDS/1
20	1798	80·26 × 89	8·8	SU HS4/2
21	996	69 × 66·6	10·5	Solex/40 PHH/2
22	1078	75 × 61	8·8	Solex/35 PDSI/1
23	1689	88 × 69·8	8·8	Solex/35 PDSIT-6/1
24	1078	75 × 61	9·2	Solex/35 PDSI-2/2
25	1492	82·5 × 69·8	8·2	Carter/1
26	1897	93 × 69·8	9·0	Solex/32 DIDTA/1
27	2490	87 × 69·8	9·5	Solex/32-35 DIDTA/1
28	2784	92 × 69·8	9·0	Solex/32 DIDTA4/1
29	1200	69·3 × 76	8·0	Solex B30 PSE1
30	1998	74·7 × 76	8·5	Zenith 1·5 CD/1
31	2498	74·7 × 95	9·5	Fuel injection
32	1159	77·7 × 61	9·0	Zenith 1·5 CD/1
33	1599	85·73 × 69·24	8·5	Zenith IV/1
34	1975	95·25 × 69·24	8·5	Zenith IV/1
35	3294	92·1 × 82·5	8·5	Zenith 42 WIAT/1

pronounced at high constant-speed than under low-speed accelerating conditions.

(c) Considering the constant-speed ratings only, the response of road performance to lead addition follows a law of diminishing returns, as shown in Fig. 133.4. (This is true of laboratory octane numbers also—see Fig. 133.1.)

(d) Again, considering the constant-speed results alone, a reduction in lead content to 0·4 g Pb/litre (which may be required in West Germany by January 1972) from 0·5 g Pb/litre (which is representative of the level currently common in German premium gasolines) will result, on average, in a KLSA loss of only 1 crankshaft degree for gasolines containing, say, 15 per cent volume olefins. However, the use of unleaded gasolines containing 15 per cent volume olefins may be expected to result in a loss of approximately 6 crankshaft degrees, compared with gasolines of the same research octane number containing 0·5 g Pb/litre.

Table 133.7. The effect upon low-speed anti-knock performance of changes in lead alkyl content

Fuel code	Delta ON, 100°C	Olefin content, % vol.	TML content, g Pb/litre	TEL content, g Pb/litre	Average knock-limited spark advance data*
L1	16·0	0·5	0·27	0·40	24·7
U1	14·8	1·0	0	0	24·1
L2	12·0	0·5	0·41	0·27	26·3
U2	12·0	1·0	0	0	24·5
L3	8·1	0	0·55	0·12	27·9
U3	8·0	0·5	0	0	24·5
L4	12·0	8·0	0·20	0·47	26·0
U4	12·0	5·0	0	0	24·5
L5	8·0	16·5	0·12	0·55	28·1
U5	8·0	14·0	0	0	27·1
L6	4·3	30·5	0	0·67	30·1
U6	3·7	30·0	0	0	28·7

* The KLSA data are relative: they refer to the distributor setting, at which knock was induced, at a speed where the gradient of the spark advance–rev/min curve is small.

Table 133.8. Lead content, FIA and octane number data for the high constant-speed knock test fuels

Reference No.	Lead alkyl Type	Lead alkyl Content, g Pb/litre (g Pb/U.K. gal)	FIA data* Aromatics, %vol.	FIA data* Olefins, %vol.	FIA data* Saturates, %vol.	Octane number data Research ON	Octane number data Motor ON	Octane number data Sensitivity
905	—	nil	53·0	0·5	46·5	97·0	87·8	9·2
906			51·0	5·0	44·0	97·2	86·8	10·4
907			48·5	10·0	41·5	97·0	86·4	10·6
908			46·0	15·0	39·0	97·0	85·8	11·2
909			43·0	20·0	37·0	97·0	85·0	12·0
910			41·0	25·0	34·0	96·9	84·6	12·3
911	TEL	0·106 (0·5)	47·5	0·5	52·0	97·0	87·4	9·6
912			45·0	5·0	50·0	97·0	87·0	10·0
913			42·5	10·0	47·5	97·0	86·4	10·6
914			39·5	15·0	45·5	97·2	85·8	11·4
915			36·5	20·0	43·5	97·0	85·4	11·6
916			33·5	25·0	41·5	96·8	85·0	11·8
917	TEL	0·212 (1·0)	44·5	1·0	54·5	96·7	87·8	8·9
918			42·0	5·0	53·0	96·8	87·0	9·8
919			39·5	10·0	50·5	96·8	86·8	10·0
920			36·0	15·0	49·0	96·8	86·2	10·6
921			33·5	20·0	46·5	97·0	85·8	11·2
922			31·0	25·0	44·0	97·0	85·0	12·0
923	TEL	0·318 (1·5)	43·5	1·0	55·5	96·9	88·3	8·6
924			41·0	5·0	54·0	97·0	87·7	9·3
925			37·5	10·0	52·5	97·0	87·0	10·0
926			34·0	15·0	51·0	97·0	86·5	10·5
927			31·0	20·0	49·0	96·8	85·9	10·9
928			28·0	25·0	47·0	96·8	85·2	11·6
929	TEL	0·425 (2·0)	42·0	1·0	57·0	96·8	88·4	8·4
930			39·5	5·0	55·5	96·9	87·7	9·2
931			36·0	10·0	54·0	96·9	87·1	9·8
932			32·5	15·5	52·0	96·9	86·8	10·1
933			29·5	20·5	50·0	96·9	86·3	10·6
934			26·0	25·5	48·5	96·7	85·8	10·9
935	TEL	0·530 (2·5)	40·5	1·0	58·5	96·9	88·8	8·1
936			37·0	5·0	58·0	96·7	88·2	8·5
937			34·0	10·0	56·0	96·7	87·6	9·1
938			30·5	15·0	54·5	96·7	87·0	9·7
939			27·5	20·5	52·5	96·7	86·4	10·3
940			24·5	25·0	50·5	96·8	85·9	10·9
941	TML	0·106 (0·5)	50·5	1·0	48·5	96·9	88·1	8·8
942			47·5	5·0	47·5	97·0	87·7	9·3
943			44·0	10·0	46·0	96·9	86·8	10·1
944			42·5	15·0	42·5	97·0	86·2	10·8
945			40·0	20·0	40·0	96·9	85·8	11·1
946			37·0	25·0	38·0	96·9	85·4	11·5
947	TML	0·212 (1·0)	48·0	1·0	51·0	97·1	89·0	8·1
948			45·5	5·0	49·5	96·9	88·3	8·6
949			43·0	10·0	47·0	97·0	87·2	9·8
950			40·0	15·0	45·0	97·0	86·8	10·2
951			37·5	20·0	42·5	97·0	86·4	10·6
952			35·0	25·0	40·0	97·0	85·9	11·1
953	TML	0·318 (1·5)	45·5	1·0	53·5	97·0	89·3	7·7
954			43·0	5·0	52·0	97·0	88·6	8·4
955			40·5	10·0	49·5	96·9	88·0	8·9
956			38·0	15·0	47·0	97·0	87·2	9·8
957			35·0	20·0	45·0	97·0	86·8	10·2
958			32·5	25·0	42·5	97·0	86·4	10·6
959	TML	0·425 (2·0)	43·0	1·0	56·0	97·0	89·4	7·6
960			40·5	5·0	54·5	97·0	88·6	8·4
961			38·0	10·0	52·0	97·0	88·4	8·6
962			35·5	15·0	49·0	96·9	87·9	9·0
963			33·0	20·0	47·0	97·0	87·2	9·8
964			30·5	25·0	44·5	97·0	87·2	9·8
965	TML	0·530 (2·5)	41·0	1·0	58·0	96·9	89·6	7·3
966			38·5	5·0	56·5	96·8	89·0	7·8
967			36·0	10·0	54·0	96·8	88·2	8·6
968			33·5	15·0	51·5	96·8	87·8	9·0
969			31·0	20·0	49·0	96·9	87·1	9·8
970			28·5	25·0	46·5	96·8	87·0	9·8

* Calculated values.

Table 133.9. Specification data for test cars

Model code	Year	Engine dimensions, bore × stroke (cm³)	Compression ratio, :1	Maximum power, kW at rev/min	Maximum torque, Nm (J) at rev/min
A	1969	76·2 × 81·28 (1485)	9·0	55 at 5500	114 at 3000
B	1966	65 × 68 (903)	9·5	39 at 6500	65 at 4000
C	1968	81·5 × 82·55 (1725)	9·2	55 at 5000	131 at 3000
D	1969	81·5 × 82·55 (1725)	9·2	65·5 at 5200	136 at 4000
E	1967	75 × 85 (2553)	8·8	85·5 at 5200	172·5 at 3600
F	1966	73·7 × 76 (1296)	8·5	45·5 at 5000	99·5 at 3000
G	1969	73·7 × 76 (1296)	9·0	56 at 6000	102 at 4000
H	1969	85·73 × 69·24 (1599)	8·5	53·5 at 5600	113·5 at 2200
I	1968	88·9 × 80 (1986)	9·5	74·5 at 5600	167 at 3500

For these reasons, as well as for reasons of refinery economics and crude oil and processing plant availability, efforts should be made to ensure that the arrival of unleaded gasolines is matched by the appearance of a new generation of motor vehicles of reduced octane requirement.

In the event of all commercially available gasolines being unleaded, the combustion chambers of car engines would not acquire lead deposits. The effect on road anti-knock ratings of unleaded and leaded gasolines in four car engines, after they had run entirely on unleaded fuel for 5000 miles, was determined with six pairs of reference fuels. The reference fuels were premium grades of a wide range of hydrocarbon types, and each pair of fuels consisted of one unleaded and one leaded to a concentration of 0·7 g Pb/litre. Techniques of accelerating the cars and running at high constant speeds revealed that leaded fuels were still less prone to knock than the unleaded ones.

The differences between the ratings of the leaded and unleaded reference fuels did not change after a further 5000 miles had been accumulated by each of the four cars on a leaded fuel. There was, however, a trend for three of the four cars to tolerate less spark advance for a knock-free operation under accelerating conditions, after accumulating 5000 miles on unleaded fuels, than after accumulating the further 5000 miles on leaded fuels.

This trend was not apparent for anti-knock ratings at high constant speeds. Investigations are continuing into car octane requirement increases and differences in KLSA after mileage build-up with unleaded gasolines.

DISCUSSION AND CONCLUSIONS

It has been demonstrated that the octane requirements of spark-ignition engines run on gasoline of low or zero lead concentration are generally higher than the requirements of gasolines with high lead contents. Further, we and other workers (6) have shown that gasolines of high lead content show superior anti-knock performance to those of low lead content, or unleaded gasolines. Therefore, even if low lead-unleaded gasolines of the future are maintained at the same research octane level as at present, some additional attention to engine design to prevent knock will be required.

The production of unleaded gasoline in Europe would involve a drop in research octane number, to obtain sufficient production at costs acceptable to the customer. There is also a loss in road anti-knock performance, as demonstrated by the results of our work. Under such circumstances it seems inevitable that a drop in compression ratio, resulting in a loss of thermal efficiency, less power, poorer acceleration, and higher fuel consumption would have to be endured, as, indeed, has been the case in North America, where unleaded gasolines have been called for by the motor industry.

Tetramethyl lead is presently used to ensure adequate anti-knock quality of the light fractions of catalytically

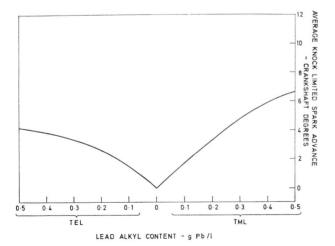

Fig. 133.4. Effect on high constant-speed anti-knock performance of changes in gasoline lead alkyl content (research ON = 97·0)

reformed gasolines. Low-speed knock caused by an absence of TML could be prevented by engine designs that encourage even fuel distribution to engine cylinders with a minimum of condensation of fuel on to the walls of the induction system. The requirements for low CO and unburnt hydrocarbon exhaust emissions are also helped by such engine design features, and these advantages might be more readily obtained by the development of fuel injection rather than carburation.

It has also been proved that increased risk of knock caused by an absence of lead alkyls in gasoline is more apparent for engines run under high loads at high constant speeds. In this respect the requirements of low oxides of nitrogen emissions and prevention of knock at high speed may both be met by controlling peak combustion chamber temperatures.

Certainly the definition of European gasoline quality for the next decade must necessarily be a compromise best reached by collaborative action on the part of the petroleum industry, motor manufacturers, and European legislative authorities.

ACKNOWLEDGEMENT

Permission to publish this paper has been given by the British Petroleum Company Limited.

APPENDIX 133.1

REFERENCES

(1) 'A.S.T.M. Designation 2699 and 2700', *A.S.T.M. manual for rating motor fuels by motor and research methods.*
(2) GREVETT, S. P. and THARBY, R. D. 'The relationship between gasoline composition and spark knock at high engine speeds', *Proc. Instn mech. Engrs* 1967–68 **182** (Pt 2A, No. 4), 104.
(3) 'Technique for determining road octane requirements', Cooperative Octane Requirement Committee.
(4) ADDICOTT, J. L. 'Some considerations on detecting and preventing spark knock in the car engine', *Proc. Instn mech. Engrs* 1967–68 **182** (Pt 2A, No. 4), 71.
(5) Coordinating Fuel and Equipment Research Committee of the Coordinating Research Council, New York, U.S.A.
(6) BELL, A. G., KEENE, J. A. and REDERS, K. 'Road anti-knock performance of low-lead and non-leaded gasoline in European cars', paper for presentation at the 1971 S.A.E. Mid-Year Meeting, Montreal, Canada.

C134/71 SWIRL—ITS MEASUREMENT AND EFFECT ON COMBUSTION IN A DIESEL ENGINE

S. OHIGASHI*   Y. HAMAMOTO*   S. TANABE*

The swirl velocity of intake air in the cylinder of a four-stroke diesel engine in a firing operation was measured successfully using a new technique which is based on the phenomenon that the path of the inductive discharge after a capacitive discharge moves downstream under the influence of the gas stream. The results of the tests suggest that the gas movement in a firing run differs from that in motoring during the intake process, but in the last stage of the compression stroke the velocities in both conditions are similar, which suggests that the swirl prior to the ignition can be estimated by a motoring test. The swirl velocity produced by the shrouded intake valve, and its effect on a mean effective pressure and on the exhaust smoke density, were also investigated. It was shown that by increasing the number of nozzle holes, good combustion giving a high power output and a low density of the exhaust smoke can be obtained when the swirl velocity is low.

INTRODUCTION

IN AN INTERNAL COMBUSTION ENGINE the movement of the charge in the combustion chamber is one of the significant factors governing the combustion process. The gas movement immediately before ignition may be one of the most significant factors in the exhaust emission problem.

Experimental methods to investigate the movement of gas in a cylinder are summarized in Fig. 134.1. In an attempt to fully understand the gas movement in a cylinder, many investigations have been carried out using scale models of cylinders (1)–(3)† or engines in test runs (4)–(7).

To measure a gas stream velocity in an engine cylinder in a test run we applied two methods: (*a*) flow visualization, using a series of high-frequency electric sparks or metaldehyde tracers, and (*b*) a hot-wire anemometer with a circuit to compensate for the change of gas temperature. These methods, however, cannot be used to measure the gas velocity in the engine cylinder during a firing run. It is expected that the velocity of the gas in the cylinder during a motoring run may differ from the velocity during a firing run. However, owing to difficulties in making measurements, no information on the gas velocity in a combustion chamber of a firing engine has been obtained.

In this study, the gas velocity in an engine cylinder was measured during a firing run by an electric discharge method (8) that was developed by the authors.

This paper presents a comparison of the gas velocities in a firing run and in a motoring run, and describes the correlation between the swirl velocity, the brake mean effective pressure (b.m.e.p.), and the exhaust smoke density in a diesel engine.

MEASURING TECHNIQUE AND SET-UP

Measuring technique

Generally, when an electric discharge occurs in a flowing gas, the ion cloud produced by the electric discharge moves downstream under influence of the gas stream, and the following inductive discharge passes continuously through the space having high ion density, i.e. the path of the electric discharge current shifts downstream with the gas flow. When the path of the inductive discharge arrives at the probe which is placed at a fixed distance downstream from the discharge electrodes, a part of the discharge current flows through the probe and the voltage of the probe rises rapidly. By measuring the time interval, τ, from the beginning of the spark to the increase in the probe voltage, the stream velocity, w, can be determined from the linear relation between w and $1/\tau$, which has been previously calibrated under the steady flow condition (8). This technique has been successfully applied to determine the velocities of pulsating flows.

Where the flame propagation is initiated by the electric discharge in the flow of the combustible mixture, such as in a cylinder of a spark ignition engine, this technique cannot be applied because the inductive discharge current passes through the ionized zone of the flame which propagates at a higher speed than the flow speed of the unburnt mixture. Therefore, the shifting velocity of the discharge path becomes higher than the stream velocity of the unburnt gas. In a compression ignition engine, however, prior to fuel injection, combustion will not be

---
The MS. of this paper was received at the Institution on 6th May 1971 and accepted for publication on 23rd July 1971. 33
* Dept of Engineering, Faculty of Engineering, Kyoto University, Kyoto, Japan.
† References are given in Appendix 134.1.

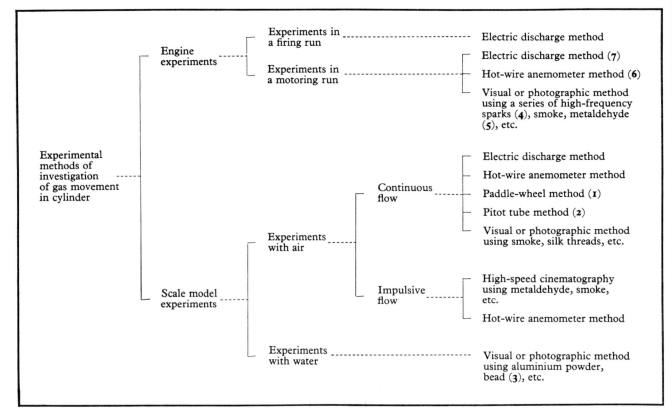

Fig. 134.1. Method of investigation

initiated by an electric discharge and, consequently, measuring technique can be easily applied.

When measuring the velocity of the gas stream in the combustion chamber of the engine, an electric discharge is produced at the desired crank angle. Since the gas velocities in the combustion chamber fluctuate from cycle to cycle, it is necessary to repeat the measurements many times. In this study the measurements were repeated about 600 times at one crank angle, and the average velocity was then obtained. A transistorized igniter for an automotive gasoline engine was used to supply the power for the electric discharge. The maximum output voltage of this equipment is 43 kV and the duration of the electric discharge is about 5 ms at an atmospheric condition when the electrode spacing of 1 mm is applied.

Test apparatus

The engine used in this study was a four-stroke cycle, single-cylinder, direct injection diesel having a stroke volume of 1·35 litres (125 mm bore/110 mm stroke). Two types of pistons were used: a cavity type, shown in Fig. 134.2a, in which the gas movement due to squish is produced in the last stage of the compression stroke, and a flat type which has no squish action. The measurements were carried out by installing the measuring plug in the cylinder head, as shown in Fig. 134.2a.

Various kinds of measuring plugs, having a different arrangement of discharge electrodes and probe (Fig. 134.2b), were provided for this study. The main difference between plug A and plug B in this diagram is in the direction of their electric discharge, i.e. vertical and horizontal, respectively. In type B the probe is located at the same depth as that of the discharge gap. In type A the probe is located at various depths from the centre of the discharge gap. To investigate the characteristics of the plugs, each plug was installed in an air duct of rectangular section (30 mm × 50 mm). The air was drawn in steadily by a blower. During the tests the measurements were made by varying the direction of the electric discharge path relative to the air flow direction.

The test results are shown in Fig. 134.3. In Figs 134.3a and b plugs A and B were set upright to the duct and made to rotate around their axis. In plug B, τ is varied with the rotating angle, θ, although the flow velocity is kept constant, and the value of $1/\tau$ attains its maximum at $\theta = 0$. This type of plug is more suitable for determining the direction of flow rather than the absolute value, because the value of τ fluctuates widely when the direction of flow fluctuates, as it does in the engine cylinder. In type A, however, since τ can only be measured in the range $\theta = \pm 30°$ and since it is hardly affected by θ, the absolute value of the flow velocity can be determined clearly although the direction of flow cannot be determined.

In Fig. 134.3c the measuring plugs of type A, having various z, were inclined at an angle of φ towards the

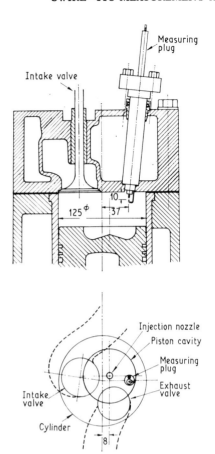

Fig. 134.2a. Schematic diagram of test engine equipped with measuring plug

direction of flow. As the probe depth (z) increases, φ_m (the maximum value of $1/\tau$) also increases. The value of φ_m indicates the direction of flow. In the engine tests, however, since the angle of the plug could not be varied, the following procedure was employed to estimate the velocity and direction of the flow which is inclined towards the cylinder axis. The measuring plugs having various z were fixed at the duct inclined by φ_m, and the relations between w and $1/\tau$ were determined by the steady flow. When these plugs are used in the engine tests the velocity is indicated by the highest K/τ value obtained (K is a constant which depends on the plug and is obtained from the above-mentioned steady flow test). Further, the direction of flow is indicated by φ_m of the plug which gives the highest K/τ value.

Preliminary test in engine cylinder

The velocity of flow inclined towards the cylinder axis was previously measured using type A plugs and by applying the procedure described in the foregoing section. It was shown that the cylinder gas moves downwards in the intake process, while it mainly moves on a horizontal plane at the last stage of the compression stroke.

As a result of the preliminary tests, the following were confirmed:

(1) During the compression stroke, the direction of flow on a horizontal plane can be determined by a type B plug, and the absolute value of the velocity can be determined by applying a type A plug ($z = 0$).

(2) During the intake stroke, the direction of the horizontal component of the velocity is determined by use of a type B plug, and the velocity is measured by fixing type A plugs (which have various values of z) so that the discharge path arrives at the tip of the probe.

GAS VELOCITY IN CYLINDER

The intake air, which may influence the velocity of the cylinder gas, was kept constant during the motoring and firing runs by operating the choke valve installed in the intake duct. The temperature of the cooling water was kept constant at 80°C in all tests.

Gas velocity in intake stroke

During the intake stroke, the velocities of the cylinder gas in a motoring and a firing run, at an engine speed of 900 rev/min, were measured. The results are shown in Fig. 134.4. The velocity of the cylinder gas during firing is higher than that during motoring, although the modes of the variation of periodic velocity are similar in both cases. This may be explained by the fact that the intake air expands owing to the transfer of heat from the intake valve, the valve seat, and the wall of the combustion chamber. However, during the intake stroke, since there can be a severe velocity gradient in the cylinder (6), it should be noted that the value measured at one point is not necessarily representative of the velocity in the entire cylinder (see Fig. 134.2a).

Fig. 134.4 shows the lift curve of the intake valve, the measured value of the air velocity in the intake duct in a motoring run, and the displacement speed of the intake air column due to the intake action of the piston, defined as $(D/d)^2 v$, where v is the piston speed at each crank angle, D the cylinder bore, and d the inside diameter of the intake duct.

Gas velocity in compression stroke

The swirl velocities over a range of crank angle degrees before t.d.c. of the compression stroke were measured in a firing and a motoring run at various engine speeds. The results are shown in Fig. 134.5. It was observed that the cylinder charge moved in a tangential direction to the cylinder wall. No noticeable differences were found between the measured values in both conditions in the last stage of compression stroke. This fact suggests that the swirl velocity immediately before ignition in a firing run can be estimated from that obtained in a motoring run of the engine.

It is shown in the diagram that the swirl velocity increases in accordance with the upward motion of the

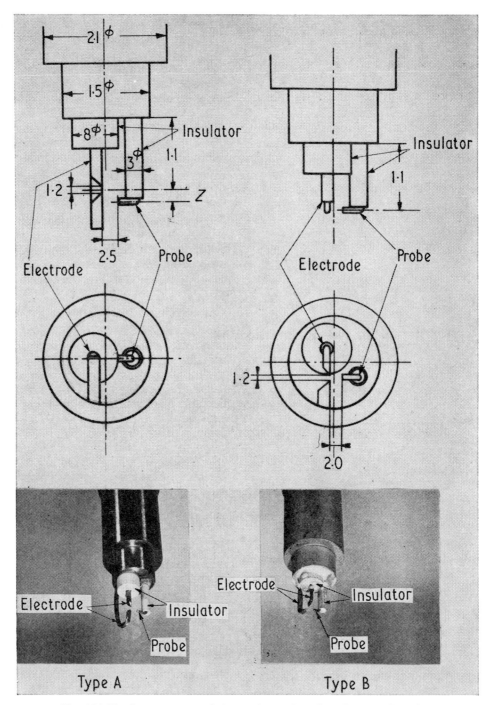

Fig. 134.2b. Arrangement of electrodes and probe of measuring plug

piston and an increase in engine speed. This may be explained as follows:

(1) The gas velocity due to the piston squish increases according to the upward motion of the piston and an increase in engine speed. As a reference, the squish velocities, estimated by the following formula (1), are shown in Fig. 134.5:

$$u = \frac{1}{a}\left(\frac{b^2-a^2}{b^2\pi x+4V}\right)\frac{V}{x}\cdot\frac{dx}{dt}$$

where u is the squish velocity at the end of the piston cavity, b the diameter of the cylinder, a the diameter of the piston cavity, x the distance from the top of the

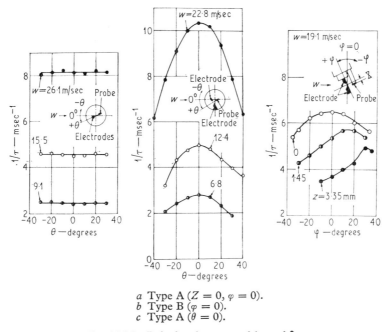

a Type A ($Z = 0$, $\varphi = 0$).
b Type B ($\varphi = 0$).
c Type A ($\theta = 0$).

Fig. 134.3. Relation between $1/\tau$ and θ, φ

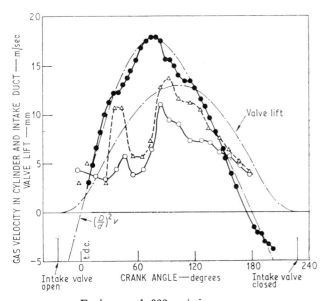

Engine speed: 900 rev/min.
Volumetric efficiency: 80 per cent.

○ Velocity in cylinder, motoring.
△ Velocity in cylinder, firing, fuelling: 0·0135 g/cycle.
● Velocity in intake duct, motoring.

Fig. 134.4. Gas velocity in cylinder and intake duct during intake stroke

If it is assumed that the total angular momentum of the cylinder charge is conserved during the compression stroke, the swirl velocity increases in accordance with the upward motion of the piston (**1**).

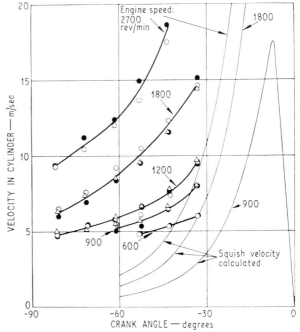

○ Motoring.
● Firing, fuelling: 0·0258 g/cycle.
△ Firing, fuelling: 0·0342 g/cycle.
Volumetric efficiency: 79 per cent.

Fig. 134.5. Comparison of swirl velocity during compression stroke in firing and motoring

piston to the cylinder head, V the volume of the piston cavity, and t is time.

(2) The mass fraction of the gas in the piston cavity to the total cylinder charge increases in accordance with the upward motion of the piston, while the radius of the piston cavity is smaller than the radius of the cylinder.

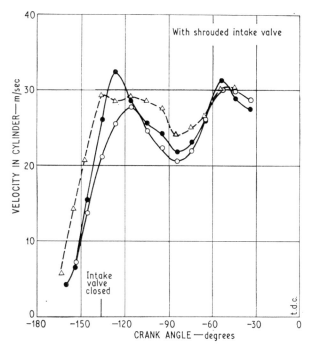

Engine speed: 1800 rev/min.
Volumetric efficiency: 81 per cent.
○ Motoring.
● Firing, fuelling: 0·0332 g/cycle.
△ Firing, fuelling: 0·0453 g/cycle.

Fig. 134.6. Comparison of swirl velocity during compression stroke in firing and motoring

(3) If the swirl velocity near the piston is higher than that near the cylinder head, as pointed out by Willis *et al.* (*3*), the swirl velocity which is measured at the neighbourhood of the cylinder head increases according to the upward motion of the piston.

When the shroud of the intake valve was fixed at the position where the strongest directional flow in the cylinder was obtained, the velocities of cylinder gas in a firing and in a motoring run were compared. The results, shown in Fig. 134.6, were obtained from tests involving higher engine outputs and higher swirl levels than were used for the tests shown in Fig. 134.5. Some differences between the gas velocity in a firing run and a motoring run are observed in the early stages of the compression stroke, but in the last stage no significant differences can be seen.

Effect of shrouded valve

The swirl velocity and its direction, which was varied by rotating the shrouded intake valve, were measured. In this test a piston without a cavity was used to eliminate the possibility of piston squish producing a secondary flow in the cylinder. To prevent the piston head coming into contact with the measuring plug, the compression ratio of the engine was reduced to 8·4:1. The measurement was made at 34° crank angle before t.d.c. by motoring the engine at various positions of the shroud. Test results are shown in Fig. 134.7. In the diagram, α indicates the position of the shroud. The flow direction, Θ, is measured anti-clockwise from the tangential direction as shown in the diagram. The velocities which have components towards the clockwise direction around the axis of the cylinder are defined as negative values.

EFFECTS OF AIR SWIRL ON COMBUSTION

In the four-stroke cycle diesel engine, the effects of cylinder air swirl on the b.m.e.p. and the smoke density of the exhaust gas were investigated. The swirl velocity was varied by rotating the shrouded intake valve as stated above. Again, to eliminate a possible secondary flow, the tests were performed with a piston without a cavity. The amount of intake air was kept constant by use of the choke valve in the intake duct and was measured by an orifice flow meter. The smoke density of the exhaust gas was measured by a Bosch smoke meter. In this experiment two fuel injection nozzles were used: one with six holes of 0·28 mm diameter, and the other with eight holes of 0·24 mm diameter. The fuel injection was initiated at 30° crank angle before t.d.c., and the injection pressure was 210 kg/cm². The compression ratio of this engine was 16:1; the engine speed was kept constant at 1800 rev/min.

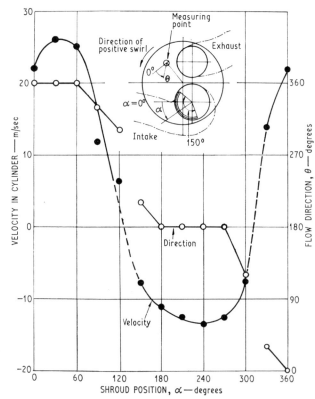

Engine speed: 1800 rev/min.
Volumetric efficiency: 83 per cent.
Flat piston.
Motoring.

Fig. 134.7. Effect of shroud position on swirl velocity at 34 degrees before t.d.c.

Fig. 134.8 shows the effects of swirl velocity on the b.m.e.p. and smoke density of the exhaust gas. In the test the volumetric efficiency and the amount of fuel supplied were kept constant at 83 per cent and 0·0427 g/cycle, respectively. It can be seen from the diagram that the swirl velocity has more influence on the b.m.e.p. and smoke density in the 8-holed nozzle than in the 6-holed nozzle.

Further, the experiments were also carried out by varying the amount of fuel supplied, keeping the intake air constant. The results are shown in Fig. 134.9. In the diagram, the b.m.e.p. and exhaust smoke density are shown against the relative air/fuel ratio, which is defined as the mass ratio of fuel to air divided by the stoichiometric ratio. The b.m.e.p. decreases as the relative air/fuel ratio increases. In the 8-holed nozzle the gradient of the curve increases as the swirl velocity decreases. In the 6-holed nozzle, however, no noticeable variation of b.m.e.p. with swirl velocity can be seen. The smoke density increases with a decrease of the relative air/fuel ratio. Using the 6-holed nozzle, it has been shown that when the relative air/fuel ratio decreases, the smoke density is high at both low and high swirl velocities. In a high relative air/fuel ratio, however, the smoke density becomes low at a high swirl velocity. In the 8-holed nozzle the smoke density of the exhaust gas decreases significantly with a decrease in swirl velocity.

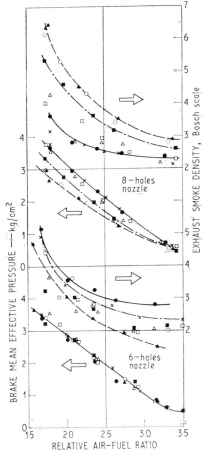

Engine speed: 1800 rev/min.
Volumetric efficiency: 83 per cent.
Flat piston.
Swirl velocity (m/s): ▲ 26·0
○ 22·0
■ 13·9
□ 13·6
△ 11·8
× 7·8
● 6·6

Fig. 134.9. Effect of swirl velocity on combustion

From the results shown in Figs 134.8 and 134.9, favourable combustion occurs in the 6-holed nozzle at a swirl velocity of about 13 m/s and in the 8-holed nozzle it occurs at about 7 m/s. This may be caused by the following: when fuel is injected into air swirling at high speed, the swirl forces the fuel spray to bend. Using the 8-holed nozzle there are more droplets near the central zone of the combustion chamber than would exist using the 6-holed nozzle since the fuel droplets from the 8-holed nozzle are smaller. Consequently, using the 8-holed nozzle, the utilization of air in the combustion chamber becomes worse than that with the 6-holed nozzle. Further, on account of the narrow space between the sprays when using the 8-holed nozzle, a burnt gas, produced by the combustion of a spray, may be supplied to a neighbouring spray if the swirl velocity is too high.

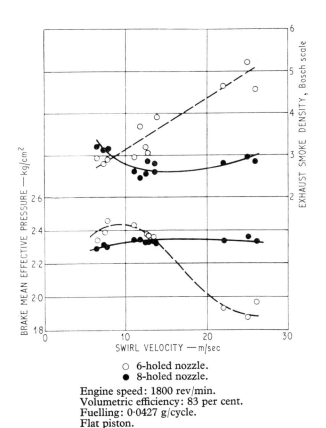

○ 6-holed nozzle.
● 8-holed nozzle.

Engine speed: 1800 rev/min.
Volumetric efficiency: 83 per cent.
Fuelling: 0·0427 g/cycle.
Flat piston.

Fig. 134.8. Effect of swirl velocity on combustion

CONCLUSION

The gas stream velocity in the cylinder of the internal combustion engine in a firing run was measured successfully by an electric discharge method. In the intake stroke and early in the compression stroke of a four-stroke cycle engine, the cylinder gas velocity in firing is higher than that in motoring.

The gas velocity at the end of compression in a firing run was almost the same as that in a motoring run. Thus, the gas velocity immediately before ignition can be estimated by a motoring test.

By investigating the effects of air swirl on output and exhaust smoke, it was shown that by increasing the number of nozzle holes, a good combustion which enables the engine to produce a high power output and a low density of exhaust smoke was obtained when the swirl velocity was low.

APPENDIX 134.1

REFERENCES

(1) FITZGEORGE, D. and ALLISON, J. L. 'Air swirl in a road-vehicle diesel engine', *Proc. Auto. Div. Instn mech. Engrs* 1962–63 (No. 4), 151.

(2) THON, L. 'Die Erzeugung einer rotierenden Luftbewegung im Zylinder von Viertakt Motoren', *Kraftfahrzeugtechnik* 1970 **3**, 67.

(3) WILLIS, D. A., MEYER, W. E. and BIRNIE, C. 'Mapping of airflow patterns in engines with induction swirl', *Trans. S.A.E.* 1967 **75** (section 1), 416.

(4) NAKAJIMA, K., KAJIYA, S. and NAGAO, F. 'An experimental investigation of the air swirl motion and combustion in the swirl chamber of diesel engines', *XIIth Congress of FISITA, Barcelona* 1968, 1–02.

(5) HIRAO, O., JO, P. D. and JO, S. H. 'Observation of the movement of gas in a two-stroke engine cylinder', *Tech. Memo. Japan Auto. Res. Inst.* 1970 (No. 1), 161.

(6) KHITRIN, L. N. *Combustion in turbulent flow* 1963, 122 (Israel Program for Scientific Translations Ltd).

(7) OHIGASHI, S., HAMAMOTO, Y. and TANABE, S. 'Gas flow velocity in inlet pipe, exhaust pipe and cylinder of two-stroke cycle engine', *Bull. Japan Soc. Mech. Engrs* 1971 **14** (No. 71), 470.

(8) OHIGASHI, S., HAMAMOTO, Y. and TANABE, S. 'A new digital method for measuring gas flow velocity by electric discharge', *S.A.E. Paper* 690180, 1969.

SECOND GENERATION CARBURETTORS FOR EMISSION CONTROL: FIRST STAGE

G. L. LAWRENCE*

This paper describes further work in the development of air valve carburettors for the purpose of exhaust emission control and attempts to analyse the fundamental flow phenomena associated with temperature change on the metering elements. The proposition advanced is that air valve carburettors enjoy an inherent advantage over conventional plain tube solutions, particularly when examined against the functional criteria demanded by the longer term legislation.

INTRODUCTION

IN AN EARLIER PAPER (1)† it was suggested that air valve carburettors, by nature of their fundamental metering system, enjoy an inherent advantage over the more conventional and widespread plain tube designs. Since that time, the legislative demands have become virtually worldwide and the standards set for the U.S.A. in 1975–76 border on what may have been considered fantasy by the established industry even in the very recent past.

Whatever opinions may be expressed regarding the lack of perspective in government circles when the total ecological situation is examined, the pressure imposed on engineers in the carburation field can cause no regrets since the spur has provided them with an opportunity for intense research and development, the more challenging because the art, having reached an advanced stage of its potential, was previously almost stagnant.

The prognosis advanced nearly three years ago appears to be even more viable today and the author's company remains dedicated to the philosophy that, given the solution of acknowledged problems associated with the flow criteria of needle/jet controlled metering, the air valve carburettor cannot be matched in its potential to contribute significantly to the reduction of exhaust emissions by control of the ingoing charge. This opinion does not deny the feasibility of practical solutions with plain tube carburettors, or with fuel injection, but suggests that either system is, inevitably, relatively complex and therefore more difficult to optimize as inherently correct and stable production units.

The difficulty of manufacturing elaborate plain tube

The MS. of this paper was received at the Institution on 8th July 1971 and accepted for publication on 30th July 1971. 33
* Zenith Carburetter Co. Ltd, Honeypot Lane, Stanmore, Middlesex HA7 1EG.
† References are given in Appendix 135.1.

carburettors with the synergistic influence of their multiple metering components has long been recognized and is now aggravated by the continuous reduction of total flow band tolerances.

Significantly, plain tube carburettors have difficulty in retaining driving flexibility during acceleration modes and stable running on steady-state road loads due to the interplay between idle and main discharge circuits complicated in the transient mode by signal loss on the metering elements attendant on inadequate gas velocities. Additional demands have now been created by the sensitivity of the more exotic treatments to achieve very low emissions, and it has been shown that reactors, both catalytic and thermal, require very close control of the induced air/fuel ratio under all conditions. Cantwell *et al.* (2) refer to this problem in their paper of 1969 and, more recently, research in the thermal reactor field by the author's company for small (European) engine capacities has underlined these findings.

Conventionally it has been the practice to submerge the acceleration defects by injecting additional neat fuel as the throttle is opened, but increased exhaust emissions reflect the difficulties of timing the discharge while changes in engine operating conditions and temperature really require a differential input. These difficulties are avoided by the progressive and relatively constant velocity function of the air valve carburettor, which also lends itself more readily to the maintenance of combustion during the deceleration mode by the simple expedient of providing for a by-passing of mixture, controlled by the single metering source, past the 'closed' throttle plate. Ideally, the introduction of an independent means to alter the air/fuel ratio in this mode is desirable to achieve optimum emission reductions.

The need to refine the transient mode function of plain

tube carburettors has resulted in significant research and the subject is well exposed in the literature. As examples, the work of Ethyl (3) (4) is well known as a practical demonstration of applied principles and, with the more theoretical analysis of Shinoda et al. (5), serves to illustrate the complicated nature of the problem.

In consideration of the standards legislated for 1975 onwards, the other dominant criteria are:

(1) consistency of performance, in production, and during the practical lifetime (which may mean 50 000 miles or five years (6));
(2) consistent mixture quality throughout the speed-load range, together with freedom from deposition problems;
(3) driving flexibility with minimum enrichment after cold start;
(4) adjustable mixture control system for deceleration;
(5) maintenance of fuel system integrity standards (7);
(6) stability of metering under all operating conditions including wide temperature variations.

In general, it is suggested that the air valve carburettor principle offers potentially the most elegant solution to these design parameters, including that of practical lifetime. Experience has shown that specification of this type of carburettor results in very reduced service sales due, it is believed, to its high tolerance of fuel-borne contamination and independence from the malfunction of power jets and acceleration devices.

The sole exception has been the inadequate flow stability of needle/jet metering systems against temperature change which has led to the need for compensatory devices. It was the object of the work described below to cure this defect.

BASIC PROBLEM

The basic problem of flow variation from the air valve carburettor with changing fuel viscosity has long been recognized and has been associated with the laminar flow criteria dictated by the metering orifice geometry of needle/jet combinations at least in the idle and low flow regimes. Prior to exhaust emission control regulations the enrichment with increasing temperature was evidenced by deteriorating idle quality—to the point of stall under adverse conditions—which caused criticism, but was commercially tolerable.

However, with the advent of regulations the dominant influence of the idling mode—the complexity of which began to be recognized—dictated the need for correcting devices to avoid unacceptable increases in CO emissions as temperatures increased over the driving cycle. The development of the Stromberg CDSE carburettor (8) incorporated such a device to qualify the needle/jet relationship automatically, and subsequently the S.U. HIF carburettor emerged with an arrangement working on similar principles. Such components are, however, subject to mechanical frailty and do not recognize the influence of physical fuel properties; nor is it easy to match precisely their boundaries of operation with the functional requirement. It was, therefore, decided to seek a more immaculate solution by the establishment of turbulent flow under all metering conditions and, thereby, inherent stability.

METERING SYSTEM DEVELOPMENT

The pilot work was carried out on a 1725 cm³ engine fitted with a single 150 CDSE carburettor—the test bed installation being suitably instrumented to monitor physical conditions and modified to control intake fuel and air temperatures under laboratory conditions.

Fig. 135.1 shows typical response curves for CO under free engine idling conditions plotted against running time.

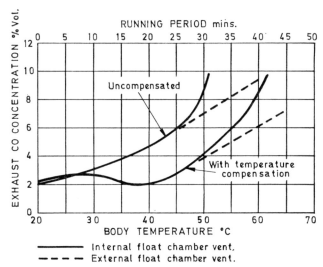

Fig. 135.1. Typical variation of exhaust CO level with engine running time

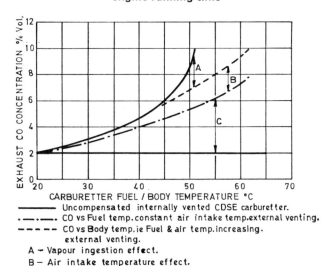

Fig. 135.2. Typical influence of fuel temperature, air intake temperature, and fuel vapour ingestion on exhaust CO level

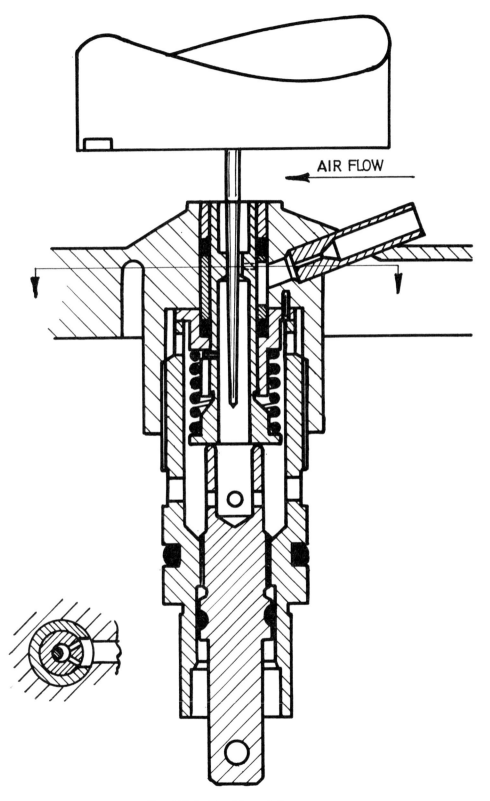

Fig. 135.3. Emulsion air jet system

Examination of the results of considerable exploratory work at this stage showed that:

(1) a dominant relationship existed between fuel temperature and CO output;

(2) a secondary relationship existed between intake air temperature, thus air density and CO output;

(3) a further relationship existed, due to increasing vapour ingestion, significantly with fuel temperatures in excess of 50°C and CO output.

Fig. 135.2 is representative of the proportional influence of these factors.

Following these findings the metering characteristics of the needle/jet orifice were investigated under stabilized conditions with external venting of the float chamber, the fuel temperature alone being varied.

It was decided at this stage to attempt the enforcement of turbulent flow criteria by the introduction of air into the fuel, thus producing mechanical interference with the flow pattern with an attendant reduction of density and kinematic viscosity of the fuel to invoke higher velocities.

A depressingly large number of arrangements were evaluated before the system shown in Fig. 135.3 conferred the sought after stability (Fig. 135.4) with consistent idle quality. Confirmatory tests were made using a completely different engine of 1159 cm³ capacity with equally satisfying results, and the performances of the engines and vehicles over the speed–load range were both demonstrated to be good.

The principle having been established attention was turned to investigating the tuneability of the system and many problems were exposed. Notably the simple total head pick-up for the emulsion air resulted in too great an interference with higher fuel flows, and difficulties were experienced with the metering needle form. Consequently, a static head arrangement was tried, and following a pressure survey of the carburettor intake condition was located in a position to limit its range of operation to low air flow rates through the carburettor. While difficulties were initially experienced due to random failures of the emulsion system to initiate flow, these problems were overcome by circuit redesign.

However, in the final analysis stability and tuning criteria were found to be best served by a return to the total head arrangement, avoiding the original problem by careful siting of a restricted pick-up tube. Similarly, the size, number, and location of the emulsion holes and tuning calibrations were optimized effectively and the production arrangement of the developed system is shown in Fig. 135.5.

STABILIZED FLOW

While, at the outset, the problem was considered to be confined by simple viscous flow parameters it became clear during related laboratory work that the flow was not truly laminar under any condition and that no established law relationship existed between mass flow and head at the metering jet. This phenomenon is probably explained by the findings of other workers (9) which show that the two states may occur in combination, especially in eccentric annuli.

Similarly, the injection of air into the fuel is subject to complex laws, and much interesting work has been pub-

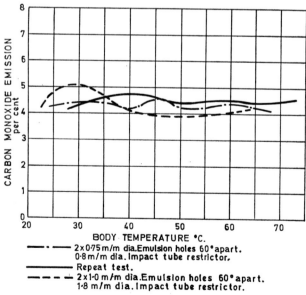

Fig. 135.4. Exhaust CO consistency test results

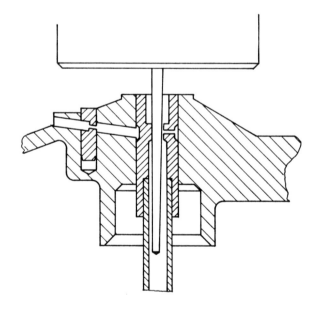

Fig. 135.5. Final emulsion air jet system

lished by Bolt et al. (**10**) in which the influence of temperature on fuel viscosity, surface tension, and two-phase flow is examined in some depth.

Nevertheless, it has been amply demonstrated that stability of metering may be achieved by the system discussed and the feasibility of compensating by this means for other physical changes may also be considered.

AIR DENSITY

The obvious significance of temperature inspired changes in air density on the metered air/fuel ratio has been masked by the characteristic performance of fixed choke carburettors which, generally speaking, have a tendency to lean out as temperature increases. The mechanics of this phenomenon are discussed later. However, the influence of air density was apparent on the air valve carburettor and attention was now turned to investigating this factor.

Theoretical calculations indicate a reduction of air mass flow of 8 per cent over a 50 degC temperature range in consideration of density, assuming sonic velocity at the throttle plate, which is consistent with the prevailing pressure conditions in the idling mode. Under the same conditions a theoretical analysis of the factors influencing air valve ride height shows this element to be unaffected, which practical testing confirmed.

To check the theory a preliminary assessment was made by running the engine, in the idling mode, over the same 50 degC temperature range, maintaining a constant CO output which necessitated reducing the fuel flow progressively from the jet by approximately 10 per cent. It could thus be concluded that the air mass flow had fallen by this amount, which agreed reasonably with the theoretical prediction. During these tests, however, some difficulty was experienced in measuring fuel flow without scatter due to vaporization effects, and a second series of tests were made to monitor air flow directly.

To achieve this the air cleaner was replaced with a capacity box fed through a sharp-edge orifice sized to achieve a pressure drop of 3 inH_2O to ensure adequate response to small changes of air flow, the depression being measured by a high sensitivity inclined manometer. Owing to the resultant high pressure drop in the intake system it was necessary to utilize internal venting of the float chamber to maintain an equivalent fuel jet area consistent with the earlier tests. Accordingly, water cooling of the float chamber was adopted to contain fuel temperatures between 25 and 30°C to avoid vapour ingestion. The results of tests in this configuration indicated reductions of the order of 12 per cent air mass flow over the same temperature range.

In consideration of the comparative results it had to be borne in mind that, because of space limitations, the air metering unit could not conform to recommended standards and also that the theoretical calculations were based on air properties alone and would be modified in practice by the entrained fuel vapour. It was concluded, therefore, that the influence of air density had been adequately substantiated and that the cure lay either in controlled air intake systems—desirable for other reasons—or the exploitation of correction by an extension of the air emulsion system influence.

FURTHER SYSTEM DEVELOPMENT

The CD-4 carburettor

At this stage it was decided to seek a second generation air valve carburettor design incorporating the emulsion metering system and other improvements indicated by the now considerable experience gained with the CDSE unit.

Design parameters were to include:

(1) Fixed needle/jet relationship.
(2) Selective venting of the float chamber.
(3) Adjustment means for idle/off-idle air/fuel ratio.
(4) Correction of idle 'dipping' problem.

Provision was to be made for:

(5) Adjustable mixture by-pass (deceleration) valve.
(6) Anti run-on device.
(7) Altitude control.
(8) E.G.R. valve.
(9) Security against roll over fuel loss (**7**).

The development of a more sophisticated design of automatic starter was also considered essential, and it is hoped that a separate paper dealing with this subject may be accepted at an early date.

The developed design, designated CD-4, is shown schematically in Fig. 135.6. Only points (2), (3), and (4) need to be discussed in the context of this paper.

Selective venting

The arrangement adopted permits external venting of the float chamber—in the emission application to a charcoal canister or other vapour storage system—during the idle and immediate off-idle modes. The point of changeover to internal venting is tuneable within the road–load range according to the specific engine application. The need for the valve is obvious from the preceding discussion, and also to avoid vapour ingestion during hot soak with attendant restart problems.

Adjustment means for idle/off-idle

Experience has shown that, apart from the very low flow regime, satisfactory flow band consistency may be achieved between production units having a fixed needle/jet relationship. It was, however, desirable to provide a ready means of adjustment at idle both for in house and vehicle production line use.

Initially an arrangement leaching air into the throttle bore via a port timed against the throttle blade angular movement was tested. However, while this demonstrated excellent control when flowed on an air box, it was found to have a deleterious effect on engine idling quality owing to its influence on mixture distribution, at least on some manifold arrangements.

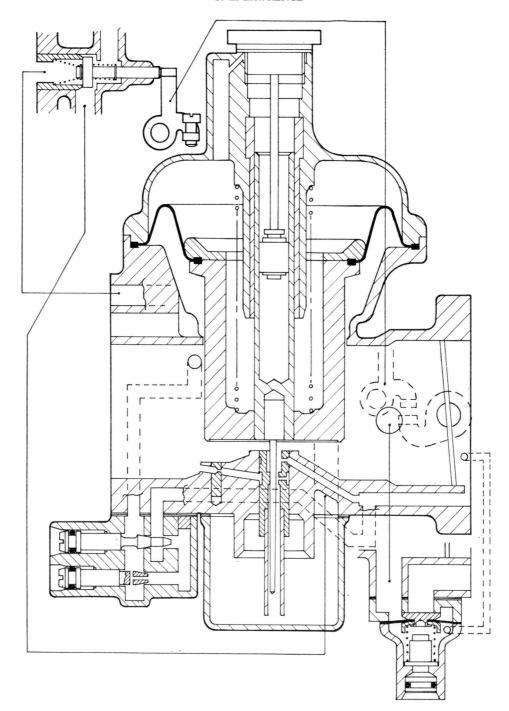

Fig. 135.6. Schematic arrangement of CD-4 carburettor

Idle 'dipping'

While this work was proceeding, parallel development of a system for correction of the idle 'dipping' problem was also in hand and the two design parameters became associated in a common solution.

The problem of idle 'dipping' is demonstrated by the failure of the engine to recover its idling speed following 'blipping' of the throttle, so that momentarily the engine speed runs down to an unstable level and in aggravated conditions may stall.

While many carburettors exhibit the fault to some degree the problem has been greater with air valve carburettors having the idling fuel discharge upstream of the throttle. Generally speaking, the quality of idling has

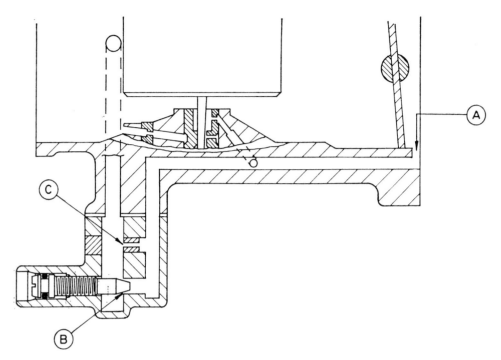

Fig. 135.7. Schematic arrangement of basic idle system

NOMINAL ENGINE CAPACITY	CARB. BODY TEMPERATURE °C.											REMARKS
	20	25	30	35	40	45	50	55	60	65	70	
	CARBON MONOXIDE per cent											
In Line 4 Cyl. 1·3 Litre (S)	2·0	2·0	1·8	1·7	2·05	1·9	2·0	1·65	1·5	1·9	2·1	Test bed
In Line 4 Cyl. 2·3 Litre (S)	2·2	2·15	2·1	2·5	2·75	2·75	2·7	2·7	2·6	2·6		Vehicle ✵
In Line 6 Cyl. 3·0 Litre (T)	2·15	2·25	2·4	2·5	2·7	2·9	3·0	3·4	3·4			Vehicle ✵
In Line 6 Cyl. 2·0 Litre (T)	2·5	2·3	1·9	2·0	2·3	2·3	2·4	2·5	2·5	2·5	2·6	Test bed
In Line 6 Cyl. 2·5 Litre (T)	1·8	1·7	1·8	1·9	2·2	2·0	1·9	2·1	2·1			Vehicle ✵
V 8 Cyl. 3·0 Litre (T)	1·4	1·3	1·5	1·6	1·7	1·9	1·9	1·9	2·0	2·2		Test bed
In Line 6 Cyl. 2·0 Litre (T)	0·9	0·8	1·2	1·2	0·9	0·5	0·5	0·8	0·6	0·5		Vehicle ✵

✵ Test terminated due to coolant water boiling.

(S) Single Carburetter Installation

(T) Twin Carburetter Installation

Fig. 135.8. Exhaust CO characteristics of different applications of CD-4 system

also been inferior to fixed choke carburettors where the discharge is downstream, the problem being associated with transport time of the fuel to the manifold.

The 'dipping' problem is aggravated by any local entrapment of the metered fuel, and this problem was explained in earlier work (8), but fundamentally a detailed investigation using a perspex carburettor and high-speed photography established that the 'dipping' or idle stagger was dictated by signal decay at the metering orifice following closure of the 'blipped' throttle. This signal decay was directly attributable to loss of air velocity caused by failure of the air valve to regain immediately its steady-state ride height due to inertia, and was therefore fundamental and inherent to the system.

To cure the defect the system shown diagrammatically in Fig. 135.7 was developed. In this arrangement a proportion of the fuel metered at the jet is taken directly downstream and discharged through the port (A) which is timed against the throttle edge to limit the range of influence. The circuit is carefully tuned to ensure uninterrupted delivery from the main metering system to retain the fully progressive driving characteristics of the air valve system. With this modification idle quality was improved and the 'dipping' problem totally eradicated.

Satisfactory idle mixture control was achieved by introducing bleed air to the circuit through a valve shown at (B) and through a fixed bleed shown at (C). This development led to dramatically improved idling quality and stable lean mixture operation on road loads, which also enabled very early 'choke' enrichment withdrawal after cold start.

Significant advantages were demonstrated with the carburettor incorporating all the features discussed. Fig. 135.8 shows typical results from a number of engines in terms of idle CO response against temperature. Fig. 135.9 tables cyclic emission results for both U.S. and E.E.C. procedures which are considered outstanding by any previous achievement. While it is not suggested that such results may readily become commonplace or typical, it demonstrates that with compatible engines very low emission levels may be achieved which may permit the use of simple carburation treatment—without air pumps—at least in the short term.

It should be recorded that the vehicles' performances were completely satisfactory and approved by their manufacturers.

FIXED CHOKE CARBURETTOR PERFORMANCE

Reference was made earlier to the phenomenon of 'leaning out' of idling mixture with increase of temperature as applied to the fixed choke (plain tube) carburettor, and from the preceding discussion it is clear that some explanation had to be sought for the apparent insensitivity to intake air density. Other workers have examined the condition: Bolt and Boerma (11) reported on the influence of intake air conditions but did not deal adequately with the specific mode, while Bier *et al.* (12) dealt with the mode and identified the influential conditions without explaining the physical cause of the phenomenon.

A series of tests were conducted using an externally vented 36 VN fixed choke carburettor modified to permit the control of temperature in the idling circuit alone (Fig. 135.10). This modification was made on the theory that vapour formation in this channel was at the root of the problem, although this theory had never been systematically examined. Care was taken to maintain all other components of the carburettor at the normal functioning

E.C.E. CYCLE RESULTS – 2 LITRE ENGINE TWIN-150 CD4 CARBS.

	CO grms/test	HC grms/test	COMMENTS
	62·3	6·04 }	cold start cycle using manually operated choke.
	60·0	4·19 }	
	30·1	6·24 }	cold start cycle without the use of choke
	34·6	6·25 }	
	35·8	6·04	hot start cycle
	12·26	7·48	hot start cycle using lean limit carburettors
E.C.E. limits	134	9·4	type test limit
	161	12·2	production test limit

U S FEDERAL CYCLE RESULTS – 2·3 LITRE ENGINE SINGLE 175 CD 4 CARB.

COLD CYCLE WITHOUT CHOKE 1972 TEST PROCEDURE.

HC grms/mile	CO grms/mile	NOx grms/mile	COMMENTS
1·6	4·48	N A	Vehicle manufacturers results.
0·972	5·66	1·231	E.R.A. results after manufacture testing.
3·4	39	3·0	U.S. Federal limits 1972/74.
0·46	4·7	3·0	U.S. Federal limits 1975.

HOT CYCLES

HC grms/mile	CO grms/mile	NOx grms/mile	COMMENTS
0·75	2·584	1·360	E.R.A. results before supply to manuf.
1·2	3·25	N A	vehicle manufacturers results.
0·596	3·485	1·231	E.R.A. results after manufacture testing

TOP GEAR CRUISE EMISSION LEVELS – 2·3 LITRE ENGINE
(VEHICLE MANUFACTURERS RESULTS)   SINGLE 175 CD4 CARB.
CARBURETTER SET AS CYCLE RESULTS.

MODE	HC ppm	CO %	CO₂ %
IDLE	88	1·87	14·92
20	26	0·21	15·68
30	23	0·12	15·19
40	23	0·14	15·13
50	20	0·13	15·01
60	20	0·14	14·83

Fig. 135.9. Vehicle emission test results using CD-4 carburettors

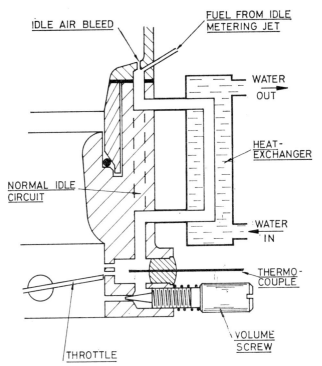

Fig. 135.10. Zenith 36 VN carburettor with cooled idle circuit

temperatures, therefore the temperature of the fuel presented to the idling passage increased in the same order.

In brief, data collected from many tests showed that when the carburettor was allowed to function normally the level of CO in the exhaust always fell with temperature. Typically, CO reduced from 4 to 1 per cent when the temperature in the idling passage rose from 11 to 32°C, at which point the idling became erratic. However, when the mixture temperature was controlled within 5 degC (16 and 21°C) the CO level in the exhaust rose from 1 to 7·7 per cent for an intake air temperature rise from 25 to 55°C, and the carburettor was clearly sensitive to air density.

To explain these findings a heat balance was conducted on the heat exchanger confining the idling circuit by measuring the coolant flow and accurately monitoring the relevant temperatures during an extended idle test. It was found that the heat abstracted was considerably in excess of that required to reduce the temperature of the mixture in the passage to maintain a constant level of 19°C at the discharge point. Thus, it followed that the remaining heat flow was utilized in changing the state of the mixture by condensing fuel vapour to liquid. It could be deduced, therefore, that the observed reduction in CO output with the normally functioning carburettor idle circuit was a direct result of volumetric increase of the mixture due to its change of state and its influence on the pressure conditions within the circuit causing a reduction of fuel flow.

To further substantiate these results the carburettor was modified to permit visual observation of the mixture within the idling passage, which clearly demonstrated the change from liquid to gaseous flow over the normal temperature range of operation.

Subsequent to this exercise, Bolt et al. (10) have reported their findings from similar research and added much valuable information.

SUMMARY

The work described traces the development of a second generation CD-4 air valve carburettor design based on the experience gained in emission control on many engines with the parent CDSE design. Functional criteria have been examined and some explanation sought for specific conditions.

It is suggested that the air valve type carburettor by definition is more adaptable to the emission control purpose than alternative designs, and that it may be more logically refined as an inherently stable unit both in production and over its lifetime.

It is accepted that alternative solutions, using plain tube carburettors, may be more commercially viable where less stringent emission levels than those associated with the U.S. standards are applied, but it is suggested that production difficulties, notably in the flow testing area, may reduce this differential.

CONCLUSIONS

(1) Air valve type carburettors are capable of contributing significantly to very low emission levels with compatible engine design.

(2) Control of intake air temperature and fuel vapour is essential to an optimized design.

(3) Air valve carburettors offer potential advantages in conjunction with thermal reactor devices.

(4) The design is inherently more suited to the sealed unit for lifetime philosophy.

ACKNOWLEDGEMENTS

The author wishes to thank the Board of Directors of The Zenith Carburetter Company Limited for permission to publish this paper, and the engineering staff of both Zenith and Engineering Research & Application Ltd for their co-operation in the research development.

APPENDIX 135.1

REFERENCES

(1) LAWRENCE, G. L. 'Emission control by carburation', *Symp. Motor Vehicle Air Pollution Control, Proc. Instn mech. Engrs* 1968–69 **183** (Pt 3E), 24.

(2) CANTWELL, E. N. et al. 'A progress report on the development of exhaust manifold reactors', S.A.E. Paper 690139, 1969.

(3) BARTHOLOMEW, E. 'Potentialities of emission reduction by design of induction systems', S.A.E. Paper 660109, 1969.

(4) ELTINGE, L. 'Potentialities of further emissions reduction by engine modifications', S.A.E. Paper 680123, 1968.

(5) SHINODA *et al.* 'Analysis and experiments on carburetor metering at the transition region to the main system', S.A.E. Paper 710206, 1971.

(6) *The Clean Air Act* 1970 (published by E.P.A.).

(7) 'Proposed rule making', *Federal Register* **35** (No. 169), 49 CFR, Pt 571.

(8) LAWRENCE, G. L. 'Mixture pre-treatment for clean exhaust; a duplex carburation system', *Proc. Auto. Div. Instn mech. Engrs* 1967–68 **182** (Pt 2A), 151.

(9) BOURNE, D. E. *et al.* 'Laminar and turbulent flow in annuli of unit eccentricity', *Can. J. chem. Engng* 1968 **46** (October).

(10) BOLT, J. A. *et al.* 'The influence of fuel properties on metering in carburetors', S.A.E. Paper 710207, 1971.

(11) BOLT, J. A. and BOERMA, M. 'The influence of inlet air conditions on carburetor metering', S.A.E. Paper 660119, 1966.

(12) BIER *et al.* 'A study of factors affecting carburetor performance at low air flows', S.A.E. Paper 690137, 1969.

C136/71 STRATIFICATION AND AIR POLLUTION

J. E. WITZKY*

This paper discusses the latest development of Southwest Research Institute's 'swirl stratified charge engine' and states the tests used to measure the exhaust emission of this engine. The same engine was operated as a carburetted engine at identical conditions, but the observed exhaust emission results cannot be compared to the stratified charge engine due to the unstable operation of the single cylinder Otto cycle engine. The bag sampling method was applied and the exhaust samples were immediately analysed in our Vehicular Emissions Laboratory.

INTRODUCTION

EXHAUST GAS EMISSION from automotive type engines has emerged as a major problem on an international level.

Air pollution in populated areas, and the overwhelming evidence that the internal combustion engine is a major contributor of those pollutants, is the concern of all of us in general and of the automotive and petroleum industry in particular.

The problem is under attack on many fronts. Legislation in all industrial countries has or will set limits on automobile exhaust emission. The limits, as well as the test procedures, will be stringent and in 1975 not much can be left in the exhaust gases to be emitted to the atmosphere. Major changes in the internal combustion engine will be required to meet the objectives for 1972 and beyond because of the significant reduction in unburnt hydrocarbon (HC), carbon monoxide (CO), and oxides of nitrogen (NO_x) emission.

The automobile manufacturers are in the process of developing systems and procedures to reduce objectionable exhaust emission from automobile engines and, of course, the problem is so serious that many other organizations are studying various means to accomplish the same results.

In general, there are two approaches to the solution of the emission problem. One consists of the treatment of the exhaust gases after the gases have left the combustion chamber. The treatment is of oxidation, either catalytically or by flame type reaction, of the exhaust products. The better approach, of course, is the reduction of harmful emissions where they are created—inside the combustion chamber. The alteration of the combustion process usually consists in the modification of the air/fuel ratio, spark timing, and chamber design. The lead components in the motor fuels clog and thus reduce the efficiency of catalytic reactors; therefore, the automotive industry is calling for the oil industry to produce unleaded fuel. Although this can be done, it immediately introduces a number of problems. Lead-free fuels tend to wear valve seats relatively rapidly. Of course, this problem can be corrected by changing the metallurgical properties of valves and valve seats. Of much greater concern is the change of anti-knock properties of unleaded fuel. The possibility of producing lead-free fuels with anti-knock properties exists; however, the expensive new refinery processing will result in higher prices. An alternative, to avoid the extra costs of premium quality lead-free fuels, would be to design new engines of lower compression ratio. The lower compression ratio means a loss of power (10 per cent) and an increase of fuel consumption (about 3 mile/gal). An increase of fuel consumption per car and miles driven or a reduction of the thermal efficiency of spark-ignited Otto engine cannot, in the light of preserving our natural resources, be accepted.

As the urban population increases, emission control regulations will become more and more severe, and it is recognized that these mentioned alterations can only be of a temporary nature. Eventually a new type of power plant must be developed, a power plant that will not introduce any appreciable quantity of pollutants to the atmosphere. The need for the development of a new engine type will exert its impact on both the engine manufacturers and the petroleum industry equally because the development of engines is inseparably connected to the available fuel.

The objective of this paper is to evaluate the stratified charge engine concept with respect to exhaust emission and types of fuels.

The MS. of this paper was received at the Institution on 29th March 1971 and accepted for publication on 21st June 1971. 34
* *Southwest Research Institute, P.O. Drawer 28510, San Antonio, Texas 78228, U.S.A.*

THE SWIRL STRATIFIED CHARGE ENGINE
Definition of a stratified charge engine

A stratified charge engine is an engine utilizing the Otto cycle with controlled ignition of a heterogeneous fuel/air mixture. Such an engine has some of the characteristics of both the carburetted engine and the diesel engine.

A carburetted engine takes in a homogeneous mixture of fuel and air on the suction stroke, compresses this mixture, and ignites it with a controlled ignition source such as a spark plug. The power output of the engine is controlled by a throttle in the intake port, which controls the quantity of mixture taken in. The fuel/air ratio of the mixture remains nearly constant for all loads. Upon ignition of the mixture, combustion proceeds by means of a flame front travelling through the combustion chamber.

The intake of the diesel engine is unthrottled, and the charge taken in on the intake stroke consists of air only. Fuel is injected near the end of the compression stroke and ignites spontaneously due to the heat of the compression. No flame front exists; rather, the fuel/air mixture at each point is burned when the conditions of temperature, mixture strength, etc., are suitable. At the time of combustion, the fuel/air mixture is highly heterogeneous. The power output of the diesel engine is controlled by varying the fuel input, which means that the fuel/air mixture is extremely lean at low engine loads.

The stratified charge engine, like the diesel, inducts an air charge through an unthrottled intake port. Again like the diesel, the fuel is injected during the compression stroke. Spontaneous ignition does not occur because compression ratios are similar to those used in gasoline engines. Like the carburetted engine, the mixture is ignited by a spark and the mixture is burned by a progressive flame front. This combination of characteristics requires that a mixture richer than the average chamber mixture be available at the spark plug since the average chamber mixture at part loads may be much too lean to be ignited by a spark. Thus, the stratified charge engine requires that the mixture of fuel and air in the combustion chamber be 'stratified' into a fuel-rich component near the ignition source and a fuel-lean component elsewhere in the chamber. As in the diesel engine, the power output of the engine is controlled by the fuel input.

Characteristics of a stratified charge engine

Why a stratified charge engine? Primarily, to operate at lean fuel mixtures at part load for improved fuel economy with controlled combustion. The stratified charge engine has the high part load fuel economy of the diesel with the light weight and quiet operation of the gasoline engine (the latter being a consequence of the controlled combustion).

Fig. 136.1 shows a theoretical comparison of a stratified charge engine with a typical carburetted engine. The data shown here were computed for a constant volume combustion process without heat losses, with an ideal four-stroke inlet process. The carburetted data were calculated using a constant stoichiometric fuel/air ratio; the m.e.p.

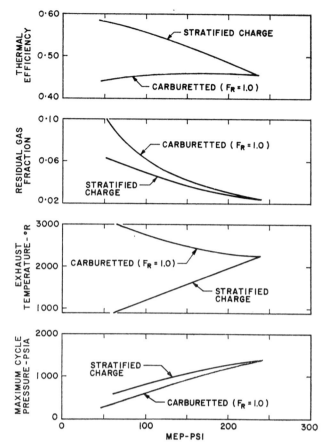

Fig. 136.1. Comparison of carburetted and stratified charge cycles

being varied by reducing the inlet density of the fuel/air mixture. To obtain the various points for the stratified charge process, the fuel/air ratio was varied from 25 to 100 per cent stoichiometric, keeping the inlet air density constant. It was assumed that the fuel charge was injected into the combustion chamber at t.d.c.

The top pair of curves compares the thermal efficiency of the two processes. At maximum power there is no significant difference in the thermal efficiencies, since both processes are operating at the same fuel/air ratio. As the load is decreased, a considerable difference in efficiency is apparent. The high efficiency of the stratified charge process is due primarily to the more efficient thermodynamic characteristics of the leaner mixture. This is not the complete story, however, since the next set of curves shows that the carburetted cycle has a consistently higher fraction of residual gases present in the fuel/air charge, which also contributes to a reduced efficiency.

The next set of curves shows how the exhaust temperature varies with the engine power output. As might be expected, the exhaust temperature of the stratified charge process is lower than that of the carburetted process. The exhaust temperature of the stratified charge engine decreases with decreasing load due to the relatively constant mass of mixture with a decreasing heat input. The

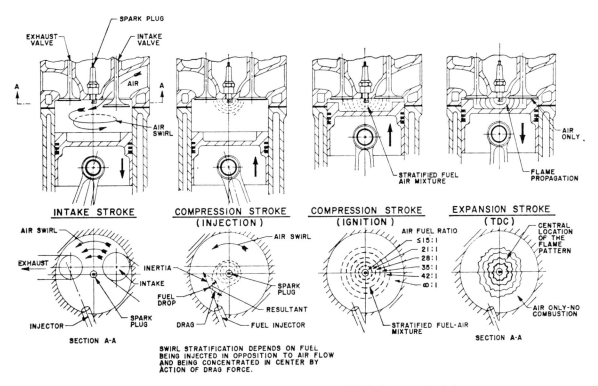

Fig. 136.2. Sequence of the swirl stratified charge principle

increase of exhaust temperature with decreasing load of the carburetted engine is a result of the throttling process.

Finally, the bottom graph shows a comparison of the maximum cycle pressure for each of the two types of processes. The stratified charge process has slightly higher peak pressures, a result of the absence of throttling and the higher compression pressures of the stratified process.

The Southwest Research Institute stratified charge engine

Fig. 136.1 applies to any stratified charge engine, and shows the characteristics obtained when an engine is operated with a variable fuel/air ratio. Fig. 136.2 is a pictorial representation of the SwRI swirl stratified charge process. The engine is a four-stroke engine with fuel injection. On the intake stroke, the incoming air charge takes up a swirling motion in the chamber due to the tangential orientation of the intake port. During the latter part of the compression stroke the fuel charge is injected against the direction of the swirling air. The forces of drag and inertia acting on the individual fuel droplets tend to force these droplets towards the centre of the chamber, providing the swirl velocities and the direction of injection are correct. This provides a fuel-rich mixture at the spark plug which effectively 'stratifies' the fuel charge. Fig. 136.2 depicts a theoretically perfect stratification of fuel.

This particular technique of stratification results in another very important characteristic; namely, a high degree of insensitivity to the octane number of the fuel. This characteristic is due to the short time duration between fuel injection and ignition. A short time period prevents extensive preflame reactions in the injected fuel and limits the quantity of fuel in the end gases at the cylinder periphery. The elimination of a long fuel residence in the combustion chamber eliminates the tendency for detonation, and makes possible the operation of the engine on a wide variety of hydrocarbon fuels.

Test engine

The engine used for the continuation of the swirl stratified charge principle development had a bore of 3·875 in and a stroke of 3 in. A cross-section of the gasoline engine (shown in Fig. 136.3) illustrates the conversion from the standard four-cylinder engine to the transparent single-cylinder engine used to photograph injection and combustion. For further development of the swirl stratified charge engine principle, a new cylinder head was designed where the intake and exhaust valve are arranged in a manner to create a high intake velocity air swirl of high volumetric efficiency and reduced exhaust gas restriction (Fig. 136.4). The technique of fuel injection and the injection nozzle location are the key for a successful operation of a stratified charge engine. It was therefore decided to build a new cylinder head out of a solid block of steel where the injection nozzle was located in a retainer ring (as shown in the figure) which provided the ability to investigate 360° of radial location of the injection nozzle around the cylinder axis. Two pistons, with a compression ratio of 9·5:1 and 12:1 respectively, were available and different straight-hole Rosa Master nozzles with hole

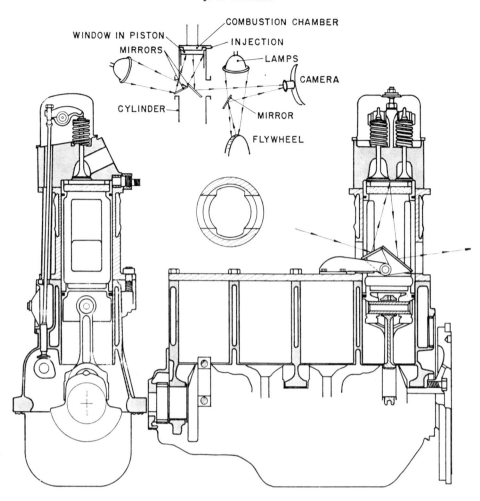

Fig. 136.3. Section and schematic of the 'picture window' engine

diameters of 0·015 and 0·020 in. The injection pump had a plunger diameter of 5 mm, tangential cam, and a delivery valve with zero pressure.

Engine test

Test runs at 1000, 2000, and 3000 rev/min were conducted with all available nozzle locations. During the entire testing period there was an offensive odour in the test cell, indicating the presence of unburnt hydrocarbons, also the performance was not as good, and previously obtained performances could not be repeated.

The next series of tests were made with the nozzle located in the cylinder head as shown in Fig. 136.5. Nozzles with different spray angles were available and the nozzle could be turned to direct the spray with or against the air swirl. With this nozzle arrangement it could be clearly established that for best performance the fuel had to be injected against the air swirl; by turning the injector to spray with the air swirl, performance deteriorated; and at an angle of 20° with the air swirl, the engine ceased to run. A vertical spray angle almost parallel with the cylinder head gave the best performance.

COMBUSTION ANALYSIS

Power output, efficiency, and combustion noise are affected by the cylinder pressure–time profile, rate of pressure rise and, consequently, the rate of heat release versus time. All the engine sees is the rate of energy release over crank angle, since this controls the maximum combustion pressure and, of course, the total work output.

Recognizing the importance of analysing a combustion cycle, a pressure–time diagram at an engine speed of 1000 rev/min and high load was investigated. The rate of heat release for this pressure–time diagram was calculated by using a digital computer program, and the rate of heat release versus crank angle degrees is shown in Fig. 136.6. As this diagram shows, the pump was timed for static injection at 60° b.t.d.c.; the actual injection occurred at 36° b.t.d.c., indicating an injection delay of 24° crank angle. The injection duration of 11° was determined by injecting gasoline into a bowl of jet fuel. The spark was timed at 11° b.t.d.c. The exposure time of the injected fuel to the air was 13° crank angle. This shows the ability of this swirl stratification system to establish a preflame reaction time of the fuel by timing the fuel injection prior

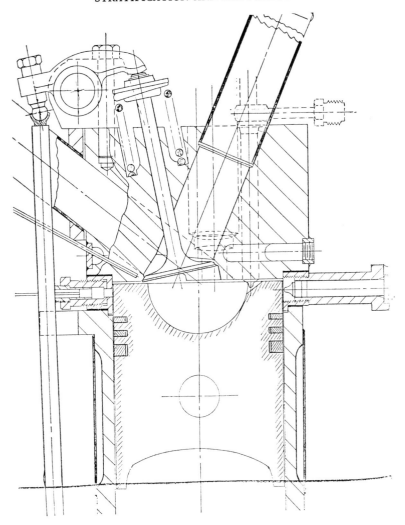

Fig. 136.4. New cylinder head (injection nozzle located in retainer ring)

to the spark discharge. By shortening the preflame reaction time, as compared to an Otto cycle combustion, it is possible to reduce the tendency of the end gas to autoignite. In the case of the stratified charge engine, sufficient preflame reaction time is provided to ensure efficient combustion, but there is insufficient time to establish the conditions necessary to auto-ignite the end gases. Fig. 136.6 further shows that the release of energy starts 4° before t.d.c., indicating an ignition delay of 7° crank angle, whereas the peak rate of heat release is reached 9° after t.d.c. and the combustion completed 20° after t.d.c. The rate of pressure rise for this diagram is 42 lb/in^2 per degree crank angle. This rate of pressure rise is slow when compared with the diesel cycle; however, it is faster than that of a conventional Otto cycle.

Fig. 136.7 shows one combustion sequence at the same speed and load as the cylinder pressure and heat release diagram.

As this high-speed photograph of injection and combustion clearly shows, the injection penetration was excessive, the tip of the injection spray was carried with the air to the periphery of the combustion chamber. This picture informed us that additional work with reduced nozzle penetration was required.

An injection nozzle was modified to reduce the penetration by reducing the L/D ratio. The penetration was reduced from 1·625 in of the standard nozzle to 1·125 in of the modified nozzle. The reduction of penetration improved the performance considerably. The single-cylinder engine in this configuration was used to determine the exhaust emission of the stratified charge engine.

PERFORMANCE AND PRESSURE-TIME DIAGRAMS

As already mentioned, the same engine was used for the carburetted and stratified charge version. For the carburetted version, a production piston and cylinder head were installed; whereas, for the stratified version, the piston number 2 and the head as shown in Fig. 136.5 were used. For each run to measure exhaust emissions, pressure diagrams and records of the engine parameters were taken. It can be assumed that the friction hp of both conversions,

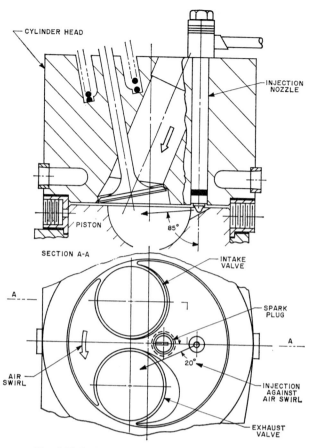

Fig. 136.5. Nozzle location in cylinder head

Table 136.1. Test conditions and operating procedure
Each test consisted of 10 conditions as shown

Run no.	Engine revolutions per minute	Load, lb
1	Low idle	0
2	1000	6·8
3	1000	11·0
4	1000	13·0
5	2000	8·0
6	2000	12·0
7	2000	16·0
8	3000	9·5
9	3000	13·0
10	3000	14·5

when operated at the same load and speed, is identical. It is, therefore, possible to compare the fuel consumption of the stratified and carburetted engine. By comparison of the two cycles, it has to be considered that the carburetted engine was unstable and difficult to control in spite of a rich air–fuel mixture. Fig. 136.8 shows the fuel consumption, the air/fuel ratio, the exhaust gas temperature, and the oxygen still present in the exhaust gases over the speed range and half power setting. It is interesting to note that the curves for O_2 are almost parallel to the air/fuel ratio curves. The fuel consumption of the stratified charge engine is lower for all loads and speeds. The pressure–time diagrams of the stratified charge engine were steadier than in the carburetted engine and showed no cycle-to-cycle fluctuations, but a higher rate of pressure rise when compared with the diagrams of the carburetted engine. The rate of heat release for the carburetted engine at 1000 rev/min and low load, and the rate of heat release of the stratified engine at approximately the same load and speed, were calculated by use of a digital computer program. The corresponding heat release rate versus crank angle degree is plotted in Fig. 136.9. The diagrams show that the rate of heat release is quite different between the two cycles. The slow burning of the carburetted engine is mostly responsible for the difference in fuel consumption. The carburetted engine yields a minimum indicated specific fuel consumption of 0·64 lb/i.h.p. h, whereas the stratified engine operates at a minimum fuel consumption of 0·376 lb/i.h.p. h.

EXHAUST EMISSION SAMPLING SYSTEM

A $\frac{1}{4}$-in stainless steel sampling probe was inserted midstream of the exhaust pipe 6 in from the engine. The sample was drawn through a filter cartridge, an ice chilled moisture trap, and was collected in a 2 ft × 3 ft Tedlar sampling bag. All sampling tubing and connections used were stainless steel or Teflon.

Table 136.2. Summary of stratified charge engine emissions
(These are averages of three stratified charge runs)

Run no.	Engine cond.	Engine revolutions per minute	Load, lb	CO, %	CO_2, %	HC/IR, p.p.m.	HC/FIA, p.p.m.	NO, p.p.m.	O_2, %
1	S.C.	1000	0	0·157	4·02	2310	3968	134·0	15·75
2	S.C.	1000	6·8	0·15	3·93	2464	5196	130·0	15·49
3	S.C.	1000	11·0	0·188	8·43	2172	3898	645·0	10·41
4	S.C.	1000	13·0	0·40	10·43	2801	5357	51·0	6·70
5	S.C.	2000	8·0	0·15	5·11	2058	3036	330·0	14·30
6	S.C.	2000	12·0	0·18	7·63	2062	2820	646·6	11·56
7	S.C.	2000	16·0	0·28	11·96	2290	4354	1508·3	5·07
8	S.C.	3000	9·5	0·15	9·01	3068	3454	991·6	9·33
9	S.C.	3000	13·0	0·35	12·00	1491	2786	1375·0	6·58
10	S.C.	3000	14·5	4·80	10·60	2100	3992	1200·0	1·80

STRATIFICATION AND AIR POLLUTION

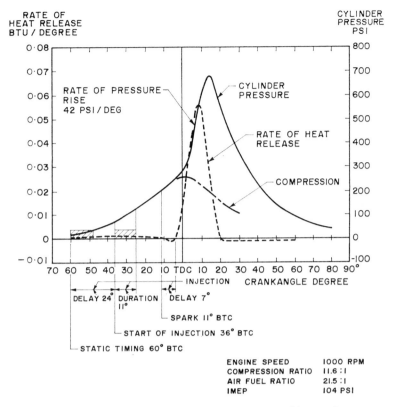

Fig. 136.6. Pressure–time diagram and rate of heat release

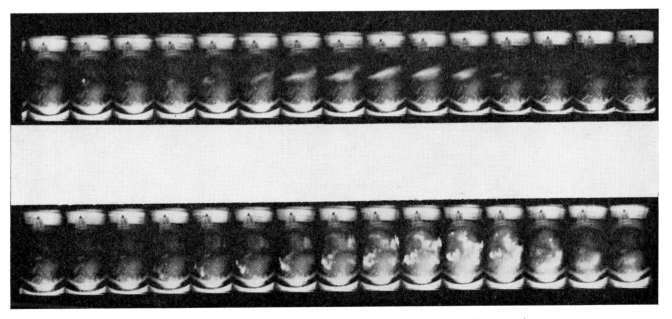

Fig. 136.7. High-speed photographs of injection and combustion (1000 rev/min)

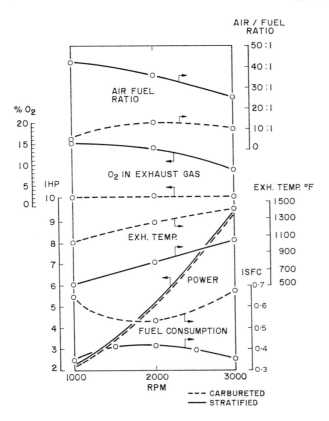

Fig. 136.8. Performance

Operating procedure

The engine was operated at each condition until temperatures were stable and a bag sample of exhaust gas was collected. This exhaust sample was immediately taken to the Vehicle Emissions Laboratory for analysis. Meanwhile, the engine was set to the next operating condition. Total elapsed time for each condition was approximately 25 min. Time between sampling and analysis was less than 3 min.

Duplicate carburettor-version runs were made. As it was extremely difficult to maintain operating conditions for any length of time, some of the bag samples were not sufficient to get an accurate analysis of all components. The lack of control in operating the single-cylinder carburetted engine is indicated by the unacceptable variation in emissions results. This variation is not a fault of the operator or emissions sampling and analysis, but is a problem stemming from operation of the carburetted version of the engine, and these results are not included in this report.

Triplicate stratified charge version runs were made. The operation of the stratified charge engine was very stable through all load and speed conditions.

Instruments used for the exhaust gas analysis were:

Carbon monoxide (CO): Beckman non-dispersive infrared analyser, low and high range—two instruments.
Carbon dioxide (CO_2): Beckman non-dispersive infrared analyser, low and high range—two instruments.

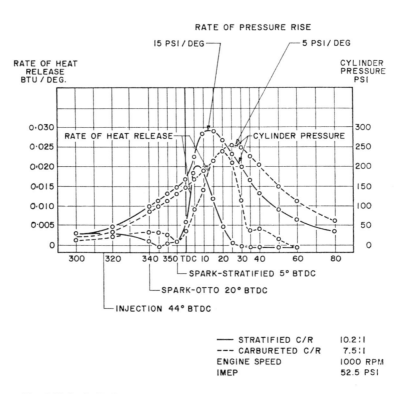

Fig. 136.9. Cylinder pressure and rate of heat release of an Otto and stratified charge cycle at the same low load and speed

Unburnt hydrocarbons (HC/IR): Beckman non-dispersive infrared analyser—0–1000 p.p.m. *n*-hexane range.
Unburnt hydrocarbons (HC/FIA): Beckman hydrogen flame ionization analyser, p.p.m. carbon.
Nitric oxide (NO): Beckman non-dispersive infrared analyser, range 0–800 and 0–4000 p.p.m.
Oxygen (O_2): Beckman polarographic oxygen process analyser.

In an attempt to summarize the exhaust emissions data, the three-run arithmetic average of the stratified charge engine is listed in Table 136.2. For all the test runs for exhaust analysis, regular gasoline was used.

RESULTS

The requirements to restrict atmospheric pollution to be imposed in the U.S.A. from 1975 onwards as it appears at the present can only be met by the use of catalytic reactors, or direct flame-type after-burners as an alternative. Afterburners will not be so effective when the cleanliness of the engine exhaust is improved in the future. However, because the lead components in the gasoline tend to clog and thus reduce the efficiency of catalytic after-burners, the automotive industry demands from the oil refiners an unleaded fuel. By simply omitting the lead compounds the anti-knock properties or octane numbers will be reduced; the compression ratio has also to be lowered (one ratio for every four octane numbers), which results in less power and an increase in fuel consumption. The insensitivity of the stratified charge engine would fit perfectly into this picture. The stratified charge engine could operate at a high compression ratio, yielding a power output and fuel consumption even superior to the present high compression gasoline engine.

The amount of carbon monoxide (CO) is drastically reduced in the stratified charge engine. By omitting run no. 10 in Table 136.2 the air/fuel ratio is 14:1 below the stoichiometric and of the remaining nine runs the average CO is only 0·22 per cent. This fact is of special interest to Europe where carbon monoxide is the main problem and oxides of nitrogen are relatively unimportant. The emission of carbon monoxide, as in all engines operated above the stoichiometric air/fuel ratio, is no problem in this stratified charge engine.

Unfortunately, the exhaust emission data of the gasoline engine cannot be validly compared with the stratified charge version of the engine due to the unstable operation of the Otto cycle engine, therefore the Otto cycle pollution data are not included. In Fig. 136.8 the oxygen still present in the exhaust gases is shown. These curves of excess oxygen are substantially parallel with the air/fuel ratio curves and are an indication that the oxygen cannot be used to reduce the unburnt hydrocarbons. The exhaust sampling probe was inserted midstream of the exhaust pipe 6 in from the cylinder head manifold interface. If the sampling probe had been installed farther downstream, and the pipe kept hot, it is possible that the excess of O_2 in the exhaust gases would have further reduced the level of unburnt hydrocarbons.

Why does the high percentage of oxygen not reduce the unburnt hydrocarbons in the stratified charge engine? In theory, a stratified charge engine should minimize their formation. Three reasons can be cited for this peculiar behaviour:

(*a*) A combustible air/fuel mixture is concentrated in the centre of the combustion chamber and the excess oxygen does not have time in one cycle to oxidize completely all the components of the fuel.

(*b*) There may still be diffused fuel in the periphery of the combustion chamber in a mixture strength too weak to be ignited by an advancing flame front (leaner than 45:1).

(*c*) The quench zone in a stratified charge engine is normally eliminated; however, a new quench zone at the interface between the combustible mixture and the air is introduced and there could be considerable mixing with a quenching effect at this interface.

CONCLUSIONS AND RECOMMENDATIONS

In conclusion, the drastic reduction of carbon monoxide (CO) and the relatively high oxygen (O_2) content in the exhaust gases of the stratified charge engine are emphasized. Further tests are recommended to find a way to use the high oxygen content for a further reduction of unburnt hydrocarbons. To reduce the oxides of nitrogen it is believed a catalytic reactor is required similar to that used in the standard Otto cycle engine.

APPENDIX 136.1

BIBLIOGRAPHY

WITZKY, J. E. and CLARK, J. M. jun. 'A study of the swirl stratified combustion principle', S.A.E. Paper 660092, 1966.

WITZKY, J. E. 'The swirl stratification process', *Automot. Engr* 1968 (February).

DIESEL ENGINE EXHAUST EMISSIONS AND EFFECT OF ADDITIVES

M. S. EL NESR*   S. SATCUNANATHAN†   B. J. ZACZEK‡

This paper presents the results of an investigation into the effects of fuels (both undoped and doped with selected additives) on the smoke levels, concentrations of carbon monoxide, formaldehyde, and oxides of nitrogen in the exhaust of a diesel engine. The results show that, in general, the levels of these contaminants are lowest for an optimum range of cetane numbers. Fuels doped with isoamylnitrate and aniline (which change the cetane number of the fuel) yield results similar to undoped fuels of corresponding cetane number. It is shown that cyclohexane as an additive, though not affecting the cetane number of the fuel, brings about significant reductions in the levels of all the contaminants under all conditions and at the same time improves the fuel consumption.

INTRODUCTION

AIR POLLUTION caused by exhausts from transport vehicles continues to be hazardous to life and property. The increase in the volume of heavy traffic dealing with the transport of goods on the highways of the world makes it imperative that ways and means be found to reduce the concentrations of the noxious constituents in engine exhaust. The constituents which cause the greatest concern are smoke, carbon monoxide, formaldehyde, and oxides of nitrogen and, in addition, oxides of sulphur.

Various methods have been tried over the years to reduce the levels of these contaminants in exhaust gases. The principal methods are:

(a) improvement of fuel;
(b) improvement of combustion conditions;
(c) removal of the contaminants from the exhaust gases before emission to atmosphere;
(d) use of afterburner devices.

Of these methods, (c) and (d) appear to be the least desirable as they would add significantly to the capital and operating costs of the unit. Furthermore, these methods seek to rectify a situation after it has occurred. It would be preferable, therefore, to concentrate more on finding some means of improving the fuel characteristics and the combustion characteristics either of the fuel or the engine in order to achieve the desired result.

Since engines vary considerably in their combustion characteristics (e.g. compression ratio, combustion chamber design, injection timing and type, speed, etc.), a common quality fuel for all engines would not appear to be feasible. The solution, therefore, while being of general applicability, must also be capable of specific variation. This type of solution can be achieved by the use of additives.

Different workers have used various substances as additives to bring about improvement in ignition quality, smoke density, and a reduction in the concentrations of the undesirable constituents in exhaust gases (1)–(8)§. Unfortunately, however, the mechanism of action of additives is still largely obscure, and hence the choice of substances for use as additives may tend to be arbitrary and their evaluation expensive and time consuming. Some guide to the choice of additives is therefore necessary.

The two main factors that govern the composition of the products of combustion are the ignition and combustion processes. An alteration in this composition may therefore be brought about by using, as additives, substances that are known to alter the ignition properties or that are suspected to have some effect on the combustion pattern. Accordingly, an investigation was undertaken in which various substances were tested for their effects on the smoke levels and the concentrations of carbon monoxide, formaldehyde, and oxides of nitrogen (9). The choice of additives was largely guided by the ignition delay characteristic of the substance itself (10) or by the effect the substance had on the ignition delays of liquid fuel droplets impinging on a hot surface (11). This paper presents the results of this investigation.

The MS. of this paper was received at the Institution on 10th May 1971 and accepted for publication on 2nd August 1971. 33
* Atomic Energy of Canada Ltd, Toronto, Canada.
† The University of the West Indies, St Augustine, Trinidad, West Indies.
‡ Enfield College of Technology, Middlesex.

§ References are given in Appendix 137.1.

APPARATUS AND EXPERIMENTAL PROCEDURE

Engine, fuels, and additives

A layout of the engine is shown in Fig. 137.1. A 40-gal reinforced oil drum was connected to the inlet pipe of the engine for cetane number determination by the 'throttling' method. The necessary electronic equipment was hooked on to obtain pressure diagrams for determination of cetane number by the 'ignition delay' method. The specifications for a Witte horizontal, single-cylinder, four-stroke engine with precombustion chamber are:

Compression ratio	17:1
Bore	5 in
Stroke	8 in
Rated power output	12 hp
Speed	720 rev/min
Ignition timing	t.d.c.
Injector	Bosch pintle type
Load	Electric dynamometer with resistor loads.

The properties of the fuels used in the investigations are given in Table 137.1.

Several additives were tested and results are presented for three representative substances—namely, isoamylnitrate, aniline, and cyclohexane. Isoamylnitrate progressively increased the cetane number with increasing dosage; aniline increased the cetane number with decreasing dosage for dosages below a certain amount and lowered the cetane number for dosages above this amount; while cyclohexane had no significant effect on the cetane number (**II**).

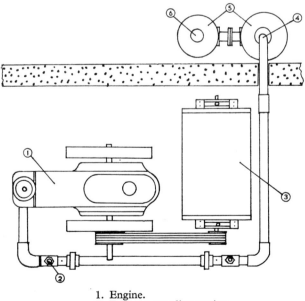

1. Engine.
2. Exhaust sampling station.
3. Electric dynamometer.
4. Inlet to silencer.
5. Silencer.
6. Exhaust outlet.

Fig. 137.1. Engine layout

Table 137.1. Properties of fuels used in investigations

Properties	Reference fuels		Commercial fuels		
	A	B	E	F	G
Specific gravity at 60°F	0·8025	0·7694	0·8440	0·8430	0·8340
Aniline pt, °F	174	187	159	148	135
API gravity	44·7	52·5	36·4	36·5	38·5
Sulphur, % by mass	0·044	0·005	0·370	0·680	0·130
Cetane number	71	22·7	57·5*	50*	40*

* Cetane numbers obtained by throttling method by linear interpolation between reference fuels.

Exhaust measurements

Smoke levels were measured with the Hartridge smoke meter. Concentrations of carbon monoxide, formaldehyde, and oxides of nitrogen were measured with a Drager gas detector. The Drager gas detector was chosen because of its simplicity of operation, though its accuracy is limited. However, the results presented in this paper are the average of several readings and give a good qualitative if not an exact quantitative indication of the concentrations of the constituents. Furthermore, the main aim of the investigation was the determination of the effects of fuels and those of certain additives on the concentrations of exhaust constituents. For these purposes the Drager gas detector measurements were considered sufficiently accurate. Spot checks of carbon monoxide concentrations were made with an Orsat gas analyser and found to be in good agreement with those from the Drager gas detector.

RESULTS AND DISCUSSION

Smoke

Fig. 137.2 shows the results obtained at full load for the smoke levels with the undoped fuels and one of the fuels doped with various dosages of the three additives, respectively, plotted against the cetane number of the fuels (both doped and undoped). The pattern over the entire load range was similar. It is evident that the smoke level is a minimum for a certain range of cetane numbers—lower or higher cetane numbers giving higher smoke values. In the present case the optimum cetane number is about 65. Acceptable smoke levels are possible with fuels of cetane number about 47 and above.

With the additives isoamylnitrate and aniline, which bring about a change in the cetane number of the fuel, the effect is substantially the same as using an undoped fuel of the same cetane number as the doped fuel. Improvement of smoke levels can therefore be brought about by using ignition improving additives. The use of ignition improving additives for smoke improvement, however, appears to be limited to changing the cetane number to the optimum range. In some instances it may even be advisable to lower the cetane number of the fuel, by the use of additives or otherwise, to reduce the smoke levels.

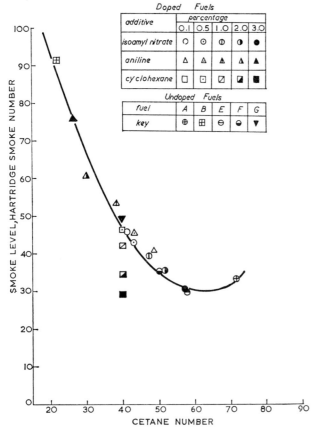

Fig. 137.2. Smoke levels for undoped fuels and fuel G doped with additives. Engine at full load

The findings for the effect of the cetane number of undoped fuels are in general agreement with those of some earlier workers (12)–(14). However, Golothan (7) found that black smoke density increased with cetane number in certain cases and was not affected in other cases. McConnell and Howells (8) found that cetane number had very little effect on the smoke levels. The latter used cetane numbers in the range 47–69, which, as can be seen in Fig. 137.2, are about the optimum range and hence the small differences in smoke levels observed.

The effect of cetane number on smoke may be viewed in terms of the ignition delay which governs both the period of uncontrolled combustion and that of controlled combustion. With longer ignition delays, characteristic of low cetane number fuels, a larger proportion of the fuel is vaporized and mixed with the air. When ignition occurs, the flame formed propagates through this mixture and lowers the general oxygen concentration, which causes the burning time of the rest of the fuel droplets to be extended. This, combined with the fact that the available burning time is reduced by the longer ignition delay, causes the unburnt fuel and carbon particles still in the process of burning to discharge into the exhaust where they are quenched, giving rise to black smoke. On the other hand, with very high cetane number fuels the ignition delay is very short and the available burning time is long. However, the fuel begins to burn as it enters the combustion chamber and, hence, the late coming fuel droplets in the spray pass through a high temperature zone which is deficient in oxygen, resulting in cracking of the fuel and carbon formation. The burning time of carbon particles is longer than that for equivalent fuel droplets. This then would result in black smoke. Between these two extremes there would exist a range of cetane number fuels which, when used in a particular engine, would provide the optimum conditions for combustion and so yield minimum smoke levels. This analysis fits the findings of the present investigation.

The delay angle response to cetane number changes varies from engine to engine and in the same engine, depending on operating conditions. In general, however, cetane numbers of fuels available commercially lie in the optimum range. It may therefore be found that the smoke levels from most engines are insensitive to cetane numbers of commercially available fuels. Hence, it would appear that only a limited advantage, from the point of view of smoke reduction, could be gained by using high cetane number fuels, or by using additives for improving the cetane number of commercial fuels.

However, since other fuel properties such as viscosity and volatility have been found to have an effect on the smoke levels (7) it may be possible to specify fuels that satisfy these requirements and then improve their ignition quality by the use of suitable additives.

The action of cyclohexane as an additive is interesting. Though having no significant effect on the cetane number of the fuel (11) it considerably improves smoke levels, particularly at dosages above 1 per cent. It was also found that when cyclohexane was added to the 71 cetane number fuel it reduced the smoke levels significantly, whereas isoamylnitrate when added to the same fuel increased the smoke levels quite appreciably. Thus it appears that further reduction in smoke levels when using fuels of optimum cetane number may be obtained by using cyclohexane as an additive.

One of the main reasons for choosing cyclohexane for testing as an additive was its steep ignition delay temperature curve as obtained for droplets impinging on a hot surface (10). It was suspected that this characteristic may affect the combustion pattern in the cylinder. This was confirmed to an extent by the fact that the pressure diagrams obtained also showed some significant departures from those obtained with the undoped fuels or with fuels doped with the other additives.

Carbon monoxide and formaldehyde

The dependence of the concentrations of carbon monoxide and formaldehyde on the cetane numbers of the undoped and doped fuels is seen in Figs 137.3 and 137.4 respectively for the worst conditions, i.e. at full load for carbon monoxide and at no load for formaldehyde. These results are parallel to those obtained for smoke. As for smoke,

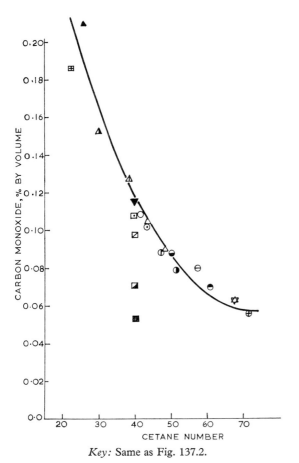

Fig. 137.3. Concentrations of carbon monoxide for undoped fuels and fuel G doped with additives. Engine at full load

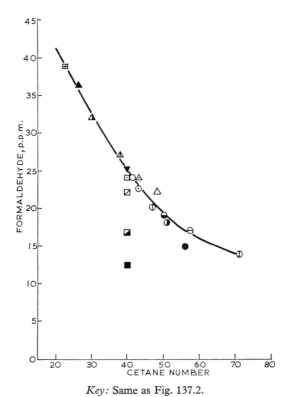

Key: Same as Fig. 137.2.

Fig. 137.4. Concentrations of formaldehyde for undoped fuels and fuel G doped with additives. Engine at no load

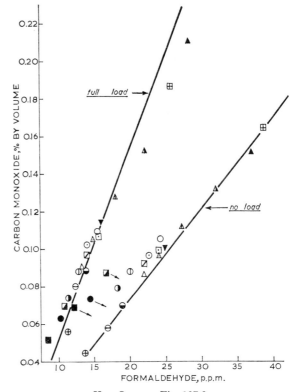

Key: Same as Fig. 137.2.

Fig. 137.5. Correlation between concentrations of carbon monoxide and formaldehyde

these concentrations decrease with increase in cetane number, the rate of decrease beginning to level off for values of cetane number above 55. The fuels doped with isoamylnitrate and aniline behave in a similar manner to undoped fuels of the same cetane number. It is again clear that no significant improvement in the concentration of these constituents can be obtained by using fuels of cetane number greater than about 55 or by using ignition improving additives.

Cyclohexane again brings about substantial reductions in the concentrations of these two constituents. It appears, therefore, that this is one method of making further improvement in the concentrations of carbon monoxide and formaldehyde when using fuels of the optimum cetane number.

Fig. 137.5 shows an interesting correlation between the concentrations of carbon monoxide and formaldehyde, the lines being drawn through the points for the undoped fuels. Apparently the two are nearly linearly related both for the undoped fuels and the doped fuels, irrespective of whether the additive changes the cetane number or not. This is a useful correlation in that it indicates that any measures taken to reduce the concentration of carbon

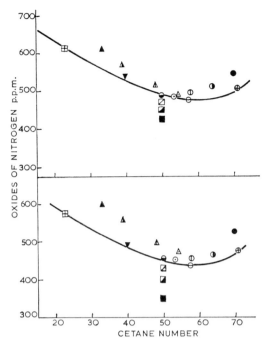

Key: Same as Fig. 137.2.

Fig. 137.6. Concentration of oxides of nitrogen for undoped fuels and fuel F doped with additives. Upper curve for 75 per cent full load and lower curve for full load

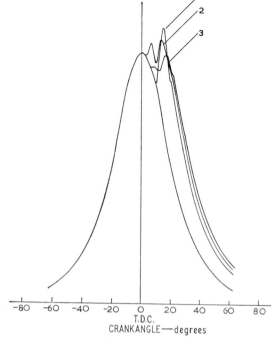

1. Cetane number 38.
2. Cetane number 68.
3. Cetane number 57·5.

Fig. 137.7. Effect of cetane number on pressure diagrams

monoxide will also bring about a proportional reduction in formaldehyde concentration.

Oxides of nitrogen

The effect of the cetane numbers of the undoped fuels and the additives on the concentration of oxides of nitrogen is seen in Fig. 137.6 where the results are given for 75 per cent full load and full load conditions, the curves being drawn through the points for the undoped fuels. Over the load range the concentration increased with load, reached a maximum at near 75 per cent full load, and then decreased slightly. The effect of load is somewhat parallel to the results obtained by McConnell and Howells (**8**) who found that for an indirect injection engine the maximum occurred nearer half full load condition. The magnitude of the concentrations here is slightly below those obtained by McConnell and Howells, probably due to the lower compression ratio and the later fuel injection timing.

For the undoped fuels, the oxides of nitrogen concentration is a minimum for a certain range of cetane numbers, and it is of interest that this range is the same as that for minimum smoke. McConnell and Howells (**8**) reason that the concentration of oxides of nitrogen should be virtually insensitive to the cetane number of the fuel. Their results, however, indicate that of the two cetane number fuels (42 and 69) tested in the indirect injection engine, the lower cetane fuel gave the higher concentration, in agreement with the present findings. Moreover, the present results are for a wide range of cetane numbers and therefore show the trends quite clearly.

The results in Fig. 137.6 should be studied in conjunction with the pressure diagrams shown in Fig. 137.7. The highest peak pressure (and therefore the highest peak temperature) is obtained with the fuel of cetane number 38 and the lowest peak pressure with the fuel of cetane number 57·5, while the fuel of cetane number 68 gave a peak pressure intermediate between the other two. It is evident, therefore, that the concentrations of the oxides of nitrogen are directly related to the magnitude of the pressure peaks.

It is generally accepted that high temperatures and high pressures are conducive to the formation of oxides of nitrogen (**15**). However, the effects of the rates of pressure rise and pressure drop (also temperature rise and temperature drop) during the combustion process do not appear to have been studied, though it has been shown that in spark-ignition engines, nitric oxide once formed in the combustion process remains fixed throughout the expansion and exhaust processes (**15**). Fig. 137.7 indicates that not only do the peak pressures vary with cetane number but also the rates of change of pressure during the combustion process. This could have a bearing on the concentration of oxides of nitrogen and should be investigated.

Isoamylnitrate, which changes the cetane number of the fuel, raised the concentrations of nitrogen oxides above that due to the base fuel and also above that of an undoped fuel of the same cetane number, though the general trend of the effect of cetane number is still the same. Aniline, on the other hand, at cetane improving dosages gave concentrations slightly below those due to the base fuel but above those for a fuel of corresponding cetane number. At other dosages the concentrations were increased. The general increase in concentrations of oxides of nitrogen by these two substances is suspected to be due to the nitrogen in the additive itself.

The effect of cyclohexane is quite interesting. It produced quite substantial reductions in the concentrations of oxides of nitrogen. As mentioned earlier, cyclohexane modified the pressure diagram quite significantly, producing lower peak pressures and slower rates of pressure changes. These may be the causes for the reduction in the oxides of nitrogen.

Fuel consumption

The use of additives automatically increases the basic cost of the fuel to the operator and hence additives should not significantly increase the fuel consumption but rather the reverse. Fig. 137.8 shows the effect of the three additives on the fuel consumption with the engine running at 75 per cent full load. The general trend over the entire load range was similar.

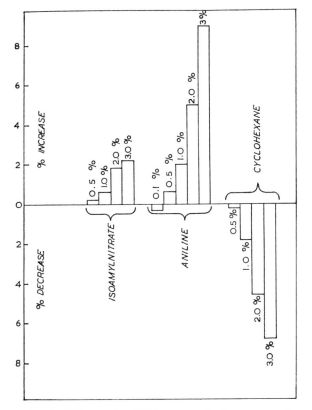

Fig. 137.8. Effect of additives on fuel consumption. Engine running at 75 per cent full load. Base fuel F

Fuel consumption depends on the heating value of the fuel and the heat release pattern in the cylinder. The heating value of isoamylnitrate is much lower than that of the diesel fuel and this may to a certain extent account for the increase in fuel consumption. However, without altering the injection timing so that ignition commences at the same point, it would be difficult to rule out the effects that isoamylnitrate may have on the heat release pattern. Similar considerations apply for aniline. It may be worth noting here that at 3 per cent dosage, aniline caused the engine to run very roughly with severe knocking. This despite the fact that even at this dosage the cetane number of the doped fuel was well within acceptable limits. The action of aniline in bringing about changes in cetane number of the fuel must therefore be through a different mechanism to that of isoamylnitrate.

Cyclohexane, whose heating value is of the same magnitude as the diesel fuel, gave significant improvements in fuel consumption. Since the effect on cetane number and ignition delay as evidenced on the pressure diagram was negligible, it appears that this additive modifies the heat release pattern in the cylinder, leading to the improved fuel consumption.

GENERAL COMMENTS

The results presented so far were obtained on a low-speed, low compression ratio engine with injection timed at t.d.c. In practice, however, injection is generally before t.d.c., and speed and compression ratios are higher. For a similar engine design (i.e. with precombustion chamber), speed and compression ratio will not alter the pattern of the results obtained but may alter the magnitudes of the values obtained. Injection timing will have an effect that needs to be examined as this will affect the heat release pattern in the cylinder and thus affect the composition of the exhaust. A test run on a 2·2-litre, four-cylinder, four-stroke marine engine fitted with a Ricardo Comet Mark III head (normally used for laboratory demonstration experiments) at 1420 rev/min and over the load range with an undoped fuel of cetane number 53 with and without cyclohexane as an additive showed improvement in smoke, carbon monoxide, formaldehyde, and oxides of nitrogen over the entire load range. The fuel consumption was only marginally improved. These results indicate that the findings reported in this paper are of general applicability.

Work has yet to be carried out on the effects of speed, compression ratio, injection timing, and type of combustion chamber, i.e. direct injection or indirect injection. The effect of using the three different kinds of additives together also needs to be investigated.

CONCLUSIONS

From the results presented in this paper, the following conclusions may be drawn:

(1) For undoped fuels and fuels doped with additives that change the cetane number: (a) smoke levels are a

minimum for a certain range of cetane numbers; (*b*) concentrations of carbon monoxide and formaldehyde decrease with increasing cetane number—the rate of decrease falling off at high cetane numbers; (*c*) concentrations of oxides of nitrogen also show a minimum for a certain range of cetane numbers—aniline and isoamylnitrate, when used as additives, tend to increase the concentrations of oxides of nitrogen above those of a fuel of corresponding cetane number.

(2) Cyclohexane, when used as an additive, improved smoke levels and concentrations of carbon monoxide, formaldehyde, and oxides of nitrogen over the whole load range.

(3) Isoamylnitrate and aniline generally increased the fuel consumption while cyclohexane improved the fuel consumption.

APPENDIX 137.1

REFERENCES

(1) BOGEN, J. S. and WILSON, G. C. 'Ignition accelerators for compression ignition engines', *Petrol. Refiner* 1944 **23** (No. 7), 118.

(2) DAVIES, E. B. and ALEXANDER, B. J. 'The use of heterocyclic tertiary amines for the control of corrosion caused by flue gases', *J. Inst. Fuel* 1960 **33**, 163.

(3) WILKINSON, J. J. and CLARKE, D. G. 'Problems encountered with the use of high sulphur content fuel oils at Marchwood generating station and experience with chemical additives', *J. Inst. Fuel* 1959 **32**, 61.

(4) NORMAN, G. R. 'A new approach to diesel smoke suppression', S.A.E. Paper 660339, 1966.

(5) GLOVER, I. 'The fuel additive approach towards the alleviation of the nuisance of diesel smokes', *J. Inst. Petrol.* 1966 **52** (No. 509), 137.

(6) MILLER, C. O. 'Diesel smoke suppression by fuel additive treatment', S.A.E. Paper 670093, 1967.

(7) GOLOTHAN, D. W. 'Diesel engine exhaust smoke: the influence of fuel properties and the effects of using barium containing fuel additive', S.A.E. Paper 670092, 1967.

(8) McCONNELL, G. and HOWELLS, H. E. 'Diesel fuel properties and exhaust gas—distant relations?' S.A.E. Paper 670091, 1967.

(9) EL NESR, M. S. M.Ph. thesis, University of London, 1966.

(10) SATCUNANATHAN, S. and ZACZEK, B. J. 'The spontaneous ignition and ignition delay of liquid fuel droplets impinging on a hot surface', *Proc. Instn mech. Engrs* 1967–68 **182** (Pt 3H), 382.

(11) SATCUNANATHAN, S. and EL NESR, M. S. 'The effects of additives on the cetane ratings of diesel fuels as related to the ignition delays of liquid fuel droplets impinging on a hot surface' (to be published).

(12) WILSON, G. C. and ROSE, R. A. 'Behaviour of high and low cetane diesel fuels', *S.A.E. Trans.* 1937 **41** (No. 2), 343.

(13) WETMILLER, R. S. and ENDSLEY, L. E. 'Effect of diesel fuel on exhaust smoke and odour', *S.A.E. Trans.* 1942 **50** (No. 12), 509.

(14) BROEZE, J. J. and STILLBROER, C. 'Smoke in high-speed diesel engines', S.A.E. Annual Meeting, Detroit, 1948 (January); abstract in *J. Soc. automot. Engrs* 1948 **56** (No. 4), 54.

(15) NEWHALL, H. K. and EL MESSIRI, I. A. 'A combustion chamber designed for minimum engine exhaust emissions', S.A.E. Paper 700491, 1970.

C138/71 SPATIAL AND TEMPORAL HISTORY OF NITROGEN OXIDES IN THE SPARK-IGNITION COMBUSTION CHAMBER

L. MUZIO*   H. NEWHALL†   E. S. STARKMAN‡

Oxides of nitrogen resulting from engine combustion are becoming progressively more important. Chemical kinetics, combined with engine thermodynamics, have been used to assess the mechanisms which control the formation of nitric oxide. Qualitative and quantitative agreements have been obtained between theoretical calculation and experimental determination. Recent investigations showed the presence of major gradients in temperature (to 600°K) and thus nitric oxide concentration (to an order of magnitude) in the burnt gases. Engine designs and operating variables affect these gradients. It also follows that the earlier kinetic models were fortuitous in the determination of absolute magnitudes corresponding to experimental measurement.

INTRODUCTION

UNTIL 20 years ago oxides of nitrogen were not considered of concern in combustion processes. Now they loom as the conceded most difficult of undesirable combustion products to control. Haagen-Smit (1)§ is credited with having established in 1952 the role of oxides of nitrogen in the formation of photochemical atmospheric pollution. At the time of his discovery there was practically no research on engine-produced oxides of nitrogen, much less a reliable measure of the magnitude of contribution.

Control measures recently adopted in the United States (2) prescribe that in the year 1976 this constituent of vehicle exhaust must be reduced to no more than 10 per cent of the level exhausted by the essentially uncontrolled 1971-model car. (While the State of California prescribes some oxide of nitrogen restriction for these 1971 models, the Federal Government does not.) The task of accommodating the projected constraint is universally accepted as formidable, even by those who are responsible for the passage and enforcement of the control legislation, particularly when compared to the control of unburnt hydrocarbons and carbon monoxide.

As indicated previously, compared to hydrocarbons and carbon monoxide, oxides of nitrogen until the 1950s had been largely neglected as products of engine combustion.

The MS. of this paper was received at the Institution on 7th June 1971 and accepted for publication on 22nd July 1971. 22
* *Asst. Professor, Mechanical Engineering, Columbia University, U.S.A.*
† *Associate Professor, Mechanical Engineering, University of Wisconsin, U.S.A.*
‡ *Professor, Mechanical Engineering, University of California, U.S.A.*
§ *References are given in Appendix 138.1.*

Partly, this was because oxides of nitrogen represent a negligible fraction of the energy of the conversion processes in the engine, compared to other species. Accurate measurements will show that, even during transient peak concentration, the oxides of nitrogen represent no more than 0·01 mole fraction of the gases in the cylinder, and the more usual concentrations range from 0·0001 to 0·001 mole fraction. The complexity of accounting for these trace constituents was thus one consideration in their being neglected in the era of engine development, when energy conversion performance was the principal criterion.

The seriousness of air pollution has impinged on the relative importance of the individual products of combustion and, as a consequence, efforts to identify the mechanisms responsible for oxides of nitrogen formation have been concerted, particularly during the last 10 years.

It is important at this juncture to identify the oxides of nitrogen in order of progressive oxidation. The possible states of oxidation are: N_2O, nitrous oxide; NO, nitric oxide; NO_2, nitrogen dioxide; and N_2O_5, nitrogen pentoxide. The first and last are now known to be of insignificant importance to engine combustion, having neither been calculated nor experimentally found to exist in measurable quantities in the processes considered. Of the remaining two, NO by far predominates as the engine-produced species (3)–(5), which, however, slowly converts to NO_2 in the atmosphere. Hereafter, therefore, when reference is made to 'oxides of nitrogen', it is almost explicit that NO is the chemical compound of consequence.

It will be the purpose of this paper to provide some of the more recent analytical and experimental evidences relating to NO formation in combustion engines, and to offer the state of understanding of the mechanisms of formation as they may be influenced by engine operation and design.

EARLY OBSERVATIONS AND EXPERIMENTAL EVIDENCE

Historically, definitive engine investigations into the formation and emission of NO appear to commence with the work of Wimmer and McReynolds (3). They utilized an electromagnetically actuated sampling valve to study the history of NO during the combustion and expansion process in a single-cylinder, spark-ignition engine. Sampling at 160° a.t.c., they concluded that NO was formed in concentrations close to chemical equilibrium at 'peak cycle temperature'. The peak cycle temperature in their study was inferred from instantaneous measurements of cylinder pressure and volume. They also speculated that there was significant decomposition for lean mixtures during the expansion stroke.

Wimmer and McReynolds made no reference to the relative position of the sampling valve with respect to the spark plug, which, with the rest of their text, strongly implied that they assumed a spatially homogeneous concentration burnt gas region.

The next major definitive study was that of Alperstein and Bradow (6), reported in 1967. A sampling valve was again used to obtain NO histories in a single-cylinder, c.f.r. engine. Although there is speculation as to the quantitative reliability of their results due to their sampling procedure (7), they were the first to show spatial variations of NO in the engine cylinder. Utilizing the distance between the sampling valve and the spark plug as the significant variable, they pointed out that higher levels of NO existed at locations near the spark plug.

In order to overcome the limitations of physical sampling from the combustion chamber, Newhall and Starkman (5) devised an infrared spectrometric approach for studying the history of NO in the last mass to burn during the expansion process in a c.f.r. engine. These results showed little or no change of NO in the end gas during the expansion process. Regarding the absolute levels of NO, they concluded that the chemical equilibrium value at the average peak cycle temperature was not an adequate indicator and that the NO must be formed at spatially localized peak equilibrium values, and subsequently frozen.

Until 1968, much of the opinion regarding NO genesis in engines was speculative. Particularly, this was the case with work on the formation process. The effect of kinetics had been given little attention.

Eyzat and Guibet (8) were early in the attempt to answer some of the remaining questions through the use of an analytical combined thermodynamic and chemical kinetic model of the combustion process. Their model assumed two zones in the combustion chamber separated by the flame front. Behind the flame was a uniform burnt gas zone, and ahead was a uniform unburnt gas zone. Both were assumed to have uniform but different temperatures.

A simplified kinetic mechanism was used for the formation of NO.

This utilized only the bimolecular reaction between oxygen and nitrogen:

$$O_2 + N_2 \rightleftharpoons NO + NO$$

Their calculations resulted in NO concentration levels lower than those of the peak equilibrium burnt gas values. They also showed that freezing occurred during the expansion process. The results were in satisfactory agreement with experiment; and although of use in predicting overall trends in exhaust NO content, more recent experiments have indicated the necessity of a more detailed model than the two-zone model of Eyzat and Guibet.

A phenomenological study, using a sampling valve, by Starkman, Stewart and Zvonow (7), was the first experimental evidence of a major spatial variation (of about 1200 p.p.m. NO) across part of the cylinder of a c.f.r. engine combustion chamber.

More recently, Lavoie et al. (9), utilizing the $NO + O \rightarrow NO_2 + h\nu$ recombination continuum to spectroscopically measure NO concentrations in a single-cylinder L-head engine, confirmed this important observation. Their results showed spatial variations of NO of the order of 0·5 mole per cent across the chamber, with differential localized freezing during the expansion process.

Once recognized, the spatial variation of NO concentration in the combustion chamber is explainable with available fundamental thermodynamic arguments. With little or no mixing in the chamber, each element of gas undergoes a different sequence of thermodynamic processes from its neighbouring gas (see Fig. 138.1). The first element of gas burns at an initial low pressure and is subsequently and successively isentropically compressed to peak pressure. On the other hand, the last element to burn is first compressed essentially to peak pressure and then is burned. These events result in the formation of a

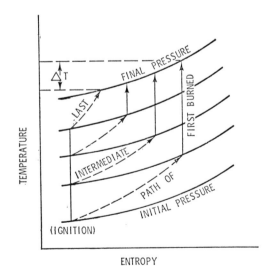

Fig. 138.1. Thermodynamic diagram of non-uniform combustion model

substantial temperature gradient across the combustion chamber, with the highest temperature existing adjacent to the spark plug.

Rasseweiler and Withrow (10) were the first to observe experimentally the temperature distribution in a combustion chamber. They used a sodium D-line technique, and spectroscopically measured an average 200°K difference in temperature across the combustion chamber of a single-cylinder, L-head engine. Recent infrared measurements confirm the results of Rasseweiler and Withrow. The gradient from one end of the chamber to the other apparently can even be as much as 600°K (11).

AVAILABLE THERMODYNAMICS, EQUILIBRIUM, AND KINETICS FOR APPLICATION TO REAL SYSTEMS

During the past 15 years a wealth of detailed and accurate thermochemical data relevant to the calculation of high-temperature chemical equilibrium properties of combustion gases have become available (12). At the same time, continued development of digital computers and related numerical calculation techniques has permitted incorporation of such data into piston-engine combustion calculations.

Chemical equilibrium considerations and effects

Thermochemical information confirms experimental observation that the principal oxide of nitrogen formed in the combustion processes is NO. Equilibrium considerations indicate particularly that at usual engine cycle temperatures the higher oxides of nitrogen, including NO_2, can exist only in a concentration at most several orders of magnitude lower than that of NO. For this reason, discussion here will be restricted to the single oxide of nitrogen, NO.

Formation of NO from the more stable molecular species nitrogen and oxygen is endothermic by some 21 kcal/mole. Hence, from the standpoint of equilibrium thermodynamics, the appearance of NO is disproportionately favoured by high temperatures.

Relatively simple chemical equilibrium calculation predicts that concentrations of this species in engine combustion gases at peak engine cycle temperatures should reach levels ranging from several hundred parts per million (p.p.m.) to as much as 1 or 2 mole per cent, depending upon temperature and fuel–air stoichiometry.

For the CHON (carbon, hydrogen, oxygen, nitrogen) system of chemical equilibrium combustion products, 11 chemical species evolve as being of significance, defined as appearing in chemical equilibrium at maximum possible engine cycle temperatures in quantities of approximately 1 p.p.m. or greater. These include H_2O, CO_2, H_2, CO, O_2, N_2, NO, OH, H, O, and N (13).

Complete quantitative equilibrium distribution of species is obtained through the simultaneous solution of 11 non-linear algebraic equations, four of which are derived from mass conservation laws and the remaining seven from equilibrium considerations. Alternatively, direct minimization of the Gibb's free energy of the combustion system may be employed with similar results. It is beyond the scope of this paper either to present detailed calculation procedures or to comment on the relative merits of alternative calculation procedures.

It is emphasized that NO appears as one of the 11 significant chemical equilibrium products of combustion, and as such is necessarily included, though perhaps implicitly, in all equilibrium engine cycle calculations of the level of sophistication considered here.

During the period 1960–65, engine cycle thermodynamic calculations, even though facilitated by the digital computer, were concerned with temporal but not spatial variation in species equilibrium. They were mostly based on the traditional constant volume Otto cycle, involving assumed instantaneous conversion of fuel–air mixture to combustion products of uniform temperature and hence uniform composition (14). Such calculations were necessarily incapable of representing many facets of real engine performance, including the effect which finite burning rate has upon consequent development of spatial non-uniformities in temperature, and thus composition, within burnt products of combustion.

These early 1960s studies were followed by several calculations in which finite burning rates have been treated, either by inclusion of empirical burning rate laws (15) or through inference from experimental engine pressure recordings (16). Such analyses were based on chemical equilibrium species calculations of the type previously described, but applied on a time-dependent local or zonal basis. No account was taken of the kinetic control of the rate of formation of chemical species. In the absence of needed experimental facts, such calculations also required commitment to assumptions regarding the state of mixedness of burnt products. Generally, either one or the other of two extremes was considered: complete mixedness of the combustion chamber, with total annihilation of spatial gradients in temperature and composition of burnt products; or no mixing, with implied total preservation of temperature and concentration gradients as developed in the burning process. It would seem reasonable to expect that in reality the situation might lie between these extremes, depending upon turbulence levels and other factors of design and operation of the particular engine under consideration.

Notwithstanding the limitations implicit through the above considerations of spatial non-uniformities in temperature and composition, it is instructive to consider the behaviour of NO in chemical equilibrium with combustion products having constant uniform thermal properties. The real combustion process might then be viewed as a temporal succession of spatially distributed states whose equilibria are represented by such values.

Results of such chemical equilibrium calculations, relevant to the appearance of NO in engine combustion

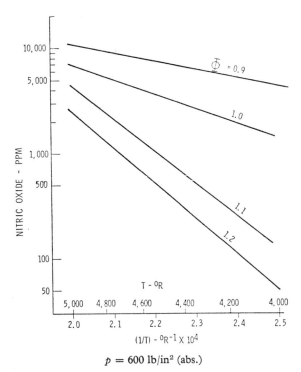

Fig. 138.2. Influence of temperature on chemical equilibrium nitric oxide concentrations

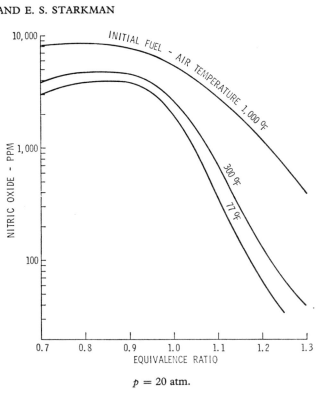

Fig. 138.4. Chemical equilibrium nitric oxide concentrations at adiabatic flame temperature

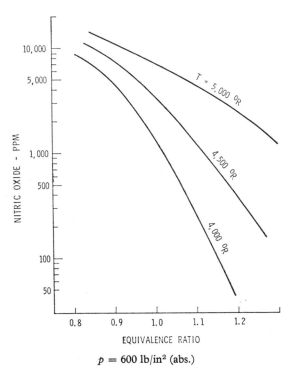

Fig. 138.3. Influence of equivalence ratio on equilibrium nitric oxide concentrations

products, are presented in Figs 138.2, 138.3, and 138.4. In these figures, NO concentrations existing in the chemical equilibrium products of combustion of uniform temperature, pressure, and stoichiometry are plotted. While these data are strictly applicable to the combustion of octane with air, they are broadly representative of combustion products associated with most hydrocarbon structures typically employed as spark-ignition engine fuels.

Fig. 138.2 illustrates the influence of temperature on chemical equilibrium NO levels for several constant fuel-air equivalence ratios. The marked sensitivity of this equilibrium to temperature is most important, and clearly demonstrated that NO increases monotonically with increasing temperature.

For Fig. 138.3 these data have been rearranged to show the influence of equivalence ratio for several constant combustion gas temperatures. The shift in equilibria favouring increased NO with increases in excess air (oxygen) is evident.

It is recognized that in reality temperature and equivalence ratio are not independent parameters. Temperatures achieved in a given combustion process are uniquely determined by the particular equivalence ratio employed. Therefore Fig. 138.4 is a supplement to the information presented in Figs 138.2 and 138.3, and includes the dependence of temperature on equivalence ratio. Here equilibrium NO concentrations are based on calculated chemical equilibrium adiabatic flame temperatures. Several initial reactant temperatures are considered. The combined effects of variation of stoichiometry with accompanying variation of temperature are clearly apparent and can be interpreted through consideration of the independent temperature and stoichiometry effects presented in Figs 138.2 and 138.3. Generally, combustion temperatures will be maximum for mixtures slightly fuel-rich of chemically correct.

Implicit in the foregoing discussion of equilibrium calculation were the assumptions that (a) sufficient time is available, during events associated with the combustion and expansion processes, to permit combustion gases (particularly NO) to reach the chemical equilibrium state; and (b) no reverse action to destroy these species would follow. If these assumptions were correct, then such calculations would be sufficient to describe completely the chemical and thermodynamic behaviour of the system. However, recent experiments (9) have demonstrated convincingly that in many cases the time required for NO to reach local equilibrium levels is significantly related to residence times characteristic of combustion systems.

In the case of the spark-ignition engine it is apparent that the cylinder combustion gases, or a portion thereof, are thermally quenched by the ensuing expansion process following combustion and before local peak equilibrium NO concentrations are achieved (17). Additionally, it is recognized that decomposition reactions, which according to the principles of thermodynamics governing equilibrium should lead to the complete destruction of NO by the end of expansion, are thermally quenched shortly after expansion begins. As a consequence, NO leaves the combustion chamber in quantities corresponding to equilibria at temperatures much higher than the exhaust temperature (3). This phenomenon has been confirmed experimentally through *in situ* infrared measurements made directly on an engine cylinder (18).

The role of chemical kinetics

It becomes apparent that the preceding chemical equilibrium description of combustion gases is in many cases incomplete, and consideration must be given to the finite rate processes governing both the formation and destruction of NO. Due largely to the needs of the aerospace establishment related to atmospheric re-entry problems arising in the late 1950s and early 1960s, considerable data related to the rates of high-temperature oxygen–nitrogen reactions have become available (19)–(21). Such data subsequently have been directed to the formation of NO within the spark-ignition engine. A summary of the pertinent reactions, together with accepted rate constants, appears in Table 138.1. Forward and reverse directions of each reaction are considered, and as a consequence both formation and destruction of NO are represented.

It is clear from an inspection of these reactions that their rates in either forward or reverse directions will, at any instant, be critically dependent upon the existing concentrations of atoms and radicals. Each of the pertinent reactions involves either radical–molecule or radical–

Table 138.1. Chemical reaction rates

Reaction	Rate constant, cm^3/mole-sec, $T - °K$, $R - 1·986$ cal/°K	Ref.
Reaction 1 f. $N_2 + ON \rightarrow O + N$ b. $NO + N \rightarrow N_2 + O$	$7 \times 10^{13} \exp(-75\,500/RT)$ $1·55 \times 10^{13}$	(20)
Reaction 2 f. $O_2 + N \rightarrow NO + O$ b. $NO + O \rightarrow O_2 + N$	$13·3 \times 10^9 T \exp(-7080/RT)$ $3·2 \times 10^9 T \exp(-39\,100/RT)$	(20)
Reaction 3 f. $N_2O + O \rightarrow NO + NO$ b. $NO + NO \rightarrow N_2O + O$	$1·42 \times 10^{14} \exp(-28\,000/RT)$ $2·6 \times 10^{12} \exp(-63\,800/RT)$	(21)
Reaction 4 f. $N + OH \rightarrow NO + H$ b. $NO + H \rightarrow N + OH$	$4·2 \times 10^{13}\star$	(22)

★ Reliability questionable.

radical interaction. Thus, equally important as specification of numerical rate constant values is the quantitative delineation of the chemical environment in which the reactions occur, i.e. the existing concentrations of reactant atoms and radicals within the particular system of interest.

Reactions 1, 2, and 3 of Table 138.1 have been studied independently by a number of investigators using diverse experimental techniques ranging from shock tubes (20) to isothermal flow reactors (21). Reasonable agreement among the observed reaction rates has been obtained, and the resulting numerical constants, as presented in Table 138.1, appear to be sufficiently trustworthy for engine work.

In contrast, the rate constant presented for reaction 4 should be considered suspect at the present time, as its inordinately large value was derived from a single set of measurements based on a flow discharge technique (22). Ordinarily, an atom–radical reaction of this type would not be expected to be competitive with atom–molecule reactions such as reactions 1 and 2, and the question of the precision of the reported rate constant would be of little consequence. However, reaction 4 cannot be dismissed. As a result of the unusually large rate constant, there exist certain combustion regimes in which reaction 4 may be dominant relative to reactions 1 and 2, particularly under fuel-rich conditions in which relatively low levels of the atomic and diatomic oxygen species requisite to reactions 1 and 2 are present. Fortunately, the fuel-rich regime produces far less NO than the lean.

Reaction 3 has generally been considered unimportant to the formation of NO since, on the basis of equilibrium calculations, levels of the reactant species N_2O in combustion gases are thought to be negligible. However, recent studies of the reaction $H + N_2O \rightarrow N_2 + OH$ (23) suggest that in the reverse reaction OH attack on N could represent a significant non-equilibrium source of N_2O molecules, which might participate as reactants in reaction 3. Further study will be required to evaluate fully the implications of this possibility.

Application of chemical kinetics to the engine process

Recently, two investigations have proposed models to account for the unmixedness in the combustion chamber and the effect on the kinetic history of NO. These are discussed in detail in references (9) and (11).

The model in reference (11) divides the chamber into a number of separated, unmixed, and adiabatic zones in which it maintains its identity throughout the combustion process and subsequent expansion. Energy balances are performed on each zone and on the entire cylinder with the assumption for a brief time frame that the C—H—O system is maintained in chemical equilibrium. The temperature histories in each of the zones could then be solved. Since the reactions pertinent to the formation of NO are slow relative to the reactions responsible for the energy release, the NO reactions are decoupled from the thermodynamic analysis and integrated separately. In this model the Zeldovich mechanism shown below was used to describe the NO formation:

$$N_2 + O \rightleftharpoons NO + N$$
$$N + O_2 \rightleftharpoons NO + O$$

A synopsis of the calculation using this model is shown in Fig. 138.5. As can be seen, the temperature difference across the combustion chamber has a marked effect on the NO formation. This figure shows typical results for the combustion of a stoichiometric mixture of propane and air in a c.f.r. engine with a compression ration of 7:1 and a 40° b.t.c. ignition.

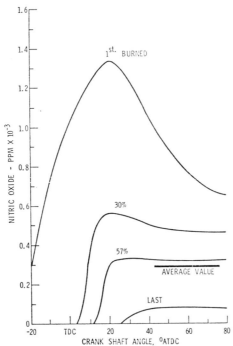

Fig. 138.5. Calculated nitric oxide
History for selected segments of the fuel-air mixture:
Propane-air, $\phi = 1 \cdot 0$.
Compression ratio = 7:1.
Ignition = 40° b.t.c.

The implications of these calculations are among the more important that have been developed to explain experimentally observed phenomena relating to engine production of NO. The results of these calculations explain the observations in experiments discussed previously (7).

Higher levels of NO are observed nearer the spark plug than at the far end of the combustion chamber. It can be noted from Fig. 138.5 that in the first elements of mass to burn, the NO is formed at near peak equilibrium levels. About 50 per cent decomposes during the expansion process. The history differs for the last mass to burn. The temperatures being lower, the formation reactions do not occur sufficiently fast to form local equilibrium quantities. Freezing occurs at values significantly lower than equilibrium.

From these calculations it can be seen that a model for application to a real system must be capable of the specification of the local thermodynamic state in the engine cylinder.

Recent experimental developments

Recent experiments (24) indicate that in high-pressure combustion processes representative of those occurring in spark-ignition engines, NO is formed primarily in post-flame combustion gases at rates fully predictable by consideration of reactions 1 and 2 of Table 138.1. This finding is based on *in situ* spectroscopic measurements of NO formation within a high-pressure combustion vessel (25).

The experimental apparatus is shown schematically in Fig. 138.6 and consists of a cylindrical pressure vessel fitted with diametrically opposed quartz windows. By means of a discharge ultraviolet lamp and grating spectrometer set to the appropriate wave-length, spectral absorption due to the formation of NO as an axially propagating flame front enters the optical path can be measured and recorded. In this manner, the formation of NO in the immediate vicinity of the flame front is observed.

Experimental measurements of NO formation in the combustion of hydrogen-air mixtures are plotted in Fig. 138.7, where the zero of time represents the instant of flame front arrival in the optical path of the spectroscopic system. Fig. 138.8 presents similar but more recent results obtained for combustion of propane-air mixtures (25). For both fuels the NO formation process is similar. Nitric oxide is formed primarily in post-flame combustion gases, and the approach to chemical equilibrium concentrations involves significant lapses of time relative to residence times of many practical combustion systems.

The solid lines of Figs 138.7 and 138.8 represent independent theoretical calculation of the rate of NO formation for conditions corresponding to the experimental vessel conditions. These calculations were based on a simple Zeldovich mechanism employing reactions 1 and 2 of Table 138.1:

$$N_2 + O \rightleftharpoons NO + N \quad . \quad . \quad (138.1)$$
$$O_2 + N \rightleftharpoons NO + O \quad . \quad . \quad (138.2)$$

SPATIAL AND TEMPORAL HISTORY OF NITROGEN OXIDES IN THE SPARK-IGNITION COMBUSTION CHAMBER

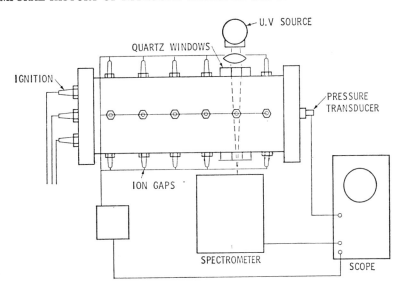

Fig. 138.6. Experimental system for ultraviolet absorption measurements

leading to the rate equation:

$$\frac{d(NO)}{dT} = K_{1f}(N_2)(O) - K_{1b}(NO)(N) + K_{2f}(N)(O_2) - K_{2b}(NO)(O) \quad (138.3)$$

Numerical rate constants employed in the integration of this equation are those presented in Table 138.1.

Since it was determined *a posteriori* from experimental results that the formation of NO is slow relative to the probable rates of relaxation of atoms and radicals to their post-flame equilibria, post-flame equilibrium concentration values for nitrogen atoms and oxygen atoms were substituted in equation (138.3). Numerical evaluation of rate constants and atom concentrations was based on local time-varying, post-flame temperatures inferred from recorded pressure traces.

It is comforting that inordinate agreement exists between experimentally measured and theoretically predicted NO concentrations. This is particularly significant in view of the fact that the predictions involved solution of equations containing no adjustable parameters, the reaction rate constants being published values derived from independent investigation of elementary reactions. This result offers strong though admittedly circumstantial evidence that at pressures and temperatures pertinent to engine combustion processes, nitric oxide is formed primarily in substantially equilibrated (with the exception

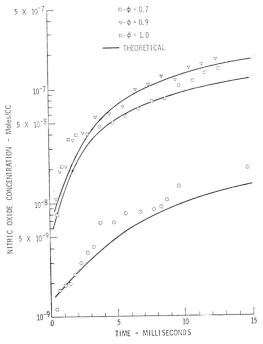

Fig. 138.7. Experimental and theoretical nitric oxide formation rates for H_2–air mixtures

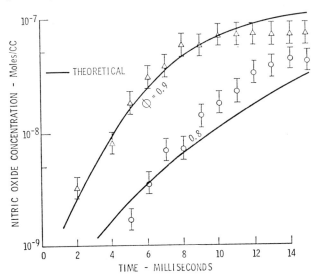

Fig. 138.8. Formation of nitric oxide in propane–air combustion

of NO) combustion gases and by a relatively simple mechanism involving reactions 1 and 2 of Table 138.1. The calculation procedure involved in the use of this mechanism is therefore straightforward and within the means of most computer facilities.

The inclusion of reaction 4 in the reaction mechanism was found to have a negligible effect on calculated results for fuel-lean and chemically correct mixture ratios. However, for fuel-rich conditions, inclusion of this reaction with the questionable rate constant value of Table 138.1 resulted in large overestimates of the rate of NO formation. This finding is consistent with the earlier observation that the reported rate constant value appears to be inordinately large.

IMPLICATIONS AND APPLICATIONS TO REAL SYSTEMS

The foregoing should serve to establish that there is an understanding of the processes which are responsible for the oxides of nitrogen produced in homogeneous mixture spark-ignited engines. It is now known that NO is the only one of the oxides of nitrogen produced in quantity sufficient to warrant consideration. It is also known that a maximum of eight reactions (counting both forward and reverse), along with appropriate rate constants, is necessary to calculate reliably the NO history in the post-combustion gases. It has also been shown that the NO appears post-flame, and therefore pre-flame and flame reactions are not necessary to the calculation procedure.

Thus, progress has advanced sufficiently far to allow rational assessment of the principal controlling engine variables. The next logical step is to undertake a comprehensive analysis of the influence of a multitude of variables. Such a comprehensive analysis is not the subject of this paper; but the simple example that follows will serve to illustrate the kind of information which can be generated, and engine design and operating variables which can be of consequence.

The example chosen will be that of a stoichiometric mixture of octane and air, burned in a cylindrical $3\frac{1}{4}$-in diameter combustion chamber of an engine operating at an assumed speed of 1800 rev/min, a 7:1 compression ratio, and 40° b.t.c. spark advance. Constant spatial velocity of the flame is assumed, and the spark plug is located at the edge of the cylinder.

Fig. 138.9 is a three-dimensional representation of the NO concentration above the piston when it has moved halfway down the stroke. By this time in the stroke, referring back to Fig. 138.5, the concentration of NO is substantially frozen at all points.

Fig. 138.9 serves to further illustrate the point previously made, that the highest concentrations of NO exist in those gases first ignited. (Note that swirl, or gross rotation of the cylinder charge, can be introduced as a variable.)

If the total content of the cylinder is discharged on each cycle, then the average concentration in the exhaust will be that of the mean ordinate of Fig. 138.9. However, some of the combustion gas is trapped as residual fraction. Thus,

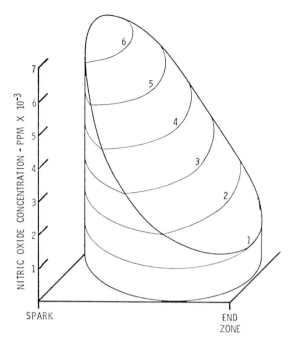

Fig. 138.9. Representation of nitric oxide concentration in cylindrical combustion chamber fired from a peripheral location

the amount remaining and its concentration are highly important. The location of the exhaust valve with respect to the concentration map is accordingly also of importance. Obviously, if the exhaust valve is so located that it first releases the gases of high NO concentration, the maximum exhaust content of NO will be the consequence, and vice versa.

Consider further the importance of spark plug location. In a previous report (26) it was disclosed that the optimum location, all other factors being equal, is for the spark plug to be at the edge of the combustion chamber rather than the centre. Considering further the influence which location of the exhaust valve may have on the exhaust concentration of NO, a centrally located spark plug is doubly undesirable because the hottest gases, and thus the maximum content of NO, will leave the cylinder first. Probably the most fruitful area for application of the approach utilized above is in combustion chamber design for minimizing temperature gradients, and thus NO emissions.

Thus far, the authors of this paper have concerned themselves only with models having combustion chambers cylindrical in shape and with peripheral and centrally located spark plugs.

CONCLUSIONS

(1) Sufficient thermochemical and chemical kinetic knowledge exists to permit reliable calculation of the time-dependent distribution of NO in the post-combustion gases of a spark-ignition engine combustion chamber.

(2) Application of the kinetic model to real systems shows that design parameters may be important to NO

levels found in the exhaust. Of particular importance as examples are spark plug and exhaust valve locations.

ACKNOWLEDGEMENT

Portions of this work were supported under research grants from the U.S. Environmental Protection Agency, Office of Research Grants, Air Programs (AP 385 and AP 582).

APPENDIX 138.1

REFERENCES

(1) HAAGEN-SMIT, A. J. 'Chemistry and physiology of Los Angeles smog', *Ind. Engng Chem.* 1952 (June) **44**, 1342.
(2) CLEAN AIR AMENDMENTS OF 1970, Public Law 91-604 1970 (31st December) (U.S. Government Printing Office).
(3) WIMMER, D. B. and MCREYNOLDS, L. A. 'Nitrogen oxides and engine combustion', *S.A.E. Trans.* 1962 **70**, 733.
(4) CAMPAU, R. M. and NEERMAN, J. C. 'Continuous mass spectrometric determination of nitric oxide in automotive exhaust', *S.A.E. Paper 660116*, 1966 (January).
(5) NEWHALL, H. K. and STARKMAN, E. S. 'Direct spectroscopic determination of nitric oxide in reciprocating engine cylinder', *S.A.E. Trans.* 1968 **76**.
(6) ALPERSTEIN, M. and BRADOW, R. L. 'Exhaust emissions related to engine combustion reactions', *S.A.E. Trans.* 1967 **75**, 876.
(7) STARKMAN, E. S., STEWART, H. E. and ZVONOW, V. A. 'An investigation into the formation and modification of emission precursors', *S.A.E. Paper 690020*, 1969.
(8) EYZAT, P. and GUIBET, J. C. 'A new look at nitrogen oxides formation in internal-combustion engines', *S.A.E. Trans.* 1968 **77**.
(9) LAVOIE, G. A., HEYWOOD, J. B. and KECK, J. C. 'Experimental and theoretical study of nitric oxide formation in internal combustion engines', *Combust. Sci. Technol.* 1970 **1**.
(10) RASSEWEILER, G. M. and WITHROW, L. 'Flame temperatures vary with knock and combustion chamber position', *S.A.E. Jl* 1935 **36**.
(11) MUZIO, L. J., STARKMAN, E. S. and CARETTO, L. S. 'The effect of temperature variations in the engine combustion chamber on formation and emission of nitrogen oxides', *S.A.E. Paper 710158*, *S.A.E. Automot. Engng Congr.*, Detroit, 1971 (January).
(12) JANAF THERMOCHEMICAL DATA TABLES, Joint Army–Navy–Air Force Thermochemical Panel, Contract AF 33(616)–6149, 1960 (Dow Chemical Company, Midland, Michigan).
(13) NEWHALL, H. K. and STARKMAN, E. S. 'Thermodynamic properties of octane and air for engine performance calculations', *S.A.E. Tech. Prog. Ser.* 1964 **7**.
(14) EDSEN, M. H. 'The influence of compression ratio and dissociation on ideal Otto cycle thermal efficiency', *S.A.E. Trans.* 1962 **70**, 665.
(15) PATTERSON, D. J. and VAN WYLEN, G. 'A digital computer simulation for spark-ignited engine cycles', *S.A.E. Tech. Prog. Ser.* 1964 **7**.
(16) PETERS, B. D. and BORMAN, G. L. 'Cyclic variations and average burning rates in a S.T. engine', *S.A.E. Paper 700064*, 1970 (January).
(17) SPINDT, R. S., WOLFE, C. L. and STEVENS, D. R. 'Nitrogen oxides combustion and engine deposits', *S.A.E. Trans.* 1956 **64**, 799.
(18) NEWHALL, H. K. 'Kinetics of engine-generated nitrogen oxides and carbon monoxide', *Twelfth Int. Symp. on Combustion* 1969, 693 (Mono of Maryland, Baltimore).
(19) KAUFMAN, F. and KELSO, J. R. 'Thermal decomposition of nitric oxide', *J. Chem. Phys.* 1955 **23** (No. 9), 1702.
(20) WRAY, K. L. and TEARE, J. D. 'Shock tube study of the kinetics of nitric oxide at high temperatures', *J. Chem. Phys.* 1962 **36** (No. 10), 2582.
(21) FENNIMORE, C. P. and JONES, G. W. 'Nitric oxide decomposition at 2200–2400°K', *J. phys. Chem.* 1957 **61**, 654.
(22) CAMPBELL, I. N. and THRUSH, B. A. 'Reactivity of hydrogen to atomic nitrogen and atomic oxygen', *Trans. Faraday Soc.* 1968 **64** (Pt 5).
(23) WOLFRUM, J., Universität Göttingen, personal communication to H.K.N. (1971).
(24) NEWHALL, H. K. and SHAHED, S. M. 'Kinetics of nitric oxide formation in high-pressure flames', *Thirteenth Int. Symp. on Combustion*, Salt Lake City, Utah 1970 (August).
(25) SHAHED, S. M. and NEWHALL, H. K. 'Kinetics of nitric oxide formation in propane–air and hydrogen–air–diluent flames', *Combust. Flame* (to be published).
(26) STARKMAN, E. S. 'Some basic considerations in formation of combustion engine exhaust products', Paper No. IId, *Internal Combustion Engine Conf.*, Buchaust 1970 (May).

C139/71 ACCOMPLISHMENTS OF THE I.I.E.C. PROGRAMME FOR CONTROL OF AUTOMOTIVE EMISSIONS

W. J. KOEHL* D. P. OSTERHOUT† S. E. VOLTZ‡

The status of the I.I.E.C. programme for control of automotive emissions and the technological achievements in some specific tasks are described. Catalysts and converter systems for the reduction of carbon monoxide, hydrocarbons, and oxides of nitrogen have been developed; some further improvements in activity and stability will be required to meet the latest proposed U.S. emission standards. A fundamental predictive mathematical model has been developed for carbon monoxide and hydrocarbon converters, and some examples of its predictive capabilities are presented. The use of thermal reactors for the control of carbon monoxide and hydrocarbons and exhaust gas recirculation for oxides of nitrogen reduction is also discussed. The performance of four concept emission control systems developed in the I.I.E.C. programme is given for actual vehicle operation.

INTRODUCTION

THE Inter-Industry Emission Control (I.I.E.C.) programme was begun in 1967 by the Ford Motor Co. and the Mobil Oil Corpn as a three-year co-operative research effort into ways of providing a virtually emission-free automobile. It now has the following 11 participants and is currently extended to 31st December 1971: American Oil Co.; Atlantic Richfield Co.; Fiat S.p.A.; Ford Motor Co.—project manager; Marathon Oil Co.; Mitsubishi Motors Corpn; Mobil Oil Corpn; Nissan Motor Co. Ltd (Datsun); Standard Oil Co. (Ohio); Sun Oil Co.; and the Toyo Kogyo Co. Ltd. (Arrangements have also been made to make I.I.E.C. research results available to Toyota Motor Co. Ltd and Volkswagenwerk A.G.)

The original emission goals of the I.I.E.C. programme, set in 1967, were aimed at achieving—at a research level—the following emission levels for hydrocarbons (HC), carbon monoxide (CO), and oxides of nitrogen (NO_x), as determined by the seven-mode, seven-cycle Federal emission test procedure (F.T.P.) (**1**)§.

Exhaust
 Hydrocarbons 0·82 g/mile
 Carbon monoxide 7·1 g/mile
 Oxides of nitrogen 0·68 g/mile

Crankcase Nil
Evaporative Nil

The emission levels were one-half those needed, according to available estimates, to restore the air quality in the Los Angeles basin (the most critical air shed in the U.S.) to that of 1940. The year 1940 is considered to be the last smog-free year in Los Angeles. At the end of 1970 the I.I.E.C. emission goals were revised to meet the requirements of the U.S. 1970 Clean Air Act Amendments for 1976 using the new non-repetitive cycle constant volume sampling (C.V.S.) emission test procedure as published in the 10th November 1970 U.S. Federal Register. At present (May 1971), the actual exhaust emission levels proposed by the U.S. are: HC, 0·46 g/mile (1975); CO, 4·7 g/mile (1975); NO_x, 0·4 g/mile (1976).

To date, 23 different projects have been initiated. Work is currently being conducted on 12 projects primarily concerned with improved catalysts for reduction of NO_x and oxidation of HC and CO, catalytic converter system design, thermal reactors, exhaust gas recirculation (E.G.R.), mathematical modelling of emission control systems, and fuel and lubricant effects on emissions. The active projects also include development and evaluation of concept emission vehicles equipped with the most promising fuel, catalytic and hardware components developed in the other I.I.E.C. projects.

Several progress reviews on the I.I.E.C. have been presented previously (**2**)–(**4**). In addition, eight papers (**5**)–(**12**) on I.I.E.C. technology and systems concepts were given at the meeting of the Society of Automotive Engineers, Detroit, Michigan, in January 1971. This paper will describe the status of the research in several of the

---
The MS. of this paper was received at the Institution on 12th July 1971 and accepted for publication on 4th August 1971. 13
* Group Leader, *Automotive and Aviation Fuels, Mobil Oil Co. Ltd, Research and Technical Service Dept, Coryton, Stanford-le-Hope, Essex.*
† Manager, *Automotive Emissions Research, Research Dept, Mobil Research and Development Corpn, Paulsboro, New Jersey 08066.*
‡ Research Associate, *Research Dept, Mobil Research and Development Corpn, Paulsboro, New Jersey 08066.*
§ References are given in Appendix 139.1.

currently active projects, namely, catalyst systems for HC, CO, and NO_x control, mathematical modelling of HC–CO converter systems, thermal reactors, exhaust gas recirculation, and concept vehicle systems.

DEVELOPMENT OF CATALYTIC CONVERTERS FOR OXIDATION OF CO AND HC

The many publications and patents on the oxidation of CO and HC over various types of catalysts provided some guidance for the development of catalysts for emission control systems. Background and experience in catalysis of petroleum-refining processes also contributed to these efforts.

Most of the development work during 1968–70 was directed to meeting the original I.I.E.C. goals with a relatively low-cost catalyst for operation in leaded (3 g Pb/U.S.G.), low leaded (0·5 g Pb/U.S.G.), and unleaded fuel. Several transition metal oxide catalysts had the required activities to meet these goals. With the advent of the 1970 Clean Air Act Amendments, emphasis was shifted to the development of low-temperature ignition catalysts (200–400°F) to assist in further reducing emissions during cold start-up and catalysts with increased high-temperature stability (1400–1800°F).

The development of a typical catalyst involved the following steps:

(1) Preparation of catalyst sample (~ 100 g) by impregnation of active ingredient (metal or metal oxide) on stable support (alumina) or other appropriate method.
(2) Drying and calcination of catalyst.
(3) Determination of physical properties of catalyst, such as surface area, particle density, and hardness. In some instances, chemical analyses are also highly desirable.
(4) Determination of initial catalyst activity (extent of conversion under fixed conditions or determination of reaction rate constants) for oxidation of CO and HC in synthetic gas mixture (CO, C_3H_6, O_2, NO, H_2O, N_2).
(5) Ageing of catalyst in engine exhaust.
(6) Determination of activity of aged catalyst for oxidation of CO and HC in synthetic gas mixture.
(7) Preparation of large catalyst sample (~ 10 lb) for vehicle testing.
(8) Repeat steps (2)–(6) with large catalyst sample.
(9) Determine performance of catalyst in converter in engine exhaust.
(10) Confirm catalyst effectiveness and durability in converter on automobile.

Some of the necessary and desirable characteristics of a catalyst for the control of CO and HC emissions are:

(1) High activity for oxidation of CO and HC, including sufficient activity at low temperatures (200–400°F) to reduce emissions during cold start-up.

Table 139.1. Ageing of HC–CO catalysts at 1400–1600°F*

Fuel	Catalyst	Equivalent ageing, miles	Activity, reaction rate constant at 750°F	
			CO	HC
Unleaded	$Pt–Al_2O_3$	0	607	427
		5 000	639	408
		10 000	405	258
Unleaded	$CuO.Cr_2O_3–Al_2O_3$	0	277	153
		5 000	177	29
		10 000	117	13
0·5 g Pb/gal	$Pt–Al_2O_3$	0	607	427
		5 000	424	146
		10 000	198	24
0·5 g Pb/gal	$CuO.Cr_2O_3–Al_2O_3$	0	277	153
		5 000	15	1
		10 000	20	2

* V-8 engine on dynamometer stand under cyclic operation simulating modified A.M.A. durability schedule.

(2) Thermal stability to withstand exposure to temperatures up to 1800°F.
(3) Long catalyst life (equivalent to 50 000 miles).
(4) Low density.
(5) Low heat capacity.
(6) Sufficient hardness and low attrition.
(7) Reasonable cost.

The effects of ageing on the activity for oxidation of CO and HC are given in Table 139.1 for platinum on alumina and copper chromite on alumina catalysts. The catalysts were aged in an engine exhaust for periods equivalent to 5000 and 10 000 miles. Samples of the catalysts were removed from the ageing unit at these equivalent mileages, and their activity for oxidation of CO and HC was determined in a synthetic gas mixture. The platinum catalyst is more stable in both unleaded and low-leaded fuels than the copper chromite catalyst at 1400–1600°F. This behaviour is typical of these two types of catalysts. The platinum catalyst has sufficient activity and stability to meet the original I.I.E.C. emission goals established in 1967, but the physical durability (attrition) has not yet been fully established.

Several catalysts have been developed which have significant activity for oxidation of CO and HC at low temperatures (200–400°F). These catalysts have maintained most of their activities in ageing tests up to 10 000 miles. These catalysts have been prepared on both low-density spheres and commercial alumina spheres. The combination of low-temperature activity and low density should provide a factory catalyst for cold start-up.

The initial screening of these catalysts is conducted by heating the catalyst sample in a furnace and determining the activity for oxidation of CO and HC at several temperatures. Some comparable data for copper chromite on alumina, platinum on alumina, and a fast warm-up catalyst are given in Table 139.2. The fast warm-up catalyst exhibits activity for oxidation of CO and HC at lower temperatures than the other two catalysts. The particular

Table 139.2. Cold start-up performance

Furnace temp., °F	Copper chromite–alumina catalyst			Platinum–alumina catalyst			Fast warm-up catalyst		
	Catalyst temp., °F	Converter, %		Catalyst temp., °F	Converter, %		Catalyst temp., °F	Converter, %	
		CO	HC		CO	HC		CO	HC
150	—	—	—	—	—	—	245	43	0
200	—	—	—	190	0	1	340	62	1
250	—	—	—	232	0	1	405	74	5
300	310	17	0	285	0	1	475	82	19
350	375	23	0	385	0	1	532	83	43
400	595	81	31	—	—	—	578	92	63
450	633	97	85	—	—	—	628	96	80

platinum catalyst in Table 139.2 is relatively inactive under these conditions.

The mathematical model for the HC–CO converter, which is described in a later section of this paper, played an important role in catalyst development. Based on catalyst activities and properties determined in laboratory tests, the performance of a particular catalyst in a complete converter system could readily be predicted with the model. The effects of different inlet conditions, converter geometries, and other parameters on catalyst performance were studied with the model.

Considerable attention was devoted to catalyst attrition. In addition to increasing catalyst strength, the actual mechanical mechanisms by which attrition occurred in converters were established, and some converter configurations were developed to minimize attrition. Various types of monolith catalysts (platinum and others) are being investigated, which have good activity and relatively low attrition.

DEVELOPMENT OF CATALYTIC CONVERTERS FOR NO_x REDUCTION

The concentrations of NO_x (NO_2+NO) in an engine exhaust can be reduced to some extent by spark retard, fuel enrichment, and E.G.R. However, a NO_x reduction catalyst may be required to meet the new emission goals. Since the literature on NO_x decomposition and reduction is somewhat limited, an exploratory research project has been conducted within the I.I.E.C. Most of the exploratory work has been carried out by G. H. Meguerian and co-workers at American Oil Co. with subsequent vehicle testing by Ford and Mitsubishi. Some of their results were recently reported in considerable detail by Meguerian and Lang (7), and Kaneko et al. (12).

Over 600 catalysts were prepared and evaluated. A pulse-flame reactor was used to generate NO_x, CO, and O_2 levels comparable to those in engine exhausts. The gaseous mixture was then passed over the NO_x catalyst being tested. Promising catalysts were subjected to an engine exhaust for 50–100 h; those which retained 80 per cent effectiveness were usually tested in actual vehicle systems. Radial flow converters (235 in³) were used to test catalysts in exhausts from Ford V-8 engines mounted on dynamometers. Similar converters were used to test catalysts on Ford vehicles which were equipped to measure exhaust gas temperature at inlet and outlet of converter, catalyst mid-bed temperature, engine manifold vacuum, and concentrations of HC, CO, NO_x, and O_2 at inlet and outlet of the converter.

No satisfactory catalyst to decompose NO_x in an oxidizing atmosphere at the high space velocities associated with a lean calibrated exhaust system has yet been formulated. Catalysts which promote the reduction of NO_x by H, CO, or HC present in the unleaded gasoline exhaust of a rich calibrated engine have been developed. Four types of catalysts were investigated.

(1) *Precious metal catalysts*—Platinum, palladium, rhodium, and rhenium are active for NO_x reduction.

(2) *Copper oxide catalysts*—Copper oxide on alumina or silica is active for NO_x reduction. The addition of chromia increases the activity, which is strongly dependent on the ratio of copper oxide to chromia. These catalysts deactivate quite rapidly in an engine exhaust environment. Attempts to stabilize copper oxide–chromia catalysts with promoters were unsuccessful; vanadia decreased the rate of deactivation under certain circumstances.

(3) *Iron oxide catalysts*—Iron oxide–chromia catalysts are active for NO_x reduction and their activity and stability are influenced by composition, support, and chemical stabilizers.

(4) *Metallic alloys*—Oxidized stainless steel reduces NO_x above 850°F. Many metals and alloys have been subsequently investigated as potential catalysts. They must usually be activated by oxidation at high temperatures or treatment with oxidizing agents such as oxalic acid. Copper, Monel, and stainless steel are quite active; copper-plated stainless steel has particularly good stability and appears promising.

Ammonia is a product in the reduction of NO_x. Under most conditions, ammonia has been observed in the outlet gases of NO_x converters. Experimental studies have also shown that the ammonia formed is oxidized to NO_x in HC–CO converters. Some catalysts have been developed which minimize the production of ammonia in NO_x

converters, although durability remains to be established.

The performance of NO_x catalysts has been studied in conjunction with HC–CO converters. The development of an improved dual-bed catalytic converter for both catalysts has increased the efficiency of NO_x reduction catalysts in emission-control systems. However, catalyst attrition has continued to be a problem, and further work will be required to develop a catalyst with sufficient activity, stability, and low attrition to meet the desired emission goals for 50 000 miles.

MATHEMATICAL MODEL FOR HC–CO CATALYTIC CONVERTER SYSTEMS

Mathematical models have been developed for many different chemical and petroleum-refining processes. They have been proved useful for design, optimization, and control of commercial processing units and have provided guidance in process and catalyst development. Two types of models have evolved. The empirical correlation model is usually based on relationships which have been derived by graphical or statistical means from operational data. In contrast, fundamental predictive models utilize engineering and other technical relationships to account for the physical and chemical changes which occur in the system being modelled. The relationships are often derived from the basic principles of chemical kinetics, thermodynamics, fluid mechanics, and related areas. Such models can be extrapolated over extremely wide ranges of operating variables. Empirical correlation models are only reliable for interpolative predictions.

A predictive model has been developed for CO and HC catalytic converter systems. For transition metal oxides and many other oxidation catalysts, the oxidation of CO is first order and the integrated rate equation in an integral flow reactor can be represented as:

$$k_{CO} = -\frac{F_C}{V_C} \ln(1 - X_{CO})$$

where k_{CO} is the apparent first-order rate constant, F_C the gas flow rate, V_C the catalyst volume, and X_{CO} the fractional conversion of CO.

Based on experimental results for catalysts such as copper chromite, CO oxidation is essentially independent of the concentrations of HC and O (for $O_2 > 2$ per cent). Lower O concentrations do effect CO oxidation.

Since the oxidation rates of HC are strongly dependent on their molecular structures, two 'kinetic lumps' were used to describe the oxidation of HC. They were labelled 'fast oxidizing' and 'slow oxidizing'. Typical of the HC in the latter group is methane, which does not take part in photochemical smog reactions. The fraction of HC oxidized in an integral flow reactor is:

$$X_{HC} = 0.8(1 - e^{-V_C/F_C k_I}) + 0.2(1 - e^{-V_C/F_C k_{II}})$$

where k_I is the apparent rate constant for 'fast oxidizing' HC and k_{II} the apparent rate constant for 'slow oxidizing' HC. The relative concentrations of 'slow oxidizing' HC in many engine exhausts is about 20 per cent.

The values of the rate constants for a particular catalyst were determined in an isothermal integral flow reactor with both synthetic mixtures of gases and engine exhaust gases. Activation energies and frequency factors were calculated for each rate constant from Arrhenius plots (ln k versus $1/T$, where T is the absolute temperature).

The above kinetic equations were combined with mass and energy balances, converter configurations, catalyst properties, fluid flow patterns, and other factors to give a completely predictive mathematical model of CO and HC catalytic converter systems.

For certain catalysts, the oxidations of CO and HC are not first order, and complex rate equations are required to describe adequately the oxidation kinetics. In addition, the oxidation rates are inhibited by CO, HC, and NO.

The input information required to the model includes converter inlet conditions, catalyst properties, exhaust gas properties, and other miscellaneous data depending on the specific system being considered. The model has been successfully applied to widely different test cycles, catalysts, and flow patterns (axial and radial).

The predictions of the model have been compared successfully with many experimental results. An example, previously discussed by Kuo et al. (5), is shown in Figs 139.1 and 139.2. These data compare the predicted and experimental performances of a HC–CO converter (392-in^3 radial flow) located in the standard muffler position. The vehicle had a 289-in^3 engine with a Thermactor (exhaust air injection) exhaust-control system and operated on non-leaded fuel. The vehicle was run on a chassis dynamometer, and the results in Figs 139.1 and 139.2 represent the performance during F.T.P. cycles after 12 000 miles of ageing. The catalyst was copper chromite on alumina, and the intrinsic activities for CO and HC oxidations at 12 500 miles were 28.5 and 39.5 per cent, respectively, of the initial activities.

The experimental inlet and outlet concentrations of CO together with the predicted outlet CO are given in Fig. 139.1. The agreement between the predicted and experimental data is very good. The corresponding HC results are plotted in Fig. 139.2. The slight difference between the predicted and experimental concentrations of outlet HC before 100 s is probably due to adsorption of some HC on the catalyst. The model predictions of the mid-bed temperatures of gas and catalyst also compared favourably with experimental results.

The advent of the Federal C.V.S. test procedure (U.S. Federal Register, 10th November 1970) with its more stringent requirements provided an opportunity to utilize the I.I.E.C. mathematical model for some scoping studies. The model was used to identify areas where changes might be required in systems hardware or engine calibration for a concept car that was tuned to meet the former I.I.E.C. emission goals with the previous F.T.P. test cycle. Typical inlet data to a catalytic converter are given in Figs 139.3 and 139.4. The solid lines are the inlet concentrations of CO and HC, respectively; the points are the predicted outlet concentrations. The inlet and predicted outlet

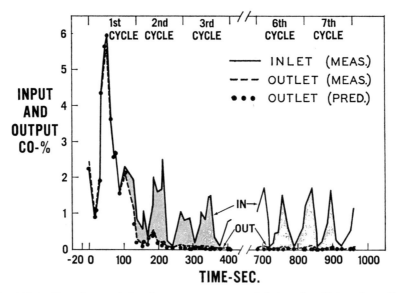

Fig. 139.1. Experimental simultaneous converter inlet and outlet and predicted outlet CO (F.T.P. cycle)

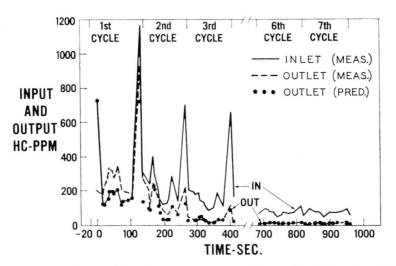

Fig. 139.2. Experimental simultaneous converter inlet and outlet and predicted outlet HC (F.T.P. cycle)

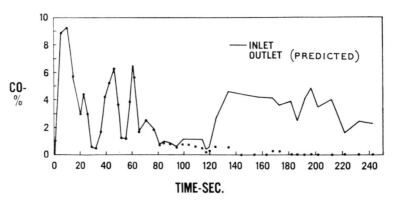

Fig. 139.3. C.V.S. CO concentration

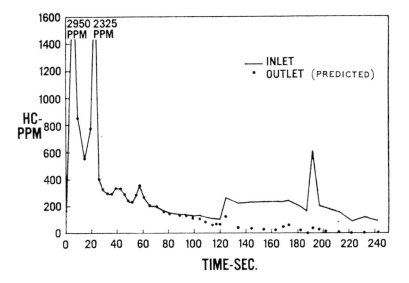

Fig. 139.4. C.V.S. HC concentration

concentrations of CO and HC during the first 100 s are approximately equal. The inlet and predicted temperatures (mid-bed) are shown in Fig. 139.5. These figures reflect the inactivity of this catalyst system at low temperatures. These model predictions prompted the development of catalysts with significant activities at temperatures between 200–400°F. After 100 s the predicted outlet concentrations become negligible. The rather high CO and HC peaks during the early portion of the test strongly contribute to the average emissions.

The total inlet concentrations for CO and HC in this test are 62·2 and 4·45 g/mile, respectively. The model predictions indicate that to meet the 1975 U.S. emission standards of 4·5 g CO/mile and 0·46 g HC/mile, the converter must be 100 per cent efficient for CO and HC conversion after 20 and 5 s, respectively.

The model was used to explore improvements which might satisfy the 1975 U.S. emission goals. The modifications included decreasing engine emissions during the initial portion of the test cycle, more rapidly heating the catalyst, increasing catalyst activity, preheating catalyst bed, and altering converter configuration. Some effects of altering inlet gas composition and increasing warm-up rate on converter performance are given, respectively, in Tables 139.3 and 139.4. Emission levels for the converter are strongly dependent on inlet gas composition and the catalyst warm-up rate. The predicted effects of catalyst activity on CO and HC emissions are illustrated in Fig. 139.6. The activity of the catalyst would have to be increased by a large factor to meet the 1975 U.S. emission goals. Engines have been tuned to reduce considerably emissions during the cold-start period, taking some of the burden off the catalyst when still cold.

Thermal reactor and exhaust pipe models have also been developed, but the details will not be discussed in

Table 139.3. Predicted effect of inlet gas composition on converter performance

Inlet composition	C.V.S. emissions, g/mile	
	CO	HC
Base case	9·2	1·5
Decrease two initial HC peaks by 50 per cent	9·2	1·1
Maximum CO = 3 per cent Maximum HC = 400 p.p.m.	5·8	0·7

Table 139.4. Predicted effect of warm-up rate on converter performance

Warm-up rate	C.V.S. emissions, g/mile	
	CO	HC
Base case	9·2	1·5
Inlet warm-up rate = 24 degF/s	5·9	1·1
Instant warm-up to 1000°F	3·4	0·8

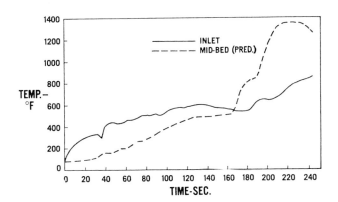

Fig. 139.5. C.V.S. temperature for fresh catalyst

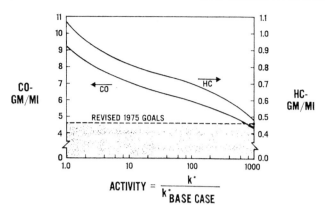

Fig. 139.6. Effect of catalyst activity on CO predicted emissions for C.V.S.

this paper. Parts of the model of HC–CO catalytic converters will be applicable to NO_x catalytic converters; detailed kinetic studies of the NO_x reactions must be obtained before a predictive NO_x model can be developed.

THERMAL REACTORS FOR HC–CO CONTROL

The thermal reactor is a relatively large volume chamber which replaces the conventional exhaust manifold and provides the temperature and residence time needed, together with added air, for virtually complete combustion of exhaust HC and CO.

In designing a reactor, provision must be made for several functions: (1) adding the air needed to burn completely the residual HC and CO, (2) mixing the added air with the exhaust gas, (3) retaining heat so that the gas temperature is high enough for rapid combustion, and (4) containing the gas mixture until combustion is virtually complete. These requirements are met in the design illustrated in Fig. 139.7, which has been evolved from tests with many design configurations. Air is added to the exhaust gas at the exhaust ports, where mixing begins. Further mixing occurs in the large central core of the reactor. The core also acts as a radiation shield to conserve heat in the gas mixture and as a baffle to lengthen the flow path through the reactor and ensure sufficient residence time for combustion.

The material used in the fabrication of a thermal reactor, particularly the internal sheet metal, is a very critical factor in determining the service life of the reactor. Reactor material must retain its strength and resist corrosion by exhaust gases during operation at temperatures which are normally in the range of 1600–1800°F but may occasionally reach as high as 2000°F under maximum output conditions or during an engine malfunction.

Candidate materials have been evaluated for resistance to oxidative corrosion by exposing them to exhaust gas in a reactor environment. Tests were conducted with alloy specimens placed in test fixtures, mounted on a Ford V-8 engine. The engine was operated in such a way that exhaust gas temperatures in the test fixtures cycled between nominal extremes of 800°F and 1700–2000°F. Details of the test procedure and specimen evaluation have been published elsewhere ([13]).

Effects of potentially corrosive fuel additives—lead, halogens, phosphorus—were studied first using a chrome steel containing 12% Cr and 3% Al. This alloy was chosen with the expectation of rapid corrosion in order to differentiate fuel component effects in relatively short duration tests. Results are illustrated in Fig. 139.8, where corrosion rate (expressed as average thickness loss per 50 h exposure) is plotted against average exposure temperature in the

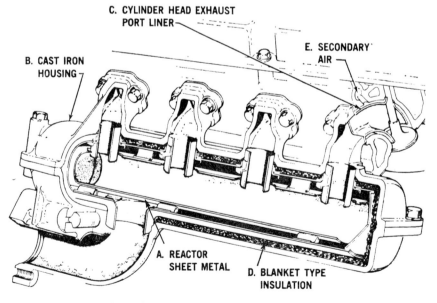

Fig. 139.7. Typical I.I.E.C. thermal reactor

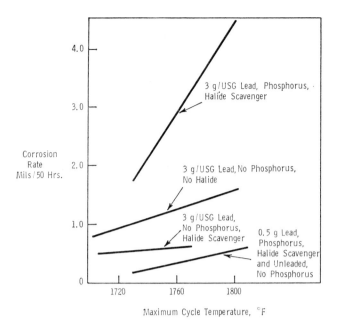

Fig. 139.8. Effect of fuel on corrosion of 12% Cr, 3% Al alloy in thermal reactor environment

Table 139.5. Thickness loss

Alloy composition, % wt (major components)		Corrosion rate estimated from weight loss, mils/50 h, at 1850–1900°F	
Chromium	Nickel	Average	Range
25 (AISI Type 310)	20	0·49	0·32–0·62
23	60	0·35	0·18–0·46
21	25	0·16	0·04–0·24
22	49	0·24	0·09–0·34

high-temperature phase of the test cycle. As expected, fuel effects were clearly differentiated. At 1800°F the corrosion rate with fuel containing 3 g Pb/U.S.G. together with Motor Mix halide scavenger and 0·2 theories of phosphorus was about 10 times the rate with fuel containing none of these additives. It should be noted that a fuel containing 3 g Pb/U.S.G. but no halide or phosphorus gave a corrosion rate midway between that of the leaded fuel containing both halide and phosphorus and one with only halide. With 0·5 g Pb/U.S.G., containing equivalent quantities of halide scavenger and phosphorus, the corrosion rates were practically the same as with no lead.

At about 1700°F the differences among the fuels tested were small, and the corrosion rates were low enough to suggest that this inexpensive 12% Cr, 3% Al alloy might be a candidate for reactor construction, especially if it could be protected against corrosion at higher temperatures by means of a coating. Hence, the second phase of the materials study concentrated on oxidation-resistant metallic coatings.

In exhaust from fuel containing 3 g Pb/U.S.G., plus halide scavenger and phosphorus, the 12% Cr, 3% Al alloy coated with a flame-sprayed Ni–Cr–Al coating was found to corrode at a lower rate than the uncoated base alloy in unleaded exhaust. Such specimens have survived 500 h test exposure in exhaust from leaded fuel with no significant visible deterioration; and another alloy (18% Cr, 2% Al, 1% Si) with the same coating has survived 1000 h exposure, also with no visible deterioration of the coating. Based on these laboratory tests, it is projected that, from the oxidative corrosion standpoint, a useful service life of at least 50 000 miles is attainable. From the manufacturing standpoint, however, these coated materials appear impractical because the coatings are prone to the formation of cracks during fabrication. At present, the most promising candidates are nickel-containing alloys.

Several alloys containing 20–60% Ni were found to have quite low corrosion rates in exhaust from low-lead fuel (0·5 g/U.S.G.) at 1850–1900°F. Data are summarized in Table 139.5 in terms of a thickness loss which is calculated from a weight change measured after descaling the specimen.

Above 1900°F, corrosion rates of the 20% Ni and 25% Ni alloys increased appreciably while that of the 60% Ni alloy remained low. With lead-free fuel, corrosion rates appeared slightly lower, but probably not significantly so in view of the inherent scatter in the test data.

The 49% Ni alloy (Hastalloy X) is included as a point of reference to reactor durability in test vehicle service where this material has been reported (**14**) to have an excellent service life. The other three alloys are appreciably less costly, and a pair of reactors made of the 60% Ni alloy has now been proved in more than 70 000 miles of vehicle service with gasoline containing 0·5 g Pb/U.S.G.

CONTROL OF NO_x BY EXHAUST GAS RECIRCULATION

The E.G.R. system functions by recirculating a portion of exhaust gas to the induction system, where it acts as an inert diluent for the fresh charge of fuel and air. Consequently, the peak combustion temperature and the amount of NO_x formed are reduced. In order to function with optimum effectiveness, the recirculation system must provide for control of the flow or recycled exhaust, and for cooling the exhaust to a temperature at which the desired NO_x reduction can be achieved. In one example, the exhaust is tapped just ahead of the muffler to achieve the desired temperature; and flow is controlled by means of a vacuum-operated on–off valve. The recirculated exhaust is introduced into the induction system via a spacer between the carburettor and the manifold. The same spacer is also used for the P.C.V. entry.

Many different E.G.R. system designs have been described in the literature (see reference (**10**) for literature references). A significant effort of the I.I.E.C. programme has been applied to studying potentially deleterious effects of E.G.R. on the engine system. Using a modification of the Ford MS Sequence VB test procedure the effect of recirculating 15 per cent of the exhaust gas was investigated. With the MS08 fuel (0·08% S) specified for the VB test, a small depreciation in overall sludge and varnish ratings was found. However, within the precision of the standard VB test, the observed differences are not considered significant. Similar results were obtained for both a premium quality 10W–40 oil and a borderline VB-quality reference oil. With a leaded, detergent–dispersant premium quality fuel, somewhat improved varnish ratings were found. An unleaded simulated MS08 fuel (0·11% S) in combination with the premium quality oil produced ratings not significantly different from the leaded MS08 base rating (**3**). These tests indicate that the E.G.R. current quality lubricants and leaded fuels should maintain cleanliness in those parts of the engine exposed to lubricating oil and normally rated in the sequence VB procedure. In all the tests with E.G.R., deposits formed at the point of entry of the recycled exhaust into the intake system, and exhaust back pressure had to be progressively increased to maintain 15 per cent recycle during the 192-h test. When the exhaust was recycled to the carburettor, deposit build-up interfered with throttle closing at idle. In these tests, with entry via the spacer, neither the quantity of deposit formed nor the extent of flow restriction was appreciably different for different fuel types such as MS08, leaded premium or unleaded fuel. Vehicle test work within I.I.E.C., as well as work by others (**15**), indicates that it may be possible to control these deposits by trapping devices in the E.G.R. system. However, other deposit problems may develop at different locations than reported here.

CONCEPT EMISSION VEHICLES

The catalyst technology and chemical, mechanical, and metallurgical engineering technology described in the preceding sections have been combined with automobile manufacturer engineering experience in concept emission control systems designed to meet all the I.I.E.C. emission goals. These are:

Concept A	Thermal reactor plus E.G.R.
Concept B	HC–CO catalyst converter plus E.G.R.
Concept C	HC–CO catalytic converter, plus NO_x catalytic converter plus E.G.R.
Concept A–B	Combinations of Concept A with Concept B.

The main features and performance at *low mileage* of these concept emission systems as applied to standard-size U.S. cars (300–400-in^3 displacement engines), and cars more typical of those found outside the U.S. (1200–1600-cm^3 engine displacement), are described below (**10**) (**12**).

CONCEPT A

This concept, illustrated in Fig. 139.9, uses a thermal reactor for HC–CO control and E.G.R. with enriched carburation for NO_x control. At the present development level, the hardware and fuel specifications being used on this package, as applied to a U.S. standard-size (4000-lb) car, include:

(1) Two 97-in^3 I.I.E.C. reactors.
(2) Reactor inlet and outlet sheet-metal liners.
(3) Cylinder heads modified with exhaust port liners.
(4) One engine-driven secondary air pump (16 in^3).
(5) Below-the-throttle E.G.R. system (E.G.R. pick-up taken before muffler).
(6) Production-type carburettor with richer calibration.

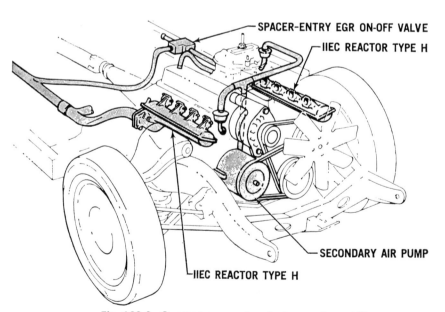

Fig. 139.9. Current concept emission package 'A'

Table 139.6. Typical emission performance

Exhaust	Emissions, g/mile		
	4000-lb car (10)		2000-lb car (12)
	F.T.P.	C.V.S.	F.T.P.
HC	0·25	0·30	0·3 –0·36
CO	5·4	9·0	6·1 –6·7
NO_x	0·62	1·4	0·40–0·65
Economy loss from base-line vehicle during emission test	24%	20%	17–25%

(7) Production distributor with modified curve (higher than production).

(8) Provision for more spark retard during the warm-up (until engine-water temperature reaches 120°F).

(9) Dual-wall exhaust pipe extending from exhaust gas outlet to muffler inlet.

(10) Prototype reactor protective system to limit maximum core temperature to 1850°F.

(11) Unleaded fuel or 0·5 g Pb/U.S.G. fuel.

Typical emission performance of test vehicles equipped with the above hardware and fuel at low vehicle mileage is shown in Table 139.6.

This package meets the I.I.E.C. emission targets and represents a significant reduction over current production levels, but it does not meet 1976 proposed U.S. Federal standards. In city–suburban driving the fuel economy loss was 18–28 per cent, while performance was not appreciably degraded.

CONCEPT B

This concept, illustrated in Fig. 139.10, utilizes a catalytic converter for HC–CO control, and E.G.R. with enriched carburation for NO_x control. Other main features of this concept include:

(1) Secondary air injection into the exhaust to ensure a sufficiency of oxygen for HC–CO oxidation over the catalyst.

(2) A programmed protection system to divert the exhaust around the main catalytic converter when the temperature of the catalyst bed exceeds the maximum temperature capability of the catalyst. (Typically 1300–1400°F for transition metal oxide catalysts, higher for noble metal catalysts.)

(3) A second HC–CO catalytic converter located near the end of the tailpipe to convert HC and CO when front-mounted converter is by-passed. (This converter may not be necessary when using catalysts capable of withstanding high temperature.)

(4) Modified distributor to provide optimum timing for fast warm-up and minimum emissions from the engine.

(5) Non-leaded fuel.

Typical emission and economy results for this concept at low mileage are shown in Table 139.7.

Table 139.7. Typical emission and economy results

Exhaust	Emissions, g/mile		
	4000-lb car (10)		2000-lb car (12)
	F.T.P.	C.V.S.	F.T.P.
HC	0·20	0·80	0·59
CO	3·7	11·0	6·8
NO_x	0·79	1·30	0·68
Economy loss from base-line vehicle during emission test	12%	8%	5%

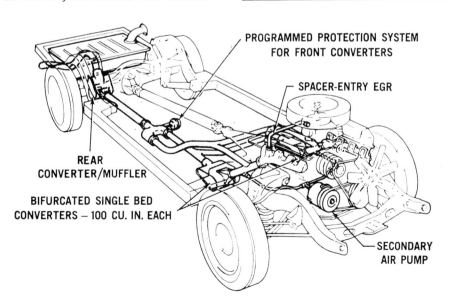

Fig. 139.10. Concept emission package 'B'

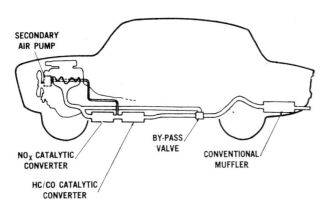

Fig. 139.11. Concept emission package 'C'

Table 139.8. Typical emission and economy data

Exhaust	Exhaust emissions, g/mile 4000-lb car (10)	
	F.T.P.	C.V.S.
HC	0·34	0·85
CO	2·6	10·0
NO$_x$	0·47	0·90
Economy loss from base-line vehicle during emission test	5%	5%

Like Concept A, Concept B meets the original I.I.E.C. goals but not the proposed U.S. 1976 emission standards.

CONCEPT C

Concept C uses a NO$_x$ reduction catalyst along with E.G.R. and enriched carburation to achieve NO$_x$ control, and an HC–CO catalyst to control HC and CO. The amount of E.G.R. and enriched carburation is less than with concepts A and B, which facilitates driveability and improves fuel economy. In addition, the NO$_x$ converter, located upstream from the HC–CO converter, lessens the thermal load on the HC–CO catalyst. Fig. 139.11 shows a schematic representation of this system.

Other than the inclusion of the NO$_x$ catalytic converter, the main features of this concept are similar to concept B. However, the secondary air is injected between the NO$_x$ catalytic converter and the HC–CO catalytic converter. The former, of course, requires a reducing atmosphere (oxygen deficient) to function.

Typical emission and economy data for Concept C at low mileage are shown in Table 139.8. Obviously, this more complex concept is attractive from a fuel economy viewpoint. However, it does not meet the proposed 1976 U.S. standards.

CONCEPT A-B

A 'maximum effort' system was devised in an attempt to meet the extremely low emission standards proposed in the 26th February 1971 and 25th May 1971 U.S. Federal Register. This concept, illustrated in Fig. 139.12, combines the main features of Concepts A and B.

As noted previously, the main difficulties are: (a) achieving rapid warm-up to minimize HC and CO emission during the first 2 min of the C.V.S. test, and (b) achieving the anticipated NO$_x$ standard. The thermal reactors serve as preheaters for the HC–CO catalytic converters which are located very close to the thermal reactor outlet, and aid the catalysts in reducing HC–CO levels. Noble metal HC–CO catalysts are employed; such catalysts are at the present state of the art more capable of withstanding exposure to high exhaust temperature.

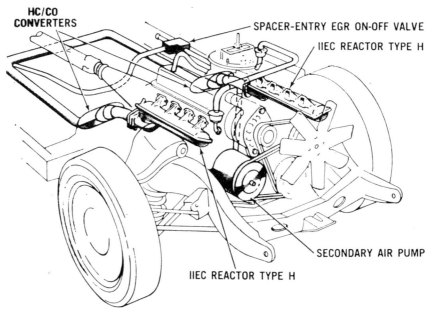

Fig. 139.12. Combined concept emission package reactor–E.G.R.–catalyst system, major hardware components

Table 139.9. Average emission results

Exhaust	Exhaust emissions, g/mile
	C.V.S.
HC	0·28
CO	3·4
NO$_x$	0·76
Fuel economy loss from baseline (city–suburban driving schedule)	27%

Maximum E.G.R. is employed, together with carburettor enrichment, to prevent NO$_x$ formation in combustion.

Average emission results for Concept A–B at very low mileage (less than 500 miles) are shown in Table 139.9. This concept meets the proposed 1975 standards for HC, CO, and NO$_x$; it does not meet the proposed 1976 NO$_x$ standards. At this level of control the only positive feature is the emission level achieved. In addition to the severe fuel economy loss, the high exhaust gas temperatures experienced caused severe functional problems.

The major factor influencing fuel economy is the control of NO$_x$. Testing at lower NO$_x$ levels is not completed, but the estimated fuel economy penalty for the 0·4 g/mile would be even higher.

MAJOR PROBLEM AREAS

While considerable success was achieved in developing individual components of these concept systems to meet the emission goals of the I.I.E.C. programme, many major problems remain, particularly with the Concept A–B system. Some of the more important problems are outlined below.

Durability

The emission results reported here were achieved at low mileage. While individual system components have in many instances been subjected to durability testing on the dynamometer or in vehicles, complete concepts have not. In the final analysis a true measure of durability can only be made by testing the complete system in typical customer service, since the major components necessarily interact with each other. Both test-track mileage and customer-type mileage are now being accumulated on the more promising versions of the various concept systems.

Fuel economy

It is apparent that there is a sizeable fuel economy penalty with any of the concept packages. Most of this economy penalty is caused by NO$_x$ control—the requirement to use richer than stoichiometric mixture ratios to minimize NO formation during combustion, and/or provide a reducing atmosphere for catalytic reduction, and to achieve reasonable driveability with E.G.R. This is particularly evident in Concept A–B, where a 25 per cent (approximate) mileage decrease is encountered at a control level of 0·7 g/mile NO$_x$ compared to a 10 per cent decrease at a 1·3 NO$_x$ control level. In an effort to minimize this penalty, further work is being carried out.

Driveability and performance

Carburettor and fuel injection calibrations and ignition timing schedules must be altered in an attempt to 'tailor' engine output emissions to the capability of the control system. This has upset driveability (hesitation on acceleration, etc.) and performance. Of necessity, both driveability and performance have been compromised to achieve minimum emissions.

Packaging and effects on other vehicle components

A number of the components require alteration of the vehicle structure to maintain road clearance, or simply to fit within the vehicle, or as a result of high-temperature problems. For example, Concept A–B produces higher-than-normal exhaust gas temperatures; this affects such components as engine transmission and braking fluids, the floor of the vehicle, and the petrol tank.

Production

The vehicle systems in this programme represent one-of-a-kind, 'best effort' systems. Much work lies ahead before any of these systems can be engineered for mass production, taking into consideration such factors as adaptation to various car makes and models, cost optimization, fabrication, and manufacturing procedures. There is a great difference between making one, two, or even 10 of a kind and making 10 000 to 100 000 or more, whilst at the same time meeting very strict emission standards.

CONCLUSION

Basic mechanical engineering, chemical engineering, mathematics, and chemical technology have all been employed by a large international group of companies, with widely diverse backgrounds, in a major effort to solve a very practical problem—air pollution from the automobile. Several solutions have been developed which show promise that stringent automobile emission requirements can be met; and research is continuing in an effort to develop and provide the system which will offer to the motoring public the best possible choice in reliability, economy, performance, and cost.

APPENDIX 139.1
REFERENCES

(1) FEDERAL REGISTER, 4th June 1968 Volume (Pt II).
(2) TAYLOR, R. E. and CAMPAU, R. M. 'The I.I.E.C.—a co-operative research program for automotive emission control', Preprint No. 17-69, Midyear Meeting, Div. Refining, Am. Petrol. Inst., Chicago, 1969 (May).
(3) OSTERHOUT, D. P., JAGEL, K. I. and KOEHL, W. J. 'The I.I.E.C. program—a progress report', presented at A.S.T.M. meeting, Toronto, 1970 (June).
(4) MEISEL, S. L. 'Exhaust emissions and control', presented at symposium of Washington Acad. Sci. and Am. Ordn. Ass., Washington, D.C., 1971 (January).

(5) KUO, J. C., LASSEN, H. G. and MORGAN, C. R. 'Mathematical modeling of catalytic converter system', S.A.E. Paper No. 710289, 1971 (January).

(6) JAGEL, K. I. and DWYER, F. G. 'HC/CO catalysts for vehicle exhaust emission control', S.A.E. Paper No. 710290, 1971 (January).

(7) MEGUERIAN, G. H. and LANG, C. R. 'NO_x reduction catalysts for vehicle emission control', S.A.E. Paper No. 710291, 1971 (January).

(8) HANCOCK, E. E., CAMPAU, R. M. and CONNOLLY, R. 'Catalytic converter vehicle system performance: rapid versus customer mileage', S.A.E. Paper No. 710292, 1971 (January).

(9) JAIMEE, A., ROZMANITH, A. I., SCHNEIDER, D. E. and SJOBERT, J. W. 'Thermal reactor design, development and performance', S.A.E. Paper No. 710293, 1971 (January).

(10) CAMPAU, R. M. 'Low emission concept vehicles', S.A.E. Paper No. 710294, 1971 (January).

(11) ROZMANITH, A. I., MIXON, L. W. and WOTRING, W. T. 'Effect of fuel and lubricant composition on exhaust emissions', S.A.E. Paper No. 710295, 1971 (January).

(12) KANEKO, Y., JURODA, H. and TANAKA, K. 'Small engine—concept emission vehicles', S.A.E. Paper No. 710296, 1971 (January).

(13) SJOBERG, J. W. and ROZMANITH, A. I. 'Corrosion resistant material for automotive exhaust thermal reactors', Nat. Ass. Corrosion Engrs Paper No. 80, 1971 (March).

(14) CANTWELL, E. N., ROSENLUND, I. T., BARTH, W. J., KINNEAN, F. L. and ROSS, S. W. 'A progress report on the development of exhaust manifold reactors', S.A.E. Paper No. 690139, 1969 (January).

(15) MIKITA, J. J. and CANTWELL, E. N. *Exhaust manifold reactors—a solution to the automotive emission problem* 1970 (April) (Natn Petrol. Refiners Ass., San Antonio, Texas).

C140/71 THE MECHANISMS OF SOOT RELEASE FROM COMBUSTION OF HYDROCARBON FUELS WITH PARTICULAR REFERENCE TO THE DIESEL ENGINE

D. BROOME*   I. M. KHAN†

Information on the formation and combustion of soot in laboratory hydrocarbon flames is reviewed. Chemical and physical factors influencing soot release are discussed, including fuel composition and additives, and effects of temperature, pressure, and turbulence. An account of the mechanisms of carbon formation follows. The paper then considers how this fundamental information on soot release relates to the diesel combustion process. Soot formation in various types of diesel engines, augmented with the work done at the authors' companies, is also discussed.

INTRODUCTION

IN ORDER to improve knowledge and understanding of the mechanisms of soot formation and release in diesel engines, the authors' companies commissioned a joint study of the whole problem, which included as its first part a review of published work on fundamentals as revealed mostly by laboratory studies. In the first part of this present paper the results of the review are summarized, while in the second part the authors' views are given on the application of the fundamental data to the diesel problem, augmented by their own work. The references on laboratory flames given in Appendix 140.1 do not comprise a complete list, and are only intended to provide an introduction.

SOOT FORMATION AND COMBUSTION AS REVEALED BY LABORATORY STUDIES

Experimental techniques

A very wide and sophisticated range of experimental techniques has been used, and these are described in the literature [e.g. (1) (2)‡]. In brief, it is to be noted that studies of conventional flames of both premixed and diffusion types are invariably carried out under steady-state conditions, the burner constructions and physical conditions being adjusted to accentuate the desired characteristics. Intermittent state experiments at very high temperatures have been carried out in chemical shock tubes and by flash photolysis (3) (4). In many cases only relative measurements of soot formation have been made, although recently more absolute methods have been favoured (5)–(7).

Nature of carbon formed in flames

Carbon in a wide range of forms may result from combustion of fossil fuels; the following main forms may be noted, although in some particular cases the distinctions are by no means clear.

Vitreous carbon

A hard, shiny black deposit formed on hot surfaces—except when detached by mechanical or thermal action it is not present in the gaseous products of combustion.

Cenospheres

Hollow or porous carbon particles of very large relative size, up to 1 mm, 10^7 Å, formed by cracking (pyrolysis) in the liquid phase. These are quite distinct from normal soot particles, and are only formed in large industrial burners.

Tarry wet gas carbon

Typically formed by longer exposure of hydrocarbons to relatively low flame temperatures, due to cracking in the vapour phase. Black in appearance and characterized by the presence of heavy hydrocarbon molecules, which may be removed by solvent extraction. As the concentration of these latter falls, the distinction between the tarry wet and dry gas carbon or soot is not at all clearly defined.

The MS. of this paper was received at the Institution on 9th June 1971 and accepted for publication on 9th August 1971. 22
* Consultant, Ricardo & Co., Engineers (1927) Ltd, Shoreham, Sussex.
† Principal Research Engineer, C.A.V. Ltd, Research Dept, Larden Road, Acton, London, W.8.
‡ References are given in Appendix 140.1.

Dry gas carbon or soot

Formed in the vapour phase at higher temperatures, grey rather than black in colour. Dry gas carbon is the concern of this paper, since none of the other types is emitted by a diesel engine under normal operating conditions.

Composition and physical properties of soot

X-ray diffraction and electron microscopy have shown that the crystallites, being the basic element of soot, may be embedded in a mass of polymeric material when the gross product is really tarry wet gas carbon. With dry gas carbon as normally found in diesel exhausts, the proportion of polymeric material is very small.

Soot contains typically 1–3 per cent by mass (i.e. 10–30 per cent on an atomic basis) of hydrogen, which may be chemically or physically bound depending on the conditions of formation.

The following is a summary of the data on the structure, size, and composition of soot particles from Schalla and McDonald (8), and Sweitzer and Heller (9).

The platelet

A hexagonal array of carbon atoms, containing typically about 90 arranged in 35 hexagonal groups. Molecular weight is about 1000.

The crystallite

A stack of platelets, normally two to five, but up to 20 have been found. The crystallite diameters are 17–30 Å; height depends on the number of platelets, but is 12 Å for a three-platelet crystallite. The mean platelet spacing is 3·55 Å (cf. that of graphite, 3·35 Å).

The particle

An agglomerate of crystallites, randomly packed but orientated with their planes generally parallel to the particle surface. Particle diameters of 20–6000 Å are possible, but 50–500 Å are more typical. For a particle of about 200 Å diameter the number of crystallites is about 1500, and the molecular weight is about 5×10^6.

Effect of chemical factors on soot release

Hydrocarbon type

This has been studied by a great many workers [e.g. (1) (10) (8)] and certain general tendencies have clearly emerged. Thus, pure hydrocarbon fuels may be classified as follows in increasing order of soot formation for a given number of carbon atoms: normal paraffins, iso-paraffins, cyclo-paraffins, olefins, cyclo-olefins, di-olefins, and aromatics. However, exceptions to this order can be found. With most fuels, increasing carbon atom number, particularly at low numbers, increases the tendency to smoke, although this is not true of cyclo-paraffins and some olefins (11). Various attempts have been made to correlate these results, e.g. based on oxygen requirements in diffusion flames, and on the C—C bond strength as defining the stability of the carbon chain (8); while general tendencies can be predicted no exact correlations have yet emerged. Trends are similar in both premixed and diffusion flames.

Additives and diluents

Additives are taken to be substances, present only in small concentrations normally in the fuel, which are active in that they affect chemical mechanisms rather than physical conditions in influencing soot release. Thus, the presence of halogen compounds can increase soot formation in diffusion and sooting premixed flames. Additions of oxides of nitrogen or compounds that readily yield them on decomposition will, however, appreciably reduce soot as well as the concentrations of poly-cyclic aromatic hydrocarbons in the burnt gases. An anomaly is sulphur trioxide, which reduces soot formation in diffusion flames but increases it greatly in premixed flames. Nickel and alkaline earth salts are very effective in reducing soot release: it would seem that electric (ionic) effects are involved, resulting in a reduction in the initial growth rate of particles (see 'Particle growth', below) and increasing their combustion. It is also to be noted that certain organic compounds, such as methanol, can as additives reduce soot formation (12).

Diluents are inactive additions that change solely physical conditions, such as gas temperature and partial pressures: in this way they can exert a marked influence on soot release and may indirectly affect the reaction route (see 'Effect of physical factors on soot release' and 'Nucleation', below).

Effect of physical factors on soot release

Equivalence ratio

The equivalence ratio (i.e. the fuel/oxidant ratio expressed relative to stoichiometric) is well known to be important in relation to soot release (6) (13), since it will control not only the balance between oxidation and the resultant pyrolytic reactions but also the temperature and partial pressure of the fuel undergoing pyrolysis. Some of these factors are discussed below. Of itself, the equivalence ratio does not have any direct effect on the reaction mechanisms or route (5).

Temperature

In premixed flames an increase in flame temperature raises markedly the equivalence ratio at which soot release first occurs and the quantity formed at any given equivalence ratio above the threshold value (6) (13). Since soot is formed only with mixtures richer than stoichiometric, only some of the hydrocarbon molecules can be involved in the primary oxidation reactions, suggesting that soot results essentially from pyrolytic reactions in the remainder, which would be expected to be markedly temperature sensitive, being controlled by chemical kinetics.

In diffusion flames, pyrolysis on the fuel-rich side of the

flame must always be of importance; and again, higher temperatures always increase the soot released. Temperature changes may exert a considerable influence on reaction routes discussed under 'Nucleation', below, and therefore by this means on soot formation.

Pressure

Only in some of the more recent work (**6**) (**14**) have pressure levels of similar order to diesel conditions been approached, most experiments being at values close to atmospheric. In general, these latter show a marked increase in soot release as pressures rise; but this cannot be maintained at higher values and, indeed, the later work indicates a levelling off.

Again, the importance of pyrolytic reactions is suggested, since pressure variations, particularly at low overall levels, will influence diffusion rates and, hence, available reaction times.

Turbulence

Turbulence is here taken to include all effects of physical mixing, micro- or macroscopic. Regrettably, very little fundamental work has been devoted to this important subject, since in most cases laminar flames have been used. However, in one case (**6**) a change from turbulent to laminar conditions in a premixed flame increases soot release by a factor of 10, almost certainly due to the increased time available for pyrolysis.

Pyrolysis of hydrocarbons

Some of the indirect evidence for the importance of pyrolytic reactions in the formation of soot has been touched on above, and this is now generally acknowledged (**5**) (**10**) (**13**).

Details of the products of pyrolysis of some of the hydrocarbons found in engine fuels will not be presented here due to lack of space, but such reactions have been extensively studied [e.g. (**2**) (**10**) (**13**)]. Suffice to say that the term pyrolysis covers the following general reaction types involving hydrocarbon molecules alone: hydrogenation; dehydrogenation; cracking, i.e. molecular splitting; polymerization, i.e. molecular addition, the product being of similar type to the original molecule; condensation, i.e. molecular addition, where very large molecules of diverse types are formed. Polymerization and condensation reactions, being generally exothermic, are favoured at lower temperatures, whereas at high temperatures molecular splitting is prevalent.

It can be appreciated that with hydrocarbons of high carbon number the products of pyrolysis can be very varied, depending on physical conditions, particularly of temperature and time. In all flames, pyrolytic reaction rates are controlled by chemical kinetics. By controlling the gas temperature the equivalence ratio will exert a marked influence on reaction mechanisms and rates.

Mechanisms of soot formation

Starting with a hydrocarbon fuel molecule with, in the diesel case, typically 12–22 carbon atoms and about twice that number of hydrogen atoms, a soot particle has finally about 10^5 carbon atoms and much fewer hydrogen atoms. In addition to physical coagulation after formation of the first particles, it is clear that chemical reactions of aggregation and dehydrogenation must occur. The problem is to define the specific reaction routes or mechanisms involved.

Essentially, the emission of a soot particle from the flame can be divided into three major phases under all normal combustion conditions nucleation or formation of precursors; growth of the nuclei into soot particles; and physical coagulation of these first particles into even bigger units. These phases will be discussed in turn.

Nucleation

Shock tube experiments (**15**) (**16**) have shown the existence of an induction period between the initiation of the reaction and the first detection of soot particles; this suggests the need for something different from the original fuel molecule before soot can be formed. In the past, much time was spent trying to identify one basic reaction mechanism involving a particular precursor; but it is increasingly evident that many routes are possible, not only dependent on different physical and chemical conditions but also at any given overall set of flame conditions. A useful diagram in (**17**) high-lighted this complexity and in particular the dependence of reaction types on physical conditions. Based on this, the authors have developed their own diagram to illustrate the overall picture, and this is given in Fig. 140.1. To be noted immediately is the great importance of temperature controlling possible reactions.

It is convenient to subdivide reaction routes in this manner. Thus, relatively low-temperature (less than about 1500°K), slow reactions are characteristic of some laboratory diffusion flames; here polymerization or even condensation reactions predominate, before dehydrogenation occurs. Under these conditions, aromatics or poly-cyclic compounds have been proposed as precursors.

Fast reactions at higher flame temperatures (say 2000–3500°K), as found in laboratory premixed flames and the turbulent diffusion flames of diesel combustion systems, involve initial decomposition or splitting of the fuel molecule. An intermediate product is often acetylene (**3**) (**18**) (**19**). Studies of concentration profiles along rich, premixed acetylene flames (**5**) (**20**) have been very instructive here, and a typical result is shown in Fig. 140.2. At first sight the maximum concentrations of the poly-acetylenes would appear to be linked to the initial appearance of soot particles; but the latter's H/C ratio is rather higher than that of the former, from which it is deduced that they are not the direct precursors. Characteristic of the soot formation zone is the presence of very heavy (150–600 atomic units, 15–20 carbon atoms), but unstable, hydrocarbon radicals, and it is suggested that at these higher temperature conditions these radicals are the

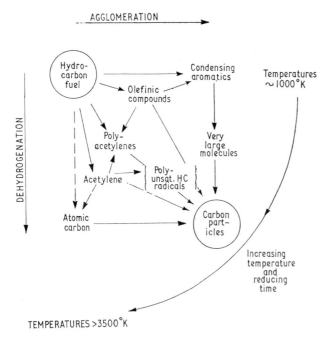

Fig. 140.1. Very simplified scheme of reactions involved in the formation of carbon from hydrocarbon fuels, showing the influence of reaction time and temperature

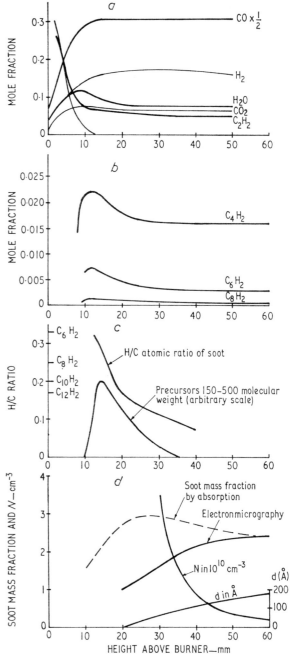

a Concentration profile in a flat acetylene–oxygen flame, $C_2H_2/O_2 = 0.95$, pressure 20 mmHg, flow velocity 50 cm/s.
b Concentration profiles of polyacetylenes in similar flame, $C_2H_2/O_2 = 1.4$, pressure 20 mmHg, flow velocity 50 cm/s.
c Variations of C/H ratio, and of soot precursors (qualitative) for similar flame.
d Concentrations of soot, mean number of particles, and their diameter in same flame as *b*.

Fig. 140.2. Formation of intermediaries, soot precursors, and growth of soot particles in a premixed flame

essential link in the chain. The aromatic compounds are thought to be by-products since they increase in concentration along with the soot; their presence is certainly not essential (20).

Other precursors have been proposed, such as hydrocarbon ions (21)–(23) and atomic carbon (1). As to the former, electric effects are known to be important, but no real link with the pyrolytic reaction chain has yet been established. On cooling from very high temperatures (very much greater than 3500°K, only realizable in plasma torches or shock tubes), condensation of carbon vapour is a possibility, but it is not considered significant under normal flame conditions.

Particle growth

Formation of the first soot particles from the precursors, and the initial aggregation of the former, are both very rapid processes occurring in the hotter zone of the flame. It is suggested (13) that the small soot particles (40 000 mass units, 40 Å diameter) exhibit radical character and grow by adding poly-acetylenes and other hydrocarbons as well as by agglomeration: this would account for the observed fall in H/C ratio of the particles along the flame (see Fig. 140.2). This phase blends into the second, slower phase where the particles have become relatively inactive, and the particle density is to an order of magnitude constant. The governing mechanisms are thought to be the decomposition of the adsorbed unsaturated hydrocarbons, requiring high activation energies. At low temperatures these remain largely unconverted, giving rise to tarry, wet gas carbon.

Coagulation

After emission from the hottest zones of the flame, the small but fully formed soot particles continue to grow by coagulation (24): this is quite distinct from initial growth, being reversible under the appropriate conditions (25) (26). Coagulation rates are directly linked to the particle number density (27). Important factors are particle size and size distribution, temperature, pressure, and turbulence all influencing the probability of collision; and the shape and structure of the particles and absorption of vapour (24) influencing their 'stickiness'. Coagulation is markedly affected by the geometry of the flame.

Kinetics of formation and growth of soot particles

From the preceding sections it may be appreciated that currently a complete quantitative treatment of the soot formation from even the simplest of hydrocarbon fuels under the most straightforward experimental conditions is not possible. All that can be said is that the gaseous phase nucleation process responsible for the first observed soot particles is kinetically a high-order process, and occurs at temperatures about 150–200 degC above those required for soot decomposition on a surface, i.e. for the growth of soot particles (28). For a given hydrocarbon the rates of particle formation and growth depend essentially on the temperature (28) (29), while for the same flame temperature the rate of particle formation is a function of the fuel type (29). A quantitative description should therefore become possible, provided (1) that the same fuel and reaction mechanisms leading to nucleation are involved, and (2) that the physical factors such as temperature and residence time are known.

Combustion and gasification of soot particles

The carbon present in the soot particles can react with available oxygen and with products of combustion wherever temperatures are high enough. The final rate of soot release from the flame is therefore the difference between the rate of formation, as described earlier, and the rate of combustion and gasification.

It is well established (30) that rates of these reactions are controlled by kinetic considerations also, since particle sizes in possible reaction zones are well below those at which diffusion into the particle becomes important: again equilibrium considerations do not apply.

Considerable work has been carried out on surface reaction rates for the carbon–oxygen combination [see, e.g., (31) and (32)] and on the combustion of soot in flames (33) (34), although the analysis is made difficult by the problem of determining the effective surface area. At all except the lowest temperatures, carbon monoxide is thought to be the primary product (32). Activation energies of the carbon–carbon dioxide and carbon–water vapour reactions are roughly similar and much higher than that of carbon–oxygen. Hence, the latter is likely to be much more significant when there is an overall excess of oxygen, as with diesel conditions, although at very high temperatures the balance differs somewhat. Again, the flame geometry is important in controlling recirculation of the soot particles and the oxygen availability at a given point.

In addition to the above, consumption of soot in the flame also occurs as a result of reaction with OH hydroxyl radicals (35), themselves generated in the flame. While the reactivity of the hydroxyl radicals is high, their concentrations are low, and direct oxidation is more important at equivalence ratios weak of stoichiometric (36).

SOOT RELEASE IN THE DIESEL ENGINE

A study of the first part of this paper will have shown how complex a problem is the formation and release of soot; therefore the difficulties in interpreting these fundamental data in relation to the special features of diesel systems will be appreciated. However, in this section some of the aspects of the matter will be considered.

It must be noted that this paper is only concerned with soot release, or in diesel parlance black smoke emissions; other forms of diesel smoke such as blue–white smoke due to emission of unburnt hydrocarbons or of lube oil (37) are not considered.

Black smoke emissions occur to some extent at all loads (overall trapped air/fuel ratios), although with some chamber forms they are at a very low level over most of the operating range. Even when subjectively very severe the carbon released as soot represents only a very small fraction of the total carbon available, e.g. a reading of 5 Bosch smoke units is equivalent to about 0·5 g of carbon as soot/m^3 of exhaust, or roughly 1 per cent of the total carbon supplied in the fuel in a typical case.

Characteristics of diesel combustion

Except with the very largest diesel engines (and then not always), diesel combustion systems are effectively of the closed type, unlike conventional laboratory burners. The closure—together with the presence of microturbulence and applied microscopic gas motion either compression, induction, injection, or combustion induced—results in vastly higher mixing rates than are possible with simple laminar flames, together with a greater or lesser degree of recirculation. This is the major reason for the complexity of the diesel combustion process, apart, that is, from its basic heterogeneous nature.

As to flame types, a part at least of the fuel injected during the ignition delay period will be evaporated and premixed with the air; note, however, that regardless of the general air/fuel ratio at any instant there will be wide variations in the actual mixture strength over the chamber. In some systems, particularly those where fuel reaches the chamber wall, part of the fuel injected during delay will remain unmixed until later in the cycle. This fuel, together with that injected after the initiation of combustion, will in general burn as a turbulent diffusion flame, although in comparison with a laboratory flame of the same title this is a considerable oversimplification due to recirculation of products of combustion and other factors.

Indeed, in some respects such diesel flames have more the characteristic of rich, premixed laboratory flames. The major point is that some degree of pyrolysis of fuel is virtually unavoidable, regardless of the overall trapped air/fuel ratio at which the engine is operating.

Although pressure levels in diesel combustion systems are far above those used, e.g. in (20), and the composition of the basic fuel differs also, it is presumed that the reaction mechanisms leading to the generation of soot are similar in both types of combustion system. This is confirmed by the similarities in the effects of different fuel types (e.g. aromatics versus paraffins (38)) and of fuel additives (39). Initial particle growth would be expected to be even more rapid in the diesel case due to the higher level of turbulence and the denser atmosphere, resulting in increased collision frequency. Although the sizes of the final particles from the different combustion systems are similar, in the diesel case they take the form of spherical coagulates (see Fig. 140.3) rather than the chain structures observed in low-pressure laboratory flames. In diesel systems the available time for soot formation is very short, typically 4–10 ms, as opposed to laboratory flames where times may be as high as 100 ms (see Fig. 140.2). It is suggested that this difference in time span accounts for the much lower conversion rates of fuel to soot in the diesel engine, notwithstanding the higher pressure conditions. Finally, it is to be noted that there is a complete correspondence between diesel and laboratory systems on the qualitative effects of physical characteristics; namely, local equivalence ratio, temperature, and pressure, as discussed in the following sections.

Operation of particular diesel combustion systems

High-speed direct-injection systems

The conventional direct-injection (D.I.) system comprises a multi-hole nozzle and an axisymmetric chamber, both being approximately co-axial with the cylinder. Neither the shape of the chamber nor the exact number of nozzle holes is critical (40)–(42). Due to fuel impingement on the chamber wall, air swirl generated during the induction stroke is essential.

A detailed study of soot release in such systems has been carried out by one of the authors and the results reported elsewhere (40) (43), but the main points may be summarized as follows. Fuel injected before ignition may become

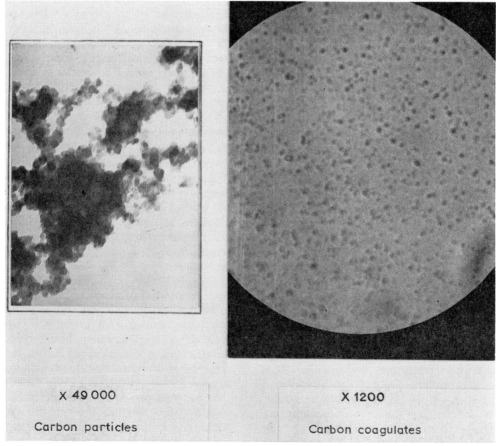

Smoke density: 0·2 g/m³.
Engine speed: 1500 rev/min.

Fig. 140.3. Photomicrographs of soot in diesel engine exhaust

fully mixed or not depending on the atomization and rate of penetration of the fuel sprays, the time required by the sprays to reach the chamber surfaces, the degree of air swirl, and the time available before ignition occurs. A greater or lesser fraction of the available air will be involved with this mixing process, depending essentially on the same factors, and will determine the local equivalence ratio and the resulting soot release (if any) when combustion of this premixed fuel takes place.

Fuel unmixed prior to the start of combustion, i.e. both that captured by the chamber wall during the delay period and that injected after ignition of the first fuel, will ultimately produce regions of over-rich mixture subjected to conditions favourable to pyrolysis: particularly where the mixing is slow, soot will form in such regions. The amount of fuel unmixed at the start of combustion will be controlled by the geometry of the chamber and its relationship to the injection characteristics, and by the relative durations of the ignition delay and injection periods; while the rate at which it mixes and burns is a function of the injection and air swirl rates. In practice then, soot release in a given engine is largely a function of the proportion of fuel injected after the start of combustion [e.g. as shown in Fig. 3.10 of reference (**43**)].

Higher overall temperatures increase soot formation far more than its subsequent combustion, other things being equal. Apart from variations in compression conditions, changes in mean cycle temperature are primarily a function of the phasing of the combustion period within the cycle, but only assume prominence in controlling soot release at retarded timings when the fuel quantity available for pyrolysis is large. Combustion of soot is a function of the temperature–time history of the soot formed and of the local oxygen availability. It is not thought to be at all significant once the main heat-release period is concluded, due to the rapidly falling gas temperatures in the expansion stroke and the lack of proper mixing of products of combustion and unused oxygen inherent in such systems.

Quiescent direct-injection systems

These D.I. systems are only found in the larger bore engines where fuel impingement is avoided by the increased free-spray path length available, together with the use of a large number of sprays. Little or no swirl necessary for limiting the thermal loading is adequate to give acceptable combustion.

The smoke-limited overall air/fuel ratio of such systems is generally higher than that of the smaller high-speed D.I. engines, due to the low rates of air entrainment and mixing within the fuel sprays. Since all combustion effectively occurs within the fuel-spray envelopes, it is clear that local equivalence ratios can be very high, even at quite modest overall air/fuel ratios, leading to significant soot formation. For the same reasons combustion of soot so formed will be minimal.

The M-system

While, as has been noted, some degree of fuel impingement on the chamber walls is inevitable in the smaller D.I. systems, in the M-system (**44**) (**45**) and others (**46**) the degree of impingement is greatly increased so that the bulk of the fuel is mixed and burnt off or near the wall. To this end very high levels of induction-generated air swirl are necessary, about twice those of conventional D.I. engines. Such systems can show a lower overall smoke-limited air/fuel ratio than the high-speed D.I. systems discussed earlier.

Contributory factors are (1) the controlling influence on the bulk fuel temperature exercised by the chamber wall, and (2) that reduced residence time of fuel-rich vapour in the high-temperature zones, due to increased mixing rates, will result from the high levels of air swirl.

Indirect-injection systems

These are here taken to include both pre-combustion chamber and swirl-chamber systems, since the information available to the authors suggests that these and their variants differ more by degree than by kind, at least as far as the basic mode of soot release is concerned. The following notes are based on experimental work carried out on some swirl-chamber systems at the authors' companies.

In all such systems the whole fuel charge is injected not only into just a part of the clearance volume (typically 20–50 per cent) but also into a region of violent air motion, swirl rates being up to six times those of conventional D.I. systems (**47**)–(**49**). In practically all indirect-injection (I.D.I.) systems (except possibly at very retarded timings) the fuel injected during the delay period can only mix with the proportion of air contained within the pre-chamber. The timing of injection will control not only the quantity of fuel introduced before ignition at high fuellings, due to its influence on the ignition delay period, but also the proportion of the total air charge contained within the pre-chamber at the ignition point. In consequence, the overall mixture strength within the pre-chamber at the start of combustion will be enriched at advanced timings, as shown in Fig. 140.4 for two variants at different compression ratios of the Ricardo Comet system. Thus, during the initial combustion phase, a greater or lesser quantity of soot will be formed from the first fuel to be injected, dependent on the equivalence ratio distribution within the pre-chamber at this time.

Fuel injected after combustion is initiated is subjected to high temperatures but is vigorously mixed with the air remaining in the cylinder by the high-velocity outflow of gas from the pre-chamber. Thus, while the initial soot formation from this latter fuel may be significant, in general excellent conditions exist for its rapid combustion so long as there is sufficient air available, i.e. at all except the lowest overall air/fuel ratios. Thus, the basic smoke characteristic with respect to injection timing is the reverse of the D.I. engine [compare the data of Fig. 140.4 with that given in references (**40**) and (**43**)]. The effective dependence of smoke emissions on the equivalence ratio in the swirl chamber at the start of combustion is again demonstrated in Fig. 140.5, which shows results for two

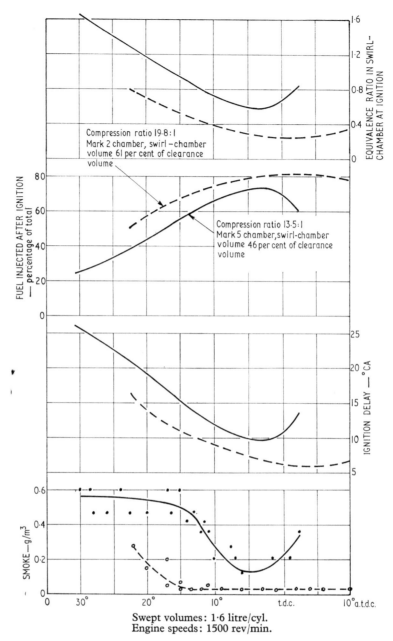

Fig. 140.4. Factors affecting smoke emissions from swirl-chamber engines

different forms of swirl-chamber system. However, it is clear that the effect on smoke emissions of changes in fuelling is not fully explained by this factor alone. In this context the other important factor, as shown in Fig. 140.5, is the amount of fuel injected after ignition.

Other factors, such as the overall rate of fuel–air mixing in the cylinder, the temperature distribution in both space and time, and the resulting extent of soot combustion, will also be important, as indeed they are in all systems. It is thought that the high rates of mixing possible in I.D.I. systems are their principal characteristic, responsible not only for the very low smoke-limited overall air/fuel ratios at which they can operate but also for the rather more retarded optimum timings for maximum efficiency that they exhibit. This is a valuable characteristic in that it not only places optimum economy rather closer to the best smoke condition but also the use of retarded timing on these type of engines, among other things, results in low emissions of the oxides of nitrogen.

Effect of fuels on soot formation in diesel engines

The preceding sections of this paper refer only to the use of conventional diesel fuels (gas oils) in different combustion systems, whereas a wide range of fuels may be burnt in a compression-ignition engine, and modes of soot release under such conditions deserve some comment.

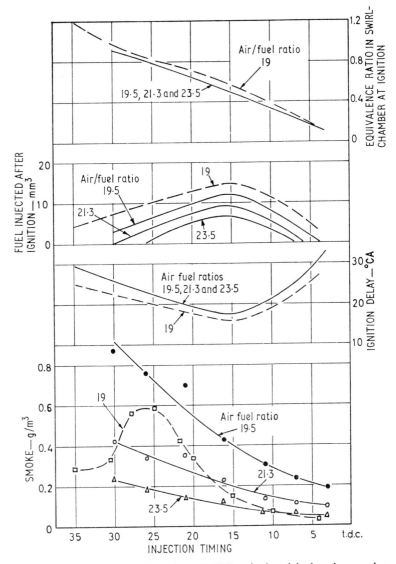

Fig. 140.5. Smoke emissions from 0·45 litre/cyl swirl-chamber engines

Many types of liquid fuels may be introduced by injection, e.g. gasoline (44) (50) and alcohols (51) (52), but atomized liquid (50) or gaseous (53) fuels may also be introduced via the induction system with only a pilot injection to act as an ignition trigger. Soot release under the latter conditions is greatly affected by the thermal stability of the fuel, and by the equivalence ratio at which it will yield soot, which factors may be interrelated, cf. the observation that an increase in octane number of the aspirated fuel reduces soot emission (50).

With aspirated fuels, particularly if gaseous, the times available for mixing result in most cases in considerable uniformity of equivalence ratio throughout the chamber. Dependent on the factors already mentioned, it is likely that such fuel mixture, being everywhere weaker than stoichiometric, will in burning produce little or no soot. Soot may, of course, be formed from the pilot spray, but if the proportion of fuel so introduced is low, smoke emissions should also be negligible. A limit to this mode of operation, i.e. in the amount of fuel aspirated, is, of course, set by detonation at the rich limit, and by failure of the flame to propagate through aspirated charge at the weak limit (54).

It must not be forgotten that changes in the chemical structure of fuels will inevitably result in changes in physical properties, so that in addition to chemical effects considerable changes in mixing patterns will result even in the more conventional diesel engine. Fig. 140.6 shows Schlieren photographs of the combustion of sprays of gas oil and gasoline in a simplified swirl-chamber combustion system. The gasoline evaporates and mixes with a much larger proportion of the available air, due to its higher volatility as well as the greater time available (long ignition delay period). As a result, soot is formed only in the still over-rich core of the gasoline spray when combustion commences. In the gas oil spray the rate of mixing

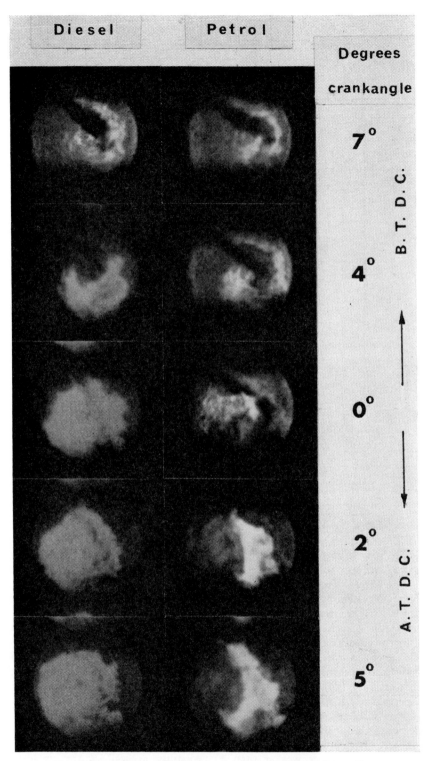

Fig. 140.6. Effect of fuel type on mixing and soot formation

is much slower, so that some soot is formed even in the premixed zones. This is clearly seen in the original colour photographs, where in the latter case the whole chamber is 'filled' with yellow flame, while much of the combustion of the gasoline is non-luminous.

Summary of variables affecting diesel smoke

Various fundamental variables affecting soot release in different types of diesel engine have been outlined in the preceding sections of this paper. These variables are themselves influenced by the directly controllable variables of chamber type, injection specification and resulting characteristics, compression conditions within the cylinder, etc. Fig. 140.7 has been drawn to summarize the whole picture, and to illustrate qualitatively the interrelation of these factors.

SUMMARY AND CONCLUSIONS

Soot, as just one type of carbon formed in the combustion of hydrocarbon fuels, is considered to result solely from vapour-phase pyrolysis of fuel molecules, occurring when insufficient oxidant is present in the high-temperature zones. Such pyrolytic reactions are kinetically high-order processes and as such are very sensitive to physical factors, of which temperature is the most important. A very large number of reaction routes may be involved in the formation of soot from the fuel molecule, but under temperature conditions typical of diesel flames the essential precursors are thought to be heavy unstable hydrocarbon radicals. Growth of the first detectable soot particles is very rapid and largely irreversible, and occurs in the hotter zones of the flame.

Not even in a single case, however, has a complete quantitative picture yet been built up. Kinetics of soot coagulation and combustion are by comparison better understood.

Coagulation occurring subsequently is a physical process. Combustion, gasification, or reaction with hydroxyl radicals after formation of the soot particles can occur if physical and chemical conditions are appropriate; but where oxygen is available, direct oxidation seems to be the more important of these reactions at normal flame temperatures.

The similar effect of fuel type and of additives in laboratory flames, as well as diesel engines, leads to the

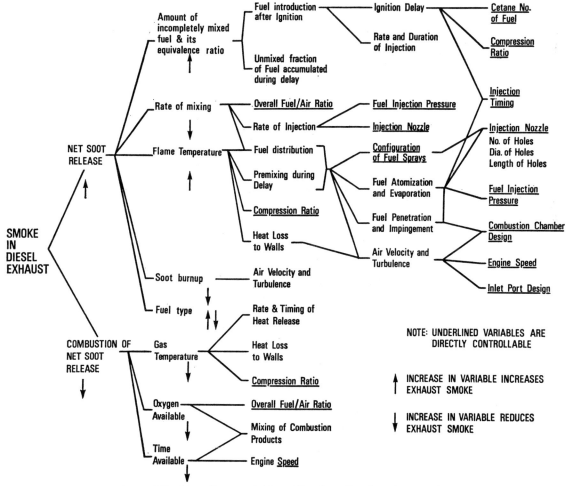

Fig. 140.7. Chart lists variables affecting diesel engine smoke

belief that similar soot formation processes are involved in both cases. The main difficulty in the application of detailed knowledge from laboratory flames (if it existed) to the diesel engine will be in defining the kinetics of the physical processes which determine the rates and levels of soot formation.

For a given fuel, local conditions of equivalence ratio and temperature over the injection and combustion periods are very important, and analysis of the history of the injected fuel elements along these lines allows a qualitative if not quantitative description of soot release in given systems to be made. Fuel variations influence soot formation via chemical properties, such as their thermal stability, as well as by the corresponding changes in physical properties influencing the mixing process.

A quantitative study of the soot formation in well-defined combustion systems, with or without the understanding of the chemical reaction mechanisms, is required to help in the design and development of the practical combustion system such as that used in the diesel engine.

In addition to the references specifically mentioned throughout the paper, further relevant data may be found in references (55)–(60).

ACKNOWLEDGEMENTS

Thanks are due to the Directors of C.A.V. Ltd and of Ricardo and Co., Engineers (1927) Ltd for permission to publish. The authors are particularly grateful to Dr A. E. W. Austen, Technical Director, C.A.V. Ltd, and Mr B. W. Millington, Joint Technical Director, Ricardo and Co., Engineers (1927) Ltd, for giving them the opportunity to carry out this study.

APPENDIX 140.1

REFERENCES

(1) GAYDON, A. G. and WOLFHARD, H. G. *Flames, their structure, radiation and temperature* 1953 (Chapman & Hall).
(2) MINKOFF, G. F. and TIPPER, C. F. H. *Chemistry of combustion reaction*, 237 (Butterworth).
(3) PORTER, G. *Combustion researches and reviews* 1955, 108.
(4) GLICK, H. S. *Seventh Symp. on Combustion* 1958, 98 (Combust. Inst., Pittsburgh).
(5) BONNE, U., HOMANN, K. H. and WAGNER, H. G. *Tenth Symp. on Combustion* 1965, 503.
(6) MACFARLANE, J. J., HOLDERNESS, F. H. and WHITCHER, F. S. E. *Combust. Flame* 1964 **8**, 215.
(7) CHAKRABORTY, B. B. and LONG, R. *Combust. Flame* 1968 **12**, 226.
(8) SCHALLA, R. L. and MCDONALD, G. E. *Fifth Symp. on Combustion* 1955, 316.
(9) SWEITZER, C. W. and HELLER, G. L. *Rubb. Wld* 1956 **134**, 855.
(10) PALMER, H. V. and CULLIS, C. F. *Chemistry and physics of carbon* 1965, 265 (Edward Arnold).
(11) CLARKE, A. E., HUNTER, T. C. and GARNER, F. H. *J. Inst. Petrol. Technol.* 1946 **32**, 627.
(12) CHAKRABORTY, B. B. and LONG, R. *Combust. Flame* 1968 **12**, 168.
(13) HOMANN, K. H. *J. Combust. Inst.* 1967 (August).

(14) MACFARLANE, J. J. and HOLDERNESS, F. H. 'Laboratory studies of carbon formation in fuel-rich flames at high pressures', *Symp. Diesel Engine Combustion, Proc. Instn mech. Engrs* 1969–70 **184** (Pt 3J), 57.
(15) HOOKER, W. J. *Seventh Symp. on Combustion* 1958, 949 (Combust. Inst., Pittsburgh).
(16) ATEN, C. F. and GREENE, E. F. *Combust. Flame* 1961 **5**, 55.
(17) STREET, J. C. and THOMAS, A. *Fuel, Lond.* 1955 **34**, 4.
(18) HOMANN, K. H., MOCHIZUKI, M. and WAGNER, H. G. *Z. Phys. Chem. N.F.* 1963 **37**, 299.
(19) RAY, S. K. and LONG, R. *Combust. Flame* 1964 (June).
(20) HOMANN, K. H. and WAGNER, H. G. *Eleventh Symp. on Combustion* 1966, 371.
(21) PLACE, E. R. and WEINBER, F. J. *Proc. Roy. Soc.* 1966 **289**, 192.
(22) PAYNE, K. G. and WEINBER, F. J. *Proc. Roy. Soc.* 1958 **250**, 316.
(23) HOWARD, J. B. *Twelfth Symp. on Combustion* 1969, 877.
(24) GREEN, H. L. and LEN, W. R. *Particulate clouds, dusts, smokes and mists*, 138 (Spon).
(25) ERICKSON, W. D., WILLIAMS, G. C. and HOTTEL, H. C. *Combust. Flame* 1964 **8**, 127.
(26) TESNER, P. A. *Seventh Symp. on Combustion* 1958, 546 (Combust. Inst., Pittsburgh).
(27) WHYTHAW-GRAY, R. and PATTERSON, H. S. *Smoke* 1932 (Edward Arnold).
(28) TESNER, P. A. *Seventh Symp. on Combustion* 1958 (Combust. Inst., Pittsburgh).
(29) TESNER, P. A. *Eighth Symp. on Combustion* 1960, 627, 801.
(30) CHAMBERLAIN, C. T. and GRAY, W. A. *Nature* 1967 **216**, 1245.
(31) FIELD, M. A., GILL, D. W., MORGAN, B. B. and HAWSLEY, P. G. W. *Combustion of pulverised coal* 1967 (Br. Coal Util. Res. Ass., Leatherhead).
(32) NAGLE, J. and STRICKLAND-CONSTABLE, R. F. *Proc. Fifth Conf. on Carbon* 1962 **1**, 154.
(33) LEE, K. B. Ph.D. Thesis, Dept Fuel Chemistry and Chemical Engng, Univ. Sheffield, 1961.
(34) TESNER, P. A. and TSIBULSEVSKY, A. M. *Combust. Flame* 1967 **11**, 227.
(35) FENIMORE, C. P. and JONES, G. W. *J. phys. Chem.* **71**, 593.
(36) LINDEN, L. H. and HEYWOOD, J. B. *Combust. Sci. Technol.* 1971 (January) **2**, Nos 5 and 6.
(37) LYN, W-T. *Lucas Engng Rev.* 1964 (July) **1** (No. 1), 10.
(38) BURT, R. and TROTH, K. A. 'Penetration and vaporization of diesel fuel sprays', *Symp. Diesel Engine Combustion, Proc. Instn mech. Engrs* 1969–70 **184** (Pt 3J), 147.
(39) SPENGLER and HAUPT. *M T Z* 1970 (March) **31** (No. 3), 102.
(40) KHAN, I. M. and WANG, C. H. T. 'Factors affecting emissions of smoke and gaseous pollutants from direct injection diesel engines'. Paper 151 of this Symposium.
(41) MILLINGTON, B. W. and BARNES-MOSS, H. W. 'Diesel engines for commercial vehicles', *Commercial vehicles, engineering and operation* 1967, 25 (Instn. Mech. Engrs, London).
(42) WATTS, R. and SCOTT, W. M. 'Air motion and fuel distribution requirements in high-speed direct-injection diesel engines', *Symp. Diesel Engine Combustion, Proc. Instn mech. Engrs* 1969–70 **184** (Pt 3J), 181.
(43) KHAN, I. M. 'Formation and combustion of carbon in a diesel engine', *Symp. Diesel Engine Combustion, Proc. Instn mech. Engrs* 1969–70 **184** (Pt 3J), 36.
(44) MEURER, J. S. 'Multifuel engine practice', *S.A.E. Int. Congr. and Exposition of Automot. Engng* 1961 (January).

(45) MEURER, J. S. *M T Z* 1966 (April) **27** (No. 4), 131.
(46) PISCHINGER, F. *M T Z* 1966 (October) **27** (No. 10), 389.
(47) LYN, W-T. and VALDMANIS, E. *J. photogr. Sci.* 1962 **10**.
(48) KHAN, I. M. and GRIGG, H. C. Paper A.18, *CIMAC Congr., Stockholm* 1971.
(49) ALCOCK, J. F. and SCOTT, W. M. 'Some more light on diesel combustion', *Proc. Auto. Div. Instn mech. Engrs* 1962–63 (No. 5), 179.
(50) DERRY, L. D., DODDS, E. M., EVANS, E. B. and ROYAL, D. *Proc. Instn mech. Engrs* 1954 **168** (No. 9), 280.
(51) NARASIMHAN, T. L., RAO, M. R. K. and HAVEMANN, H. A. *J. Indian Inst. Sci.* 1956 **38** (No. 4).
(52) MUHLBERG, E. *ATZ* 1963 (January) **65** (No. 1), 16; 1963 (March) **65** (No. 3), 73.
(53) LYON, D., HOWLAND, A. H. and LOM, W. L. 'Controlling exhaust emissions from a diesel engine by LPG dual fueling'. Paper 126 of this Symposium.
(54) KARIM, G. A., KLAT, S. R. and MOORE, N. P. W. 'Knock in dual-fuel engines', *Proc. Instn mech. Engrs* 1966–67 **181** (Pt 1, No. 20), 453.
(55) BARTHOLME, E. and SACHSE, H. Z. *Elektrochem* 1949 **53**, 326.
(56) PARKER, W. G. and WOLFHARD, H. G. *J. Chem. Soc.* 1950, 2038.
(57) WOLFHARD, H. G. and PARKER, W. G. *Proc. phys. Soc. Lond.* 1949 **62A**, 102.
(58) KNIGHT, B. E. 'Similarity considerations in assessing diesel engine fuel spray requirements', *Proc. Instn mech. Engrs* 1965–66 **180** (Pt 3N), 10.
(59) KHAN, I. M., GREEVES, G. and PROBERT, D. M. 'Prediction of soot and nitric oxide concentrations in diesel engine exhaust'. Paper 142 of this Symposium.
(60) ELLIOTT, M. A. *Ind. Engng Chem.* 1951 (December).

C141/71 RECENT AUTOMOTIVE AIR POLLUTION CONTROL LEGISLATION IN THE U.S. AND A NEW APPROACH TO ACHIEVE CONTROL: ALTERNATIVE ENGINE SYSTEMS

J. J. BROGAN*

The Congress of the United States passed the 1970 Clean Air Act Amendments because of concern over our air environment. Among other things, this legislation requires a further 90 per cent reduction in major pollutant emissions from automobiles sold in model year 1975 and after. To meet this requirement, clean-up of the conventional Otto cycle engine is being attempted. Another approach, supported by our Federal government, is to sponsor a programme to develop alternative engine systems such as Rankine systems, the gas turbine, and the stratified charge engine. The programme is described and status of the work presented.

INTRODUCTION

IN 1969 THE AUTOMOBILE contributed nearly half of the total pollutants to the air environment of the U.S.A. A summary of major air pollutants nationwide from the automobile as well as those from all other sources appears in Fig. 141.1. These data were derived by the Environmental Protection Agency based on estimates of annual mileage, fuel consumption, and pounds of pollutants per gallon of fuel consumed. From inspection of this figure it can be seen that the automobile plays a significant role in producing carbon monoxide, hydrocarbons, and nitrogen oxides and a minor role in producing sulphur oxides and particulate matter. Although the population had, in the past, been enamoured with the automobile, its role in deteriorating the air quality has created great concern.

As a result, in 1970 Congress passed the most stringent legislation ever for the control of air quality in the United States. The legislation, entitled the Clean Air Act Amendments of 1970, was signed into law by President Nixon on 31st December 1970.

Broadly stated, these amendments extend the authority of the Federal government to become directly involved in all activities where significant pollution arises. The amendments provide the blueprint for ensuring that clean air will be restored to the U.S.A. within this decade.

LEGISLATIVE REQUIREMENTS

Specifically, with regard to motor vehicles, the law requires that new light duty vehicles must achieve 90 per cent

The MS. of this paper was received at the Institution on 15th June 1971 and accepted for publication on 16th August 1971. 33
* *5 Research Drive, Ann Arbor, Michigan 48103.*

reduction from 1970 vehicle emissions of carbon monoxide and hydrocarbons by the 1975 model year and 90 per cent reduction from 1971 vehicle emissions of nitrogen oxides by the 1976 model year. The exhaust emission standards from 1971 to 1976 are presented in Table 141.1. The 1970 standards are identical to the 1971 levels shown; the 1974 levels are the same as those for 1973; the 1975 and 1976 levels shown are those mandated by this recent legislation.

The Environmental Protection Agency may extend the 1975 and 1976 deadlines by one year where certain conditions are met. These conditions are: (*a*) if it is essential to public interest or public health and welfare; (*b*) good faith effort to meet the standards has been made; (*c*) the necessary technology is not available or has not been available for a sufficient time to meet deadlines; and (*d*) the independent review by the National Academy of Sciences has indicated that means of meeting the standards are not available.

Lastly, the Clean Air Act Amendments require that, beginning with the 1972 model year, manufacturers warrant vehicles to be designed, built, and equipped to

Table 141.1. Federal motor vehicle emission standards, g/mile

Substance	1971	1972	1973	1974	1975	1976
HC	4·1	3·0	3·0	3·0	0·41	0·41
CO	34·0	28·0	28·0	28·0	3·4	3·4
NO_x	—	—	3·1	3·1	3·1	0·4

Standards shown through 1974 are equivalent standards based on the 1975 Federal test procedure (CVS 3-bag technique).

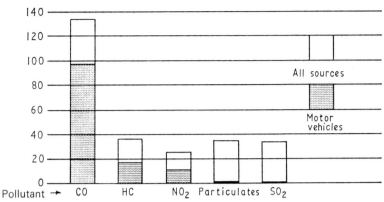

Fig. 141.1. Nationwide emissions estimates, 1969

conform at the time of sale with applicable emission standards and to be free of defects in materials or workmanship that would lead to non-conformity during a period of either five years or 50 000 miles.

Automobile firms worldwide which anticipate selling in the United States are conducting intensive research on means to clean up automobiles to meet these forthcoming standards and to maintain the required durability. Since the present U.S. automobile market consists almost entirely of spark ignition, gasoline fuelled, internal combustion engines, all of the 'clean-up' efforts have essentially focused on this type of engine. The 'clean up' has required development of new fuel delivery systems, engine modifications, and exhaust treatment. At this time we do not know of an automobile manufacturer who has demonstrated meeting all three pollutant levels simultaneously with the conventional internal combustion engine without severe degradation of driveability, road performance, and fuel economy. The technical difficulties inherent in the control of the intermittent combustion process, coupled with the requirement for 50 000 mile durability, dictate that different control approaches also must be considered seriously. One promising approach is to develop alternative power sources which offer the potential of meeting the emission standards without marked reductions in performance.

BASES FOR THE DEVELOPMENT PROGRAMME

In anticipation of forthcoming stringent exhaust emission standards, a panel of scientists, originating in the Executive Office of the President, was formed in 1969 with the purpose of evaluating efforts by industry in the U.S.A. to develop alternative power systems for the automobile. The conclusion reached by the panel was that industry was not making a sufficient effort to support such developments. Lacking sufficient industrial interest, a Federal research and development programme was recommended. This programme was announced by President Nixon on 10th February 1970, and it was under way by July. In addition, the President announced another Federal programme to stimulate industry by providing financial incentive to groups who independently develop their own alternative power systems. This is known as the Clean Car Incentive programme. The total Federal effort consisting of the Research and Development programme and the Incentive programme is called the Advanced Automotive Power Systems (AAPS) programme.

GOAL OF THE AAPS PROGRAMME

What is the goal of the AAPS programme? The original goal, as announced by President Nixon in 1970, was to develop a virtually pollution-free power system for the automobile within five years. The programme was structured to provide demonstration of several types of engine systems before that date. In addition, a longer term development activity was initiated and aimed at system demonstrations beyond the 1975 date. However, with passage of the recent legislation, the need for Federal demonstration of an alternative power system achieving the 1976 standards in the shortest time was intensified. Thus, while the original goal remains, more emphasis is now placed on developing only systems which offer the potential of early demonstration. Lastly, the programme provides incentive for industry to meet the 1976 emission standards with conventional engines, by seeking to provide a demonstrated technology for alternative power systems that can—if industry should fail in its efforts—be offered for adoption by industry because they meet the standards.

Goal achievement can be accomplished by demonstration of a new power system either developed directly under Federal government sponsorship or by similar demonstrations whose development is sponsored by those in industry, perhaps as a result of stimulation brought about by the existence of the Research and Development programme or of the Incentive programme, or both.

The remainder of this paper will be devoted to a description of the Research and Development programme followed by a description of the Incentive programme.

RESEARCH AND DEVELOPMENT PROGRAMME

What are candidates for development in the R & D programme? Five types of power systems were initially part of the programme when it began in July 1970. These included Rankine cycle power plants, the gas turbine, two kinds of hybrids—heat engine/electrics and heat engine/flywheel systems—and the all-electric system. Restructuring of the programme and re-ordering of development priorities to provide early demonstrations were implemented in July 1971. It was decided to accelerate the development of the most promising near term systems such as the gas turbine and the Rankine cycle systems and defer further work on longer range projects such as the hybrids and all-electric systems. In addition, the programme added another near term candidate system, the stratified charge engine. Thus, the programme began with five types of systems; we now have three and one of these three, the stratified charge engine, is new to the programme.

Very little development work for the automotive application had been conducted on most of the original candidate systems; therefore, it is worthwhile to discuss the status of the work conducted for each candidate, original and current, in the AAPS programme. The main body of the paper treats the current candidates and the discussion of the remainder of the systems appears in Appendix 141.1.

INDIVIDUAL CANDIDATES AND THEIR STATUS

Rankine cycle engine

Consider the Rankine cycle engine as illustrated in Fig. 141.2. In this engine there is an external combustor and an enclosed working fluid which is heated, expanded to do work, then condensed into a liquid, with the fluid being continuously recycled. The three types of Rankine systems which are presently in the system design and component test phase are:

(1) Organic working fluid—reciprocating expander.
(2) Water base working fluid—reciprocating or rotary expander.
(3) Organic working fluid—turbine expander.

As we see them, the problem areas associated with a practical design of the Rankine cycle are as follows:

Condenser size weight	Freezing (water)
Boiler size weight	Lubrication
Control complexity	Feedpump
Engine efficiency	Seals (non-water)
Minimize emissions	Valving design

The problems appear mainly in the inefficiency of components and complexity of the control system and, of

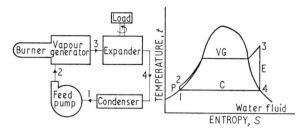

Fig. 141.2. Basic Rankine cycle engine

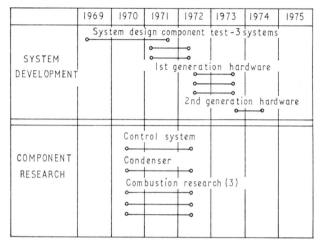

Fig. 141.3. Schedule of the planned work on the Rankine cycle

course, with exhaust emissions. Fig. 141.3 summarizes a schedule of the planned work on the Rankine cycle.

Complete system development for the three systems listed above, coupled with parallel research on the most critical components, is shown in Fig. 141.3. We point out that the three systems contractors are designing complete systems including all components. An integral part of this effort is to define the design requirements that each component must meet. For example, the temperature of the condensate, maximum and minimum flow rates and heat release rates are part of the condenser design requirements. Initially, each system contractor was designing a condenser to meet his specific requirements. In addition, an independent contractor was conducting research on improved condenser surfaces to meet the design requirements of each system contractor.

Recently, Garret AiResearch in Los Angeles, the independent condenser research contractor, demonstrated a new heat exchanger surface which is at least 20 per cent more efficient than surfaces presently in use in aerospace applications and 45 per cent more efficient than heat exchanger surfaces used in automobile radiators. The surface is of the slotted fin family and can be fabricated at low cost. This new surface lends itself to use on all three Rankine systems; therefore, the system contractors will apply this surface in all future condenser designs.

Configuration selection of a single type of combustor and vapour generator for each of the three systems is also under consideration. The approach to develop a practical low emission combustor and vapour generator is similar to that employed in the development of the condenser; namely, three system contractors and an additional team of specialists working parallel to arrive at competing component designs.

At the time of writing the scheduling for the Rankine cycle systems was that shown in Fig. 141.3. Prototype hardware delivery is expected in 1973. This schedule will be refined in the near future to show earlier delivery of hardware and system test, thereby showing the effect of a more intensified effort to develop this candidate.

Gas turbine

More work has been conducted by industry in the U.S.A. on the gas turbine than on any other candidate as an unconventional engine for the automobile application. For the gas turbine we are focusing our efforts on solving some of the problems which have plagued past attempts to get the gas turbine on the road. These problems include the need for reducing the nitrogen oxide emissions in the exhaust, developing manufacturing techniques and materials for facilitating mass production of turbines inexpensively, improving the part-load fuel economy, and increasing system reliability.

Use of this problem-solving approach, if successful, is intended to stimulate industry to apply the results of this research to their own turbine designs. As we see it, the demonstration of the turbine in the automobile probably will be performed by industry. It is because industry is so close to practical hardware on this system that our approach here emphasizes problem-solving rather than a government sponsored system demonstration.

Contract work on this candidate was initiated only recently. Cost comparisons are being made on production costs and operating of several turbine configurations ranging from the low pressure ratio regenerative system through to a high pressure ratio non-regenerative system. The result will be available by the November meeting.

Stratified charge engine

As mentioned earlier, the new candidate brought into the programme is the stratified charge engine. This engine is a spark-ignited, gasoline-fuelled, internal combustion engine with many hardware characteristics of the conventional engine. Differences appear mainly in the combustion chamber design, use of fuel injection, and in the resulting combustion process. In one version of this engine the fuel is injected into the cylinder as the spark appears and, therefore, the fuel burns as it enters. The fuel burns initially in an over-rich condition with fuel injectors designed to permit fluid swirl at the top of the cylinder. The burning expands into the lower portions of the cylinder and, overall, the resultant combustion products are similar to those obtained using a very lean mixture. Lean burning is very desirable from the emissions viewpoint. Much of the initial work on this engine in the U.S.A. was sponsored by the U.S. Army Tank–Automotive Command in Warren, Michigan. Our work on this engine has emphasized achievement of reductions in emissions of nitrogen oxides. Testing of the latest modifications made to this type of engine will be completed in October 1971 and results will be reported in the oral presentation of this paper.

The goals of the R & D programme, the candidates, and our approach to further development of each have been discussed. As stated earlier, another important programme is under way to develop a virtually pollution-free automobile using an alternative power source. This work consists of efforts on the part of industry with financial incentive provided by the government to inspire independent work. That is an apt description of the Federal Clean Car Incentive programme.

CLEAN CAR INCENTIVE PROGRAMME

The programme provides a market for large and small auto and non-auto manufacturers who often possess new and unique approaches toward engine designs. However, in the past, they have lacked incentive to further independent development.

The programme description is summarized in Fig. 141.4. After successfully passing stringent emissions and performance testing, first on a leased prototype car, then on 10 purchased copies of the prototype for demonstration, the successful engine system will be further tested after procurement of up to 500 vehicles. If the low emissions levels are maintained and the road performance satisfactory, the car is then eligible for certification as a low-pollution vehicle.

Lastly, the Incentive programme is expected to provide a valuable source of information from actual vehicles from which to judge the capability of the industry to meet 1975–76 emission standards.

This programme began in January 1971 with approximately 20 proposals received from industry to enter the prototype phase. Ten different vehicle systems have been accepted into the programme. While contract negotiations are not as yet complete, as we see it now, delivery of the first prototype of low emission cars into this programme will be made before August 1971.

The Incentive programme has been planned at $26.5 million over a three-year period. The rate of use and extent of use of these funds is dependent on the rate at which selected candidates proceed through the programme. At each test stage, any given candidate can be eliminated. The period of major costs should be in 1973 and 1974 since a number of candidates would be in the demonstration test stage in 1973 and the fleet test of one or more candidates should begin in 1974.

WORLDWIDE PARTICIPATION IN BOTH PROGRAMMES

Both programmes are open to firms residing outside the U.S.A. Most contractual work in the R & D programme is awarded on the basis of competitive proposals by

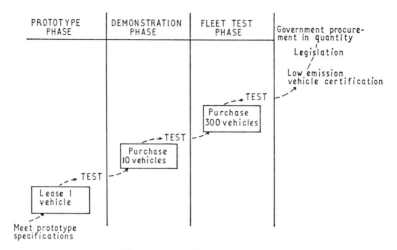

Fig. 141.4. Elements of the Federal Incentive plan

industry in response to Statements of Work which are well publicized and are then sent to firms requesting them. Multiple awards or contracts are common to the R & D programme. For example, we now have three different firms under contract, working on three different approaches to develop high-efficiency combustors for the Rankine cycle system. Four parallel approaches will be considered for the gas turbine combustor design. As a step towards improving lines of communication between the programme and industry in other countries, we have provided the Statement of Work for all contract work under way to interested industrial firms world wide. In addition, the final reports from completed contracts will be made available to all interested parties in any country.

We welcome participation from other countries in both programmes. Any firm or government representative who wishes further information on technical work under way, completed, or on participation should contact the Environmental Protection Agency in Washington, D.C.

SUMMARY OF THE ADVANCED AUTOMOTIVE POWER SYSTEMS PROGRAMME

I have defined the R & D and the Incentive programmes. Together they form the Advanced Automotive Power Systems programme. Although their goals are similar—to produce a virtually pollution-free automobile—their methods differ. We feel confident that the ultimate goal will be achieved within the five-year time frame. If the conventional internal combustion engine can be modified adequately by industry, so much the better, but through these two programmes alternatives will be made available if emissions from the internal combustion engine cannot be satisfactorily controlled.

APPENDIX 141.1

HYBRID AND ALL-ELECTRIC POWER SYSTEMS, FORMER CANDIDATES IN THE AAPS PROGRAMME

This appendix describes the technical work conducted up to July 1971 for three power systems which originally were part of the AAPS programme. The intensified effort required to develop more near term power systems required that further work on these three systems be deferred at the present time.

Hybrids

The hybrid candidates included the heat engine/electric and the heat engine/flywheel. The heat engine/electric hybrid consists of a small size low-powered engine (80–100 hp) and an array of inexpensive lead–acid batteries. These two power sources can be arranged in series or parallel configurations such as illustrated in Fig. 141.5. The same series versus parallel configurations can be applied where the flywheel replaces the battery system. The hybrid system is designed to extract power from the heat engine alone, or from the heat engine and the battery at the same time. The parallel configuration appears attractive where the conventional internal combustion engine is used as the heat engine. This combination can provide a practical system at reasonable development cost within two years. However, the heat engine requires the same types of after-treatment that the conventional engine requires where it is the only power source as in contemporary automobiles. Although relatively lower exhaust emissions appear likely with use of the parallel hook-up, we are not confident that the emissions can be made low enough to meet the 1976 standards. High manufacturing cost complexity and large system volume requirements were the other factors in deciding to defer further work on this system. The series configuration shown in Fig. 141.5, or versions thereof using a small gas turbine, has much technical merit. A high-speed alternator coupled directly to the turbine shaft can be used for charging the battery.

In either configuration we find that the system operates best by running the heat engine at a constant speed—say, equivalent to 40 mile/h road speed—and, where vehicle accelerations are required, the additional power comes

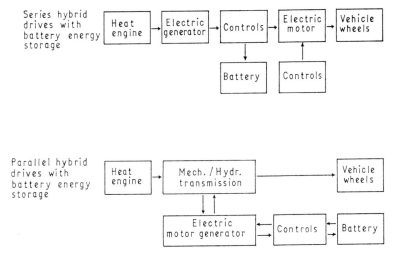

Fig. 141.5. Different arrangements for hybrids

from the battery system. For the parallel configuration, above this 40 mile/h road speed, the engine speed slowly converges on the new steady-state level demanded by the driver with additional power required during this slack period provided directly from the battery system. For the series hook-up the heat engine probably would run at fixed speed all the time.

There are several potential adventages which the hybrid concept offers. One is that the engine speed range is relatively small with the attendant ease of control of exhaust emissions under such conditions. Another advantage is that the very high road performance for a standard size U.S. automobile, of approximately 4000 lb, can be demonstrated using a relatively small and inexpensive heat engine.

As stated earlier, the heat engine/flywheel system would function in similar manner to the heat engine/electric with the battery replaced by a mechanical storage device, namely, the spinning flywheel. We had progressed on this system from parametric analysis of many practical flywheel materials such as shown in Table 141.2 and numerous practical configurations as shown in Fig. 141.6 to the design and fabrication of specific flywheels for cars.

Two flywheels have been tested to verify predicted energy and power densities. Since the important parameters for this application are the levels of power and energy densities, testing was conducted to verify that our expectations of practical densities up to 5000 W/lb and 12 W-h/lb, respectively, were safely achievable for the flywheel. Two full size flywheels were tested. One of the flywheels was of fibreglass and the other fabricated of 4340 steel. To give you a feel for some characteristics of the flywheel for a full-size family car, a flywheel made of 4340 steel and in a constant stress configuration weighs 42 lb and operates at a maximum speed of 24 000 rev/min. Detail designs of the key component of this system, the transmission, were completed prior to dropping this candidate.

All-electric car

Alkali-metal battery research and development, with emphasis on lithium/sulphur, supported by the AAPS programme for the automobile application, was under way for more than a year at Argonne National Laboratories prior to deferring further work on the all-electric automobile as a candidate.

Table 141.2. Flywheel materials

Material	Density, ρ, lb/in^3	Poisson's ratio, ν	Ultimate tensile strength, F_{tu}, kip/in^2	Yield tensile strength, F_{ty}, kip/in^2	Rec. working stress, σ, kip/in^2	σ/ρ ($\times 10^6$)	Material cost, $/lb	Normal cost
18Ni–400 (maraging steel)	0·289	0·26	409	400	260	0·900	2·25	4·20
18Ni–300 (maraging steel)	0·289	0·30	307	300	200	0·692	2·25	5·47
4340 steel	0·283	0·32	260	217	130	0·459	0·60	2·20
1040 steel	0·283	0·30	87	58	36	0·127	0·30	3·97
1020 steel	0·283	0·30	68	43	25	0·088	0·30	5·73
Cast iron	0·280	0·30	55	37	20	0·071	0·30	7·10
2021–T81 (aluminium)	0·103	0·33	62	52	26	0·252	0·53	3·54
2024–T851 (aluminium)	0·100	0·33	66	58	35	0·350	0·50	2·40
6 Al–4 V (titanium)	0·160	0·32	150	140	82	0·512	4·00	13·13
'E' glass	0·092	0·20	130	—	65	0·706	0·42	1·00
S-1014 glass	0·067	0·20	250	—	128	1·910	2·40	2·11

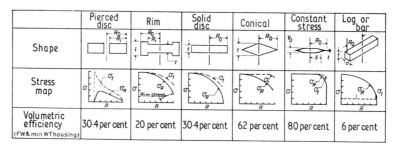

Fig. 141.6. Flywheel geometries

The lithium/sulphur cell programme was essentially a laboratory programme to determine if these cells show promise for use in a high specific energy, high specific power battery. At this stage in its development many of the problems that must be solved to make the lithium/sulphur battery both technically and economically attractive have been identified.

Laboratory cells of less than 10 cm² electrode area have demonstrated capacity densities (at the 1 h rate) of 0·4 A-h/cm² and lifetimes of over 800 cycles and 1100 h (at reduced capacity densities). No cell operated to date, however, has achieved the required high-performance levels over an extended lifetime. The sulphur electrode limits cell performance—the capacity per unit volume and sulphur utilization must be improved to provide a viable battery. Sulphur electrode design modifications, additives to the sulphur, and the proper choice of electrolyte are all capable of increasing the cell performance.

The cells operated up to this time have been relatively small (<10 cm² electrode area) and unsealed. The full-scale cells will be much larger (~350 cm²) and must be sealed. Initial scale-up efforts are necessary to provide information on problem areas peculiar to the sealing, design, construction, and operation of larger cells. Materials are available for laboratory and first generation scaled-up cells, but the applicability of low-cost lightweight construction materials that are necessary for a high energy density, high power density, low-cost battery is yet to be demonstrated. Finally, the assembly of cells into batteries will present new problem areas of cell matching, safety, reliability, long-term hermetic sealing, and temperature control.

The programme has until now been a laboratory effort to demonstrate the technical feasibility of lithium/sulphur cells. A great deal of research and development remains to be done to reach the stage of a reliable battery of 20 kW originally planned for demonstration in 1978.

C142/71

PREDICTION OF SOOT AND NITRIC OXIDE CONCENTRATIONS IN DIESEL ENGINE EXHAUST

I. M. KHAN* G. GREEVES* D. M. PROBERT*

The paper discusses the development of models to predict soot and nitric oxide formation in direct-injection-type diesel engines. The soot formation model has been applied to three engines and found to predict the effects of injection timing, injection rate, fuelling, air swirl ratio, and engine speed. The nitric oxide formation model has been applied to one engine and found to predict the effect of all the above-mentioned variables, though the effect of engine speed is overestimated. The validity of the models and their contribution to the understanding of the processes involved in soot and nitric oxide formation are also discussed.

INTRODUCTION

FUEL INJECTION EQUIPMENT (FIE) application to diesel engines and development of diesel combustion systems generally remains a time-consuming experimental process; thus, there is much interest in developing methods for predicting diesel engine behaviour. The need for this is further accentuated by the necessity to develop combustion systems to satisfy an ever-growing set of often conflicting requirements: efficiency, specific power output, visibility of exhaust, emission of invisible pollutants, cold-starting capability, and noise.

Work to date has gone some way in enabling certain aspects of the diesel engine performance to be calculated (I)–(3)†. This paper reports recent work on the development of models for the prediction of the concentration of soot and nitric oxide (NO) in the diesel engine exhaust. A stage has been reached where the models can be used for predicting with a fair degree of accuracy the effects of FIE and engine operating variables on emission of soot and NO. Work on the further refinements of models and their application to a larger number of engines is required; hence, this paper is to be regarded as an interim report.

Notation

A — Mass of air entrained, g.
A_t — Total mass of air in cylinder, g.
a — Mass of air consumed, g.
C_s — Soot formation rate coefficient, mg/Nm s.
C_z — Rate coefficient for Zeldovich mechanism, degK/atm$^{1/2}$ s.

D — Diffusivity constant, cm^{-1}.
D_n — Nozzle hole diameter, cm.
E_r — Air entrainment ratio.
E_s — Activation energy for soot formation, cal/mole.
E_z — Activation energy for Zeldovich mechanism, cal/mole.
H — Lower calorific value of fuel, cal/g.
K — Equilibrium constant for NO.
M_a — Mass of micromixed air, g.
M_f — Mass of micromixed fuel, g.
[NO] — Current mole fraction of NO.
[NO]$_e$ — Equilibrium mole fraction of NO.
[N$_2$]$_0$ — Original mole fraction of N$_2$ available for NO formation.
n — Equivalence ratio exponent for soot formation.
[O$_2$]$_0$ — Original mole fraction of O$_2$ available for NO formation.
P — Cylinder pressure, atm.
P_i — Cylinder pressure at ignition, atm.
P_j — Injection pressure, atm.
P_u — Partial pressure of unburnt fuel, kN/m^2.
R — Universal gas constant, cal/g mole degK.
r — Radius of wall jet front from axis, cm.
S — Soot formation per unit volume, g/m^3 at NTP.
T — Thermodynamic mean temperature in cylinder, degK.
T_a — Temperature of air zone, degK.
T_e — Mean temperature in jet, degK.
T_i — Temperature at ignition, degK.
T_p — Local temperature for NO formation, degK.
T_u — Local temperature for soot formation, degK.
t — Time since beginning of injection, s.
t_0 — Initial reference time in wall jet, s.
t_w — Current reference time in wall jet, s.

The MS. of this paper was received at the Institution on 8th June 1971 and accepted for publication on 13th August 1971. 23
* C.A.V. Ltd, Research Dept, Larden Road, Acton, London, W.3.
† References are given in Appendix 142.2.

t_Y	Time for impingement, s.
U	Free jet front velocity, cm/s.
U_Y	Free jet impingement velocity, cm/s.
V_f	Jet front velocity, cm/s (= U or W).
V_{NTP}	Volume of cylinder contents at NTP, m³.
V_u	Volume of soot formation zone, m³.
W	Jet front velocity on the wall, cm/s.
W_0	Initial velocity on the wall, cm/s.
X	Fuel injected, g.
x	Fuel prepared for heat release neglecting chemical kinetics, g.
x'	Fuel combusted, g.
Y	Impingement distance, cm.
y	Penetration of free jet front, cm.
Z	Ratio of gas density relative to air at NTP.
γ	Isentropic index for air zone.
δ	Jet thickness on wall, cm.
δ_0	Initial jet thickness on wall, cm.
θ	Cone half angle.
ρ_a	Gas density in cylinder, g/cm³.
ϕ	Local original fuel/air equivalence ratio.
ϕ_o	Overall equivalence ratio.
ϕ_p	Local products equivalence ratio for NO formation.
ϕ_u	Local unburnt equivalence ratio for soot formation.
$\phi_{u\ mean}$	Mean unburnt equivalence ratio in jet.

MECHANISMS AND KINETICS OF SOOT AND NITRIC OXIDE FORMATION

Soot

The details of the mechanisms leading to soot formation are not known even in the case of the simplest laboratory flames (4). In the diesel exhaust there is evidence that the diameter of soot particles is fairly uniform irrespective of engine load and speed (5), so that the rate of soot formation is practically controlled by the rate at which the numbers of particles increase, i.e. by the rate of nucleation. In addition, it appears that for a given fuel undergoing pyrolytic reactions under the temperatures encountered in diesel combustion systems, the rate of soot particle nucleation is governed by the formation of certain species of gaseous precursors.

In an elemental volume, the formation of precursors from unburnt or partially oxidized fuel vapour will be controlled by chemical kinetics. Also, the amount of unburnt fuel relative to the unused oxygen present is likely to affect significantly the race between soot precursor reactions and oxidation reactions.

This is illustrated in Fig. 142.1, which shows the nature of the dependence on equivalence ratio for a rich hexene–air premixed flame, a continuous kerosine spray, and a diesel combustion system (6). It can be seen that the soot formation is proportional to

$$(\text{overall equivalence ratio})^n$$

The values of the exponent, however, cannot be used directly because these have been derived (in two cases)

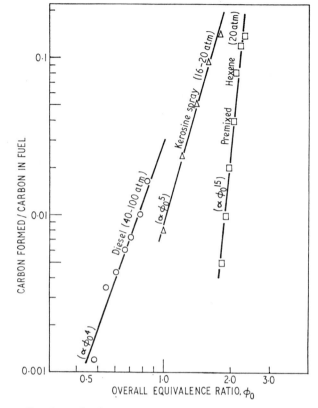

○ Based on diesel exhaust soot results for variable fuelling, 7° b.t.d.c. injection timing.
△ Taken from reference (6).
□ Taken from reference (6).

Fig. 142.1. Effect of overall equivalence ratio on carbon formation

from heterogeneous systems, and also because the temperature (in all three cases) varies with equivalence ratio.

It follows from the preceding considerations that the rate of soot precursor formation, and therefore soot formation, in a partially oxidized element of fuel–air mixture can be characterized by an Arrhenius-type equation with an equivalence ratio function; the following equation is proposed:

$$\frac{dS}{dt} = C_s \frac{V_u}{V_{NTP}} \phi_u^n P_u\, e^{-E_s/RT_u} \quad . \quad (142.1)$$

The activation energy E_s in the equation is derived from previously reported work on the effect of inlet air temperature on exhaust soot (7) and is found to be close to 40 000 cal/mole. Values of the constants C_s and n are not known and are estimated by comparing predictions with experimental results.

Nitric oxide

It is fairly well established that in hydrocarbon–air combustion systems the rate of NO formation is controlled by the Zeldovich mechanism (8)–(10). A simplified reaction scheme for lean mixtures is as follows:

$$O_2 \rightleftharpoons 2O$$
$$O + N_2 \rightleftharpoons NO + N$$
$$N + O_2 \rightleftharpoons NO + O$$

which leads to the following equation for the rate of NO formation (9):

$$\frac{d[NO]}{dt} = C_z [O_2]^{1/2} [N_2] \left(1 - \frac{[NO]^2}{[NO]_e^2}\right) \frac{P^{1/2}}{T_p} e^{-E_z/R\, T_p} \quad (142.2)$$

where
$$[N_2] = [N_2]_o - [NO]/2$$
$$[O_2] = [O_2]_o - [NO]/2$$

The coefficient for rate of NO formation C_z and activation energy E_z in equation (142.2) are given as $C_z = 5\cdot1 \times 10^{15}$ degK/atm$^{1/2}$ s, and $E_z = 137\,000$ cal/mole. The equilibrium value $[NO]_e$ required for equation (142.2) is the concentration of NO that would prevail under equilibrium conditions and is calculated as follows:

$$K = \frac{[NO]_e^2}{[N_2]_e [O_2]_e} = 21\cdot12 \, e^{-43\,260/R\, T_p} \quad (142.3)$$

which can be transformed into:

$$[NO]_e = -\frac{b + \sqrt{(b^2 - 4ac)}}{2a} \quad (142.4)$$

where $a = 1 - K/4$, $b = K([N_2]_o + [O_2]_o)/2$, $c = -K[N_2]_o[O_2]_o$, $[N_2]_o = 0\cdot79$, and $[O_2]_o = 0\cdot21(1 - \phi_p)$.

Equilibrium values of NO calculated from equation (142.4) are independent of pressure, since there is no change in the total number of moles involved ($N_2 + O_2 \rightleftharpoons 2NO$). Complex equilibrium calculations, including the species N_2, O_2, H_2, H_2O, CO_2, CO, NO, H, O, OH, and N (11) (12), are in good agreement with the simple calculations of equation (142.4), except for equivalence ratios close to and above unity (Fig. 142.2).

GENERAL CONSIDERATIONS ON MODELLING THE DIESEL COMBUSTION PROCESS

The main aspects of the fuel–air mixing process are clarified by considering a diesel-fuel spray burning in quiescent air. Fig. 142.3 shows the schematics of mixing and burning within such a spray, along with the consequences regarding soot and NO formation. It has to be

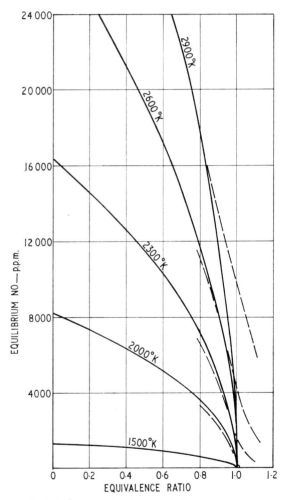

——— Equilibrium NO by simple calculation.
- - - - Equilibrium NO by complex calculation at 40 atm. (11) (12).

Fig. 142.2. Equilibrium NO as a function of temperature and equivalence ratio

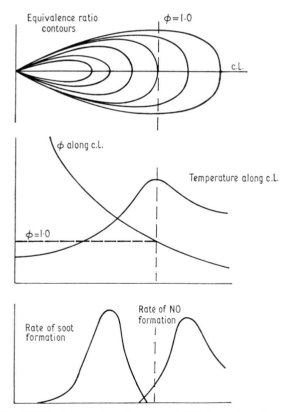

Fig. 142.3. Schematic of soot and NO formation in a fuel spray

emphasized that burning takes place throughout the spray, but at any instant the rich zones contain a high fraction of unburnt fuel (responsible for soot formation), and the lean zone contains the bulk of the unused air (responsible for NO formation).

This picture, as far as soot formation is concerned, is confirmed by high-speed Schlieren films of diesel combustion taken at the authors' laboratory.

Fig. 142.3 illustrates the equivalence ratio and temperature distribution at one instant during the combustion of an airborne spray. Ideally, a model should be able to predict such a picture at very short intervals of time throughout the combustion period while taking into account air velocities, turbulence, and spray impingement on the chamber walls. With such a model, soot and NO formation in a diesel combustion chamber could be calculated by using the data available on the kinetics of the mechanisms and by comparing with experiments.

The lack of detailed knowledge of the mixing process and of the conditions affecting it makes it practically impossible to construct such a model at the present stage. A relatively simple approach, embodying the main physical characteristics as already outlined, is desirable and has been adopted in this paper.

BASIS OF THE MODELS

The model chosen for development is based upon the fuel spray jet model of Grigg and Syed (2). This model predicts the rate of heat release mainly by calculating the rate of air entrainment (macromixing of fuel and air), and the rate of micromixing of fuel and air. Times for droplet evaporation are assumed to be negligible.

This model has been extended to include the impingement of the jet on the wall and subsequent air entrainment by the wall jet. Furthermore, account is taken of the effect of air swirl ratio on the amount of air entrainment by the jet, and on the diffusivity constant which controls the rate of micromixing of fuel and air within the spray. The main features of this free jet and wall jet model are illustrated in Fig. 142.4a; the details are given in Appendix 142.1.

DEVELOPMENT OF SOOT PREDICTION MODEL

The model, as described in the preceding section, does not predict the equivalence ratio distribution within the jet. The value of the exponent n of the equivalence ratio is also not known. The soot formation model is developed by comparing the predictions with the experimental data available on three direct-injection engines (see Table 142.1).

Assumptions are made to define the stratification of the fuel charge within the jet, and the numerical value of the exponent n in equation (142.1) is chosen to give the right trends of variation of exhaust soot with the FIE and engine operating variables. Once the values of all the other model constants are obtained, the value of rate coefficient C_s in equation (142.1) can be chosen by simple scaling to equate predictions with experimental results.

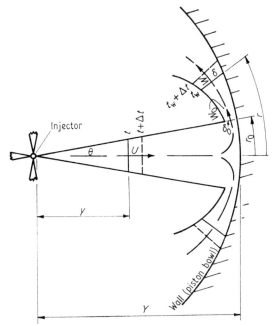

Fig. 142.4a. Fuel spray free jet and wall jet air entrainment model

Table 142.1. Some engine details

Detail	Engine A	Engine B	Engine C
Bore, mm	98	100	130
Stroke, mm	127	101	142
Swept vol./cylinder, cm³	966	797	1885
Compression ratio	16:1	17:1	16:1
Bowl diameter, mm	58	58	72

Fuel charge stratification and temperature in the jet

Account of the fuel charge stratification which exists in a diesel spray is taken by assuming that all the unburnt fuel within the spray, at a particular instant during the combustion process, is confined in the centre of the spray and on the wall (in zone c of Fig. 142.4b) and that the equivalence ratio in this zone is given by:

$$\phi_u = \frac{X}{x} \phi_{u\,mean} \quad . \quad . \quad (142.5)$$

where

$$\phi_{u\,mean} = \frac{(X-x)15}{A-a} \quad . \quad . \quad (142.6)$$

and

$$\frac{X}{x} \quad \text{or} \quad \frac{\text{fuel injected}}{\text{fuel mixed}}$$

represents the extent of fuel charge stratification within the jet.

The temperature in the soot formation zone (T_u) is taken to be the mean temperature in the jet (T_e), which is calculated from heat release and is given by:

$$T_u = T_e = \frac{A_t T - (A_t - A)T_a}{A} \quad . \quad (142.7)$$

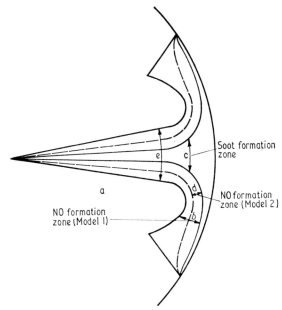

Fig. 142.4b. Models for soot and NO formation

where the temperature T_a in the air zone (zone a, Fig. 142.4b) is given by:

$$T_a = T_i \left(\frac{P}{P_i}\right)^{(\gamma-1)/\gamma} \quad . \quad . \quad (142.8)$$

Value of the exponent n

It was found that the prediction of the soot concentration in the engine exhaust for different injection timings was markedly influenced by the numerical value of the exponent n. This is shown by Fig. 142.5, where the predictions of soot concentrations in the exhaust of engine A, for a range of injection timings and for $n = 0, 2, 3,$ and 4, are compared with the experimental results. From this, and from results of similar comparisons at other engine speeds, $n = 3$ was found to be satisfactory and was used for all subsequent predictions.

Diffusivity constant, rate coefficient, and entrainment ratio

The entrainment ratio [defined by equation (142.17)] characterizes the relative variation in air entrainment with changes in the engine speed, as well as with changes in swirl ratio at a given engine speed. The diffusivity constant [equations (142.18) and (142.19)] controls the rate of micromixing of fuel with air in the jet and is varied to take into account the changes in mixing rate with changes in air velocity and associated turbulence occurring in the diesel combustion chamber. These changes occur with variation in the engine speed as well as with a variation in the swirl ratio. These effects can also be appreciated on the high-speed Schlieren films. There is, however, no previous quantitative information available on these aspects.

Analysis of the effects of entrainment ratio E_r and diffusivity constant D showed that variations of E_r markedly affect soot release, while those of D have a similar effect on the rate of heat release. Thus, the values of D for engines A, B, and C were determined by matching predicted heat release with the experimentally observed heat release for a range of engine speeds while using an

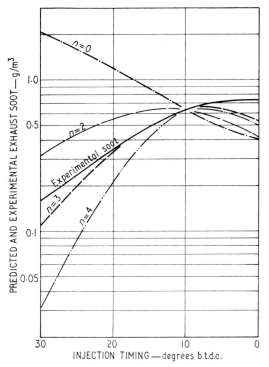

Engine A, 1100 rev/min, 60 mm³/stroke. Soot formation coefficient (C_s) chosen for correct prediction at 10° timing.

Fig. 142.5. Effect of equivalence ratio exponent on prediction of soot in exhaust for a range of injection timings

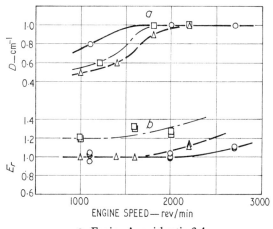

○ Engine A, swirl ratio 3·4.
△ Engine B.
□ Engine C.

Fig. 142.6. Values of diffusivity constant to give heat release match and of entrainment ratio to give match of predicted and experimental exhaust soot

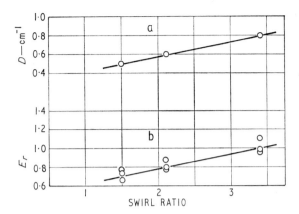

○ Engine A, 1100 rev/min, swirl varied by shrouded valve.

Fig. 142.7. Values of diffusivity constant to give heat release match and of entrainment ratio to give match of predicted and experimental exhaust soot

entrainment ratio of unity. These values are given in Fig. 142.6a. See below for values plotted in Fig. 142.6b.

An exercise was carried out to determine diffusivity constant (D) and entrainment ratio (E_r) as a function of swirl ratio for the engine A where experimental results at a number of swirl ratios and an engine speed of 1100 rev/min were available. The values of D and the E_r curves versus swirl ratio are given in Fig. 142.7a and 142.7b. The extremely important effect of air swirl on entrainment ratio is evident from Fig. 142.7b.

An entrainment ratio of unity was assumed for engine A at 2000 rev/min, and the rate coefficient C_s determined as 4.68×10^5 for best match of predicted and experimental exhaust soot (see 2000 rev/min results, Fig. 142.9, below). This value of C_s was retained for all subsequent predictions. Using an entrainment ratio of unity, exhaust soot predictions were carried out for other speeds of engine A and also for engines B and C; they are presented in Fig. 142.8. Only a limited number of plotted points are shown for engine A as the rest of the results are presented elsewhere. The important effect of entrainment ratio can be inferred from the separate correlations in Fig. 142.8. The values of E_r required to make predicted and experimental exhaust soot agree in Fig. 142.8, have been plotted in Fig. 142.6b.

Details of predictions

The predictions of exhaust soot with the model described in the preceding sections have been carried out for a wide range of injection timings, injection rates, fuellings, and engine speeds for all the three engines A, B, and C. To simplify the presentation only the results for engine A are given in Fig. 142.9, and prediction is compared with the experimental exhaust soot. There is good agreement, and similar agreement is obtained for engines B and C.

It is of interest here to look at the details of soot prediction with two injection rates at a number of injection timings obtained for the engine A. Fig. 142.10 shows that

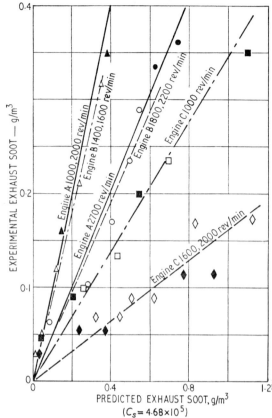

▲ Engine A, 1100, 2000 rev/min } varying timing.
● Engine A, 2700 rev/min
△ Engine B, 1400, 1600 rev/min } varying fuelling.
○ Engine B, 1800, 2200 rev/min
□ Engine C, 1000 rev/min } varying timing.
◇ Engine C, 1600 rev/min
■ Engine C, 1000 rev/min } varying injection rate.
◆ Engine C, 1600, 2000 rev/min

Fig. 142.8. Experimental exhaust soot versus predicted values for engines A, B, and C, using an entrainment ratio of 1·0

the effects of both injection rate and injection timing are correctly predicted. The instantaneous values of equivalence ratio (ϕ_u) and temperature (T_u) in the soot formation zone, and the soot concentration in the cylinder referred to NTP conditions, for three conditions chosen to illustrate the effects of a change in injection rate and also in injection timing, are given in Fig. 142.11 as a function of degrees crank angle after the beginning of injection. A retard in timing has little or no effect on the ϕ_u curves, but the decrease of ignition delay gives an earlier temperature rise and therefore under conditions where ϕ_u is much greater gives greatly increased soot formation.

An increase in injection rate at the same timing, however, has no effect on ignition delay, but the equivalence ratio ϕ_u curve is shifted towards the left of the figure due to a higher rate of air entrainment and mixing. Consequently, when temperature rise occurs, the ϕ_u is much lower, giving a substantial reduction in soot formation.

This is in spite of higher peak temperatures in the jet for an increase in injection rate (Fig. 142.11).

The agreement of the predicted effect of fuelling on exhaust soot with experimental values is shown in Fig. 142.12. The marked increase in exhaust soot with fuelling is explained by the increase in both ϕ_u and T_u.

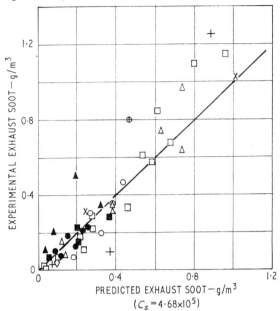

Injection timings of 35°, 30°, 20°, 10°, 5°, and 0° b.t.d.c. are included.

- ○ 2700 rev/min, 51·5 mm³/stroke ⎫
- □ 2000 rev/min, 30, 50, 60, 70, 75 mm³/stroke ⎬ normal rate injection.
- △ 1100 rev/min, 60 mm³/stroke ⎭
- ● 2700 rev/min, 51·5 mm³/stroke ⎫
- ■ 2000 rev/min, 60 mm³/stroke ⎬ fast rate injection.
- ▲ 1100 rev/min, 60 mm³/stroke ⎭
- + High swirl ⎫
- × Medium swirl ⎬ 1100 rev/min, 60 mm³/stroke.
- ⊕ Low swirl ⎭

Fig. 142.9. Experimental exhaust soot versus predicted values for engine A. Diffusivity constants and entrainment ratios were taken from curves of Figs 142.6 and 142.7

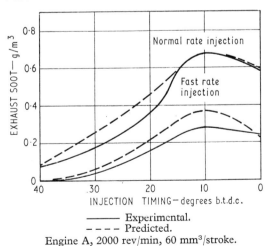

——— Experimental.
– – – Predicted.
Engine A, 2000 rev/min, 60 mm³/stroke.

Fig. 142.10. Predicted and experimental exhaust soot versus injection timing for normal and fast rate injection

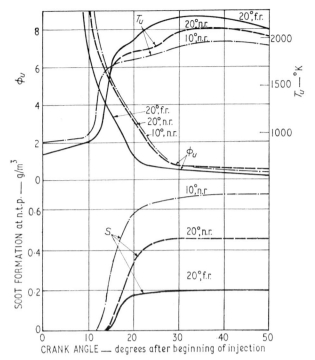

Engine A, 2000 rev/min, 60 mm³/stroke.
10° b.t.d.c. timing, normal rate injection (10°, n.r.).
20° b.t.d.c. timing, normal rate injection (20°, n.r.).
20° b.t.d.c. timing, fast rate injection (20°, f.r.).

Fig. 142.11. Effect of injection period and injection timing on soot formation, showing instantaneous values of equivalence ratio, temperature, and soot concentration

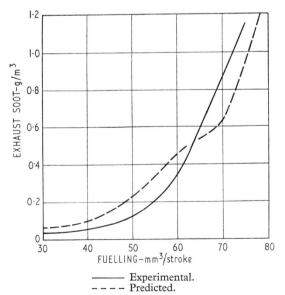

——— Experimental.
– – – Predicted.
Engine A, 2000 rev/min, 20° b.t.d.c. injection timing.

Fig. 142.12. Predicted and experimental exhaust soot versus fuelling

It is appropriate at this point to summarize the development of the model to show clearly the extent to which equation (142.1) has been fitted to experimental results by choice of constants, and the extent to which it has been used to predict other experimental results. The values of E_s, n, and C_s in equation (142.1) should be invariant. Values of $E_s = 40\,000$ cal/mole and $n = 3$ have been determined from experimental data. The value of C_s has been determined by direct comparison of predicted with experimental exhaust soot once all other model parameters are determined.

The soot formation zone variables, i.e. ϕ_u, T_u, P_u, and V_u of equation 142.1, are completely defined by the model, except for the diffusivity constant D and entrainment ratio E_r. The value of D has been determined by experimental heat release data. There is no knowledge as to the absolute magnitude of E_r; but for a given engine, engine speed and swirl ratio E_r has been assumed to be constant. E_r has been taken as unity for the 2000 rev/min speed of engine A, and a value of $C_s = 4.68 \times 10^5$ determined on the basis of this assumption, by direct comparison of predicted and experimental exhaust soot. For other engines, engine speeds, or swirl ratios, the value of E_r has been deduced by fitting predictions to experimental values of exhaust soot.

Thus, except for D and E_r, the model has been completely defined. Further analysis of the type shown in Figs 142.6 and 142.7 will be required before D and E_r can be determined for any direct-injection engine at the design stage.

DEVELOPMENT OF NO PREDICTION MODEL

The same basic fuel jet model (see under 'Basis of the models') was used for NO prediction model development. This was done by comparing prediction with experimental results obtained on engine A. Two models were used to quantify NO formation.

Model 1

The physical basis of this model is essentially the same as that for the soot formation model (Fig. 142.4b). NO formation occurs in the lean zone of the jet (zone b) where the equivalence ratio, ϕ_p, (model 1) is given by:

$$\phi_p \text{ (model 1)} = \frac{15x'}{A} \quad . \quad . \quad (142.9)$$

The temperature in this zone (T_p, model 1) is assumed equal to the mean temperature in the jet (T_e) and is given by equation (142.7).

The rate equation given above [equation (142.2)] was used in conjunction with model 1 to predict the effect of injection timing and fuelling on the concentration of NO in the engine exhaust. It was found that the variation in NO emission levels with changes in injection timing could be correctly predicted by choice of the activation energy E_z and rate coefficient C_z. These values, along with the

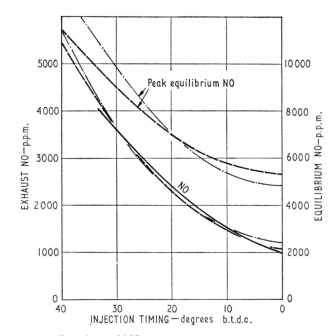

——— Experimental NO.
—·—· NO prediction (model 1), $C_z = 1.32 \times 10^6$, $E_z = 30\,000$ cal/mole.
— — — NO prediction (model 2), $C_z = 3.5 \times 10^6$, $E_z = 35\,000$ cal/mole.

Engine A, 2000 rev/min, 60 mm³/stroke.

Fig. 142.13. Effect of injection timing on NO concentration in exhaust. Rate coefficient and activation energy chosen for prediction to fit experimental curve

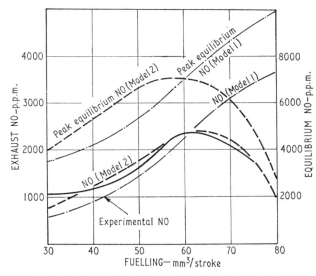

Engine A, 2000 rev/min, 20° b.t.d.c. injection timing.

Fig. 142.14. Effect of fuelling on predicted NO concentration in exhaust for model 1 and model 2

prediction and experimental results, are given in Fig. 142.13. But, as shown in Fig. 142.14, model 1 does not correctly predict the effect of fuelling on NO in the exhaust.

Model 2

It is obvious from the results of model 1 that this model underestimates the values of equivalence ratio in the NO formation zone due to the assumption of uniform equivalence ratio and temperature in the outer jet (zone b, Fig. 142.4b). In model 2 an attempt was made to delimit the NO formation zone (zone d, Fig. 142.4b) within the outer jet. This was done by assuming that the micromixed fuel and micromixed air calculated from the fuel jet model (Appendix 142.1) together constitute the NO formation zone. The equivalence ratio in this zone (ϕ_p, model 2) is given by:

$$\phi_p \text{ (model 2)} = \frac{15x'}{M_a} \quad . \quad . \quad (142.10)$$

The predictions with model 2 were carried out and values of C_z and E_z were chosen to give the best fit to the experimental data. It was found that model 2 correctly predicts both the effect of injection timing and fuelling, as shown in Figs 142.13 and 142.14. The effect of injection timing and fuelling on the peak equilibrium concentrations of NO during the cycle for both model 1 and model 2 are also given on Figs 142.14 and 142.15. The peak equilibrium values, though representing trends similar to the rate-controlled NO concentration of the corresponding models, are much higher than the experimental NO in the exhaust. Model 2 was used for all subsequent predictions of NO emissions using the values of C_z and E_z given in Fig. 142.13.

The predictions for a range of injection timings, injection rates, fuelling, engine speed, and swirl ratios for engine A are compared with the experimental observations in Fig. 142.15. The NO formation model gives good predictions of the effect of all the variables except engine speed. For correct prediction, all the points on Fig. 142.15 should lie on the line drawn for 2000 rev/min, for which speed the values of C_z and E_z were determined. Lines are also drawn through points for 1100 rev/min and 2700 rev/min. Detailed examination shows that the predicted values are higher than the experimental at 1100 rev/min, and the reverse occurs at 2700 rev/min.

It is of interest here to consider the effect of injection timing on the ϕ_p, T_p, and NO concentration histories within the engine cylinder (Fig. 142.16). According to model 2, the bulk of NO formation occurs during the main burning period, which follows the initial burning phase of rapid heat release. The increase of NO concentration due to advance in injection timing essentially results from the rise in temperatures. Similar plots of instantaneous values for a range of fuellings show that the variation of NO with fuelling is due to the combined and opposing effect of equivalence ratio and temperature in the NO formation zone.

The development of the NO prediction model (model 2) will now be summarized, to show the extent to which the model has been fitted to experimental results, and to what extent the model has been used to predict other experimental results. The values of E_z and C_z in equation

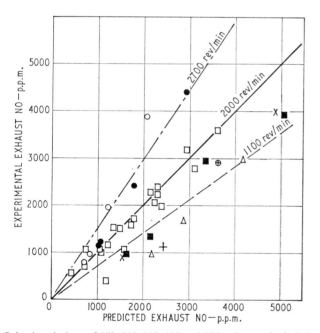

Injection timings of 35°, 30°, 20°, 10° and 0° b.t.d.c. are included.

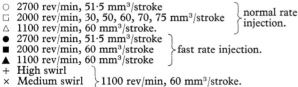

Fig. 142.15. Experimental NO concentration in exhaust versus predicted NO (model 2) for engine A

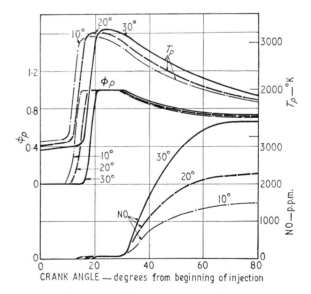

Engine A, 2000 rev/min, 60 mm³/stroke.

Fig. 142.16. Instantaneous temperature, equivalence ratio, and NO concentration (model 2) for 10°, 20°, and 30° b.t.d.c. injection timings

(142.2) should be invariant, and values of $E_z = 35\,000$ cal/mole and $C_z = 3.5 \times 10^6$ have been determined by matching to experimental results. The NO formation zone variables, i.e. ϕ_p, T_p, and the dilution of the NO zone contents to the total cylinder contents, are completely defined by the model, except for values of the diffusivity constant D and entrainment ratio E_r. The value of D has been determined by experimental heat release data, and the values of E_r have been taken from Figs 142.6b and 142.7b, i.e. E_r has been determined by the considerations outlined in the development of the soot prediction model.

As with the soot prediction model, the NO prediction model is completely defined, except for D and E_r. Further analysis is required to determine these quantities for any direct-injection engine at the design stage.

DISCUSSION ON THE VALIDITY OF THE MODELS

Soot prediction model

The relatively simple model for predicting soot formation does not take into account fuel droplet evaporation (all the unburnt fuel present in the combustion chamber is assumed to be in the vapour phase) or the detailed nature of the equivalence ratio and temperature distributions within the jets emanating from the injection nozzle. Consequently, all the unburnt fuel present in the chamber at any one time is assumed to contribute to soot formation.

High-speed Schlieren photography of diesel combustion shows that, at the beginning of combustion, soot formation takes place in the region of the spray where ignition occurs. This region is some distance away from the nozzle, the actual distance depending upon the rate of fuel jet penetration and the length of the ignition delay. Thus, at the beginning of combustion the unburnt fuel confined in the much richer regions upstream of the ignition zone does not yield soot because it is at a much lower temperature. Within a very short time, however, this is no longer true, as the flame engulfs the whole of the spray.

It is likely that during this main phase of burning, heat release occurs in the zones where the equivalence ratios are close to unity, with the rest taking place in the fuel-rich zones. Furthermore, this main phase of burning occurs with more or less all the air in the chamber within the jets (see Fig. 142.17); and due to the presence of air swirl and turbulence, the temperatures within the jets are much more uniform than at the beginning of combustion. The regions of rich mixture, at the centre of the airborne section of the jet and on or near the wall in the wall section of the jet, continue to exist, however, during and after injection, since time is needed to disperse all the unburnt fuel.

The above discussion shows that during the initial stage of combustion the model is at variance with what is actually observed, i.e. under this condition the assumption of uniform temperature throughout the unburnt fuel is incorrect. The soot formation during this phase, however, is a relatively small fraction of the total, as indicated by

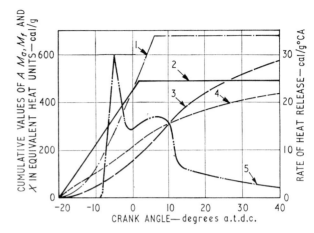

1 Air entrained, $A\left(\dfrac{H}{15A_t}\right)$.

2 Fuel injected, $X\left(\dfrac{H}{A_t}\right)$.

3 Micromixed air, $M_a\left(\dfrac{H}{15A_t}\right)$.

4 Micromixed fuel, $M_f\left(\dfrac{H}{A_t}\right)$.

5 Rate of heat release.

Engine A, 2000 rev/min, 60 mm³/stroke, 20° b.t.d.c. injection timing.

Fig. 142.17. Computed instantaneous values showing air entrainment, micromixing, and rate of heat release versus crank angle

high-speed photography. The bulk of the soot formation occurs at the beginning of the main combustion phase and lasts about 10° crank angle. For these conditions the model is fairly correct in assuming that the temperatures are uniform within the jets.

The method of specifying the stratification of the unburnt fuel charge by multiplying the mean equivalence ratio of the unburnt fuel in the jet by

$$\frac{\text{fuel injected }(X)}{\text{fuel intimately mixed }(x)}$$

is also quite plausible. This stratification ratio, however, does not give the details of the unburnt fuel equivalence ratio field within the rich zone where soot formation occurs (zone c, Fig. 142.4b). The value of the exponent n of the unburnt fuel equivalence ratio ϕ_u in equation (142.1), found by matching prediction with experiment, is closely linked to the choice of the stratification ratio; both, while quantifying the magnitude of the effect of the processes which occur, are empirical.

During the development of the soot formation model no account was taken of soot combustion. Soot, as it is transported from the rich zone where it is formed to the outer zones of the jet, is expected to be partly consumed either in the main oxidation zone (zone d, Fig. 142.4b) by the hydroxyl (OH) radicals (**13**) or in the outermost zone (zone b, Fig. 142.4b) by the oxygen. This state of affairs exists at the beginning of the main burning phase, during which practically all the soot formation takes place (see Fig. 142.11).

During the remainder of the burning phase the soot consumption occurs by its reaction with oxygen, as oxygen is present in all parts of the chamber, whereas the hydroxyl radicals will be expected to be present only in the main reaction zones (equivalence ratio close to unity). Soot combustion also continues after the main heat release period and during the early part of the expansion stroke.

It has, however, been shown that the soot mass concentration involved in the soot consumption reactions throughout the cycle is relatively unimportant in comparison with the soot release at the end of combustion (5). It follows that an initial evaluation of the soot formation model, with which this paper is concerned, need not include soot combustion. Leaving out soot combustion will obviously lead to an over-estimation of the numerical value of the rate coefficient C_s [see equation (142.1)]. It may also be that the scatter of the comparisons of soot prediction with experimental values (see Fig. 142.9) may be reduced even further when soot combustion in the jet is taken into account.

The success of the soot formation model in predicting the effects of injection timing, injection rate, fuelling, air swirl ratio, and engine speed would seem to indicate that both the physical characteristics of the model and the soot formation rate equation [equation (142.1)] are essentially correct. Further refinements of the model may be required by taking into account the history of the temperature and equivalence ratio distribution in the unburnt fuel, and by including soot combustion. It may also be necessary to look in relatively greater detail at the fuel oxidation mechanism by allowing for the formation of carbon monoxide under rich mixture conditions.

As the model becomes more and more complex, however, detailed comparisons with experiments, i.e. with the experimental values of soot concentration and temperature in the chamber during the cycle, become essential for its development. The authors are attempting to make such measurements.

The model as it stands can be used for predicting soot formation for any direct-injection engine, given the relevant engine dimensions and test conditions, where there are data on the rate of heat release, and a value of the entrainment ratio E_r can be estimated. The latter can be deduced if the soot emission from the engine for at least one experimental condition is known for each speed. Further development of the heat release model discussed in Appendix 142.1—and detailed data on the nature of air flow in engines and its effect of jet mixing with a range of nozzle hole sizes, injection pressures, etc.—are required before the model can be used for predicting soot from an engine at the design stage.

NO prediction model

Model 2, used for the prediction of the NO emission, also correctly predicts the effect of the various FIE and engine variables, though a greater reduction of NO emission with increase of engine speed was predicted than was experimentally observed. The Zeldovich NO formation equation [equation (142.2)] was used in conjunction with model 2, but it was found that the activation energy E_z required to fit the experimental results was 35 000 cal/mole, as compared to 137 000 cal/mole in the original Zeldovich equation. In addition, the value of the rate coefficient C_z was found to be different to the original value in equation (142.2).

This would seem to be due, in part, to the simplified equilibrium calculation [equation (142.4)], involving only NO, N_2, and O_2, that was used for the values of $[NO]_e$ in equation (142.2). The differences between the simple and complex calculations of equilibrium NO concentration become very noticeable for equivalence ratios close to unity (see Fig. 142.2). The differences in activation energy and rate coefficient, and the inability of the model to predict the effect of speed, may also be due to inadequacies of the model in defining the NO formation zone.

Further work is required, which must include the various species in the equilibrium calculation. In addition, the more detailed model proposed above for soot formation prediction would provide a more adequate description of the NO formation zone. Nevertheless, the model as it stands can provide useful predictions for exploring engine operating conditions of particular interest regarding reduction of NO emission.

CONTRIBUTION TO THE UNDERSTANDING OF SOOT AND NO FORMATION

Experimental observation leads to qualitative understanding of the processes involved and enables a model to be formulated in the first instance. The much more detailed and quantitative data that the model provides, in turn, contribute to a deeper understanding. The model for soot formation in diesel engines presented and discussed in this paper confirms that soot formation for a given fuel is controlled by the amount of insufficiently mixed fuel (i.e. confined at the centre of the free jet and near the wall in the wall jet), its equivalence ratio, and temperature. Earlier work showed that the percentage of fuel injected after ignition markedly influenced exhaust smoke (7). This is well explained by the model presented here, since a fuel element injected after ignition will experience temperature rise early in its mixing history, i.e. at high values of ϕ_u, giving high soot formation.

With normal rates of injection the marked reduction in exhaust soot with an advance in injection timing in the case of the direct-injection engine (see Fig. 142.10) is, according to the model, due to a reduction in both the amount of insufficiently mixed fuel and its equivalence ratio, ϕ_u, during combustion (Fig. 142.11). The effect of the injection timing is observed to be less important with a higher rate of injection (Fig. 142.10). Here, also, there are variations in the amount of incompletely mixed fuel and its equivalence ratio. Not only are these variations less important but the values of ϕ_u during combustion are lower, due to the high rates of air entrainment and mixing associated with the fast rate of injection.

The model also shows that the rate of air entrainment within the jets and the rate at which the rich zone is diluted are extremely important factors in determining the soot formation. In the model these have been characterized by the rate of travel of the front of the free jet and wall jet (see Appendix 142.1), the entrainment ratio E_r, and the stratification factor X/x.

The soot concentration in the exhaust varies greatly from engine to engine having the same compression ratio and operating with the same overall equivalence ratio and injection timing. The soot prediction model data indicate that these variations are essentially due to differences in the air entrainment ratios from engine to engine, due to the differences in their air swirl levels.

The basic variables that affect NO formation are temperature and oxygen concentration. The importance of the localized nature of the NO formation zone was made particularly evident during development of the NO model. Consequently, the entrainment ratio, through its effect on the mixing history, has a significant influence on NO formation.

CONCLUSIONS

Relatively simple considerations have led to the development of models for the prediction of soot and NO concentration in the exhaust of direct-injection diesel engines.

The soot formation model, together with the proposed rate equation, are shown to be successful in predicting the effects of injection timing, injection rate, fuelling, air swirl ratio, and engine speed. The model improves the understanding of the soot formation process within the engine cylinder. It is probable that the proposed soot formation equation will be applicable to other combustion systems, e.g. the indirect-injection engine and the gas turbine, provided adequate models representing the mixing processes are formulated.

The NO formation model has given fair predictions of the effects of injection timing, injection rate, fuelling, and air swirl ratio, though the effect of engine speed was overestimated. The Zeldovich equation was used for this model, but changes in the values of its numerical constants are required. This is believed to be due to the limitations of the model in defining the NO formation zone and to the simplified method of calculating equilibrium concentrations.

ACKNOWLEDGEMENTS

The authors wish to thank the Directors of C.A.V. Ltd for permission to publish this paper. In particular they are grateful to Dr A. E. W. Austen, Technical Director, for his enlightening discussions. Thanks are also due to Mr H. C. Grigg, Chief Research Engineer, for his support and discussions.

APPENDIX 142.1

FUEL JET MODEL FOR AIR ENTRAINMENT, MICROMIXING, AND HEAT RELEASE

The free jet conical spray model of reference (2) was extended to account for impingement of the free jet on the piston bowl wall, and for the subsequent air entrainment by the wall jet, as illustrated in Fig. 142.4a. The position of y of the jet front in the free jet is determined by the Schweitzer spray tip penetration equation.

$$y^2 = 1.042 \times 10^6 \frac{D_n P_j^{1/2} t}{Z} \quad . \quad (142.11)$$

Application of the momentum equation along the free jet axis gives the cone angle θ.

$$\tan^2 \theta = 1.09 \times 10^{-3} Z \quad . \quad . \quad (142.12)$$

As the jet front advances in the free jet the increment in air entrainment during time Δt is given by:

$$\Delta A \text{ (free jet)} = \frac{\tau}{3} \rho_a \tan^2 \theta \left(\frac{1.042 \times 10^6 P_j^{1/2} D_n}{Z} \right)^{3/2}$$
$$\times \{(t+\Delta t)^{3/2} - t^{3/2}\} \quad (142.13)$$

For transition of the jet front from a free jet to a wall jet, the transition time and loss of kinetic energy in the direction of flow during transition are neglected. The equations of reference (14) are applied to describe the velocity W and jet thickness δ on the wall.

$$W = W_0 \left(\frac{r}{r_0} \right)^{-1.06} \quad . \quad . \quad (142.14)$$

$$\delta = \delta_0 \left(\frac{r}{r_0} \right)^{1.006} \quad . \quad . \quad (142.15)$$

Equations (142.14) and (142.15), together with the assumptions for transition from a free jet to a wall jet, can be solved to give the position and volume flow of the jet front on the wall in terms of the time since impingement. The corresponding increment in air entrainment ΔA caused by advance of the jet front on the wall in time Δt is given by:

$$\Delta A \text{ (wall jet)} = \pi \rho_a \frac{Y^2 U_Y^2 \tan^2 \theta}{1.459 t_0^{0.459}} \{(t_w + \Delta t)^{1.459} - t_w^{1.459}\}$$
$$\quad . \quad . \quad . \quad (142.16)$$

where $t_0 = \dfrac{r_0}{2.06 W_0}$ and $t_w = t + t_0 - t_Y$

For any time t since the beginning of injection, equation (142.11) is used to determine whether the jet front has impinged; equation (142.13) or equation (142.16) gives the increment in air entrainment. After termination of fuel injection the jet front is assumed to proceed as before, but the truncation forms a back edge to the jet. The increment in air entrainment is then calculated by the net effects of advance of the jet front and advance of the jet back.

Regarding change of cylinder volume, it is assumed that the jet expands or contracts with volume and that the increment in air entrainment is a result only of advance of the jet front and jet back.

The above model assumes a stagnant charge, so in order to take some account of the effects of air charge motion caused by port-induced swirl, or by piston motion, an entrainment ratio E_r is defined as the ratio of actual air entrained to that entrained in a stagnant charge.

$$\Delta A \text{ (actual)} = E_r . \Delta A \text{ (stagnant)} \quad (142.17)$$

The macromixed quantities of fuel and air within the jet boundaries, as determined by fuel-injected X, and air-entrained A, are assumed to micromix by turbulent diffusion according to the following equations (2).

$$\dot{M}_a = D.V_f.(A-M_a) \quad . \quad . \quad (142.18)$$

$$\dot{M}_f = D.V_f.(X-M_f) \quad . \quad . \quad (142.19)$$

Heat release is computed from the micromixed quantities of fuel M_f and air M_a, account being taken of chemical kinetics during the first phase of combustion. Experimental values of ignition delay are used for the engine in question, and the rate of heat release from the fuel and air premixed during delay is assumed to follow a triangle rate law with 6° c.a. base. After rapid heat release, chemical kinetics are neglected. Fig. 142.17 shows the resultant computed curves for fuel-injected X, air-entrained A, micromixed fuel M_f, micromixed air M_a, and rate of heat release versus crank angle for a test condition of engine A.

The heat release model described above was applied to engines A, B, and C, using an entrainment ratio of unity, and gave good agreement of predicted with experimental heat release, by suitable choice of the value of the diffusivity constant D. The results of this work are to be published elsewhere. The thermodynamic mean temperature, T, is calculated from the heat release, the thermodynamic properties of the cylinder gases, and by deducting the heat loss to the walls. Variations of heat loss with changes in engine speed, injection timing, and fuelling were taken into account by using appropriate experimental data.

APPENDIX 142.2
REFERENCES

(1) AUSTEN, A. E. W. and LYN, W-T. 'Relation between fuel injection and heat release in a direct-injection engine and the nature of the combustion processes', *Proc. Auto. Div. Instn mech. Engrs* 1960–61 (No. 1), 47.

(2) GRIGG, H. C. and SYED, M. H. 'The problem of predicting rate of heat release in diesel engines', *Symp. Diesel Engine Combustion, Proc. Instn mech. Engrs* 1969–70 **184** (Pt 3J), 192.

(3) BIDDULPH, T. W., LYN, W-T. and VALDMANIS, E. 'Engine starting and ignition delay', *Proc. Instn mech. Engrs* 1966–67 **181** (Pt 2A, No. 1), 17.

(4) BROOME, D. and KHAN, I. M. 'The mechanisms of soot release from combustion of hydrocarbon fuels with particular reference to the diesel engine'. Paper 140 of this Symposium.

(5) KHAN, I. M., WANG, C. H. T. and LANGRIDGE, B. E. 'Coagulation and combustion of soot particles in diesel engines', Paper to be published.

(6) MACFARLANE, J. J. and HOLDERNESS, F. H. 'Laboratory studies of carbon formation in fuel-rich flames at high pressures', *Symp. Diesel Engine Combustion, Proc. Instn mech. Engrs* 1969–70 **184** (Pt 3J), 57.

(7) KHAN, I. M. 'Formation and combustion of carbon in a diesel engine', *Symp. Diesel Engine Combustion, Proc. Instn mech. Engrs* 1969–70 **184** (Pt 3J), 36.

(8) ZELDOVICH, YA. B., SADOVNIKOV, P. YA. and FRANK-KAMENETSKII, D. A. *Oxidation of nitrogen in combustion* 1947 (Acad. Sci., U.S.S.R., Moscow–Leningrad).

(9) CLAYTON LAPOINTE, discussion of the paper 'The formation and control of nitric oxide in a regenerative gas turbine burner', by CORNELIUS, W. and WADE, W. R., *S.A.E. Farm, Construction and Industrial Machinery Meeting*, and *Powerplant Meetings* and *Production Forum*, Milwaukee, Wisconsin, 1970 (14th–17th September).

(10) LAVOIE, G. A., HEYWOOD, J. B. and KECK, J. C. 'Experimental and theoretical study of nitric oxide formation in internal combustion engines', *Combust. Sci. Technol.* 1970 **1**, 313.

(11) WIMMER, D. B. and MCREYNOLDS, L. A. 'Nitrogen oxides and engine combustion', *S.A.E. Trans.* 1962 **70**.

(12) MARTENEY, P. J. 'Analytical study of the kinetics of formation of nitrogen oxide in hydrocarbon–air combustion', *Combust. Sci. Technol.* 1970 **1**, 461.

(13) LINDEN, L. H. and HEYWOOD, J. B. *Smoke emissions from jet engines* (Mech. Engng Dept, Massachusetts Inst. Technol., Cambridge, Massachusetts).

(14) GLAUERT, M. B. 'The wall jet', *J. Fluid Mech.* 1956 (December).

C143/71 THE SAMPLING AND MEASUREMENT OF EXHAUST EMISSIONS FROM MOTOR VEHICLES

J. P. SOLTAU* R. J. LARBEY†

The last few years have seen sizeable improvements in exhaust gas analysis instrumentation, but probably not in measurement accuracy. The science of exhaust pollutant evaluation is still in its infancy and the technology is hard pressed to keep up with regulations. Four different sets of rules have now been established and cover most of the western world where there is a high density vehicle population. As emissions from the petrol engine are lowered, the diesel exhaust becomes suspect and regulations have been proposed to cover these lower emitters. The B.T.C. Analytical Methods Committee continues to serve a useful purpose in correlating gas analysis activities and in providing an atmosphere where uninhibited discussion on techniques and analytical apparatus can take place.

INTRODUCTION

THIS PAPER IS A SEQUEL to one of the same title presented in 1968 at the London Symposium on exhaust pollution (1)‡. Like its predecessor, it deals briefly with the recent trends in measurement techniques and with specific aspects of the work covered by the B.T.C. (British Technical Council for the Petroleum and Motor Industries) Analytical Methods Committee.

The interest shown in this relatively new chemical engineering field has magnified due to commercial pressures and the general awareness of the part played by the internal combustion engine in global air pollution. The Committee has remained an indigenous one, but rapport with similar foreign organizations has increased. Instead of strengthening existing methods and obtaining a high degree of expertise, the bulk of the work carried out by the various laboratories since the last report has been to develop new sampling and analysis methods as required by legislators. At the time of writing there exist four main test procedures, the U.S. Federal, the Californian, the Japanese, and the European. The analytical method best known and most commonly used in emission control development is the Californian set of rules which have been in use for nearly a decade.

Until recently, the interest in exhaust pollution was restricted to engineers in the petrol engine field. With the advent of stricter regulations, the diesel engine also becomes a recognized atmospheric polluter and numerous projects are now in hand to investigate the problem. To meet the emerging demands for discussion on diesel exhaust analytical test procedures and instrumentation, a subcommittee of the B.T.C. Analytical Methods Committee was formed to deal solely with these topics.

The continuous recording gas analysis method has been generally dropped in favour of bag sampling of one type or another. Carbon monoxide, and where necessary carbon dioxide, are still measured by the non-dispersive infrared method, but for hydrocarbons there has been a swing from infrared instruments towards flame ionization detectors. There is still no unanimity on the best way to measure nitrogen oxide concentration in vehicle exhaust gases. The reference techniques remain the Saltzmann or Phenol Disulphonic Acid (PDSA) methods, but a chemiluminescence technique has been proposed for low concentration sampling. Infrared is still considered a reliable and handy method in Europe.

The present recommended, proposed, or operative standards and method of test for all countries where legislation on vehicle exhaust exists is given in Tables 143.1 and 143.2. An important point is that U.S.A. and California have mass emissions measurement standards in which the same limit applies to all cars irrespective of weight, while European standards are classed according to vehicle size, with heavier models permitted to emit a greater total mass of pollutants.

The MS. of this paper was received at the Institution on 12th July 1971 and accepted for publication on 16th August 1971. 33
* *Joseph Lucas Ltd; Chairman B.T.C. Analytical Methods Committee.*
† *The Associated Octel Company Ltd; Secretary B.T.C. Analytical Methods Committee.*
‡ *References are given in Appendix 143.1.*

THE SAMPLING AND MEASUREMENT OF EXHAUST EMISSIONS FROM MOTOR VEHICLES

Country and Standards	1969	1970	1971	1972	1973	1974	1975
U.S.A.							
Test procedure	7-Mode cycle, continuous NDIR analysis of HC and CO			Non-repetitive cycle, variable dilution sampling, FID analysis for HC, NDIR analysis for CO, chemiluminescence for NO_x			
g/mile HC	275 v.p.m.††	2.2	2.2*	3.4*	3.4*	3.4*	0.41*
CO	1.5% CO††	23	23*	39*	39*	39*	3.4*
NO_x	—	—	—	—	3.0*	3.0*	3.0*
State of California							
Test procedure			7-Mode cycle: NDIR analysis for HC, CO, and NO_x			Non-repetitive cycle, variable dilution sampling, FID analysis for HC, NDIR analysis for CO, and NO_x‡‡	
g/mile HC	275 v.p.m.††	2.2	2.2*	1.5† or VDS 3.2	1.5†	15†	
CO	1.5% CO††	23	23*	23† 39	23†	23†	
NO_x	—	—	4.0*	3.0† 3.2	3.0†	1.3†	
Canada							
Ontario	As U.S. Federal						
Japan							
Test procedure	4-Mode cycle, continuous NDIR analysis of HC and CO				Not specified		
					g/km		g/km
HC	—	—			1.7‡		0.3‡
CO	3%	2.5%*			11‡		7‡
NO_x	—	—			3.0‡		0.6‡
Australia					4.5% CO at idle‡		
Europe							E.C.E. Standards (types I and II‡)
E.C.E. test procedure	3-Phase cycle, total exhaust gas collection, single NDIR analysis of HC and CO		From September* E.C.E. Standards types I and II				
Belgium	4.5% CO at idle*						
Denmark	4.5% CO at idle*						
France			New type engines vehicles from October* E.C.E. Standards (types I and II)	All vehicles from September*			
Spain	5% CO at idle*						
Sweden			g/km*		g/km†		g/km†
HC	—		2.2 (8.9)**		1.8 (7.3)**		1.5 (6.1)**
CO	4.5% at idle		45 (183)**		30 (122)**		23 (93)**
Switzerland			4.5% CO ±1% at idle*				
West Germany	4.5% CO at idle				E.C.E. Standards* (types I and II)		
Netherlands					E.C.E. Standards* (types I and II)		

* Operative.
† Proposed.
‡ Recommended.
** Figures given are the maxima (with equivalent values in terms of g/test in parentheses).
†† Figure for vehicles in excess of 140 in³. For vehicles of displacement 100–140 in³ limits were 350 v.p.m. HC and 2·0% CO and for 50–100 in³, 410 v.p.m. HC and 2·3% CO.
‡‡ From 1972 U.S. Federal may have to be adopted. This will mean different standards.

Table 143.2. E.C.E. test procedure—standards for type I testing*

Vehicle weight† (vw), kg	Mass of carbon monoxide (per test), g	Mass of hydrocarbons (per test), g
vw ⩽ 750	100	8·0
750 < vw ⩽ 850	109	8·4
850 < vw ⩽ 1020	117	8·7
1020 < vw ⩽ 1250	134	9·4
1250 < vw ⩽ 1470	152	10·1
1470 < vw ⩽ 1700	169	10·8
1700 < vw ⩽ 1930	186	11·4
1930 < vw ⩽ 2150	203	12·1
2150 < vw	220	12·8

* Type II testing is a check on CO at idling. Standard is 4·5% CO.
† Vehicle unladen weight + 120 kg.

INSTRUMENTATION FOR EMISSION TESTING

Hydrocarbons

For the 1971 U.S. Federal, Californian, European, and Japanese methods, *n*-hexane sensitized non-dispersive infrared (NDIR) analysers are employed. The instrument calibration is carried out with standard mixtures of *n*-hexane in dry nitrogen.

Deficiencies of such a measuring technique for hydrocarbons (HCs) are:

(1) *Hexane sensitized NDIR has poor sensitivity to classes of HCs other than paraffins.*

The position is somewhat worsened by the fact that it is those very HCs involved in photochemical smog-producing reactions that are not quantitatively detected. The relative response of some classes of HCs to NDIR is given in Table 143.3. A factor has to be introduced into the calculation of results to allow for this poor sensitivity.

(2) *Interference, due to CO_2 and water vapour in the exhaust gas, has to be corrected for, either by the use of optical filters or a gas filter cell.*

Table 143.3. Relative response of NDIR and FID for various hydrocarbons (12)

Hydrocarbon	Relative response	
	Hexane sensitized NDIR	FID
Paraffins		
Methane	6	95
n-Hexane	100	105
Olefins		
Ethylene	6	100
2-Me-2 Butene	51	98
Aromatics		
Benzene	6	97
Toluene	8	97
Acetylenes		
Acetylene	2	110

For the 1972 U.S. Federal testing, the analysis of HCs is by the use of a flame ionization detector (FID). Here, the HC is burnt in a hydrogen flame across which is applied a potential difference. An ionization current is set up across the electrodes which is proportional to the concentration of the HCs. Calibration is usually carried out with mixtures of propane (0–200 volumes per million (v.p.m.)) in pre-purified air.

The technique is much more sensitive to HCs, a necessity when measuring the lower levels encountered in the air diluted exhaust, and is not selective in its response to different classes of compounds.

Carbon monoxide

All proposed or existing regulations stipulate the use of NDIR analysers for the measurement of CO, as no real problems or deficiencies exist with this technique. Ranges vary from 0–10 to 0–0·1 per cent according to the test method employed. Standard gases of CO in dry nitrogen are used for calibration. For low ranges, cells of increased length are fitted to the analysers. Some instruments have dual cell lengths.

Carbon dioxide

Although not a pollutant whose concentration must be controlled, measurement of carbon dioxide (CO_2) is necessary in most test procedures to enable corrections to be made for any air dilution of the exhaust gas stream. NDIR analysis is used universally over the range 0–16% CO_2. Calibration is carried out on standard mixes of CO_2 in dry nitrogen. Measurement of CO_2 is not required for 1972 Federal testing as the degree of pollution is related to the actual gas mass exhausted by the vehicle.

Oxides of nitrogen

Measurement of oxides of nitrogen (NO_x) to the 1971 Californian procedure is by NDIR analysis. Nitric oxide (NO) sensitized analysers are specified because NO is the main oxide of nitrogen emitted in the raw exhaust.

The detection range required is 0–4000 v.p.m. NO. Calibration is by means of standard gas mixtures of NO in dry nitrogen.

A shortcoming in this technique is that water content of the gas gives a cross response so that this has to be removed or corrected for.

1973 U.S. Federal testing includes measurement of NO_x in the diluted exhaust. The oxides of nitrogen present in the bag sample can be both nitric oxide (NO) and nitrogen dioxide (NO_2).

NDIR is not sensitive enough to measure the low levels of NO and does not detect NO_2. To overcome these problems a chemiluminescent technique has been chosen. As it measures only NO, any NO_2 present has to be converted to NO by heating the gas to 600°C. The NO is then reacted with ozone at low pressures to form NO_2 and oxygen. Some of the NO_2 is in the excited state and it undergoes transition to the ground state with the emission

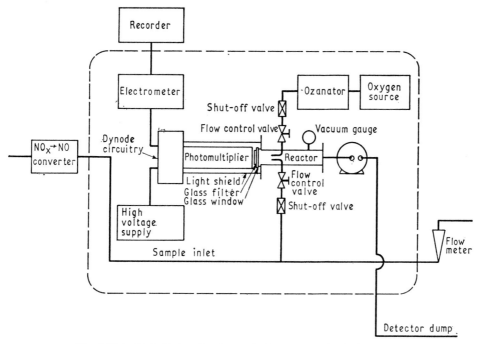

Fig. 143.1. U.S. Federal oxides of nitrogen analytical system

of light [590–630 nm region]. The quantity of light emitted, measured using a photomultiplier tube, is proportional to the original concentration of NO in the gas entering the reactor (Fig. 143.1).

The range involved is 0–100 v.p.m. NO. Calibration of this range is effected using NO in dry nitrogen standards. Converter efficiency is checked utilizing gases in the range 0–100 v.p.m. NO_2 in pre-purified air.

METHOD OF TEST FOR THE U.S.A.

1971 model year vehicles under 6000 lb gross vehicle weight (g.v.w.)—all States except California

The specified test procedure is laid down in the Federal Register (2) and is based on the familiar Californian 7-mode driving cycle, continuous sampling, and NDIR measurement of HC and CO as developed and introduced by the Californian Motor Vehicles Pollution Control Board (M.V.P.C.B.), later renamed the Californian Air Resources Board (A.R.B.), in the mid-sixties.

The 7-mode cycles (Fig. 143.2) are driven from a cold start on a dynamometer equipped with a power absorption unit and various inertia flywheels which are adjusted to represent the weight of the vehicle under test. The brake load is chosen to reproduce road load at 50 mile/h true speed on the level. Although analysis is continuous, only certain portions of the cycle are recorded for the subsequent calculations. For a complete exhaust emission test seven such cycles are driven, but only the first four and the last two are logged. A test fuel is specified for approved

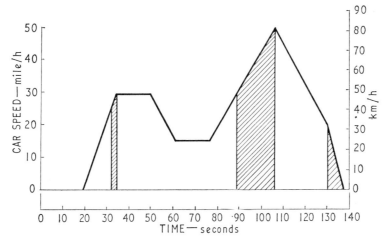

Fig. 143.2. California driving cycle

testing. Sampling is undertaken on the raw exhaust gas and is continuous, with particulate matter and water vapour content greatly reduced by filtration and cooling, respectively.

Calculations of the emission levels for the vehicle under test are normally carried out with the aid of a computer. The chart traces are integrated for the required portions of each cycle. The calculated mode data are multiplied by a 'weighting' factor indicative of the mode's importance in actual town journeys. Summation of the first four 'cold' cycles and last two 'hot' cycles is then carried out employing further fixed factors. The results are converted to a grammes per mile basis using density of component and an equation which gives exhaust volumes per mile as a function of dynamometer inertia wheel weight.

1972 and subsequent model year vehicles under 6000 lb g.v.w.—all States except California

In view of the limitations presented by the old test technique, work was undertaken by the N.A.P.C.A. (National Air Pollution Control Association) to find a better exhaust gas analysis method. An improved sampling system was developed by Broering et al. (3).

The principle is one of variable dilution of the exhaust gas with ambient air such that the total volume of gases is constant, and for this reason it is known as constant volume sampling (CVS). A small aliquot of the diluted exhaust is taken and pumped into a plastic bag. At the end of test the content of the bag is analysed for HC, CO, and NO_x (the latter for 1973 model year cars) so that a figure for the masses emitted can easily be calculated. In addition, a new

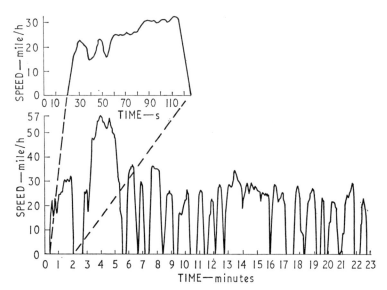

Fig. 143.3. U.S. Federal driving schedule (11)

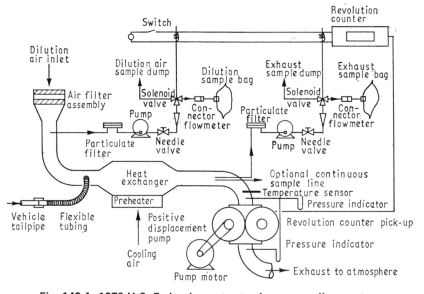

Fig. 143.4. 1972 U.S. Federal constant volume sampling system

THE SAMPLING AND MEASUREMENT OF EXHAUST EMISSIONS FROM MOTOR VEHICLES

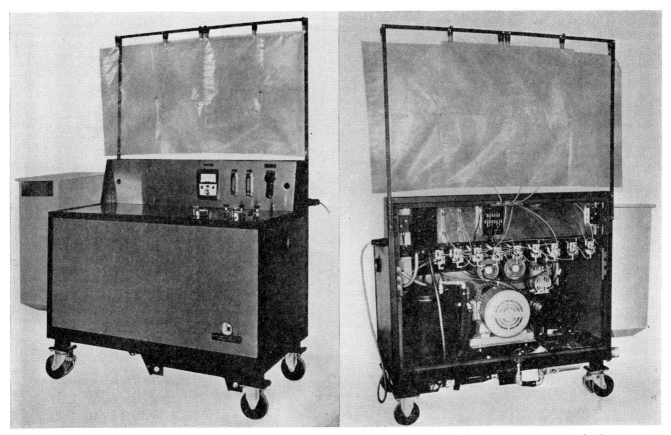

Fig. 143.5. Olson constant volume sampling rig for the 1972 Federal exhaust gas sampling method

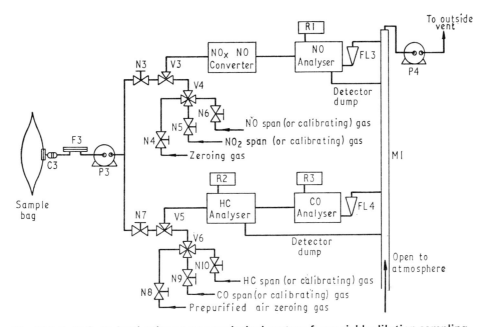

Fig. 143.6. U.S. Federal exhaust gas analytical system for variable dilution sampling

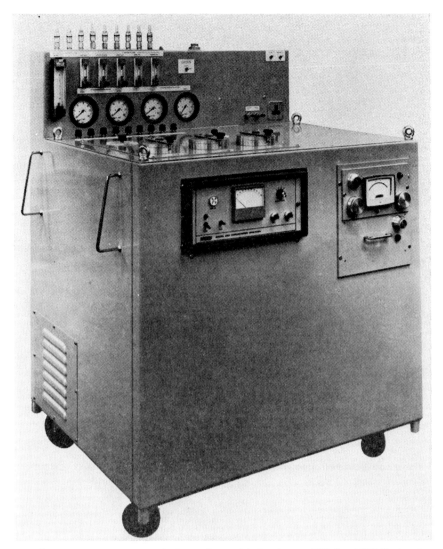

Fig. 143.7. Grubb Parsons gas analysis trolley with Beckman FID for 1972 Federal testing

driving cycle (Fig. 143.3) was introduced. The adopted standards and test producedure were published in November 1970 (4).

The chassis dynamometer setting procedure is similar to that in the 1970–71 method. From a cold start the vehicle is driven over the non-repetitive cycle.

A CVS rig, the schematic layout of which is illustrated in Fig. 143.4, is used to sample the entire exhaust gas and dilute it with a variable quantity of air to ensure no water vapour condensation occurs. Trolleys containing all the necessary CVS hardware are commercially available (Fig. 143.5). Two continous samples of gas are taken at a fixed rate, one of the dilution air and the other of the diluted exhaust, and pumped into separate Tedlar (polyvinyl fluoride) bags.

At the end of the test, the content of the plastic collection bags is analysed for HC and CO by FID and NDIR respectively. A recommended layout of the necessary filters, pumps, analysers, etc., is given in the Federal Register (HC and CO circuit in Fig. 143.6). Again trolleys incorporating both the analysers and ancillary equipment are commercially available (Fig. 143.7). The final pollutant concentration in the sample is found by subtraction of the dilution air values from those of the diluted exhaust. For HCs the equivalent volumes per mile carbon contents are found by multiplying the volumes per mile propane (C_3H_8) concentration by 3.

In the calculation, total diluted exhaust volumes per mile passing through the pump is found by knowing the number of revolutions during test, the pump capacity, and driving cycle mileage. Corrections are made to allow for pressure drop across the pump, and the final volume is adjusted to standard condition, i.e. 68°F and 760 mmHg. Thus, using this volume in conjunction with the concentration and density of pollutant in the sample, the final mass emission level is calculated as grammes per mile.

As can be seen from the flow diagrams, facility is available for continuous analysis of the diluted exhaust. This is necessary for design and development of pollution control systems.

Introduction of NO_x measurement for 1973 and subsequent model year vehicles

The Federal Authorities at the Environmental Protection Agency (E.P.A.) have given notice (5) of the proposed introduction of NO_x measurement for 1973 and subsequent model year cars. Proposed standards are given in Table 143.1. As a result, the analytical procedure has had to be altered to include the measurement of NO_x. (In raw exhaust the oxides of nitrogen contain over 90% NO.)

After considering other feasible types of NO_x detectors —e.g. Whittaker Cell—the E.P.A. finally decided to adopt the chemiluminescence technique developed for vehicle exhaust gas by the Ford Motor Company (6), shown in Fig. 143.1.

To include NO_x measurement the analytical sampling train has been modified. The new sampling system (Fig. 143.6) and procedure is given in the Federal Register. The calculation is carried out in the same manner as for HC and CO. As NO_x emissions vary with atmospheric water content, results are corrected to a standard humidity.

The Register specifies that any other approved technique which gives results equivalent to chemiluminescence may be used. Phenol Disulphonic Acid (PDSA) and modified Saltzmann are two 'wet' methods which could be employed (1).

Method of test for California (vehicles under 6000 lb g.v.w.)

California was the first body to introduce exhaust emission legislation and has its own pollution standards. Up to and including 1971 model year cars and light trucks, the California test producedure for HCs and CO is identical to the 1970–71 U.S. Federal procedure. For 1971 and subsequent model years the California A.R.B. gave notice (7) of additional NO_x standards being imposed. NDIR techniques are used for the analysis, and the results are corrected to a standard humidity.

The procedure is operational for 1971 but the requirements for 1972 model year testing are different. As the U.S. Federal legislation covers the whole of the U.S.A., any individual State not wishing to conform has to apply for a waiver. The Californian A.R.B. was granted a waiver from the 1972 U.S. Federal testing, submitting and being allowed to use either the Californian 7-mode cycle test for CO, HC, and NO_x, or Federal 1972 testing for HC and CO—in the latter case with three hot Californian cycles being driven at the end of the test with analysis for NO_x by NDIR on the last two of the three supplementary cycles. A correction is made for the 'hot' analysis.

For 1973 model year vehicles the problem is resolved, as the U.S. Federal testing includes NO_x measurement.

Federal and Californian method of test for gasoline-powered heavy duty vehicles

The only country with emission standards for heavy duty vehicles is the U.S.A. According to the Federal Register (4) the definition of such a vehicle is one whose gross vehicle weight is greater than 6000 lb.

Standards and the test procedure for these vehicles were introduced into California for 1969 models and adopted by the rest of the U.S.A. for 1970 models. Exhaust emission standards were then, and still are for 1972 model year, 275 v.p.m. HC and 1·5% CO.

The method of test is substantially different from that for light duty vehicles. In fact, it is not the actual vehicle which is tested but its engine, and the testing is carried out on an engine dynamometer.

Fig. 143.8 gives the emission test cycle. It is intended to represent typical driving patterns in urban areas. There are nine modes, run at a constant speed of 2000 rev/min, to a complete cycle; four cycles from a cold start constitute a complete test. To eliminate the need to change engine speeds between cycles, the idle modes may be driven at the beginning and end of each test. The idle mode result at the beginning of the first cycle is used for calculation of the second cycle emissions, and that at the end of the fourth cycle for calculation of the third cycle emissions.

Continuous NDIR analysis of HC, CO, and CO_2 is carried out on the raw exhaust using exactly the same instrumentation as for 1970–71 light duty vehicle testing.

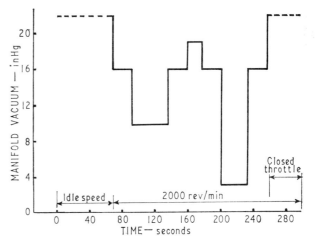

Fig. 143.8. Petrol engine heavy duty vehicle test cycle

Table 143.4. Heavy duty engine test cycle and weighting factors

Mode	Man. vac., inHg	Time in mode, s	Weighting factors
1. Idle		70	0·036
2. Cruise	16·0	23	0·089
3. P.T. Accel.	10·0	44	0·257
4. Cruise	16·0	23	0·089
5. P.T. Decel.	19·0	17	0·047
6. Cruise	16·0	23	0·089
7. Full load	3·0	34	0·283
8. Cruise	16·0	23	0·089
9. Closed throttle		43	0·021

Calculations of the results are as follows: the continuous traces are integrated over the last 3 s for all open throttle and idle modes, and totally for the closed throttle mode. A factor is incorporated to allow for any dilution taking place due to control devices. Individual mode HC and CO emissions are weighted (Table 143.4) and summed for the complete cycle. After averaging the results from the first two (warm up) and last two (hot) cycles, 35 per cent of the former plus 65 per cent of the latter gives the overall test emissions on a volumetric basis.

Despite the changeover to CVS and mass emission standards for light duty vehicles, the continuous sampling and volumetric emission standards have been retained for heavy trucks in 1972. Work has been carried out (**8**) investigating the feasibility of various types of testing of vehicles on chassis dynamometers with comparison of continuous, CVS, and servoproportional sampling. It could well be that future legislation will involve testing of vehicles instead of engines, constant volume sampling and mass emission standards.

COMPARISON OF 1972 AND 1970–71 FEDERAL TESTING

Driving cycle

The non-repetitive 1972 cycle represents more closely a downtown rush hour driving schedule than does the 7-mode Californian cycle. Also, the 1972 cycle is self-weighting in its application, whereas for the old cycle weighting is necessary during calculation. All of the new cycle is sampled and analysed which eliminates 'cycle beating', a possibility on the old technique. Purely considering development work, the 7-mode cycle probably has the advantage because of its short time span.

Sampling

Variable dilution of the exhaust has two effects: first, it reduces the rate of reaction between HC and NO in the exhaust gas and, second, it prevents condensation of water vapour and heavy hydrocarbons, which can occur with continuous sampling of raw exhaust gases. Also, it represents more closely the actual dilution of the exhaust gas emanating from the tailpipe on the roadway.

Calculation

The calculation procedure is much simpler for the 1972 test, and because no empirical factors are involved the emission levels are much more accurate as they are directly calculated on a main basis.

Emission levels

It might appear that there is a relaxation in limits for the U.S.A. between 1971 and 1972 (see Table 143.2), but this is not so as the new method gives higher values than the old one—in fact, by a factor of about 2. In hard facts, the 1972 limits represent a reduction of levels over that of 1971 of about 25 per cent for HC and 17 per cent for CO.

Accuracy of results

Indications are that the repeatability of the new technique is only marginally better than that obtained with the original Californian method.

METHOD OF TEST FOR EUROPE (E.C.E. PROCEDURE)

Some European countries under the membership of the Economic Commission for Europe (E.C.E.) have introduced legislation following an agreed test procedure (**9**). Accompanying standards are listed in Table 143.2. There are two test types concerning vehicle exhaust pollutants. Type I is the full test covering HC and CO emissions, described below; type II is a simple check of idle CO emission levels and needs no further explanation.

Dynamometer set-up procedure is similar to that used in Federal testing. The driving cycle (Fig. 143.9) is somewhat reminiscent of the old California cycle inasmuch as it is repetitive, four cycles from a cold start constituting a test. It has been devised to represent typical driving patterns in European towns with European cars.

The total raw exhaust gas over the four cycles is passed through a heat exchanger to remove most of the water vapour and then into one or more large plastic bags. At the end of test the contents of the bag or bags are analysed for HC, CO, and CO_2 by the same NDIR equipment as used in the current Californian and Federal testing.

Total exhaust gas volume in the bag is measured with a

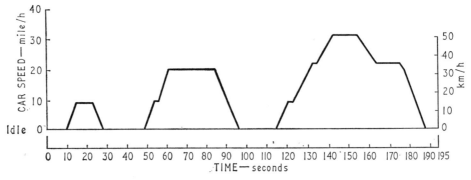

Fig. 143.9. European E.C.E. driving cycle

dry gas meter, making the necessary corrections for temperature and pressure. The volume concentration of HC and CO in the bag is converted to a total weight using the appropriate density and total exhaust gas volume. Thus a weight of HC and CO per test is found.

Standards laid down vary with the weight of the vehicle undergoing test. The test is not considered to be as severe as the U.S. Federal test.

All the points previously made apply concerning the repetitive driving cycle, sampling of raw exhaust gas, and measurement by NDIR.

METHOD OF TEST FOR JAPAN

The chassis dynamometer setting procedure is similar to that of the U.S. Federal but the brake load values may be calculated from a given equation. The car, which is thoroughly warmed up, is driven through a cycle of 17 individual modes with continuous NDIR analysis of HC, CO, and CO_2. However, only four of the modes are actually used in the calculation of the exhaust emissions. These four modes are given in Fig. 143.10, the shaded portions being the parts not used in the calculations. In practice, three of these 4-mode cycles are driven and the average taken, although this is not actually specified. Presumably the cycle has been designed to suit the domestic situation in Japan.

Having integrated the chart traces over the portions of the modes specified, the concentrations are weighted, i.e.

Operating condition	Weighing factor
Idling	0·11
Accelerating 0–40 km/h	0·35
Cruising 40 km/h	0·52
Decelerating 40–0 km/h	0·02

The weighted values are summed and the final figure represents the emission levels of the vehicle on a volume concentration basis.

Existing and recommended standards are given in Table 143.1. Again, standards are not as severe as for the U.S.A. For 1973 onwards new limits are given on a weight basis but no details of the test procedure have yet been published.

DIESEL ENGINED VEHICLE TEST METHODS

The current trend is to distinguish between the gaseous emissions and particulate emissions such as smoke. Only the former are discussed here as the latter are not included in the terms of reference of the B.T.C. Analytical Methods Group.

During the great upsurge of interest in gasoline emissions in California, the diesel engine was until recently rarely mentioned, mainly because diesel engined vehicles account for a mere 1 per cent of the total registrations.

The only legislation applicable to diesel gaseous emissions is the recently introduced Californian legislation: this is not specific to diesel engines, but applies to all vehicles over 6000 lb g.v.w. A recent paper (**10**) describes the development of the test cycle and procedures for diesel engines. Basically, the engine manufacturer must submit test data for each family of engines having the same design feature. Engine models in a family must operate on the same cycle (two or four stroke), and must have the same basic combustion and aspiration systems: models with different bore and stroke may be included if they use the same basic cylinder block. One engine in each family must be submitted for approval.

The test cycle developed for the 1973 Californian legislation (Table 143.5) is based on a 13-mode cycle, to be carried out on a test bed or chassis dynamometer. For the test bed work, a preconditioning run of 50 h at a nominal rated power is specified. However, due to the 10 per cent tolerance on load and speed, the 50-h run can be carried out at about 70 per cent rated power.

Two test speeds are specified, namely rated speed and an intermediate speed (defined as 60 per cent rated speed, or peak torque speed, whichever is higher).

The 13-mode cycle must be carried out twice, with an 8-h shutdown between tests. The overall cycle emissions are calculated by dividing the sum of the weighted mass emissions by the sum of the weighted horsepowers in each mode: weighting factors of 0·2 and 0·08 are used for the idle and power modes respectively.

Table 143.5. Test conditions for 1973 Californian diesel emissions legislation

Engine speed	Load (±2%), %*
Low idle	0
Intermediate	2
Intermediate	25
Intermediate	50
Intermediate	75
Intermediate	100
Low idle	0
Rated	100
Rated	75
Rated	50
Rated	25
Rated	2
Low idle	0

* 10 min in each mode.

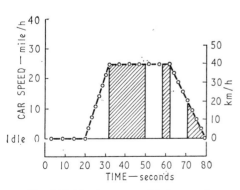

Fig. 143.10. Japanese driving cycle

The legislative limits for 1973 and 1975 are (in grammes per horsepower-hour):

	HC+NO$_2$	CO
1973	16	40
1975	5	25

NDIR analysers are specified for NO and CO: for HC measurement, a heated flame ionization detector (heated FID) must be used. The output from the analysers must be recorded on chart recorders for the final minute in each mode. A test fuel is specified for approval testing.

As for petrol engines, the principal oxide of nitrogen present in diesel exhaust is nitric oxide (NO). Since the diesel cycle tends to have higher pressures and lower temperatures than the spark ignition engine, a higher percentage of NO$_2$ is found in the exhaust: however, this is normally less than 10 per cent and is generally ignored due to measurement difficulties. Therefore, an NDIR, NO sensitized analyser is employed.

CO analysis requires little comment. Diesel engine CO levels are very much lower than for petrol engines, being in the range 0·01–0·1 per cent. For normal diesel work a CO analyser is not really necessary since, provided that the exhaust condition is kept to B.S. AU 141 levels, CO levels are considerably below those set for the 1975 Californian legislation.

Owing to the high molecular weights of the HC content of diesel exhaust, infrared analysers cannot be used due to condensation problems. The most satisfactory solution is to use some form of heated FID analyser. Essentially, the system maintains the exhaust sample at a temperature such that loss by condensation is balanced against loss by oxidation.

Fig. 143.11 shows the various analysers mentioned above incorporated in a mobile trolley, together with the required calibration gases, filters, pumps, etc.

THE WORK OF THE B.T.C. ANALYTICAL METHODS COMMITTEE

The aims of the main Committee, which has been in existence for seven years, were quoted in an earlier paper (1). The new diesel group which was formed in November 1970 to correlate analytical activity in the U.K. and maintain technical links between the industry and legislative bodies, has the following terms of reference:

(1) To act as a clearing house for literature and legislation in order to keep members up to date. Hence, reliance is placed on members who have preferential access to early drafts and information, and on M.I.R.A.

(2) To exchange information on analytical techniques and equipment and to keep a dossier of members'

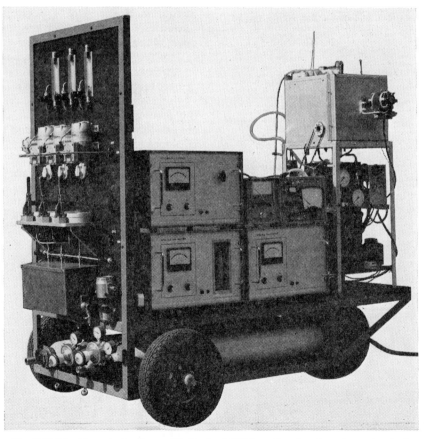

Fig. 143.11. Ricardo trolley with heated FID for diesel exhaust gas analysis

equipment to allow those with the same equipment to discuss problems.

(3) To maintain contact with instrument manufacturers and calibration gas suppliers in order that technical assistance may be given to facilitate manufacture and expedite instrument supplies to B.T.C. members.

(4) To give facilities to instrument and gas suppliers to explain and defend their activities. This has already been shown to be valuable by the main Committee.

(5) To organize interlaboratory correlation programmes in which gas bottles and test engines are circulated and the results compared, particularly with reference to total hydrocarbons as this has already been shown to be subject to a considerable scatter in different laboratories in the U.S.A.

(6) To consider whether the U.K. diesel scene needs different analytical and test procedure treatment from that of the U.S.A., bearing in mind their differing patterns of usage, and to give advice on any future legislation in the U.K. as far as the above topics are concerned.

The group is open to all firms vitally interested in the automotive diesel, e.g. vehicle and engine manufacturers, oil companies, accessory manufacturers, diesel research and development laboratories. As with the parent Committee, analytical and test equipment suppliers may not sit on the panel but may address the members when required to present information on their products.

So far 16 members and about six fully operational laboratories are engaged in this type of work. Much of the equipment has been adapted from petrol emission analysis apparatus and the total hydrocarbon analysers have been 'home-made', although moves have been started to encourage instrument manufacturers to provide suitable alternatives.

Instrument specifications

Liaison has been maintained with instrument manufacturers in the U.K. and abroad. There has been a continuous improvement in technical performance of gas analysis equipment, but development progress has not kept up with changes in sampling methods and tightening of exhaust quality standards. The stage has not yet been reached where gas analysis instruments can be taken for granted. Breakdowns are frequent and skilled operators are still required to operate testing installations. For rapid CO checks, the Committee is still of the opinion that only a specific instrument like the infrared analyser will give valid answers. Several compact apparatuses of this type are available; unfortunately none is manufactured locally and none at a realistic price considering their eventual widespread use.

Analysis apparatus designed to meet the new 1972 Federal specification is now available.

Gas and car exhaust correlation programmes

Both cross-check services mentioned in the previous paper (1) have been continued. The 'four times a year' unknown gas mixture which circulates to the Committee members has grown to a three tier arrangement. From this year each subscriber receives a bottle of a master mix of the following: CO, CO_2, and either hexane or propane in dry nitrogen to suit the Californian test; CO and propane in pre-purified air in concentration in step with the Federal 1972 analytical technique; and NO in dry nitrogen in the proportion found in raw and diluted exhaust.

Interlaboratory repeatability is much the same today as when the scheme was inaugurated, and leaves much to be desired. The coefficient of variation has been found to be greater when the concentration is lower (Fig. 143.12). If repeatability is poor on master blends of gases, then there is little hope for good vehicle correlation.

The car exhaust quality cross-check scheme has been somewhat of a failure inasmuch as repeatability between laboratories has not improved over the years. Table 143.6 shows that the coefficient of variation has deteriorated over five vehicle checks using the U.S.A. test method. It should be noted that this may partly be due to the same trend as with the gas correlations, i.e. the coefficient of variation tends to increase as emission concentration decreases. Clearly, as emission standards become more stringent and test cars have lower emissions, it becomes more difficult to maintain a low coefficient of variation. The disappointing trend of repeatability has raised the question as to whether the problem is due to variations between laboratories or merely variations of the car. The results have, therefore, prompted the Analytical Group to carry out tests where

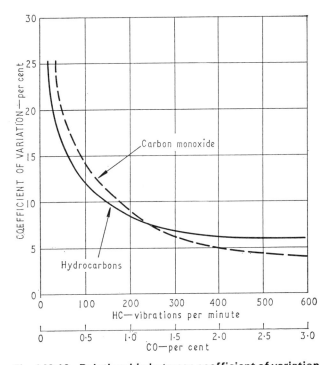

Fig. 143.12. Relationship between coefficient of variation and concentration. Hydrocarbons and carbon monoxide in B.T.C. gas correlation scheme

Table 143.6. B.T.C. exhaust gas correlation results for cars 1 to 5

Test series	Hot cycle results, 1970 Federal test					
	HC, p.p.m.			CO, per cent		
	Mean	Standard deviation	Coefficient of variation*	Mean	Standard deviation	Coefficient of variation*
1	441	54	12	1·59	0·20	12
2	188	27	14	0·64	0·15	23
3	202	26	13	1·03	0·31	30
4	205	23	11	1·44	0·41	28
5	157	33	21	0·60	0·20	33

* Coefficient of variation = $\dfrac{\text{Standard deviation} \times 100}{\text{Mean}}$.

the car exhaust quality is checked each time at the same laboratory, but where the driving between tests is representative of the driving when ferrying the car from laboratory to laboratory in the normal cross-check exercise. The coefficients of variation shown in Table 143.7 are somewhat in the same order of magnitude as when the test vehicle is analysed in different laboratories. This would indicate that the variations are caused by differences in engine condition and by basic inadequacies of the testing procedure. The laboratory cross-checks carried out following the European testing method have marginally less scatter. Car 5 gives a coefficient of variation of 22 per cent for CO and 18 per cent for HC as compared with 33 and 21 per cent respectively for the Federal test results.

The interlaboratory cross-checks have included measurement of ambient conditions and checks are kept on idle speed differences, but none of these explains emission variations. On the basis of present repeatability the emission advantage of the scheme is to identify laboratories which might be at a great variance from the mean.

Work carried out following the new 1972 Federal procedure gives hope for greater analysis accuracies. Table 143.8 shows results obtained by M.I.R.A. on behalf of the

Table 143.7. B.T.C. exhaust gas correlation scheme

Vehicles tested in one laboratory only with representative delivery journeys between each test.

Car 1—1600 cm^3, lean tuned carburettors

Test number	g/mile		
	HC	CO	NO
1	1·10	17·4	3·43
2	1·15	12·2	3·50
3	0·90	14·25	3·34
4	0·95	15·4	3·25
5	1·10	14·0	3·29
6	1·22	16·75	3·54
7	1·16	16·5	3·04
8	1·36	10·3	3·45
9	1·43	9·2	3·30
10	1·40	8·5	3·10
Mean	1·18	13·4	3·326
Standard deviation	0·18	3·2	0·165
Coefficient of variation (%)	15·00	24·5	5·00

Car 2—1600 cm^3, fuel injection

Test number	g/mile	
	HC	CO
1	1·61	16·5
2	1·40	12·2
3	1·36	14·8
4	1·35	10·4
5	1·30	13·5
6	1·36	13·5
7	1·52	14·5
8	1·36	17·6
9	1·66	19·4
10	1·42	17·1
11	1·50	17·6
12	1·56	17·0
13	1·50	17·6
14	1·40	20·25
15	1·51	20·8
Mean	1·54	16·18
Standard deviation	0·100	2·98
Coefficient of variation (%)	6·5	18·4

Table 143.8. B.T.C. exhaust gas correlation scheme. Comparison between the VDS and Californian vehicle test procedure

1700 cm^3 automatic transmission car with lean tuned carburettors.

Test number	VDS test		Californian hot cycle test		
	g/mile		g/mile		
	HC	CO	HC	CO	NO, p.p.m.
1	2·01	22·3	1·4	6·2	1587
2	2·03	25·2	2·1	6·8	1491
3	1·97	21·5	2·15	6·2	1375
4	2·36	22·6	1·34	5·25	1466
5	1·58	20·0	1·34	5·8	1500
6	1·82	20·0	1·60	5·5	1389
7	1·77	22·3	1·66	4·5	1587
8	2·15	21·4	1·50	4·0	1377
Mean	1·96	21·9	1·63	5·5	1471
Standard deviation	0·24	1·66	0·33	0·93	87·0
Coefficient of variation (%)	12·00	7·6	20·00	17·00	5·9

B.T.C. on the current correlation vehicle which will be tested by both VDS (Variable Dilution Sampling) and raw exhaust methods.

Miscellaneous activities of the Analytical Methods Committee

Calibration and span gases for analysis instruments can still pose certain problems, the principal one being the quality of the mixtures. To minimize the possibilities of inaccurate blending, a committee member is regularly in touch with the suppliers, and representatives are invited to discuss problems with members at meetings. The gas blends which should be stocked by the manufacturers have been stipulated by the Committee and have been given code letters and numbers for ease of ordering by Telex.

With the advent of the new testing techniques there is a need for calibration gases at very low levels of concentration. These have been added to the list of requirements and after some initial difficulties gases of the required accuracy are being supplied. Some laboratories choose to manufacture their own 'gold standard' calibration gases, and a code of recommended practice has been issued by the Committee.

Although all members of the Analytical Methods Committee take part in the gas cross-check scheme, some also join foreign projects and their findings add to the laboratory correlation experience.

To date the Committee has failed to find an indigenous plastic fabricator willing to manufacture Tedlar (polyvinyl fluoride) gas sampling bags of various sizes. M.I.R.A., after having obtained guarantees from members, has purchased a sealing machine and stocks of material to supply the requirements of the member firms. The project is working well and containers suitable for the various test methods are now readily available at a reasonable cost.

CONCLUSIONS

It has been shown that the time when emission exhaust standards become stable and when exhaust gas quality checks are but routine is a long way off. The analysis of vehicle effluents has in no way reached the exact science which was prophesied a decade ago. The fault does not lie so much with the analyst but with the lack of understanding of the internal combustion engine and the too rapid tightening of emission regulations. The situation will not ease in the near future as new procedures will follow new procedures before they are fully developed. The B.T.C. Analytical Methods Committee's aims are still valid and it continues to act as a forum where gas analysis problems can be aired and experience compared. Rapport with similar bodies is increasing with a beneficial gain in experience.

ACKNOWLEDGEMENTS

The authors are indebted to the B.T.C., Joseph Lucas Ltd, and the Associated Octel Co. Ltd for permission to publish this paper, and thank the members of the Analytical Methods Committee for their help and support.

APPENDIX 143.1

REFERENCES

(1) SOLTAU, J. P. and CAMPBELL, K. 'The sampling and measurement of exhaust emissions from motor vehicles', *Symp. Motor Vehicle Air Pollution Control*, *Proc. Instn mech. Engrs* 1968–69 **183** (Pt 3E), 81.

(2) *Federal Register* 1968 **33** (4th June), 108.

(3) BROERING, L. C., WERNER, W. J. and ROSE, A. H. 'Automotive mass emission analysis by a variable dilution technique', Paper presented at the Air Pollution Control Association Annual Meeting, Cleveland, Ohio, 1967 (June).

(4) *Federal Register* 1970 **35** (10th November), 219.

(5) *Federal Register* 1971 **36** (27th February), 40.

(6) NIKI, H., WARNICK, A. and LORD, R. R. 'An Ozone–NO chemiluminescence method for NO analysis in piston and turbine engines', S.A.E. Paper 710072, 1971.

(7) *California exhaust emission standards and test procedure for 1971 and subsequent model year cars* 1968 (State of California Air Resources Board).

(8) OLSEN, R. W. and SPRINGER, K. J. 'Exhaust emissions from heavy duty vehicles', S.A.E. Paper 690764, 1969.

(9) INTEREUROPE. E/ECE/324; E/ECE/TRANS/505; Regulation No. 15.

(10) BASCOM and HASS. 'A status report on the development of the 1973 California diesel emissions standards', S.A.E. Paper 700671, 1970.

(11) FUSSEL, D. R. 'Air pollution and the motor vehicle—what next?', *Petrol. Rev.* 1971 (March).

(12) PATTISON, J. N. 'Motor vehicle control news', *J. Air Pollut. Control Ass.* 1970 **20**, 834.

C144/71 CATALYTIC REDUCTION OF ATMOSPHERIC POLLUTION FROM THE EXHAUST OF PETROL ENGINES

C. D. HAYNES*   J. H. WEAVING*

The work described in this report was carried out at a British Leyland Motor Corporation research and development laboratory set up to investigate the long-term aspects of atmospheric pollution control. In particular, the work described is aimed primarily at developing a system to meet the 1975–76 U.S. Federal motor vehicle exhaust pollution legislation. The system described is a dual catalyst system in which two catalysts in series are used. The first catalyst is a reduction catalyst working in a net reducing atmosphere to reduce the nitric oxide to nitrogen. The second catalyst is an oxidizing catalyst which, together with added air, oxidizes the carbon monoxide and unburnt hydrocarbons to carbon dioxide and water. The report deals mainly with the problems encountered with the reduction catalysts, and in particular deals with the effectiveness of different catalyst types, the production of ammonia, sulphur poisoning of the catalyst, and general endurance tests. Some 1971 Californian-type cycle results are given; these, together with the other results, show that the catalysts described are neither active enough nor sufficiently durable to provide a suitable system to meet the 1975–76 U.S. Federal legislation.

INTRODUCTION

THE WORK REPORTED in this paper was carried out in a research and development laboratory set up by the British Leyland Motor Corporation to investigate long-term aspects of atmospheric pollution control. In practice, the work of the laboratory has been focused on meeting the extremely severe U.S. Federal Regulations which come into force in 1975–76 and which will require 97 per cent reduction in carbon monoxide (CO), 98 per cent reduction in hydrocarbons (HC), and 90 per cent reduction in oxides of nitrogen (NO_x) compared with an untreated vehicle. The laboratory has investigated or is investigating, to a greater or lesser extent, the known methods that stand any chance of meeting these regulations.

There seem to be three possibilities: (1) exhaust gas recirculation (for NO reduction) and a thermal reactor (for CO and HC oxidation); (2) exhaust gas recirculation plus an oxidation catalyst; (3) a dual catalyst system, i.e. an NO_x reduction catalyst followed by a CO and HC oxidation catalyst; and (4) as (3) except that exhaust recirculation is used to supplement the NO reduction catalyst.

Exhaust recirculation at steady-state conditions can give up to 80 per cent reduction in NO, but even in these conditions no published work claims to achieve the 90 per cent reduction required by the 1975–76 legislation. Poor driveability and excessive fuel consumption make 75 per cent a more likely maximum NO reduction over a driving cycle. This may be a sufficient reduction with vehicles that have low baseline NO_x levels.

This paper deals only with catalytic methods. The oxidation reactions are clear and simple, namely: CO and unburnt HC with injected air give carbon dioxide (CO_2) and water (H_2O):

$$2C_8H_{18} + 25O_2 = 16CO_2 + 18H_2O \quad (144.1)$$
$$2CO + O_2 = 2CO_2 \quad . \quad . \quad (144.2)$$

Reactions for NO removal, however, are not so simple as was originally supposed. It was known (1)† that the reaction between NO and CO will only take place providing that the amount of oxygen present is strictly limited as the latter has a preference for oxidizing CO, thereby removing the reducing agent:

$$2CO + 2NO \rightarrow 2CO_2 + N_2$$

However, because of the presence of a large quantity of water in the form of steam in the exhaust, the so-called water–gas shift reaction takes place:

$$CO + H_2O \rightarrow H_2 + CO_2 \quad . \quad . \quad (144.3)$$

and in fact NO may be reduced by hydrogen to nitrogen:

$$2NO + 2H_2 \rightarrow N_2 + 2H_2O \quad . \quad (144.4)$$

† References are given in Appendix 144.1.

The MS. of this paper was received at the Institution on 24th June 1971 and accepted for publication on 11th August 1971. 23
* British Leyland Motor Corporation Ltd, Atmospheric Pollution Control Research Laboratory, Browns Lane, Coventry CV5 9DR.

However, reduction does not stop here and a percentage of ammonia (NH_3) is formed:

$$2NO + 5H_2 \rightarrow 2NH_3 + 2H_2O \quad . \quad (144.5)$$

This ammonia, when passed with air to the oxidizing catalyst, is unfortunately partly reformed into NO:

$$4NH_3 + 5O_2 \rightarrow 6H_2O + 4NO \quad . \quad (144.6)$$

The problem is therefore to design a catalyst which produces a minimum amount of NH_3 or to devise a means of oxidizing the NH_3 to N rather than to NO.

This paper reports the investigation of a large number of catalysts and the reactors that have been devised for this purpose. Although there have been recent claims for successful control of all three pollutants with one catalyst, this is at present not considered a practical method. The main difficulty is that the exhaust gas composition must be maintained within such close limits as to be impractical with known carburation or fuel-injection systems. For this reason, dual-catalyst systems only are described here.

TEST METHODS

Catalysts have been tested in one or more of the following four ways.

A small-scale furnace test

In this test method a small quantity of catalyst is placed in a container in a furnace. Vacuum pumps are used to draw a small proportion of the engine exhaust gas through the sample. This system enables catalysts to be tested rapidly at different space velocities, and by varying the engine air/fuel ratio, at different CO and O_2 levels.

Space velocity is

$$\frac{\text{exhaust volume flow at NTP}}{\text{catalyst volume}}$$

and has units of h^{-1}. Tests were carried out at space velocities at 12 000 to 96 000 h^{-1}.

For oxidation catalyst tests, air, either from an engine-driven pump or the air line, can be added. A diagrammatic layout is given in Fig. 144.1.

An engine testbed test

In this test the whole exhaust gas flow from an Austin America engine (1300 cm^3) passes through the catalyst. Tests were usually carried out with the engine set to the 50 mile/h road load condition. This speed was used

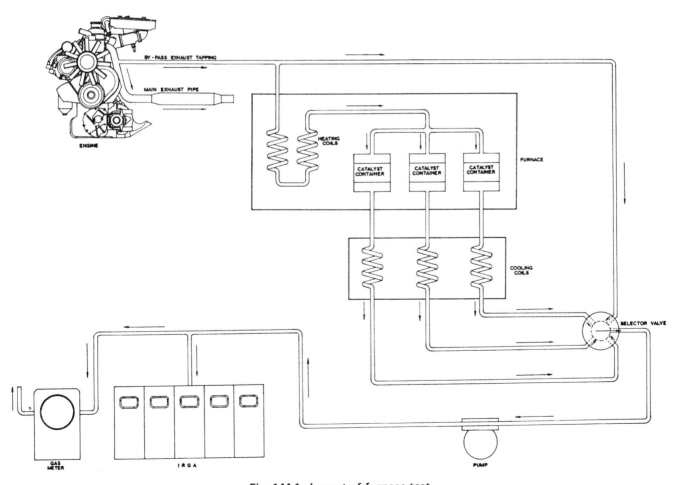

Fig. 144.1. Layout of furnace test

because it gave approximately the same catalyst temperature as the average temperature achieved over the 1971-type U.S. cycle tests. The dynamometer was equipped with an inertia wheel so that it was also possible to perform 1971-type cycle tests.

This test method made it difficult to make changes in space velocity and catalyst temperature independently, but tests for NO reduction were conducted over a range of air/fuel ratios. Oxidation catalysts were also tested over a range of air/fuel ratios and secondary air-injection rates.

Vehicle tests

A few of the more promising catalysts were tested in a vehicle (Austin America). Tests were of two forms: (a) system development tests to determine optimum carburation, optimum secondary air flow rate, and catalyst converter position; (b) endurance tests where the vehicle was driven on the road to determine the possible life of catalysts.

Fig. 144.2 shows the type of catalyst box used for both types of test, with the exception that for some endurance tests a welded, rather than bolted, construction was used to join the top and bottom halves. Fig. 144.3 shows the underside of the car fitted with a twin-catalyst system, and Fig. 144.4 is a diagrammatic layout of the exhaust pipework and bypass system.

The vehicle is fitted with a catalyst overtemperature protection system which causes the exhaust gas to be bypassed around the catalyst in a separate pipe system. In a more practical production system it is likely that the bypass valves and gas passageways would be self-contained within the catalyst converters, rather than separate external units.

To speed up the testing of catalysts whilst bypass valves were developed, the test vehicle shown uses diesel engine turbocharger bypass valves. These are of the correct size and are designed for the temperatures and conditions found in this application. Since operation is by engine oil pressure, they have a large operating force which prevents jamming. In the application shown they are triggered by a thermocouple measuring the mid-bed temperature of the catalyst.

A testbed endurance test

A 2-litre Triumph engine was rigged up so that it could be left without attention and run 24 h/day. Six catalysts were tested at once in a container, shown in Fig. 144.5. In all, 3 litres of catalyst were used; since this was ×1·5 as much catalyst as was used in the vehicle, the engine was set to provide ×1·5 the 50 mile/h road load air flow of the vehicle.

The design of box ensures that all catalysts receive exhaust gas of the same composition and operate at the same temperature. It has the disadvantage that equal flow rates can only be assured if all catalysts are of the same physical form and size, and deteriorate physically at the same rate and in the same manner.

This disadvantage was unimportant because the catalysts deteriorated chemically long before there was any significant change in catalyst volume as a result of attrition.

GENERAL TEST METHODS

Conventional infrared gas analysers were used to analyse the CO, CO_2, unburnt HC, and NO. A paramagnetic analyser with a comparatively slow response time was used to analyse the O_2 content. Ammonia levels were measured either by Draeger tubes or by wet chemical methods. It is probable that the higher NH_3 levels were underestimated because of the difficulty of obtaining a sample without loss of NH_3 into water condensate.

Table 144.1 gives the inspection data of the test fuel. The test vehicle was a standard Austin America (Austin 1300) as sold on the U.S. market with no major engine or body alterations.

TEST RESULTS AND DISCUSSION

Pelleted catalysts

The work described in this report deals mainly with work carried out on NO reduction catalysts in which the main active constituent was iron, nickel, or copper. These

Table 144.1. Inspection data of test fuel

RON	91·7
MON	82
R.v.p.	9·08 lb
Aromatics	31·2 per cent
Olefines	11·3 per cent
Saturates	57·5 per cent
Lead	0·0022 g Pb/U.S. gal
Sulphur	0·016 per cent (weight)

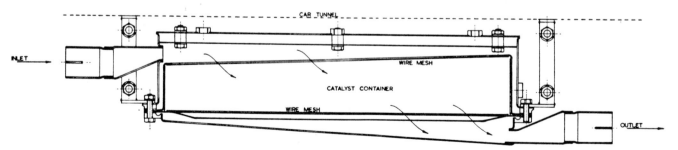

Fig. 144.2. Layout of catalyst box (underfloor)

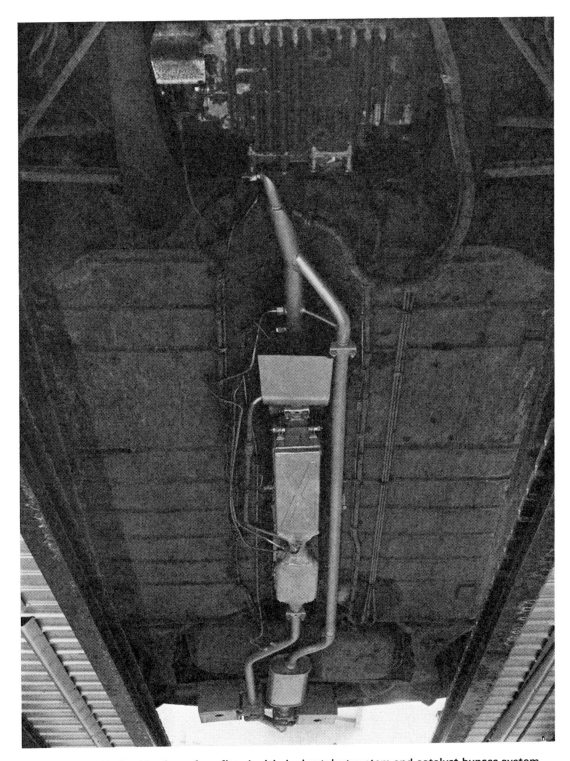

Fig. 144.3. Underside view of car fitted with dual catalyst system and catalyst bypass system

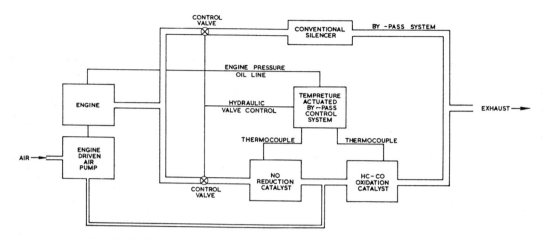

Fig. 144.4. Diagrammatic layout of exhaust system and bypass valves

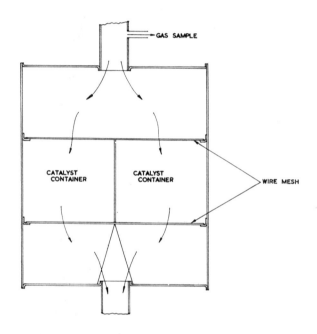

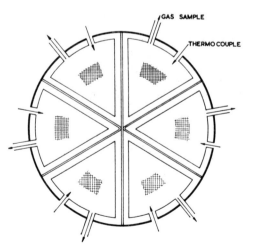

Fig. 144.5. Layout of catalyst endurance box

catalysts, together with platinum, were also tested for oxidation and this work is also described, though in less detail. Some work was also carried out with Monel metal in the form of wire and in the form of 'saddles' of 0·040-in material, as described by Esso (**2**).

Early tests at 500°C on iron, nickel, and copper catalysts were very encouraging, and NO reductions of virtually 100 per cent were achieved in furnace tests at a space velocity of 12 000 h^{-1} and 90–95 per cent in some engine tests at 25 000 h^{-1}. Typical results are shown in Fig. 144.6, and in addition results with an inlet exhaust level of 2% CO are given in Table 144.2. Results are given at 2% CO, since to obtain sufficient NO reduction it is likely that the engine would have to be set to give approximately 2–3% CO during the cycle. The results in both Fig. 144.6 and Table 144.2 are only typical and are not necessarily the best obtained with a particular type of catalyst. The values in Table 144.2 and Fig. 144.6 are for different tests. It is interesting to compare these results with similar tests reported in the Interindustry Emission Control Programme, figure 23 (**3**).

Table 144.2. Performance of catalysts at 2% CO and 500°C

Catalyst material	Percentage oxidation or reduction							
	Furnace test*				Engine test†			
	Without secondary air		With secondary air		Without secondary air		With secondary air	
	NO	CO	CO	HC	NO	CO	CO	HC
Iron .	100	80	—	—	93	60	50	50
Iron .	93	50	—	—	—	—	—	—
Nickel .	97	80	—	—	—	—	—	—
Nickel .	100	80	—	—	90	45	75	80
Copper .	100	95	100	80	90	45	80	25
Copper .	88	75	—	—	—	—	80	70
Platinum.	90	50	93‡	65‡	—	—	—	—

* Space velocity 12 000 h^{-1} except where noted.
† Space velocity 25 000 h^{-1} except where noted.
‡ Space velocity 36 000 h^{-1} except where noted.

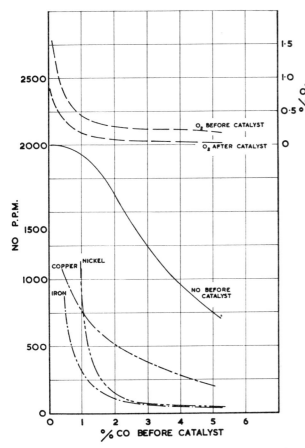

Fig. 144.6. Iron, nickel, and copper as NO reduction catalysts

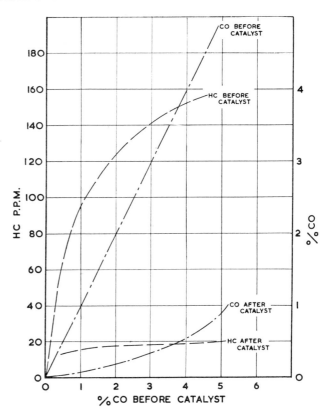

Fig. 144.7. Oxidation of HC and CO by platinum catalyst supplied with additional air

A feature of the NO_x reduction catalysts is the drop in effectiveness as the CO level falls and the O_2 level increases. In general, this takes place well before the exhaust gas becomes net oxidizing rather than net reducing, although with the more active catalysts at low space velocities the catalyst remains active until the exhaust gas is neither reducing nor oxidizing. With some catalysts which undergo transition between metal and metal oxide the catalyst may temporarily remain active in net oxidizing exhaust gas.

In addition, Table 144.2 gives results of oxidation tests for some of the same catalysts. In Fig. 144.7 results are shown for tests on a platinum catalyst with additional air. The results given in Table 144.2 are for the maximum oxidation percentage obtained over the range of secondary air flows used.

It will be noted that some CO removal takes place when the catalyst is being used as an NO reduction catalyst without the addition of air. This is probably the result of several reactions, including oxidation with remnant oxygen, NO reduction, and reaction with water.

The first difficulty encountered was the considerable loss in effectiveness as the space velocity was increased. Table 144.3 gives the results at 2% CO for an iron catalyst.

With the Austin America vehicle a catalyst capacity of between 2 and 2½ litres was required to keep the maximum space velocity down to 50 000 h^{-1} over the 1971 U.S. procedure. This comparatively large volume of NO reduction catalyst was undesirable mainly because of its effect on the warm-up rate of the second, or oxidation, catalyst.

The next difficulty encountered was the now well-known problem of NH_3 production by the NO reduction catalyst. Fig. 144.8 shows the NO reduction and NH_3 formation over a range of air/fuel ratios. The NO inlet level is as Fig. 144.6. Peak levels of up to 1000 p.p.m. NH_3 were measured with some catalysts, and with rich mixtures approximately 60 per cent of the NO was being reduced to NH_3. More generally the peak levels were around 500–800 p.p.m.

It is interesting that the NH_3 was discovered not because of its odour, although 20 p.p.m. is reputed to be detectable in air, but by the increase in NO levels across

Table 144.3. Effect of space velocity

Space velocity, h^{-1}	Percentage NO reduction	Percentage CO oxidation (without added air)
12 000	93	50
24 000	85	43
48 000	50	31
96 000	33	30

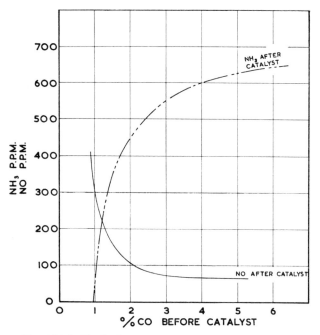

Fig. 144.8. Typical ammonia production by catalyst

the oxidation catalyst. The NO increase was in some cases up to 600 p.p.m., and in extreme cases there was sufficient catalyst dust present to cause an NO increase along an apparently plain portion of exhaust pipe into which secondary air was injected. Initial reaction was to disbelieve the NO analysis rather than to believe that the NO increases were real.

Initially, two methods were investigated to overcome the NH_3 problem. The first was to raise the operating temperature of the catalyst, with the intention of reducing the equilibrium NH_3 concentration. In practice this was ineffective because the increased reaction rate raised the NH_3 levels even at the maximum catalyst temperature, which was limited by the onset of sintering or area loss. This effect is shown in Table 144.4.

Table 144.4. Effect of temperature on NH_3 formation

Temperature, °C	NO reduction, %	NO p.p.m. before catalyst	NH_3 p.p.m. after catalyst
300	29	2000	60
500	91	2000	500
600	100	2000	1000

The second and more promising line of work was found to be that of limiting the secondary air flow rate. It was found that if the oxygen level after the second catalyst was limited to approximately 0·25 per cent, then approximately 80 per cent of the potential CO and HC oxidation was achieved without any NO formation (see Table 144.5 and Fig. 144.9). This practical difficulty of controlling the secondary air flow in close relationship to the mass of CO and HC to be oxidized should not be minimized.

The secondary air flow control would have to take into account engine speed, throttle position, engine stiffness, and carburettor tolerances. Another difficulty was the positioning of the catalysts in the exhaust system. To obtain a good emission result over the cycle it is necessary to ensure a rapid warm-up, and yet prevent excessive temperature and, hence, bypassing of the catalyst. The former requires that the catalyst is near the engine and the latter requires it to be more distant. A further complication is that the temperature of the catalyst changes with time as the catalyst deteriorates.

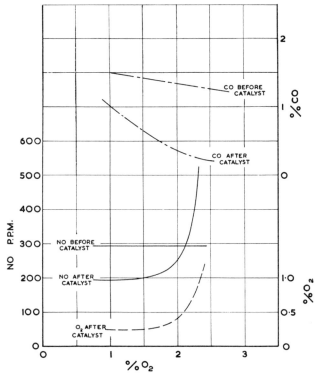

Fig. 144.9. Effect of secondary air on CO oxidation and NO_x production across the second catalyst

Table 144.5. Examples of NO formation in oxidation catalyst

Catalyst		Reduction catalyst						Oxidation catalyst					
		Before			After			Before			After		
Reduction	Oxidation	NO	CO	O_2	NO	CO	O_2	NO	CO	O_2	NO	CO	O_2
Copper	Copper	2020	1·9	0·6	200	1·3	0·2	200	1·2	2·9	810	0·25	2·2
Copper	Copper	2020	1·9	0·6	200	1·3	0·2	200	1·3	0·7	200	0·40	0·2

Despite these difficulties, some tests to the 1971 California procedure were carried out with an iron catalyst for NO reduction and nickel for oxidation. Careful control of the air flow and the position of the catalysts gave levels of 11 g/mile CO, 0·5 g/mile HC, and 0·6 g/mile NO (2·3, 0·22, and 0·4 are the approximate equivalents of the 1975–76 standards when the old procedure is used). The poor oxidation performance is a result of inadequate warm-up rate and is not indicative of the warmed-up performance of the oxidation catalyst. In addition, the need to keep the CO levels from the engine at a minimum of 2 per cent to ensure good NO_x reduction performance adversely affects the CO levels emitted when the oxidation catalyst is below its effective operating temperature.

Later tests with the warm-up of the oxidation catalyst improved by additional insulation, and repositioning of the catalyst boxes have given the following results: 0·38 g/mile HC, 3·5 g/mile CO, and 0·6 g/mile NO_x to the 1971 California procedure, and 0·57 g/mile HC, 8·4 g/mile CO, and 0·74 g/mile NO_x when tested to the 1973 Federal procedure. These tests are not directly comparable with the previous tests because of modifications made to the catalyst at the same time as the converter position was altered for quicker warm-up. It is stressed that little endurance work has been done on these modified catalytic reactors.

Another difficulty which arose fairly early on was the rapid deterioration of engine exhaust-valve seats. This deterioration caused the O_2 level in the exhaust to rise but was generally not otherwise indicated, i.e. compression pressures and cylinder leakage rates were often unaffected. This is probably due to the use of unleaded fuel in a vehicle which in normal driving was often operated at full throttle. Indeed, the 1971 driving cycle in parts required full throttle. At present, no effort has been made to adapt the vehicle to lead-free fuel, other than to reduce the compression ratio from 8·8 to 8·0.

It was found that the NO_x reduction catalyst fitted to the car suffered a fairly rapid deterioration in performance; therefore, catalysts were tested on the endurance system previously described. The endurance rig soon showed up similar deterioration, and chemical analysis showed this was due to sulphur poisoning. In the case of copper and nickel, a sulphide rapidly formed, which proved to be a poor catalyst. The iron catalyst was less rapidly affected, and since iron sulphide was also a catalyst the deterioration was less severe. The results in Fig. 144.10 show the deterioration in catalyst performance at 2% CO.

The main physical deterioration in the catalysts experienced in vehicle tests was sintering, with the resultant loss of catalyst surface area. Reductions of from 200 to 30 m²/g were experienced even though the maximum catalyst temperature was limited by the previously described overtemperature protection system.

Attrition of a catalyst has only been a problem in experimental rigs where the catalyst was deliberately subjected to the upward flow of exhaust gas from a single cylinder with no damping volume between the engine and

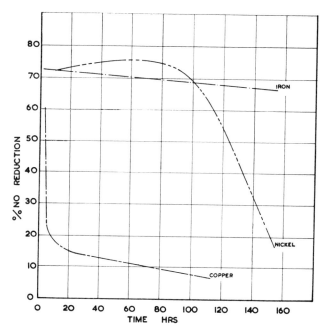

Fig. 144.10. Effect of time on nickel, iron, and copper as NO-reducing catalysts

the catalyst. In the vehicle tests, where the gas from four cylinders passes downwards through the catalyst, attrition has presented no problem in approximately 10 000 miles. There is some apparent loss in volume in the catalyst immediately on use, but it is difficult to determine whether this is real, because the catalyst changes its chemical form and as a result changes both its weight and volume.

Monel metal

Monel metal wire and 'saddles', as suggested by Esso (2), were tested at temperatures of around 800°C. Although extremely effective for NO reduction, for instance an NO reduction of 75 per cent was obtained at a space velocity of 135 000 h^{-1} and 90 per cent reduction at 25 000 h^{-1} was obtained, Monel metal did not seem practical because of the rapid physical deterioration. Its effectiveness is thought to be due to the oxidation of the copper, which increases the surface area of the metal. Unfortunately, the oxide layers formed are not adherent to the parent metal and flake off, thus restricting the life of the catalyst. An advantage of the catalyst was the comparatively low NH_3 levels produced, i.e. 50 p.p.m. NH_3 for NO reduction of 1000 p.p.m.

CONCLUSIONS

The work so far carried out is far from complete, and the catalysts tested would not result in systems able to meet the 1975–76 U.S. Federal legislation. Since it still appears that catalysts ate the most probable method of meeting the 1975–76 legislation in the U.S., the following factors are under active investigation:

(1) Reduction catalysts that produce little or no NH_3. Although there is no specific legislation against NH_3, the

U.S. Federal Register is presumed to cover this in its restriction on the production of any additional obnoxious gas as a result of fitting any device or pollution reduction system. In any case, the reformation of NO in the oxidation catalyst limits the overall NO reduction capabilities of any system producing NH_3.

(2) Catalysts that tolerate sulphur.

(3) Engine valve-seat materials that tolerate unleaded fuel.

(4) Oxidation catalysts that become active at lower temperatures.

(5) Reduction catalysts that are effective with weaker mixtures. Even if the operation with net oxidizing mixtures is only temporary, it would greatly assist the use of reduction catalysts over cycle tests where transient mixtures are sometimes difficult to maintain in a narrow range. The lower the CO level required by the reduction catalyst, the easier is the work of the oxidation catalyst. In addition, the CO emission in the period before the oxidation catalyst becomes effective may be critical.

ACKNOWLEDGEMENTS

The authors wish to thank I.C.I. for their excellent co-operation in this project, including development of catalysts and the conducting of research into many chemical problems that have arisen during the investigation. They acknowledge the careful experimental work performed by Ian Hey and Harold Dermott. Finally, they would like to thank the Directors of the British Leyland Motor Corporation for permission to present this paper.

APPENDIX 144.1

REFERENCES

(1) SHELEF, M., OTTO, K. and GANDHI, H. 'The oxidation of CO by O_2 and by NO on supported chromium oxide and other metal oxide catalysts', *J. Catalysis* 1968 **12**, 361.

(2) BERNSTEIN, S. L., KEARBY, K. K., RAMAN, A. K. S., VARDI, J. and WIGG, E. E. 'Application of catalysts to automotive NO_x emissions control', S.A.E. Preprint 710014, 1971.

(3) MERGUERIAN, G. H. and LANG, C. R. 'NO_x reduction catalysts for vehicle emission control', S.A.E. Preprint 710291, 1971.

C145/71 PETROL INJECTION CONTROL FOR LOW EXHAUST EMISSIONS

J. P. SOLTAU* K. B. SENIOR*

Only a few spark ignited internal combustion engines are currently fitted with petrol injection. This situation is due partly to the high cost compared with basic carburettors and partly to the absence, until recently, of a real need for a more sophisticated fuelling system. With the introduction of strict antipollution laws the factors are changing and technical requirements may dictate the use of an exact petrol injection arrangement. However, a fuel metering system can only be as good as its controlling computer. This paper discusses various ways by which the correct fuel requirement information can be fed to the controller, and the limitations of different systems. It is suggested that the optimum arrangement is a full three-dimensional electronic control with various corrections for ambient and engine variations from standard. The best position of the interface between the electric and hydraulic functions is still open to conjecture.

INTRODUCTION

INDIRECT OR DIRECT PETROL INJECTION for spark ignited internal combustion engines is by no means novel: it has been used successfully in car racing since the late 1930s. In the early 1960s petrol injection systems became normal equipment on all Grand Prix cars and on the majority of other competition vehicles. However, the first use of such equipment on series-produced automobiles did not appear until the mid-1950s and the increase in applications has been slow. Currently only about 1 per cent of the western world's car production is fitted with petrol injection. The cautious increase in acceptance of a system which is by definition more exact than carburettors has been due principally to high cost, and the lack of a real need for a very accurate fuel system.

With any automotive fuel system the control tolerance acceptance for good driveability and maximum power is relatively large; thus quite satisfactory power units can be built with elementary forms of fuelling devices. With the advent of regulations governing the exhaust gas emission levels the situation has radically changed. For any engine condition the difference in fuel requirement between that giving a 'clean' exhaust and that producing unstable running can be very small indeed, or even non-existent. This means that the fuelling system for a modern petrol engine must be able to give the exact requirement with virtually no tolerance. It has been the authors' experience that such exact needs can best be met with an injection apparatus, and the following discussion looks at the available controlling and correcting parameters.

The MS. of this paper was received at the Institution on 12th July 1971 and accepted for publication on 16th August 1971. 22
* *Joseph Lucas Limited, Great King Street, Birmingham 19.*

INJECTION SYSTEMS

Petrol engine fuel injection equipment can inject continuously or intermittently and the quality of the spray can vary from a plain jet to a finely atomized mist. Also, injection can take place in the intake manifold or in the cylinder itself. Direct injection has been found inferior to manifold injection, the reason being that adequate time is not available to provide a physically homogeneous mixture throughout the charge. An interesting fact is that when injecting directly into the engine the combustion process is essentially of the diesel type with burning occurring on the surface of droplets, while intake manifold injection produces true vapour phase combustion. The differences in the burning process are shown in Fig. 145.1.

When injecting into the manifold, the quality of spray is of secondary importance, as it would appear that the fuel break-up occurs mainly on impact with the manifold walls and the inlet valve. Although extensive tests have been carried out in an endeavour to find the minimum degree of atomization for satisfactory petrol injection functioning, no measurable difference in engine efficiency or exhaust pollution has been recorded with wide variation in spray quality. Test rig work has shown that the size of fuel droplets at the nozzle can influence the amount of liquid carburant which appears behind the inlet valve (Fig. 145.2) but in an actual working engine little difference in performance can be observed. Generally a cone spray angle between 60° and 20° is acceptable, although in some instances a single jet spraying at a suitable target will produce adequate mixture formation. The nozzle siting would seem to be more important than the degree of atomization, but this factor varies from engine to engine. On some units little difference can be recorded even with

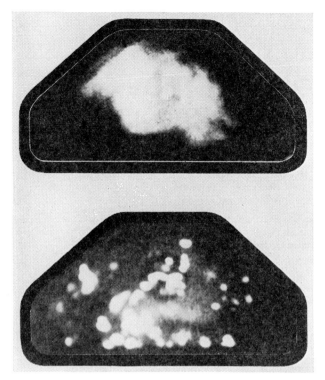

Lucas 500-cm³ research engine with quartz wall in 'flat' hemispherical combustion chamber.
Mechanical fuel injection, nozzle either in the top of combustion chamber or in intake tract.
Air/fuel ratio 14:1. Engine speed 2000 rev/min.

Fig. 145.1. Comparison between intake injection (*top*) and direct injection (*lower*) combustion

widely varying emplacements. As a rule, it is essential for a clean exhaust and acceptable driveability to inject fuel into a volume upstream of the intake valve which is no greater than the displaced volume of a cylinder.

If continuous injection is employed, the difference between idle fuel and full-load full-speed fuel will be in the region of 1 to 60. This puts a heavy burden on the control system, which has to be extremely accurate at the low end of the scale, and a dual or even triple controlling arrangement may have to be used. The authors found that the control restrictions with a continuously injecting nozzle were such as to make the arrangement not viable for clean exhaust applications. If one considers a continuous fuel injection system, with fixed orifice nozzles, a fuel line pressure controlled over a range of 1–3000 is required.

Intermittent injection can be related in time or to crankshaft position. With a 'time' system injection can be of fixed duration or quantity and the time interval between injections altered to satisfy the engine needs, or the injection pulses or injected volumes can be altered according to requirements. With such a system one has to deal with a control range of 1–60 as with continuous injection, which makes any form of control difficult, especially in that to pass the American and European exhaust quality requirements inaccuracies at idle and low load must be no

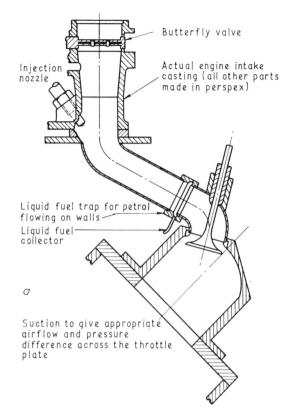

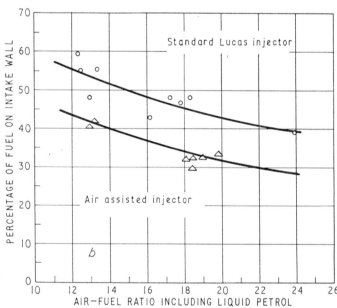

One cylinder of 4700-cm³ V8 engine. Graph shows test at idle conditions.

Fig. 145.2. Influence of nozzle spray quality on intake tract wall wetting

larger than ±2 per cent of the ideal needs. The most satisfactory type of timed injection is one where the injection takes place every cycle of the engine. With such an arrangement the speed effect of 1–10 or more is taken out of the control system and one is left with the control range of 1–4 or 1–5. This is considered satisfactory, and is in fact the injection arrangement used universally in production automobiles with mechanical or electrical fuel injection systems.

Whether the injection system is timed at a particular part of the cycle, at a similar point for all cylinders, or whether several injectors are triggered simultaneously, must be dictated by the interrelation of cost, driveability, and exhaust pollution needs. If one considers steady engine conditions there is little to be gained from a correctly timed injection for each cylinder, i.e. each cylinder intake could be injected when the inlet valve is opening. However, differences do occur during transient conditions. In an engine where injection takes place while the inlet valve is shut, it is possible for fuel appropriate to a closed throttle condition to be injected behind a closed inlet valve and to be subsequently sucked into the cylinder under full throttle conditions. This will produce hesitation which may not be acceptable from the driveability point of view although exhaust emission will normally not be significantly affected by such an arrangement unless there is a misfire. Thus, with mistimed injection, acceleration fuel enrichment will have to be provided to overcome the driveability problems. Conversely, injection during the closed inlet valve period improves the time available for adequate evaporation and can in some cases diminish marginally the exhaust hydrocarbons emitted during steady-state motoring.

Fig. 145.3 shows a plot of hydrocarbons (HC), carbon monoxide (CO), and nitric oxide (NO) against injection timing for a vehicle driven over the 1971 U.S. Federal cycle. The engine was fitted with an intake manifold depression controlled mechanical petrol injection arrangement. It will be observed that injection timing has, on this particular vehicle, an effect on exhaust emissions, especially in the case of HC and NO. The CO somewhat follows the expected pattern where the lowest values are recorded when the most propitious conditions occur for best fuel and air mixing, i.e. at the beginning of the intake stroke and in the dormant period when the inlet valve is closed. The emission of HC is fairly constant with a low value when injecting into a fast moving air stream. The NO is lowest when the charge is the coolest at the point of ignition, this condition being produced either by evaporation in the cylinder or complete evaporation in the intake system before admission to the cylinder. Similar trends are apparent in steady-state running at various road loads. The injection timing effect on driveability is most marked, and with no acceleration overfuelling device in use only fuel injection at the beginning of the intake stroke is viable.

CONTROL PARAMETERS

Currently over 80 per cent of injection equipment sold in the world makes use of indicated inlet manifold pressure as the sole or principal fuel control parameter. The reason for the wide acceptance is that the signal has a small

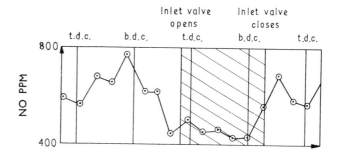
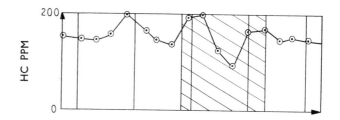
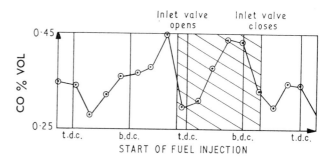

Duration of injection between 5° and 60° crank angle.
1700-cm³ four-cylinder car.
Single-throttle engine inlet manifold with intake depression controlled timed mechanical fuel injection.

Fig. 145.3. Hot 1971 Federal emission tests. Relation between NO, HC, and CO exhaust emissions and fuel injection timing

manageable range and is capable of doing considerable work, thus avoiding the need for servo assistance.

Although often criticized for its inadequacies, intake depression control is employed by each of the three leading manufacturers of petrol injection equipment on at least one of their systems on the grounds of simplicity and, hence, cost. Whether the rather elementary intake vacuum control arrangement can survive in a world of more and more stringent fuelling requirements is open to doubt, and it will depend on the type of engine produced in the future. Nonetheless, it is felt that such a control warrants some detailed analysis.

Several methods of engine tests are available to obtain the relationship between inlet manifold pressure and fuel requirements. Until recently one of the more popular ways was to plot constant torque and speed 'fish hook' curves as described by Downing* and shown in Fig.

* DOWNING, E. W. 'Practical emission control systems: petrol injection', *Symp. Motor Vehicle Air Pollution Control, Proc. Instn mech. Engrs* 1968–69 **183** (Pt 3E), 47.

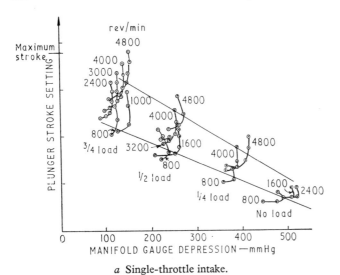

a Single-throttle intake.

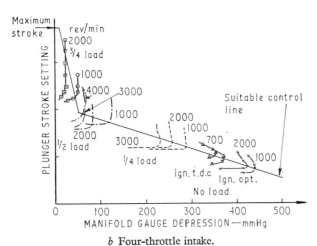

b Four-throttle intake.

2000-cm³ four-cylinder engine with mechanical fuel injection system.

Fig. 145.4. Engine speed effect on manifold pressure fuel control line

145.4. Unfortunately, such curves do not indicate the degree of exhaust pollution and are thus of limited value for current engine development work. A more satisfactory method is to obtain the fuel requirements at constant intake manifold depression and speed for, say, 0·5% CO exhaust content and also for a weaker point where either incipient misfire begins or where the exhaust oxygen content reaches 3 per cent. Such a plot covering the whole engine load and speed spectrum is shown in Fig. 145.5*a*, with the chosen lines replotted as points on a fuel versus manifold pressure graph (Fig. 145.5*b*). The maximum power fuel requirement band is calculated from a 2 per cent power drop on rich and lean settings.

With an inlet manifold of infinite volume and at moderate crankshaft speeds, the measured intake pressure indicates directly the fuel required per stroke to obtain satisfactory engine running. Unfortunately, in 'real' engines the manifold is always of a finite volume, and in some cases, where there is one throttle for each cylinder, the volume can be no greater than the cylinder volume. The fuelling line then departs appreciably from the ideal fuelling curve and this is shown in Fig. 145.5*c*.

If it were possible to measure accurately and continually the instantaneous pressure in the manifold during the intake stroke, it would be possible, by registering the pressure over this period and relating it to valve opening and cylinder position, to obtain the true requirement of fuel per engine stroke. Fig. 145.6 shows plots of instantaneous pressure against crank angle. The recorded average values are also shown on the diagram. It can be seen that the pressure is by no means uniform, varying in some cases by as much as 4 inHg. In the example shown—a four-cylinder single-throttle engine—the repeatable pressure trace pattern occupies 180° crankshaft. In a six-cylinder single-throttle it would occupy 120°, and in any installation with a throttle per cylinder, 720°. As the inlet valve opening period is usually longer (four cylinders or more per throttle) or shorter (one or two cylinders per throttle) than the pressure wave cycle, the average recorded intake manifold pressure will bear little relationship to the average pressure during the actual inlet stroke. Further complications arise due to exhaust and inlet gas reverse flow.

The influence of the manifold arrangement on the fuel control line is clearly shown in Fig. 145.4 when the 'fish hook' loops are compared on one engine with two intake alternatives. The single-throttle installation has a marked speed effect which nearly vanishes with a four-throttle arrangement.

When one throttle feeds only one cylinder, most of the fuelling control will occur at a near-closed throttle position and the range of butterfly motion will be small; thus there will be a practically constant intake configuration and relationship between the indicated and effective manifold pressure over the speed range. On the other hand, with a single throttle feeding a number of cylinders, the whole opening range of throttle tends to be employed. Engine features such as compression ratio and valve overlap also tend to have a marked effect on the fuelling curve and on the relationship between true intake manifold pressure and indicated manifold pressure.

Some of the shortcomings of vacuum pressure control can be overcome by combining a pressure signal with an air flow signal. It can be done in various ways by using a venturi, a baffle, or pitot tube in the intake tract. Fig. 145.7 shows different pressure pick-up configurations and their effect on the control line. Even with such modification a pressure signal is not ideal under all conditions. At wide open throttle where depression is virtually absent the control cannot differentiate between different engine conditions and a particular power unit may have widely different fuel requirements at, say, 2000 rev/min full load and 6000 rev/min full load. Also, at the low load end of the scale it is possible to have an idle depression which is less than the light load running depression at a higher speed.

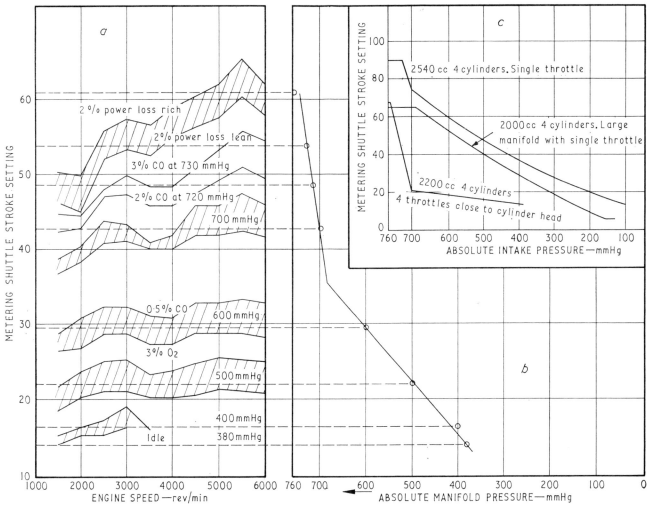

Manifold pressure controlled mechanical petrol injection engine.
1600 cm³, four cylinders, four throttles.

a Fuelling tolerance bands.
b Fuel control line based on tolerance bands.
c Typical manifold pressure fuel control lines.

Fig. 145.5

The pressure in the manifold can be measured either as depression below ambient pressure or as an absolute pressure. Generally the former gives the better control and it will be further discussed under the heading of 'Correcting factors'.

The second main control parameter which can be employed to control the fuel going into an engine is the throttle opening angle. Unfortunately, as such a single parameter control is related to the product of intake manifold pressure and engine speed, it can only be employed satisfactorily over a rather limited engine speed range. Otherwise gross fuel/air mismatch, unacceptable for clean exhaust road vehicles, will occur.

However, throttle angle or slide position fuel control has been found satisfactory for many years on racing engines operating over a narrow speed band. Currently the majority of racing units employ such a controlling system.

On Formula 3 cars where there is a restriction in the intake tract, the control, being incapable of sensing engine speed, cannot be correct when the breathing of the engine is impaired.

A third control parameter is the correctly measured intake air flow as opposed to open throttle area. In the case of the throttle arrangement the area of the opening does not represent directly the air flow, as the pressure differential across the valve varies from nil to a maximum of three-quarters of an atmosphere. Without the use of a large air box, the air mass flow to an engine cannot be accurately measured due to the highly pulsating nature of the air stream. However, a close estimate, accurate for some purposes, can be obtained from the position of a constant depression valve. Such an arrangement consists of an air valve controlled in such a way that a constant pressure drop exists across it. The position of the valve

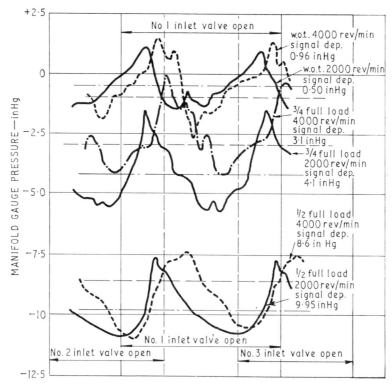

1800-cm³, four-cylinder engine: single-throttle manifold.

Fig. 145.6. Instantaneous manifold depression plotted against crank angle

will then indicate directly the air flow (assuming a constant coefficient of discharge and no reversal flow) to the engine per unit time. The principle is that used in C.D. carburettors. With a constant flow injection system, provided the nozzles inject in a constant air pressure environment, a simple control can be devised relating fuel needs to air flow. With timed injection, the only system considered in this paper, the air meter reading will have to be divided by the engine speed to obtain the requirements per engine stroke. Thus the control becomes compound, which will be further discussed under 'Combined control parameters'.

It would appear that some form of 'feedback' hunting control would be feasible. Such a control would modify the fuelling until maximum performance was obtained. It could be done using engine speed or engine torque, but the control would not be able to give the best compromise result for lowest exhaust emissions, and for this reason some form of searching exhaust gas analyser would have to be combined with the fuelling control arrangement. It is unlikely that the above system will be available for many years as rapid gas analysis or accurate torque measurement on an actual vehicle is not currently available. There would also be some insurmountable driveability problems and a form of memory would have to be included in the control, which would let the system know or remind it of the previous running condition. The great advantage of an optimized control is that the control would suit any engine/car combination without tailoring, and would also take into consideration engine wear and external conditions such as differences in barometric pressure or in ambient air temperature.

COMBINED CONTROL PARAMETERS

A better fuel control is obtained by employing a surface indicating the fuel requirements and using two control parameters to indicate the point required on the surface. Such an arrangement is usually referred to as a three-dimensional control system. The required surface is obtained from extensive tests on the engine dynamometer. The two control signals can be manifold pressure and throttle angle, or manifold pressure and engine speed, or throttle angle and engine speed, or air flow and engine speed, or air flow and throttle angle, or air flow and manifold pressure.

Fig. 145.8 shows the throttle angle/manifold pressure surface obtained on one particular engine for 1% CO at part load and for maximum power. It will be noticed that a rather small area of control ensues and, obviously, unless the two signals act simultaneously the sensor will 'look' for information outside the control surface area. The arrangement is nevertheless a possibility and will have to be constructed in such a way that one control always leads the other: the area outside the true controlling surface will have to be designed in such a way that some indication of

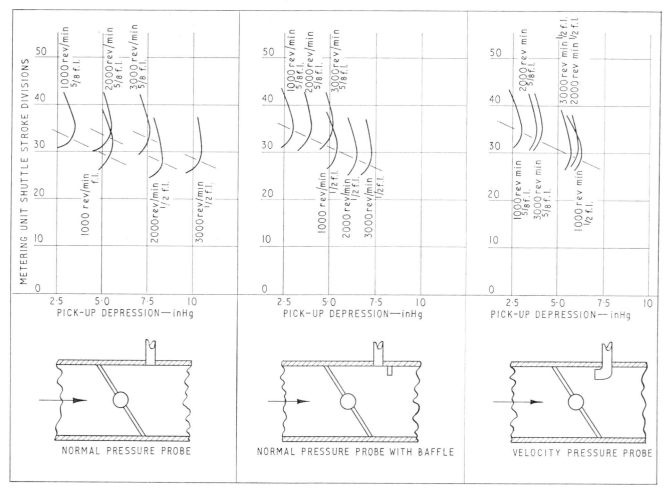

2500-cm³, six-cylinder engine; single-throttle arrangement; mechanical petrol injection.

Fig. 145.7. Effect of intake pressure probe configuration on fuelling control lines

the future fuel requirement will be given. The lines AB and BC show the path that should be followed when the engine conditions are changed rapidly. In such an instance the directly coupled throttle signal will always lead the manifold pressure signal. Fig. 145.9 shows a plot of fuel requirements as controlled by manifold pressure and engine speed. The usable control area in this case is much larger than that shown in the first instance, but unfortunately there can remain some ambiguity in the low-load and high-load areas where it is possible to have several fuel requirements at one engine speed with one indicated manifold pressure.

Fig. 145.10 shows the surface obtained when the fuel requirements are plotted against true air flow and engine speed. The controlling area is not large and there will be an inherent lag of response unless acceleration enrichment is practised. Similar fuelling control surfaces can also be built with information of air flow and throttle angle or manifold pressure. The air flow values used in the construction of the model were obtained with a viscous flow air meter and not from the position of a secondary constant depression valve as mentioned earlier.

The two control parameters of air flow and engine speed can also be employed to give a line, not a surface. To obtain this a division must be performed so that the air flow is related to engine cycle and not to time. Fig. 145.11 shows the fuelling requirements plotted against true air flow divided by engine speed. The air/fuel ratio was set to give 1% CO increasing to 4% CO at maximum torque.

A unique and fully adequate control surface is that defined by throttle angle and engine speed (Fig. 145.12). There are no ambiguities and the surface covers the whole of the rectangular area formed by the full range of throttle angle and engine speed.

The added fuelling accuracies obtainable with such a system by comparison with that of a simple manifold pressure arrangement are shown by the following example obtained with a car following the 1971 Federal test. Exhaust emissions with the full three-dimensional control were: CO 3·5 g/mile, HC 0·8 g/mile and NO_x 0·8 g/mile, as against 9·0, 1·3, and 1·87 g/mile respectively for the basic manifold depression control.

The primary control is the throttle angle and it will move at the same time as the driver alters the accelerator pedal

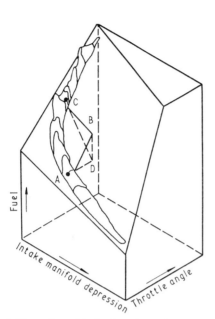

2700-cm³, V6 engine with six throttles and mechanical petrol injection.

Fig. 145.8. Fuel control surface—intake manifold depression and throttle position

1500-cm³, four-cylinder engine. Single-throttle manifold. Timed mechanical petrol injection.

Fig. 145.10. Fuel requirement surface—air flow and engine speed

3000-cm³, eight-cylinder engine. Twin-throttle manifold. Electronic petrol injection.

Fig. 145.9. Fuel requirement surface—manifold pressure and engine speed

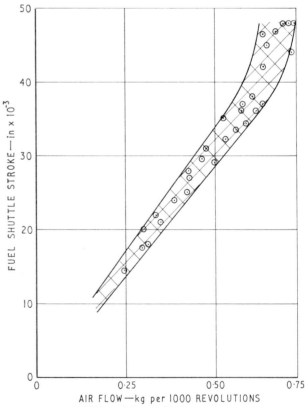

1500-cm³, four-cylinder engine. Single-throttle manifold. Mechanical timed petrol injection.

Fig. 145.11. Plot of fuel requirements against air flow divided by speed

2500-cm³, six-cylinder engine. Single-throttle manifold. Electronic petrol injection.

Fig. 145.12. **Fuel requirement surface—throttle angle and speed**

position. The speed control is secondary and the fuelling corrections will be made as the engine speed increases or decreases following the conditions. Like most other forms of control, the three-dimensional throttle angle and speed arrangement is purely a memorizing system and the instructions for the fuel quantity based on the position of the throttle and the engine speed are obtained on the engine dynamometer with production type power units. The basic control cannot take into consideration any variations from engine to engine or any variations due to engine wear or other mechanical idiosyncrasy. Therefore, extremes of engine to engine variations must be obtained so that tolerance bands can be established and the control aimed at satisfying the mean.

CORRECTING FACTORS

Any memorizing fuel control system is only correct for the engine used in the calibration tests and for the ambient conditions prevailing at the time. Any deviation from this situation must be corrected by the use of a suitable multiplication factor.

AIR DENSITY

The main correcting factor is that which modifies the control according to ambient air density, that is, atmospheric pressure and temperature. The correction should be such that it applies both when there are significant variations in barometric pressure and also when the car travels at altitude. The rectification should be one of proportionality, that is, one which multiplies the fuel requirements by a factor which takes into consideration the difference between actual pressure and temperature and datum air density for which the memory was obtained. Manifold absolute pressure, when used as the control parameter, does not correct for ambient pressure adequately. At low load with small deviations from standard pressure the correction is reasonable, but near and at full load very severe underfuel can occur. Fig. 145.13a gives a typical fuelling line. It will be seen that the steep rise in the curve occurs over a small pressure range. If the car is driven under conditions of very low barometric pressure or at altitude when full load fuel will be required, the actual fuel supplied will be as point a when in reality point b shows the amount of petrol needed by the engine under these conditions.

The deficiency in absolute pressure control can be overcome by using two separate control arrangements as shown in Fig. 145.13b: (A) absolute pressure line, and (B) a high-load curve governed by manifold depression and not absolute pressure. If the double control line is not employed, very great errors will occur especially at high altitude where the ambient pressure can be lower or equal

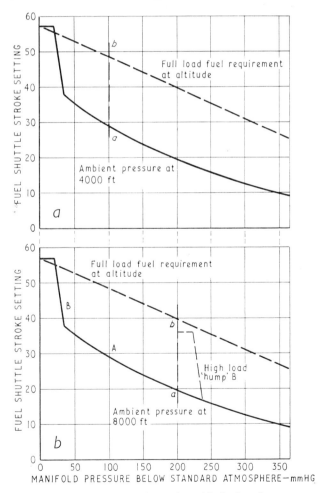

2500-cm³, six-cylinder engine with six throttles.

Fig. 145.13. **Altitude effects on fuelling control lines**

to sea level idle pressure. In the example shown in Fig. 145.13b, without the manifold depression line, at 8000 ft the engine would receive idle fuel on full throttle when, in reality, it requires about twice that amount. The secondary control has not necessarily to be operated by manifold vacuum, but can also be worked by the throttle position. A better type of intake manifold pressure control is one depending purely on manifold depression and having barometric and temperature correcting devices.

With the combined control composed of throttle angle and speed, the barometric or density correction has to be on a proportionality basis and each point on the surface has to be multiplied by a constant factor.

RAM EFFECT

The ram effect in an engine producing various peaks at either full load or part load is related to the local acoustic velocities in the intake system and the exhaust pipes. Since the inlet acoustic velocity is dependent on the square root of the temperature of the gas inside the ram pipes or intake manifold, the speed at which filling is carried out most efficiently will vary with intake charge temperatures. Therefore, variations in inlet temperature from that at which the calibration tests were carried out should have an effect on the correspondence between the engine speed scale and the volumetric efficiency curve as represented by the fuel requirements to obtain best performance. Fig. 145.14 shows two full throttle curves of b.m.e.p. plotted against engine speed at different intake air temperatures. It can be seen that the density effect overwhelms any shift in the ram peaks, which can thus be ignored.

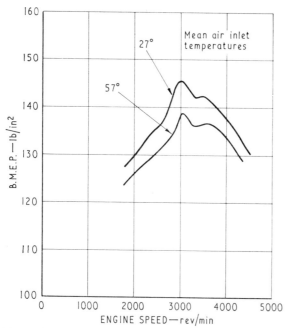

2500-cm³, six-cylinder engine with six intakes and mechanical petrol injection.

Fig. 145.14. Plot of b.m.e.p., i.e. fuel requirements against engine speed with different air temperatures

ENGINE CONDITION

As an engine wears and the valves and piston rings start to leak, the basic characteristics—especially at low load and low speed where exhaust pollution is so important—will vary from that on the test unit used to obtain the basic engine fuelling data. A small correction to the fuelling line will have to be incorporated and it will be least significant with control mechanisms using air flow as a parameter.

A car with a 1600-cm³ engine fuelled by a petrol injection system controlled from inlet manifold pressure gave the following Federal test result when functioning normally: 31 g/mile CO and 1·85 g/mile HC. With impaired breathing the values became 52 g/mile CO and 4·15 g/mile HC. A worn engine will tend to idle at a lower depression than a new one, and the apparent higher air consumption will cause the fuel system control to give too rich a mixture.

TRANSIENT CONDITIONS

When accelerating, unless the control is directly connected to the throttle mechanism, there will be a delay in adequate fuelling and a special device is required to supply additional fuel during such a mode. If it is not done there will be an appreciable flat spot in the driveability, especially with a clean exhaust engine which is running lean with no reserve in hand. The problem is not so noticeable with a normally tuned engine which will be on the rich side of stoichiometric, and the leanness during acceleration is acceptable. It is not strictly correct to say that there should be a richening of the mixture during acceleration—one injects additional fuel to prevent the mixture going lean due to deficiencies in the control system. The need for extra fuel during acceleration is not unique to non-throttle controlled petrol injection arrangements, and it is also required for carburettor installations to mask delays in the fuel to follow the air flow.

There are two ways of keeping deceleration exhaust emissions at a low level: one is to maintain good burning, the other is to cut off the fuel completely.

During deceleration any fuel present on the walls of the intake is first stripped off, leading temporarily to a rich mixture. Further, due to low charge pressure and consequent vitiation of the mixture by exhaust gas during the valve overlays period, combustion is poor. Manifold vacuum limiting techniques are used to provide enough working mixture so that adequate combustion can be maintained even at speed. A disadvantage of this technique is the tendency towards inadequate deceleration of the vehicle upon closure of the throttle without use of the brakes.

The more usual form of deceleration fuel control in petrol injection applications, although it gives higher HC emission than overrun burning, is that of fuel cut-off. An arrangement is made so that above a chosen engine speed and below a certain throttle opening no fuel is supplied to the engine. Apart from the emission bonus, the system also helps to give a good fuel consumption. There are disadvantages with fuel cut-off in that when the fuel flow

PETROL INJECTION CONTROL FOR LOW EXHAUST EMISSIONS

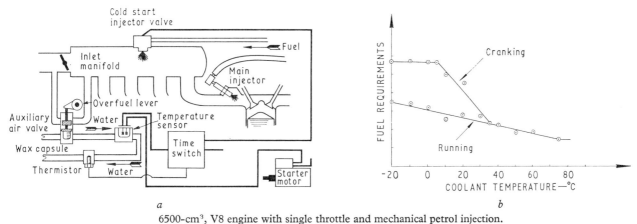

a   *b*

6500-cm^3, V8 engine with single throttle and mechanical petrol injection.

Fig. 145.15. Cold starting circuit and fuelling curve

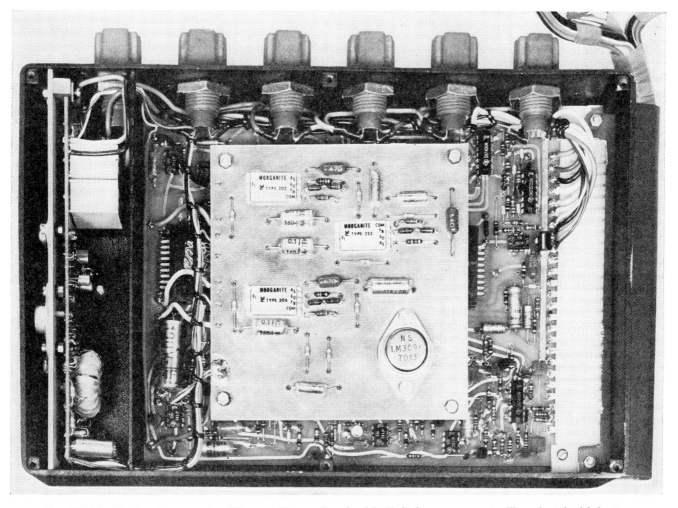

Fig. 145.16. Electronic computer (25 cm × 15 cm × 5 cm) with digital memory controlling electrical injectors to give fuelling requirements as shown in Fig. 145.12

is stopped or reinstated, HC peaks appear in the exhaust. In the first instance it is caused by fuel being trapped in the intake system and in the second by a slight delay in the resumption of good combustion. The first part of the injected fuel will tend to deposit itself against intake walls and not go into the combustion chamber. There is also a driveability problem when reinstating the fuel, especially with lean mixtures. It is often noticed that reinstatement does not take place until an appreciable load is put on the engine and it provides a shock to the vehicle with a surge forward. The problem can be diminished by controlling the reinstatement process automatically, but it adds considerable complications to the control.

COLD START AND WARM-UP

In order that an automotive fuel system can deal adequately with cold start and warm-up, especially during exhaust pollution tests, it must give a sufficient excess of fuel during cold cranking to provide a vapour phase air/fuel ratio in the combustible range. This excess is dependent on the physical characteristics of the fuel (whether winter or summer grade), the degree of atomization and mixing of the fuel, and the extent of redeposition as the very wet mixture navigates the inlet tract. After 'first fire' the excess requirement rapidly reduces since, as the engine accelerates up to its idling speed, the mixing in the manifold improves, as also does fuel evaporation within the now warmer combustion space. In order to maintain a self-sustained idle speed the charge has to be supplemented by either opening or bypassing the throttle to compensate for loss of cycle efficiency by excessive heat loss to the cylinder walls and high mechanical friction. As the load or speed of the engine changes, the evaporation and mixing within the inlet manifold and the combustion space change, as also do thermal and mechanical cycle efficiencies. With an injection system, as with carburettors, only a compromise can be maintained during the warm-up period to minimize control complexities.

In a petrol injected engine the above events are controlled by signals from a combination of wax capsules, thermal and time switches, and thermistors. A typical circuit is shown in Fig. 145.15a where the additional fuel for 'first fire' is supplied by an extra nozzle to keep the working range of the main injecting system within easily manageable limits. The starting nozzle, or gross overfuelling on the main nozzles (the choice is governed by physical and control characteristics), is triggered by the starter solenoid circuit if the water jacket temperature is below a set level, and the additional fuel is maintained after start of fire for a period of time also depending on engine temperature. The warm-up extra fuel and air are both controlled by engine temperature. Fig. 145.15b shows a typical plot obtained during the establishment of an automatic warm-up system for an engine fuelled by a mechanical system. The above work is carried out in cold chambers under strictly controlled conditions. The air/fuel ratio is measured through CO exhaust gas monitoring.

IDEAL SYSTEM

From what has been shown it would appear that the best system possible with today's technology is a system using, as main control parameters, throttle angle and engine speed. The combination will give a completely unique control, but will only be correct as long as atmospheric and engine conditions are similar to those pertaining during the original engine test and establishment of the control surface. Corrections must be made for any deviation from the test conditions, and although this can be done accurately it may involve more complication than the original control if the tolerance bands are narrow.

To construct the best control with mechanical means represents a formidable task. This has been attempted on a commercial scale in the past, but the compromises needed to keep the system reasonably simple are not acceptable with the new projected exhaust pollution regulations. There is only one way to build such a control system in a small enough package and at an acceptable cost, and that is where the main computer is electrical. Electronic techniques now enable the whole system to be digital in concept and accurate to any required degree. Integrated circuits provide a control unit of small physical dimensions and will enable the system to be manufactured cheaply in large quantities. Fig. 145.16 shows the control unit (25 cm × 15 cm × 5 cm) required to provide a system which will completely match the three-dimensional surface shown in Fig. 145.12.

Several electronic distribution and metering systems are available, but to date it has been found that the electro-mechanical constant pressure nozzle provides the simplest and most reliable hydraulic arrangement. An electronic computer could also control, via a suitable interface, a piston pump or a shuttle type pump. If there is a demand to control very accurately the moment of injection and the length of injection, such a system may well be the answer, even though the complications involved would impose a heavy cost penalty.

ACKNOWLEDGEMENTS

The authors are indebted to the Directors of Joseph Lucas Limited for permission to publish this paper.

C146/71 EXHAUST EMISSION CONTROL: FURTHER EXPERIENCE IN THE APPLICATION OF EXHAUST THERMAL REACTORS TO SMALL ENGINES

J. W. WISDOM*

This paper summarizes some of the work undertaken on the design, development, and endurance testing of thermal exhaust reactors to reduce specific exhaust pollution from European sized engines to near those levels required by the Federal legislation in the U.S.A. for 1975. The work described forms an extension to that published in May 1970 and covers in the main the application of thermal reactors to two vehicles, a Vauxhall Viva and a Saab 99.

INTRODUCTION

SINCE THE PUBLICATION of our initial studies in the use of thermal exhaust reactors on small engines for the control of exhaust emissions (1)†, the field of investigation has been widened to encompass engines of greater capacity and of increased specific outputs. The original project was undertaken in the belief that the use of a thermal reactor offered the possibility of providing a system capable of achieving very low hydrocarbon (HC) and carbon monoxide (CO) emission levels. It was anticipated that the performance available would create a considerable margin between the legislative demands and attainable values, and hence provide a basic system capable of having a substantial production lifetime, reducing at the same time the demands for involved Development and Application to tolerable levels. In the event, the legislative demands have anticipated the development time period required to produce fully developed reactors, and the Federal exhaust emission regulations in the U.S.A. for 1975 require a refinement in technique which would have been considered impossible to achieve, even in experimental form, two or even less years ago.

Following discussions with the Du Pont Company at Wilmington, U.S.A., it was agreed that the author's company would investigate the function of Du Pont reactor designs on the smaller European engines with a view to establishing technical viability and to educate the design staff on future carburettor requirements.

Early work indicated the need for designs aimed specifically at the small engine, and two factors were significant. First, the combustion loading of the developed reactor system would be greater and, second, it was believed that air valve carburettors possessed inherent advantages in the control of transient modes and thus offered distinct advantages in terms of vehicle driveability. Accordingly, constant depression Stromberg C.D. carburettors have been used throughout the investigations.

The declared policy of the U.S.A. for increasingly stringent legislation reduced the earlier scepticism concerning thermal reactors as a commercial proposition and following the publication of reference (1) approaches were made to Zenith, and several projects concerned with the development of reactors were commissioned. Where possible, data from these studies are described in this paper. In particular, data obtained from a Vauxhall Viva and a Saab 99 vehicle are presented.

DESIGN

All the work so far undertaken by Zenith on thermal reactors has been concerned with in-line four-cylinder engines. Engine capacities have ranged between approximately 1 and 3 litres. The basic layout of the reactor cores used has consisted of a tube into which the exhaust gases enter via four ducts, one in line with each exhaust port. A high degree of mixing of the exhaust gases takes place within the reactor core and also to some extent during the passage of the gases from the core to the exhaust pipe of the vehicle.

Out of a number of variations, two basic designs have been utilized for the greater part of the work. These are

---
The MS. of this paper was received at the Institution on 28th June 1971 and accepted for publication on 16th August 1971. 44
* *Engineering Research and Application Ltd (a subsidiary of Zenith Carburetter Co. Ltd), London Road, Dunstable, Bedfordshire.*
† *References are given in Appendix 146.1.*

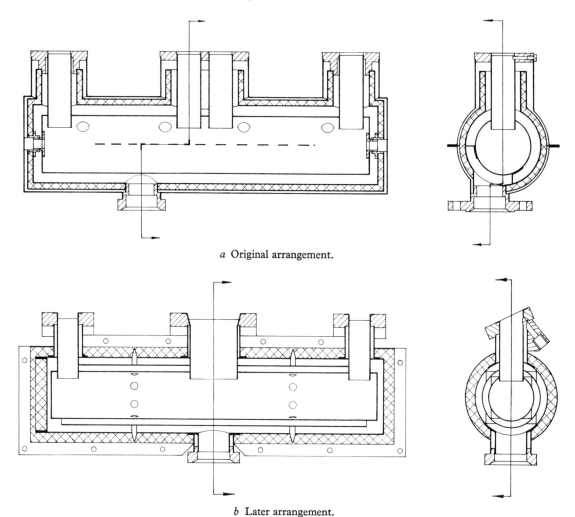

a Original arrangement.

b Later arrangement.

Fig. 146.1. Arrangements of exhaust reactor

illustrated in Fig. 146.1. While the inlet and mixing within the core remained virtually unchanged, the outlet path of the gases in the second design was modified to give more similar flow lengths to the gases from each cylinder as they negotiate the reactor.

Relatively few installation problems have been encountered, since the vehicles used have had sufficient clearance available in the region of the exhaust manifold to ensure adequate installation and service accessibility. The addition of an air pump and its appropriate driving pulley has proved more difficult. These problems will become greater in small cars and in some sports type vehicles.

Heat insulation of the reactors has been achieved by the use of Triton Kaowool, while air cooling, by passing the air from the exhaust injection pump through the reactor outer casing annulus prior to injection into the exhaust ports, is a possible alternative.

Reactor outer casings have been fabricated from mild steel with the core, liners, and the casing inner wall formed in Incaloy 800. To reduce weights and to minimize ther-

mal lag, the gauge of the Incaloy was reduced to 22 s.w.g. (0·028 in) and the mild steel to 20 s.w.g. (0·036 in). In this form the total weight of one typical installation was 10 lb, which compares with 6·5 lb for the standard exhaust manifold.

PERFORMANCE—VAUXHALL VIVA

The work undertaken on this vehicle can be divided into two phases. The first phase consisted of a brief investigation aimed at establishing if a simple, low cost, large volume, non-heat-insulated manifold was capable of attaining a satisfactory performance level. The second phase was concerned with an extension of the project described in reference (**I**), and its object was to optimize the emission control performance of the basic type of reactor design used in that work. The Viva engine and ancillaries are listed below:

Engine: Vauxhall Viva four-cylinder OHV push-rod. Compression ratio 7·5:1, capacity 1159 cm^3 (70·7 in^3).

Crankcase ventilation: Sealed system—rocker cover vented via oil trap to air cleaner elbow and also to inlet manifold through 2-mm diameter restriction.
Distributor: Standard A.C. Delco unit for low compression ratio Viva incorporating a vacuum retard capsule.
Carburettor: 150 CDSE Spec. No. 3237 with modified metering needle.
Exhaust with reactor: Single pipe $1\frac{1}{4}$ in inside diameter of same length as standard system without silencer or expansion chamber.
Air pump: A.C. Delco No. 7975580 vane type pump with a $1.0 \times$ engine speed drive ratio, 13 inHg relief valve setting.
A.I.R. system: Gulp valve No. 5558DC.

Phase I

Fig. 146.2 shows the arrangement of the large volume exhaust manifold used in the first phase. The particular vehicle available for the investigation utilized a twin downpipe exhaust system, hence the need for two outlet pipes on the reactor. Three sizes of manifold were made equal in volume to approximately 1, $1\frac{1}{2}$, and 2 times the engine swept volume.

It was envisaged that the unit giving the lowest combustion loading, i.e. the one having the greatest volume, was most likely to achieve the lowest emission levels and this arrangement was used for the initial tests. These investigations were followed by measurements using the manifold of a size equal to the swept volume of the engine. The levels measured with both manifolds were disappointingly high, and although, in general, the use of the large manifold gave improved results the gain was only marginal. For these reasons the manifold having a volume equal to $1\frac{1}{2}$ times the engine swept volume was not used.

During the tests the results indicated that the increase in mixing length for the gases from some cylinders, brought about by the use of a single outlet from the manifold, was beneficial. Subsequent investigations were therefore concerned with establishing if the lengthening of the flow path within the manifold, by means of a simple baffle, would improve the performance level. For this purpose the large volume manifold was modified by the inclusion of a central baffle, as shown in Fig. 146.2.

Temperature measurement at steady-state conditions over a range of mixture settings, as produced by the carburettor variable fuel metering jet, indicated that the temperature of the gases was insufficient to sustain efficient combustion, and hence, significantly reduce emission levels. Typical values are shown in Fig. 146.3. These findings were confirmed by vehicle tests in accordance with the 1970 U.S.A. Federal procedure using NDIR equipment, in which the composite emission levels given by the large volume manifold, with central baffle of 246 p.p.m. HC, 1·7% CO, and 881 p.p.m. NO_x (oxides of nitrogen) were no better than those obtained from the standard A.I.R. (Air Injection Reaction) treated car of 261 p.p.m. HC, 1·25% CO, and 855 p.p.m. NO_x.

Phase II

The second stage of these investigations was concerned

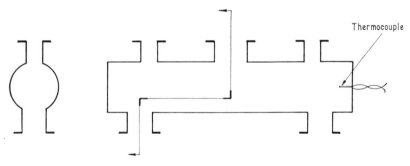

a Original arrangement with twin outlets.

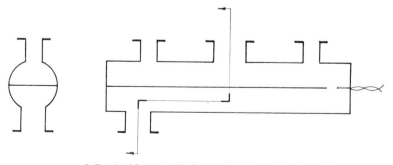

b Revised layout with internal baffle and single outlet.

Fig. 146.2. Large volume non-insulated exhaust manifold

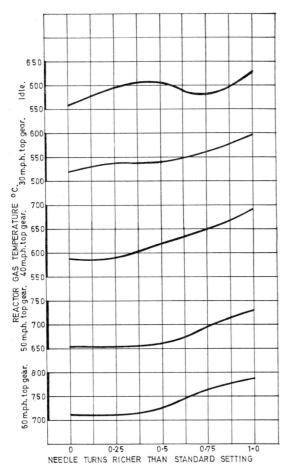

Fig. 146.3. Large exhaust non-insulated manifold. Effect of mixture setting upon reactor gas temperature

baseline testing of the vehicle in accordance with 1970 Federal procedure resulted in levels of 361 p.p.m. HC, 0·9% CO, and 1231 p.p.m. NO_x being measured.

A reactor, based on the configuration shown in Fig. 146.1b, was fitted to the vehicle. The core volume was approximately 60 per cent of the engine swept volume, and the overall length of the system was approximately 16 in and the diameter was 4 in. The drive to the air pump was not modified from the standard ratio, nor was it found necessary to modify the ignition timing.

Some of the early tests were carried out without the thermal insulation material being installed in the reactor. As might be expected, although the exhaust emission levels were acceptable, during road tests it became obvious that excessive underbonnet temperatures were occurring which caused vapour locking in the fuel system. Fitted with this arrangement, emission levels of the order of 95 p.p.m. HC, 1·23% CO, and 360 p.p.m. NO_x were measured when the vehicle was tested in accordance with the 1970 Federal procedures, using NDIR equipment. Some detail redesign of the reactor was undertaken to simplify the manufacturing and assembly processes. With the revised, fully insulated system, encouraging results were obtained for the HC and NO_x emissions, but the CO levels were greater than desirable. Investigation revealed that the CO levels tended to increase considerably after completion of the fifth cycle. It was found, by testing without the air pump being operative, that the carburettor (which was sited directly over the reactor) had inadequate temperature compensation. An asbestos heat shield was used to reduce the float bowl temperature, and with this arrangement overall values of approximately 15 p.p.m. (0·15 g/mile) HC, 0·6% (9·0 g/mile) CO, and 400 p.p.m. (1 g/mile) NO_x were obtained. Table 146.1 summarizes a typical 7-mode cycle test result.

The driveability of the vehicle was good and although no detailed performance figures were obtained, brief tests established that the changes in maximum speed, acceleration, and fuel consumption values from those of the vehicle in standard A.I.R. form were of the same order as those previously measured and described in reference (1).

with the continued development of the thermal reactor along the lines of the design principles already established in earlier work. For this work, in keeping with the general trends within the industry towards the possible use of non-leaded fuel and the control of NO_x, the compression ratio of the engine was reduced from 9:1 to 7·5:1. Initial

Table 146.1. Summary 1970 Federal test cycle—Viva vehicle

Cycle no.	Conc. (as determined)			Weighting factor	Weighted		
	HC, p.p.m.	CO, %	NO_x, p.p.m.		HC, p.p.m.	CO, %	NO_x, p.p.m.
1	133·2	1·9	556·8		11·6	0·17	48·4
2	14·9	0·35	571·0	$\dfrac{0·35}{4}$	1·3	0·03	44·7
3	9·3	0·20	475·7		1·8	0·02	41·4
4	3·4	0·23	448·5		0·3	0·02	39·0
5*							
6	2·2	0·52	276·9	$\dfrac{0·65}{2}$	0·7	0·17	90·0
7	3·7	0·59	252·8		1·2	0·19	82·1
Sum (trip composite)					16·9	0·60	350·6

* Not read.

PERFORMANCE—SAAB 99

The work on this vehicle represented something of a departure from the starting point available on the Viva, since the emission form of the engine utilized a CAP (Clean Air Package) system, i.e. no exhaust air injection was used. In addition, the front wheel drive layout of the vehicle posed a more demanding packaging problem. The Saab engine and ancillaries are listed below:

Engine: Four-cylinder in-line four-stroke OHC. Compression ratio 9:1, capacity 1708 cm^3 (104·2 in^3).

Crankcase ventilation: Sealed system—rocker cover vented via oil filter to air cleaner and also to inlet manifold via Smith's type valve.

Distributor: Standard A.C. Delco unit to emission vehicle specification incorporating ignition retard capsule.

Carburettor: 175 CDSE built to Specification No. 3301. Experimental metering needle fitted.

Exhaust system: A standard chassis exhaust system was used throughout the testing programme.

Air injection pump: A.C. Delco No. 7975580 valve type pump. Pump drive ratio 1·0 × engine speed.

A.I.R. equipment: Check valve and air injection manifold.

Evaporative loss equipment: None fitted.

To incorporate the A.C. Delco air pump some rearrangement of the alternator position and drive arrangements had to be undertaken, but, as shown in Fig. 146.4, a suitable layout was not too difficult to obtain. A similar design of reactor to that used in the later stages of the work on the Viva was used in this work and a general view of the unit is shown in Fig. 146.5. Test bed work was initially directed towards the optimization of the air pump delivery rate and engine mixture requirements at idle and steady-state road load conditions using the standard CAP ignition settings. From these data a profile for the carburretor metering needle was derived and it was found that

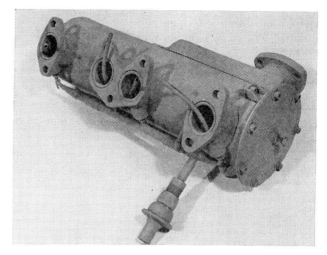

Fig. 146.5. General view of reactor for Saab vehicle

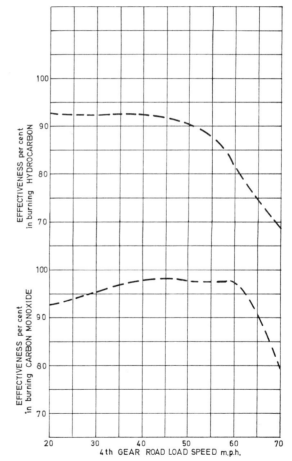

Fig. 146.6. Reactor effectiveness at constant road load conditions

Fig. 146.4. Reactor installation in Saab vehicle

an air pump drive ratio of 0·625 times engine speed would satisfy the reactor requirements.

Some measure of reactor efficiency at road load conditions was made by testing with air injection supply

effect of mixture air/fuel ratio upon core gas temperatures and exhaust emission levels at engine idle and 50 mile/h cruise conditions are shown in Figs 146.9 and 146.10 respectively.

It was not possible to directly measure absolute values of air/fuel ratio, but the range covered could be expressed as a function of the number of turns from a given datum, on the adjustable fuel metering jet used in the carburettor. At idle it can be seen that the maximum gas temperature recorded on the unshielded chromel/alumel thermocouple was 800°C, and at this condition the minimum values of approximately 1% CO and 50 p.p.m. HC were recorded with a peak CO_2 value of 14 per cent.

At the 50 mile/h condition a peak gas temperature of approximately 1000°C was measured under the richer conditions used. The minimum exhaust emissions of 15 p.p.m. HC and 0·1% CO were recorded with slightly lower gas temperatures. Measurements made during further testing indicated, as shown in Fig. 146.11, that the reactor effectiveness (*2*) in burning CO was lower during the acceleration phase. To overcome this deficiency the output

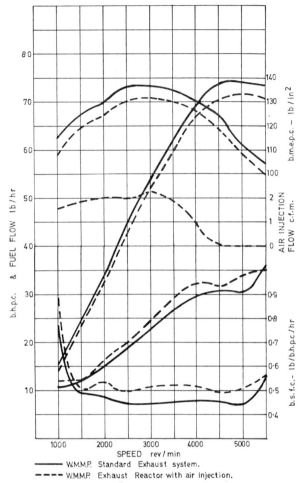

Fig. 146.7. Comparison of full throttle performance. Standard CAP system versus reactor with standard exhaust system

disconnected, and using the methods described in an S.A.E. Paper (*2*). The results are shown in Fig. 146.6.

The full throttle performance at weakest mixture for maximum power (W.M.M.P.) conditions of the engine both in standard form and when fitted with a reactor and suitable exhaust air supply is shown in Fig. 146.7. Fig. 146.8 shows the exhaust back-pressure levels for both full throttle and the road load range. From these results it can be seen that the reactor and air injector pump installation gave some 3 per cent loss in maximum power and 3·5 per cent loss in maximum torque. The loss in power output could be attributed in roughly equal proportions to (1) the power requirements of the air pump and (2) the effects of increased back pressure.

Vehicle work

When the system was installed in the vehicle and tested in accordance with 1970 procedures it was found that the composite levels measured were unacceptably high.

Investigations to examine the steady-state performance of the vehicle were undertaken. Typical results of the

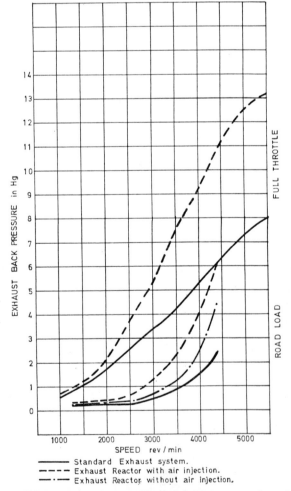

Fig. 146.8. Comparison exhaust back pressure. Standard exhaust system versus exhaust reactor with and without air injection

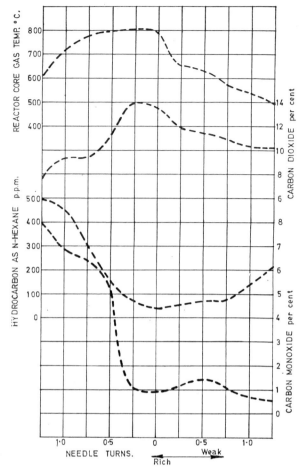

Fig. 146.9. Reactor—Saab vehicle. Effect of mixture setting at idle upon reactor gas temperature and emission performance

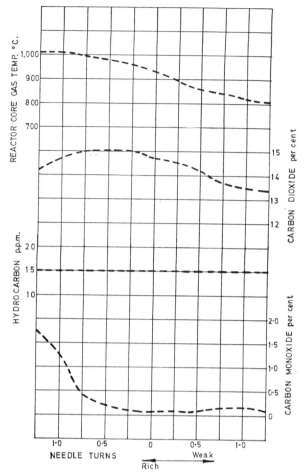

Fig. 146.10. Reactor—Saab vehicle. Effect of mixture setting at 50 mile/h cruise upon reactor gas temperature and emission performance

of the air pump was uprated by increasing the air pump/engine drive ratio from 0·625:1 to 1:1. This modification, together with the use of a slightly richer carburettor setting, resulted in improving the overall exhaust emission from 74 p.p.m. HC, 0·99% CO, and 410 p.p.m. NO_x to 47 p.p.m. HC, 1·05% CO, and 433 p.p.m. NO_x. When the ignition advance tapping was removed, composite levels of 20 p.p.m. HC, 0·97% CO, and 370 p.p.m. NO_x were recorded. A summary of the cycle results is shown in Table 146.2. This vehicle has also been tested in accordance with the 1972 Federal test procedure. Typical results obtained are given in Table 146.3. These results indicate the influence of cold start conditions upon emission results.

Table 146.2. Summary 1970 Federal test cycle—Saab vehicle

Cycle no.	Conc. (as determined)			Weighting factor	Weighted		
	HC, p.p.m.	CO, %	NO_x, p.p.m.		HC, p.p.m.	CO, %	NO_x, p.p.m.
1	89·8	1·8	566·7		7·8	0·15	49·3
2	79·8	1·6	488·9	$\frac{0·35}{4}$	6·9	0·14	42·5
3	18·4	0·9	347·1		1·6	0·08	30·2
4	5·5	0·9	312·7		0·5	0·08	27·2
5*							
6	5·9	0·8	313·6	$\frac{0·65}{2}$	1·9	0·25	101·9
7	4·2	0·8	367·3		1·4	0·26	119·4
Sum (trip composite)					19·9	0·96	370·5

* Not read.

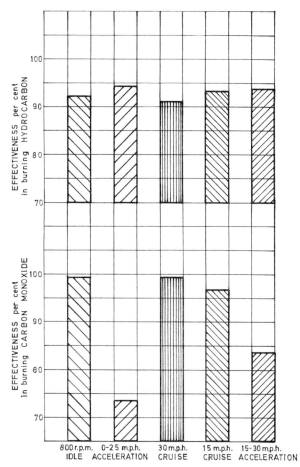

Fig. 146.11. Reactor—Saab vehicle, reactor effectiveness

Table 146.3

Hydrocarbon, g/mile	Carbon monoxide, g/mile	Starting condition
1·42	9·49	Cold
0·46	8·15	Cold
1·04	15·7	Cold
0·02	5·95	Hot
0·33	9·03	Cold
0·16	3·52	Hot

NO_x emission levels have only been measured in accordance with the 1970 test procedures using NDIR equipment. Cold start values of approximately 1·0 g/mile were obtained without the vacuum tapping being connected. When tested by an independent laboratory the cold start emissions obtained from the car equipped with the reactor were as given in Table 146.4.

Road performance

The driveability of the vehicle was assessed as being acceptable and equal to that of the standard CAP equipped car. The overall road performance levels obtained from a vehicle fitted with the reactor and a vehicle fitted with a 1971 CAP system are given in Table 146.5.

Table 146.4

Hydrocarbon, g/mile	Carbon monoxide, g/mile	Nitrogen oxides, g/mile
0·56	6·88	1·5
0·44	7·92	0·87
0·27	8·02	1·29

NO_x analysis was by modified Saltzmann method.

Table 146.5

Speed, mile/h	Fuel consumption, mile/gal	
	Reactor	CAP
20	51·5	55·7
30	50·6	52·8
40	47·3	51·0
50	41·4	45·0
60	36·3	40·5
70	33·1	37·1
Overall fuel consumption	27·9	Not available
Maximum speed, mile/h (timed lap)	90·6	91

Endurance testing

Mileage accumulation is proceeding on a vehicle fitted with automatic transmission. Initially, the tests were undertaken on a programmed chassis dynamometer, but more recently open road testing has been used. To date, a total of some 10 000 miles has been covered, using a leaded fuel, without significant deterioration in performance being measured.

During the open road testing some 6000 miles have been covered at an average speed of about 30 mile/h with a fuel consumption of 29·4 mile/gal.

Examination of the core and deflector shield, which are shown in Fig. 146.12, revealed that no significant dimensional or apparent material changes had taken place.

FLOW VISUALIZATION STUDIES

Examples of the flow patterns obtained from models are shown in Fig. 146.13. From these studies it became apparent that some of the exhaust gases were not entering into the main turbulent region in the centre of the core and were escaping into the annulus, between the core and the deflector shield, via the gap between the inlet port tubes and the reactor core. In addition, on some designs, static flow regions within the core and annuli configuration have been observed. Design modifications are currently being undertaken to overcome these deficiencies.

Generally the flow models offer a means of ready comparison between the existing and new, more novel designs, in terms of space utilization within the reactor, general and local velocity levels and system overall pressure loss.

CONTROL OF OXIDES OF NITROGEN

A notable feature of the results obtained with the use of thermal reactors has been the low levels of NO_x recorded. This somewhat unexpected bonus appears to be linked with the use of high exhaust back pressure which enables some dilution of the incoming charge to occur, probably during the period of valve overlap.

While the values of NO_x so far obtained will not meet the 1976 legislation demands, they are of the same order as those currently given by many exhaust recirculation systems.

It should be noted that apart from the work on the Viva installation during Phase II of the project, no attempt was made to utilize the silencing potential of the reactor. The standard vehicle exhaust system was used throughout the work on the Saab car. Obviously the use of high back pressure could ultimately place severe restrictions upon the output of engines with stringent breathing requirements; however, the data obtained to date suggest that it may be possible to use this method of NO_x control on some vehicle installations.

REACTOR DURABILITY

Mention has already been made of the endurance testing being undertaken on the Saab vehicle. It is apparent that this aspect of reactor performance is both critical and difficult to evaluate fully under all driving conditions.

Difficulties have been experienced with overcoming the differential thermal expansion problems which, during development, caused gaps to occur in the joints of the double skinned outer casing through which gases could escape and thus cause the loss of some thermal insulation. Revised designs based on these experiences are in hand and are giving encouraging results.

Without question, many aspects of reactor performance require extensive investigations to ensure adequate life and safety. However, it is believed that sufficient data are now available to justify some optimism with regard to the possibility of meeting the U.S.A. Federal emission regulations of 1976.

CONCLUSIONS

(1) Thermal exhaust reactors are capable of providing very low exhaust emissions. Values of the order of 0·4 g/mile HC and 8·0 g/mile CO have been consistently recorded using the 1972 U.S.A. Federal test procedure.

(2) Mileage accumulation tests have shown no significant deterioration in emission performance after completing 10 000 miles using leaded fuel. Examination of the reactor core and radiation shield has shown that the mileage accumulation has had little or no effect on the Incaloy used for their construction.

(3) NO_x levels, as measured by NDIR gas analyser in accordance with Californian 1971 legislation, were of the

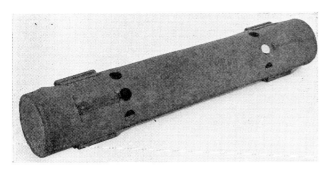

a Reactor core.

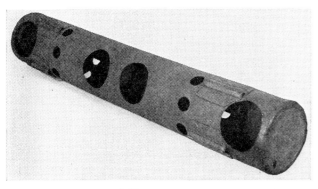

b Reactor core.

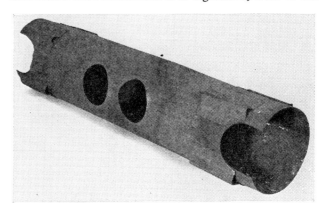

c Reactor deflector shield.

d Reactor deflector shield.

Fig. 146.12. Condition of Saab core and deflector shield after 10 000 miles

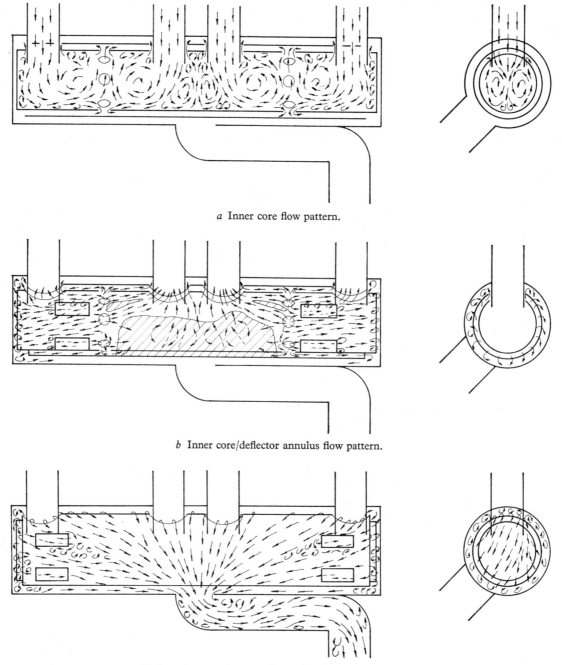

a Inner core flow pattern.

b Inner core/deflector annulus flow pattern.

c Deflector/outer casing annulus and exhaust duct flow pattern.

Fig. 146.13. Reactor flow patterns

order of 350 p.p.m. (approximately 1 g/mile). These values were obtained without the use of exhaust gas recirculation.

(4) It is believed that the low levels of NO_x were produced by the utilization of relatively high back pressure to induce inert exhaust gases into the engine cylinders. Measurements indicated that less than 2 per cent loss in maximum power and torque were experienced due to back-pressure effects.

(5) The use of a simple non-heat-insulated large exhaust manifold combined with an exhaust air injection system is unlikely to provide a satisfactory emission control system.

(6) Flow visualization studies have indicated that turbulence levels are high within the reactor core. This water analogue technique shows promise of making a significant contribution to the optimization of reactor design.

(7) Reactor systems are dependent for adequate func-

tion on close control of the ingoing air/fuel ratio and demand regulation of the metering flow band at least as precise as that required by air injection and CAP treatments.

ACKNOWLEDGEMENTS

The author would like to thank the Directors of the Zenith Carburetter Co. Limited for permission to publish this paper. In addition the author would like to express his appreciation of the help given to him and the efforts made by his colleagues at E.R.A.

APPENDIX 146.1

REFERENCES

(1) LAWRENCE, G. and WISDOM, J. 'Emission control by exhaust manifold reactor—an initial study for small engines', *Proc. Auto. Div. Instn mech. Engrs* 1969–70 **184** (Pt 2A), 249.

(2) 'Design and development of the G.M. air injection reactor system', S.A.E. Paper 660106, 1966.

C147/71 THE DESIGN, DEVELOPMENT, AND APPLICATION OF EXHAUST EMISSIONS CONTROL DEVICES

A. S. DE FORGE-DEDMAN*   J. A. HOWARD*

The purpose of this paper is to present information relating to the design, development, and application of exhaust emissions control devices which are in production, or under consideration, for gasoline engines within the 90–180 in^3 (1·5–3 litre) swept volume range, and which are mass produced in great numbers for passenger cars. The development problems and individual performance of the components of the Ford Motor Company IMCO system, as applied to the above range of engines, are discussed in detail and the ways in which the problems were solved are given. These components are a dual diaphragm distributor, a new carburettor, an intake manifold deceleration control valve, and a thermostatically controlled air cleaner. The research being undertaken on the individual components to improve this system, while retaining acceptable driveability to meet the new Federal legislation, is debated. It is also intended to present information relating to the control of crankcase emissions. The emissions control packages fitted to cars being exported to European countries will also be briefly described.

INTRODUCTION

SINCE THE U.S.A. first brought in mandatory emissions legislation in 1966, the Federal levels have progressively been lowered, the future levels being so low that there are serious doubts that sufficient knowledge exists at present to enable car manufacturers to meet these new levels at an economic price. As the mandatory requirements have recently been changed, the latest levels are shown in Table 147.1.

In 1965 Ford of Britain commenced a research programme to develop the first emissions 'bolt-on' package for the 1·5-litre and 1·6-litre Kent series of push-rod engines, which was based on the injection of air into the exhaust port, adjacent to the exhaust valve head. This package was first attached to Ford Cortinas exported to the U.S.A. in 1967. The thermactor package, as it was known, was progressively improved and used until recently when it was replaced by a new system. This new system does not require the relatively expensive air pump or the air injection control system that was necessary to avoid high exhaust pipe temperatures and explosions in the exhaust pipe during overrun conditions. The reduction in emissions levels was achieved by providing tighter production tolerances over the carburettor air/fuel ratio flow bands and closer control of spark timing. Thus a saving in production cost has been realized and more power is available at the flywheel. This new system is much quieter than the former air pump package.

This new package, the improved combustion (IMCO) system, has a deceleration valve which injects a rich mixture into the inlet manifold on engine overrun at high inlet manifold vacuums. This permits combustion to continue without misfire and so reduces unburnt hydrocarbon carry-over. While developing this package it was found that closer control of the spark timing is necessary and this has been achieved by incorporating a second diaphragm in the distributor to retard the spark at predetermined conditions. The air cleaner has been modified to include an air inlet temperature control capsule which allows hot air, heated by the exhaust manifold, to be drawn into the engine during the warm-up period. A new carburettor with an automatic choke which achieves rapid engine warm-up has been developed.

PREVIOUSLY PUBLISHED RESEARCHES

There is a vast quantity of published research work which has been completed on single- and multi-cylinder petrol engines with the object of reducing exhaust emissions. Although these engines have many different forms of combustion chambers, the experiments have enabled some general fundamental conclusions to be established about exhaust emissions. On reading the literature one can conclude that engine combustion chamber design geometries do not have a great effect on the way exhaust emissions

---

The MS. of this paper was received at the Institution on 25th June 1971 and accepted for publication on 16th August 1971. 34
* Ford Motor Co. Ltd, Research and Engineering Centre, Laindon, Basildon, Essex.

Table 147.1. U.S.A. emissions levels

Model year	Test procedure	Mandatory legislation*, g/mile			Ford production targets, g/mile		
		HC	CO	NO_x	HC	CO	NO_x
1970 California	FTP†	2·2	23	—	1·65	17·3	—
Federal and Canada	FTP	2·2	23	—	1·65	17·3	—
1971 California	FTP	2·2	23	4·0	1·65	17·3	3·0
Federal and Canada	FTP	2·2	23	—	1·65	17·3	—
1972 California	FTP	1·5	23	3·0	1·1	17·3	2·25
	CVS‡	3·2	39	3·2	2·4	29·2	2·4
Federal and Canada	CVS	3·4	39	—	2·55	21·8	—
1973 California	CVS	3·2	39	3·0	2·4	21·8	2·25
Federal and Canada	CVS	3·4	39	3·0	2·55	21·8	2·25
1974 California	CVS	3·2	39	1·3	2·4	21·8	0·97
Federal and Canada	CVS	3·4	39	3·0	2·55	21·8	2·25
1975 California	CVS	0·46	4·7	1·3	0·345	3·53	0·97
Federal and Canada	CVS	0·46	4·7	3·0	0·345	3·53	2·25
1976 Nationwide and Canada	CVS	0·46	4·7	0·4	0·345	3·53	0·3

* For mass production a development level at least 25 per cent below mandatory levels has to be aimed at.
† FTP—Federal test procedure.   ‡ CVS—Constant volume sampler.

are produced (1)*. This indicates that while changes in the exhaust emissions can be produced, the changes are so small that there is no confidence that the emissions legislation of 1976 can be met by basic engine development. The only way to achieve these low levels of emissions is to develop 'bolt-on' items. This fact is lucidly brought home by the researches of Dodd and Wisdom (2), and although that paper reported their work at M.I.R.A. in 1968 on three chambers, Dodd extended their programme to include the testing of 11 different combustion chamber designs, and indicated that there was hardly any difference in the level of exhaust emissions. All their published exhaust emissions data were plotted to a base of air/fuel ratio, and the four different methods of mixture preparation used—vaporization, injection, drip feeding, and atomization—showed no significant effect on the magnitude and distribution of the hydrocarbon (HC), carbon monoxide (CO), and oxides of nitrogen (NO_x) exhaust emissions.

Other parameters were briefly examined. For a stabilized engine condition, inlet mixture temperature was varied, and it was found to have no effect on HC and CO; the effect on NO_x was not established. The surface area to volume ratio of the combustion chambers tested was varied and found to have only a marginal effect on HC, CO, and NO_x emissions. One can summarize by saying that for the 11 different chambers, the distribution of emissions with air/fuel ratio always had a characteristic shape, thus indicating that chamber shape is not of prime importance. The peak value of NO_x always occurs at about 16:1. At this point the HC and CO emissions were a minimum. Thus to avoid producing nitric oxide (NO), one has a choice of operating the engine on a rich air/fuel ratio where unburnt HC and CO would exist, or to run excessively lean and accept the risk of severe combustion misfire and poor driveability.

It should be remembered that the researches of Dodd and Wisdom had been completed under conditions of steady state. However, with steady-state conditions it is possible to see that the maximum NO_x emissions occur at an air/fuel ratio slightly leaner than the stoichiometric air/fuel ratio where the b.m.e.p. is almost at a maximum value for a particular engine speed. When transient tests such as the Federal test procedure are completed on any engine, it can be seen that the level of NO_x is highest during the acceleration modes where the combustion chamber is developing high pressures and temperatures. In reference (3) this conclusion is illustrated by work done on transient conditions by the Toyota Motor Company, using three different engines. Although the three engines all had wedge-shaped combustion chambers, the trace shown in Fig. 15 of reference (3) is typical of the NO_x NDIR traces recorded for the 2-litre S.O.H.C. engine and 1·6-litre push-rod bowl-in-piston engine made by Ford Motor Company Ltd. We have found that chamber shape does not play an important part in controlling emissions during transient tests. In reference (3) Toyota also show the effect of overrichening the mixture on a transient test. This is in agreement with previous experience that overrichness does not alter the fact that the greatest concentration of NO_x is still produced under conditions of high combustion chamber loads where the peak combustion chamber temperature and b.m.e.p. are highest.

Two of the most active researchers who have been trying to determine which part of the combustion process produces NO_x are Newhall and Starkman (4). They have been studying the formation of NO in the combustion chamber through small quartz windows by using spectroscopic techniques. They have also been measuring the production rate of NO_x during the power stroke by taking gas samples from the combustion chamber (5). Starkman illustrates the levels of production of NO_x during the combustion process in Fig. 4 of reference (5) for different engine speeds. Newhall (6) has carried out a theoretical study of the expansion stroke using a range of chemical reaction rate values of the combustion process, and has attempted to show why the manufactured rate of NO_x is

* References are given in Appendix 147.1.

much higher than the destruction rate by disassociation. Lavoie (7) has also completed a theoretical study and some spectroscopic experiments to support his conclusion that the bulk of NO production occurs in the burnt gas and not in the high-temperature flame zone. However, more experimental work needs to be done before a comprehensive mathematical model can be developed which predicts the experimental results accurately, and thus reveals the manufacturing process of NO_x.

The general conclusion at present is, therefore, that NO_x is produced under conditions of high chamber loads where the peak pressures and temperatures are highest. Duke et al. (8) further proved this point by running a single-cylinder engine under varying degrees of spark induced detonation conditions for compression ratios of 8:1, 9:1, and 10·5:1, and noted at stoichiometric air/fuel ratios that while HCs were reduced almost linearly with increase in knock intensity, the NO_x, after rising to a maximum when slight knock occurred, remained at this value for the complete range of knock intensities investigated. CO and CO_2 were also unaffected by knock. Figs 5–8 of reference (8) illustrate this, and show the effect of change of air/fuel ratio.

The steady-state tests of Dodd and Wisdom (2), and the steady-state and transient tests undertaken by Matsuma et al. (3), Turner and Martin-Flaven (9), and ourselves, illustrate that despite the large differences of design between the engines used in these tests, the distribution of CO and HC exhaust emissions followed a familiar pattern when these engines were tested according to the 7-mode Federal test procedure.

THE FORD IMCO SYSTEM

It is against this background of research that in 1969 the Ford IMCO system was devised. The Turner and Martin-Flaven research work on double diaphragm distributors showed that spark retard during the deceleration modes greatly reduced HC emissions. However, prior to commencing tests on our own engines it was decided to reduce the variation of cylinder to cylinder air/fuel ratio to less than 1:1. Thus, the inlet manifolds were developed to achieve this at the average engine speed and throttle opening recorded during the 15–50 mile/h 'accel' mode of the Federal test procedure (FTP).

DOUBLE DIAPHRAGM DISTRIBUTOR AND SPARK MODULATION

When the double diaphragm distributor shown in Fig. 147.1 was tested on the S.O.H.C. 2-litre engine, it was

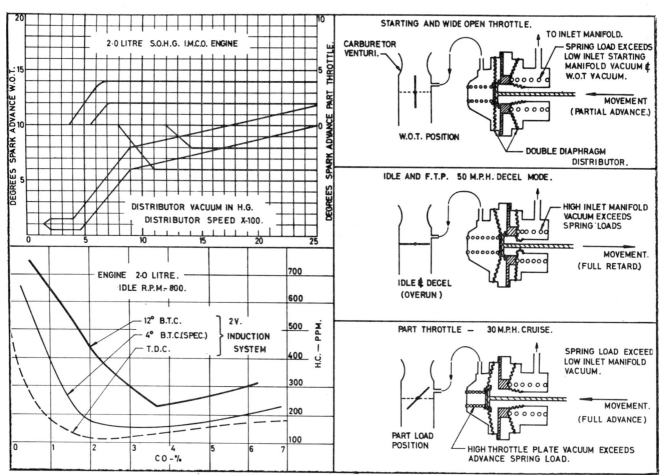

Fig. 147.1. Double diaphragm distributor

found that a considerable reduction in NO_x could be achieved if the ignition was retarded at W.O.T. conditions as well as retarding ignition on the deceleration modes. As the three diagrams show, the advance diaphragm has no vacuum at idle but experiences maximum vacuum at minor part loads. This vacuum progressively drops off as the throttle plate is opened farther. Thus this provides maximum advance during the early part of the acceleration modes of the Federal test procedure. However, to avoid producing large quantities of NO_x at high chamber b.m.e.ps a spring rate is chosen which allows the advance diaphragm to return to its original position as the carburettor approaches W.O.T. conditions, thus retarding the ignition from the optimum power value which avoids maximum cycle pressures and temperature being developed.

Under conditions of closed throttle engine deceleration, the throttle plate pressure tapping connected to the advance diaphragm is at atmospheric pressure. Thus the advance diaphragm is unloaded and the spring returns it to its seat. The retard diaphragm is, however, experiencing maximum vacuum. This is especially so at the start of the 50-0 deceleration mode of the FTP cycle. Thus the ignition is retarded and both diaphragms move together until the inlet manifold vacuum drops to 12 inHg where the retard diaphragm spring returns both diaphragms to the central position. Under idling conditions a vacuum exists behind the retard diaphragm while the pressure at the face of the advance diaphragm is atmospheric. Thus the ignition is retarded at idle. The overall characteristics of the ignition timing are shown in Fig. 147.1. Fig. 147.1 also illustrates the reduction in the level of CO and HC emissions that have been achieved at idle, when the spark is retarded from 12° b.t.d.c. to t.d.c.

In order to meet the tighter 1972 emission legislation, a more sophisticated spark control system has been developed which is modulated to be sensitive to road speed and ambient temperature. The system is fitted to automatic transmission Cortina and Capri cars exported to the U.S.A. and Canada (see Fig. 147.2a). As shown in Fig. 147.2a, a small alternator has been positioned on the speedometer drive cable and an electronic system has been developed which can switch off the throttle plate vacuum to the advance diaphragm for certain ambient temperature conditions and vehicle speeds. When the ignition is switched on, the electronic control module shuts off the carburettor vacuum to the advanced diaphragm. As the vehicle accelerates a signal is generated which is proportional to the vehicle speed. When the vehicle speed reaches

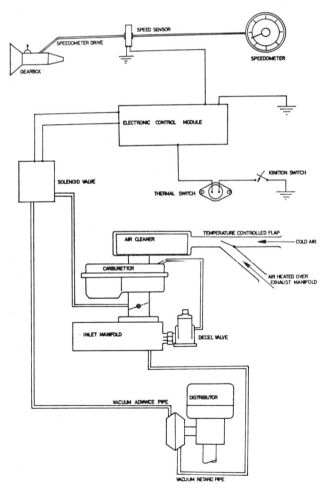

Fig. 147.2a. Automatic transmission spark modulation system

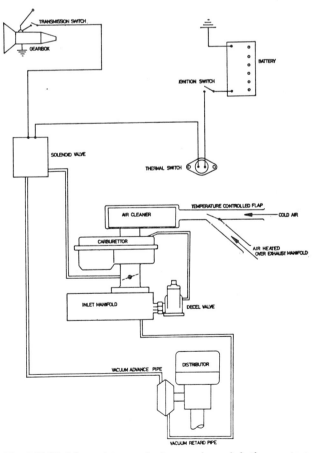

Fig. 147.2b. Manual transmission spark modulation system

25 mile/h, a signal with sufficient strength is generated to open the solenoid valve and permit spark advance to occur. On deceleration, the voltage signal decreases with the vehicle speed, and at 18 mile/h the electronic control amplifier in the modulator responds to close the solenoid valve. Thus the vacuum line to the advance diaphragm is switched off, and the ignition remains retarded unless there is a great change in ambient temperature. If the outside ambient temperature is lower than approximately 55°F, the thermal switch closes to override the signal from the alternator. The electronic control amplifier in the module responds accordingly to keep the solenoid valve open. Therefore, regardless of vehicle speed, spark advance is achieved. The switch is mounted near the driver's door away from the hot engine. Thus this system controls the maximum combustion cycle temperature with the aid of the ambient thermal switch.

In the interest of simplicity, for cars fitted with manual transmission exported to the U.S.A. and Canada, a simplified version of the above system can be employed. The system is shown in Fig. 147.2b. When the vehicle is being driven in first and second gear there is no spark advance as the carburettor vacuum has been shut off by the solenoid valve. When third gear or fourth gear is selected, the transmission switch on the gearbox opens the solenoid valve and permits spark advance to occur. However, if the ambient temperature is below 55°F the thermal switch operates and the solenoid remains open all the time, thus permitting spark advance to occur normally as experienced with the double diaphragm distributor previously described. An advantage in terms of driveability is gained with this system as ignition advance is available in the gears most frequently used. The spark modulation systems are to be part of the IMCO package for 1972.

INTAKE MANIFOLD DECELERATION CONTROL VALVE

Fig. 147.3 illustrates the deceleration control valve which is mounted on the inlet manifold adjacent to the carburettor. When the manifold vacuum exceeds 22 inHg (as experienced during the 50–0 deceleration mode of the FTP) the diaphragm lifts the valve off its seat and lets in a mixture of fuel and air from the carburettor. As the vehicle deceleration proceeds, a sufficient quantity of fuel and air is made available through the deceleration control valve to maintain combustion under high inlet manifold vacuum conditions. The manifold vacuum drops from approximately 26 inHg while the valve progressively closes, and at a vacuum of 18·5 inHg the valve is returned to its seat. Thus, in practice a quantity of extra air/fuel mixture is supplied from the carburettor for a manifold vacuum range of 26–18·5 inHg. This greatly reduces the HC emissions during deceleration modes.

THE SINGLE AND DOUBLE VENTURI CARBURETTORS

Morgan and Stojek (10) have described in detail the different types of carburettors that are to be used on the new range of S.O.H.C. four-cylinder in-line Ford engines. The Weber twin venturi type of carburettor used on the 2-litre S.O.H.C. low-compression engine exported to U.S.A. for the Pinto car is equipped with a water heated automatic choke. For the European Cortina 2-litre cars and for the Federal 2-litre Capri cars exported to North America, it was decided to develop the Weber D.G.A.V. carburettor which had the primary venturi repositioned on the opposite side of the secondary venturi in order that this carburettor could be fitted to the European 1·6-litre push-rod Kent in-line four-cylinder engines. If the Pinto carburettor was used, the automatic choke fouled the rocker cover.

In all other aspects the carburettor design is the same as the Weber D.F.A.V. carburettor used in the American 2-litre Pinto car. To achieve rapid warm-up and so reduce the CO and HC emissions on the cold cycles of the FTP, these two carburettors incorporated an automatic choke with a modulated choke plate pulldown system to lean off the mixture at intermediate ambient temperatures. This was achieved by allowing the manifold vacuum to act on a rubber diaphragm and spindle assembly incorporated in the carburettor, as shown in Fig. 147.4. A general view of the water heated bimetal strip is shown in Fig. 147.5. Thus the choke plate pulldown modulation is achieved by the manifold vacuum acting upon the diaphragm against the bimetallic spring load which varies with temperature. Therefore, the choke plate position varies during the cold cycles of the FTP for variation of manifold vacuum (combustion chamber load) and the cooling water temperature. This automatic choke control system is also fitted to the single venturi Autolite carburettor which is used on the Kent 1·6-litre bowl-in-piston push-rod engine exported to the U.S.A. and Canada.

A power valve enrichment system is employed on both single and double venturi carburettors and provides additional fuel (via the primary barrel on the twin venturi carburettor) at manifold vacuums below 6 inHg. On the Weber carburettor this is diaphragm operated, and on the single venturi Autolite carburettor this is operated by a piston.

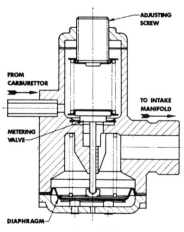

Fig. 147.3

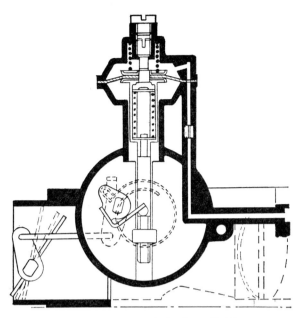

Fig. 147.4. Automatic choke (manifold vacuum modulation)

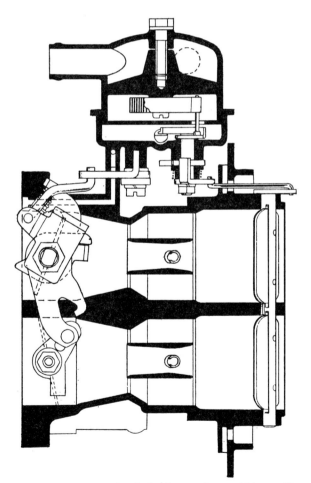

Fig. 147.5. Automatic choke (water heated bimetallic spring mechanism)

Another modification carried out on the Weber twin venturi carburettor was the introduction of a slotted main discharge nozzle in place of the rectangular form of discharge nozzle. It was adopted because it provided improved control of the main discharge spill point and increased the atomization of the fuel leaving the nozzle. Closer control of the carburettor flow band tolerances was achieved, principally in the off-idle region of the flow curve. This was reduced from ± 6 to ± 4 per cent for the Federal application and ± 5 per cent for the Economic Commission of Europe (E.C.E.) and to meet the Swedish regulations. The idle mixture control screw limiter was set to limit the CO emissions at idle to 3 per cent in the maximum rich position.

IMCO AIR CLEANER PACKAGE

Air cleaners for the new 2-litre S.O.H.C. engine are fitted with an automatic warm air control system which assists rapid warm-up and controls the air temperature at the carburettor entry to 100°F on a warmed-up engine. Control is achieved by a temperature sensitive wax capsule which operates a flap valve in the inlet tube of the air cleaner. Warm air is taken from a position adjacent to the hot cast-iron exhaust manifold. This is fed into the air cleaner where it mixes with the cold air drawn in from a relatively cool part under the bonnet. If the temperature drops, the wax capsule operates the valve to permit more warm air to be drawn from the vicinity of the exhaust manifold. In this way the air temperature in the air cleaner is controlled to 100°F±20 degF for the majority of the operating conditions. This is also necessary to maintain acceptable driveability with the leaner mixture settings used.

COMPRESSION RATIO FOR EMISSION CONTROLLED ENGINES

The effect of changes of compression ratio on the level of NO_x emissions is shown very clearly in reference (**3**). The Toyota Motor Company tested a 1075-cm³ four-cylinder engine with compression ratios of 6·9:1, 7·4:1, 8·5:1, and 9:1 for air/fuel ratios of 12:1, 14:1, and 16:1. The engine speed for these conditions was held at 2000 rev/min for a wide open throttle setting to simulate the 15–50 mile/h 'accel' mode of the FTP, and the spark timing was held at 30° b.t.d.c. As they show in Fig. 10 of their paper, it is evident that the increase of compression ratio results in an increase of combustion peak temperature, and hence the level of nitric oxide increases as the compression ratio is raised. This increase is not so large for overrich conditions such as is shown when the air/fuel ratio is held constant at 12:1. As mentioned previously, other people (**8**) increased the compression further to 10·5:1 and showed that the NO_x level rose to a maximum level and remained constant when detonation of the end gas commenced. It is thus certain that as NO_x increases almost linearly with increase in compression ratio (for the normal air/fuel ratios of 14:1 to 18:1), then to reduce NO_x the compression ratio must be reduced. The limiting factors on lowering compression ratio will be the fall-off in performance.

On test runs on Ford 2-litre S.O.H.C. engines built with compression ratios of between 9·2:1 and 8·2:1, the measured mean reduction in NO_x and HC emissions was 5 and 4 per cent per ratio reduction, respectively. It was therefore decided that a compression ratio of 8·6:1 was sufficiently low for the North American markets for the 1971 model year engines, and 8·2:1 for the 1972 model year.

Further work has been done on the 1·6-litre push-rod Kent engines being developed for the 1972 model year. The compression ratio was varied from 7·45:1 to 9·2:1 on an engine. The air/fuel ratio was held constant at 13·5:1 for a constant b.m.e.p. of 57 and a spark timing of 36° b.t.d.c. For these conditions, and for a speed of 2800 rev/min, the mean steady-state NO_x level varied by 250 p.p.m. This value is in agreement with the work of the Toyota Motor Company, who recorded a change of some 200 p.p.m. for a similar range of compression ratios. The same results were achieved at 2000 rev/min. It appears that very large reductions in compression ratio must be made if large reductions in the level of NO_x are to be achieved.

Further work was done to determine the effect on HC, CO, and NO_x emissions of various degrees of squish intensity. A 1·6-litre bowl-in-piston engine was used. We discovered that squish had only a minor effect on the HC and NO_x levels and that this effect could only be detected at higher speeds. Therefore, it was not considered important to reduce the squish intensity. No change in the level of CO was recorded. It was decided that for the 1972 model year the compression ratio should be reduced from 8:1 to 7·5:1.

OVERALL PERFORMANCE OF IMCO PACKAGE

Figs 147.6a–f show the typical overall performance of the

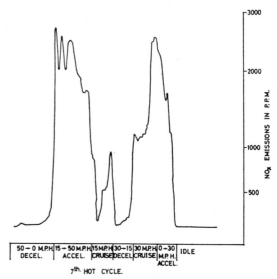

Fig. 147.6c. A 2-litre S.O.H.C. IMCO engine, automatic three-speed gearbox, Federal Capri

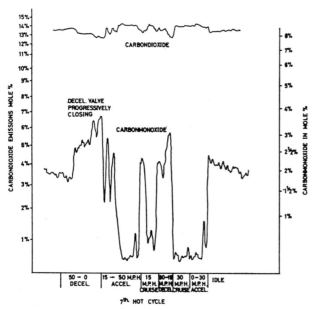

Fig. 147.6a. A 2-litre S.O.H.C. IMCO engine, automatic three-speed gearbox, Federal Capri

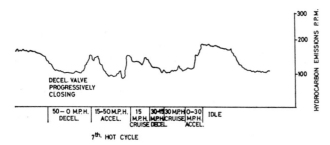

Fig. 147.6b. A 2-litre S.O.H.C. IMCO engine, automatic three-speed gearbox, Federal Capri

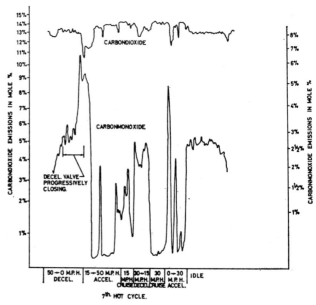

Fig. 147.6d. A 2-litre S.O.H.C. IMCO engine, manual four-speed gearbox, Federal Capri

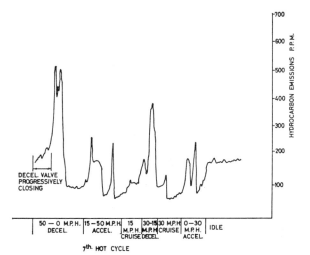

Fig. 147.6e. A 2-litre S.O.H.C. IMCO engine, manual four-speed gearbox, Federal Capri

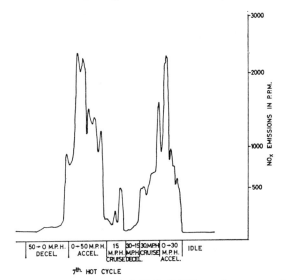

Fig. 147.6f. A 2-litre S.O.H.C. IMCO engine, manual four-speed gearbox, Federal Capri

IMCO package on a 2-litre S.O.H.C. engine fitted with a four-speed manual and a three-speed automatic gearbox in Federal Capri cars, when tested against the 7-mode cycle. The composite analysis of all seven cycles for the two cars is given in Table 147.2.

PARTICULATE EMISSIONS

Emissions of oxides of nitrogen

As other manufacturers have found, the highest levels occur during the accelerating modes when the combustion peak pressures and temperatures rise. The level of NO_x exhaust emissions drops slightly at the cruise modes and is at a minimum during deceleration and idle.

Carbon monoxide emissions

Whereas the NO emissions are high when high combustion chamber peak pressures and temperatures exist (when accelerating the vehicle), the level of CO emitted is low. This is in agreement with that experienced by other manufacturers. Under W.O.T. conditions, the combustion chamber pressure ratio prior to ignition is a maximum, as is the turbulence, squish, and swirl. Thus the chamber combustion efficiency is high and therefore the level of CO emissions is a minimum. The level of CO emissions rapidly increases during the part-open throttle cruise modes when the combustion chamber b.m.e.p. is reduced and the combustion efficiency begins to fall. The CO emissions rise to a maximum during the deceleration modes. The deceleration valve operates at high manifold vacuums and maintains combustion during the deceleration modes. However, the CO emissions are still highest during this part of the FTP cycle, thus indicating how inefficient combustion is under these conditions.

Hydrocarbon emissions

The HC emissions increase when the vehicle is being decelerated. The explanation for this is similar to that given above for the CO emissions. The operation of the 'decel' valve which injects a supply of air and fuel mixture for manifold vacuums greater than 22 inHg does reduce the level of emissions on the 'decel' modes to some extent, but for the above reasons the level of HC emissions is always a maximum during the deceleration modes.

For the same reasons the level of HC emitted during idle is high, as incomplete combustion takes place at low peak pressures and temperatures.

Control of crankcase emissions

On both the 1·6-litre push-rod engine and the new 2-litre S.O.H.C. engine, the blowby gases and oil mist in the sump are drawn off into an oil separator mounted on the side of the engine above the oil level. This separator has a vacuum applied to it by a rubber pipe connected to the inlet manifold, the location being downstream of the throttle plate. The flow through the separator is controlled by the spring loaded valve shown in Fig. 147.7, which also shows the mass flow characteristic of the crankcase p.c.v. valve.

A continuous supply of clean filtered air is supplied to the sump via the oil filler cap which is connected to the air cleaner by a rubber pipe. During part throttle openings, the manifold vacuum is higher than that developed in the air cleaner, and therefore the blowby gases and oil mist are drawn from the sump to the inlet manifold via the oil separator baffle plates. The oil collects in the base of the

Table 147.2
Gross vehicle weight, 1365 kg.   Inertia flywheel, 2500 lb.

Emission	Manual four-speed	Automatic three-speed
HC	123·8 p.p.m. (0·981 g/mile)	157·8 p.p.m. (1·391 g/mile)
CO	1·18% mole (17·556 g/mile)	1·13% mole (18·651 g/mile)
NO_x	798·5 p.p.m. (1·944 g/mile)	1055·1 p.p.m. (2·857 g/mile)

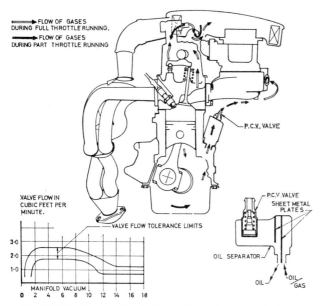

Fig. 147.7. Generalized view of crankcase emissions

separator and returns to the sump. During full throttle running, the blowby gas flow increases. However, the sump vacuum also reduces because the inlet manifold vacuum has decreased. Therefore, under these conditions some of the blowby gases are vented to the air cleaner via the oil filler cap and connecting pipe.

Fig. 147.8 shows a general view of the IMCO package as it is used on the Federal 1·6-litre and 2-litre engines.

CONCLUSIONS AND THE FUTURE

From our present experience gained by operating the spark modulation ignition systems with the IMCO package, we expect to be able to meet the 1973 Federal emissions regulations with this combined package. For future years, an exhaust gas recirculation package is under development, together with a chemical catalyst package. A thermal reactor development programme has also been started.

At present it appears that for 1974 we are likely to add the exhaust gas recirculation (EGR) system to the spark modulated IMCO package. The effectiveness of EGR for reducing NO_x has been recorded by the Toyota Motor Company (3) who claim a 50 per cent reduction of NO_x emissions for their steady-state tests. Our own experiments are at present encouraging.

For 1975 and onwards, it appears that we shall require to use every means at our disposal to meet proposed legislation, and we will probably combine a thermal reactor and chemical catalyst package to the present spark modulated EGR/IMCO system. However, it should be appreciated that it is very difficult to estimate how effective these packages will be when combined, and a great deal of further investigation is required.

APPENDIX 147.1
REFERENCES

(1) OLIVER, W. T. 'Practical emissions control systems—air

Fig. 147.8. General view of the IMCO package

injection into exhaust manifold', *Symp. Motor Vehicle Air Pollution Control, Proc. Instn mech. Engrs* 1968–69 **183** (Pt 3E), 4.

(2) DODD, A. E. and WISDOM, J. W. 'Effect of mixture quality on exhaust emissions from single-cylinder engines', *Symp. Motor Vehicle Air Pollution Control, Proc. Instn mech. Engrs* 1968–69 **183** (Pt 3E), 117.

(3) MATSUMO, K., TODA, T. and NOHIRA, H. 'Oxides of nitrogen from smaller gasoline engine', S.A.E. Paper 700145, 1970.

(4) NEWHALL, H. K. and STARKMAN, E. S. 'Direct spectroscopic determination of nitric oxide in reciprocating engine cylinders', S.A.E. Paper 670122, 1967.

(5) STARKMAN, E. S. 'Formation of exhaust emissions in the combustion chamber', Paper 15.3.D, *XIIIth FISITA Congress* 1970 (May).

(6) NEWHALL, H. K. 'Kinetics of engine generated nitrogen oxides and carbon monoxide', *12th Int. Symp. Combustion*, Poitiers (France), 1968.

(7) LAVOIE, G. A. 'Spectroscopic measurement of nitric oxide in spark-ignition engines', Fluid Mechanics Laboratory Publication 704, 1970 (March) (M.I.T.).

(8) DUKE, L. C., LESTZ, S. S. and MEYER, W. E. 'The relationship between knock and exhaust emissions of a spark-ignition engine', S.A.E. Paper 700062, 1970.

(9) TURNER, W. T. and MARTIN-FLAVEN, M. 'Some methods of controlling exhaust emissions during deceleration', *Symp. Motor Vehicle Air Pollution Control, Proc. Instn mech. Engrs* 1968–69 **183** (Pt 3E), 55.

(10) MORGAN, J. A. and STOJEK, D. J. 'A new range of 4-cylinder in-line S.O.H.C. engines by Ford of Europe', S.A.E. Paper 710148, 1971.

C148/71 THE INFLUENCE OF VALVE OVERLAP ON OXIDES OF NITROGEN EXHAUST EMISSIONS

C. HENAULT*

Exhaust gas recirculation is accepted as a technique for the reduction of emission of oxides of nitrogen. It can be achieved by increasing the amount of exhaust gas left in the combustion chamber. This paper describes tests made on a 1600-cm³ engine with engine dynamometer using five different camshafts giving valve overlap values varying from 0°–80°. Test results as confirmed by road tests are discussed and conclusions reached that in some cases low oxides of nitrogen emissions can result from a delay in the closing of the exhaust valve without the need for exhaust gas recirculation.

INTRODUCTION

THE REDUCTION of oxides of nitrogen (NO_x) from the exhaust effluent of motor cars, to comply with the proposed 1973–76 American limits, is one of the main problems affecting the motor industry. The maximum allowable NO_x values will dictate the depolluting systems which will be used on future vehicles. By the same token, the NO_x reduction techniques will affect the choice of systems to cut down the exhaust carbon monoxides and hydrocarbons. It is proposed in this paper to examine the influence of valve overlap on the formation of NO_x.

The influence of exhaust gas recirculation on the emissions of NO_x has been the subject of numerous papers and is therefore well documented; apart from the use of a reduction catalyst, it will be the technique employed for many years to come. Exhaust gas recirculation can be obtained in two ways: (1) by introducing part of the exhaust gas stream into the intake manifold, and (2) by increasing the mass of residual exhaust gas left in the combustion chamber. The second solution depends on the characteristics of the camshaft and has the great advantage that it can reduce the degree of recirculation with engine speed. One of the drawbacks is that it will have a marked influence on carburation.

INCREASING QUANTITY OF RESIDUAL EXHAUST GAS

Examination of the function of the engine near top dead centre (t.d.c.) at the end of the exhaust stroke shows that two ways of increasing the quantity of residual exhaust gas are available (Fig. 148.1). First, the blowing of exhaust gas into the intake manifold during the period when the inlet valve opens and the piston has not reached the top of its stroke. Second, the breathing of exhaust gas from the exhaust manifold between t.d.c. and the point where the exhaust valve closes. Calculations confirm the above two possibilities and yield values of trapped residual gas compatible with that of actual engine experience. The recirculating amounts are given in Figs 148.2 and 148.3.

ENGINE TESTS

Engine tests were made on a 1600-cm³ engine fitted to an engine dynamometer. All types of road conditions were tried on the bench, and confirmatory road tests were carried out in an actual vehicle. Gas analyses were made with non-dispersive infrared analysers for carbon monoxide (CO), carbon dioxide (CO_2), hydrocarbons (HC), and nitric oxide (NO). Engine physical properties were: 1565 cm³; four cylinders—bore 77 mm, stroke 85 mm, the compression ratio being 8·4:1.

To obtain information at every possible running mode, the mixture ratio was varied to obtain CO values between 0 and 2 per cent. Ignition timing was also varied near the optimum value. Five camshafts were used, and information regarding valve opening timing is shown in Table 148.1.

It will be seen that the different shafts gave overlap values from 0–80° crankshaft. The variations in valve timing have modified the engine characteristic causing differences in optimum performance.

To minimize the effect of these changes the following test methods were followed: (1) only road load values were employed; (2) these loads and engine speeds were

---

The MS. of this paper was received at the Institution on 8th July 1971 and accepted for publication on 18th August 1971. 24
* Régie Renault, 112 rue des Bons Raisins, 92 Rueil Malmaison, France.

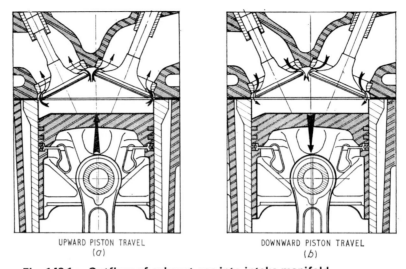

Fig. 148.1*a*. Outflow of exhaust gas into intake manifold

Fig. 148.1*b*. Inflow of exhaust gas through the opened exhaust valve

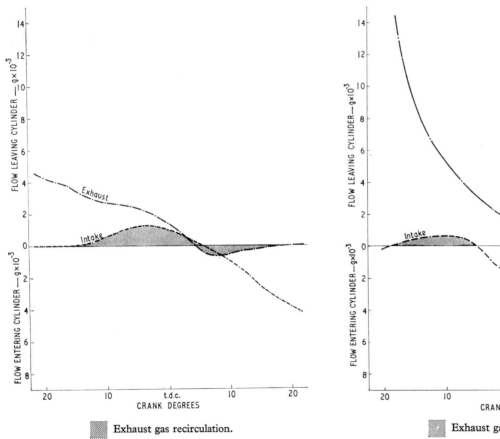

▨ Exhaust gas recirculation.

Fig. 148.2. Calculated flow through valves; crankshaft speed, 1000 rev/min

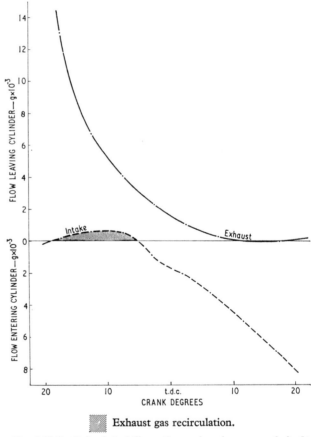

▨ Exhaust gas recirculation.

Fig. 148.3. Calculated flow through valves; crankshaft speed, 3000 rev/min

Table 148.1. Valve opening timing data

Inlet opens, °b.t.d.c.	Inlet closes, °a.b.d.c.	Exhaust opens, °b.b.d.c.	Exhaust closes, °a.t.d.c.
0	52	56	0
16	52	52	22
20	52	58	18
35	65	65	35
40	72	72	40

similar for all camshafts; (3) in all cases the starting point mixture richness was fixed at 2% CO and the ignition at optimum setting; (4) from the above setting CO, HC, and NO were measured for various mixture ratios and ignition timing values with a fixed throttle opening. The values of NO have not been corrected for humidity as only slight variations in ambient conditions were observed.

TEST RESULTS

Results are given for two speeds, i.e. 1000 and 3500 rev/min. The conclusions drawn from the work are also applicable to other speeds not mentioned. Fig. 148.4 shows engine performance with the various camshafts. The performance depends on the valve timing and is not truly relevant to the NO_x problem considered here. Figs 148.5 and 148.6 show the emissions of HC and NO_x in relation to the mixture richness in percentage CO. The figures are given for the speeds 1000 and 3500 rev/min, with engine loads corresponding to road loads for the speeds in question in top gear.

The results show conclusively that there is a very marked effect on NO emissions with valve overlap. The NO decreased rapidly with increase in valve overlap, while the emissions of HC were roughly constant. Nevertheless, at low speeds there is a trend for an increase in the HC emissions, probably due to a direct flow of air–fuel mixture from the intake to the exhaust caused by the various valve motions in the gas streams.

Examination of the ignition advance influence depicted in Figs 148.7 and 148.8 shows that optimum timing is not significantly altered by camshaft peculiarities. This is so

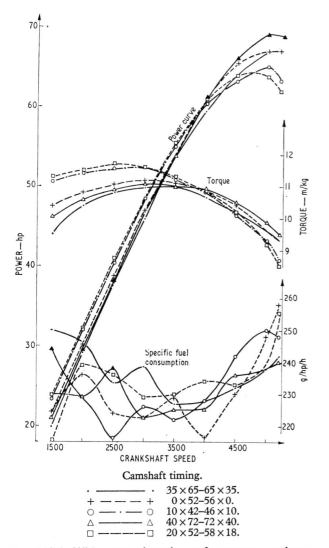

Fig. 148.4. Wide-open throttle performance; optimum A/F and ignition advance

Camshaft timing.
· ——— · 35 × 65–65 × 35.
+ – – – + 0 × 52–56 × 0.
○ —·—·— ○ 10 × 42–46 × 10.
△ ——— △ 40 × 72–72 × 40.
□ – – – □ 20 × 52–58 × 18.

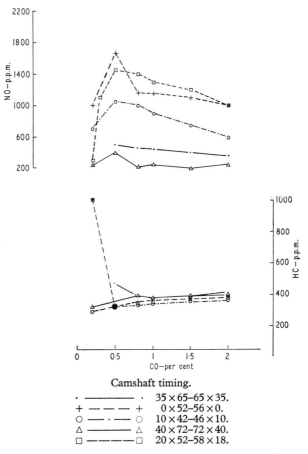

Fig. 148.5. NO and HC emissions as a function of A/F represented by exhaust CO content; 1000 rev/min; 3 hp; optimum ignition timing at 2% CO

Camshaft timing.
· ——— · 35 × 65–65 × 35.
+ – – – + 0 × 52–56 × 0.
○ —·—·— ○ 10 × 42–46 × 10.
△ ——— △ 40 × 72–72 × 40.
□ – – – □ 20 × 52–58 × 18.

at constant power and air/fuel ratio, and it will be noticed that the slope of the NO_x emission curve plotted against ignition timing is least when the camshaft gives the highest degree of valve overlap. The slopes become identical at 2500 rev/min, the lines being shifted vertically by an amount related to the combustion chamber residual gas quantity.

It has not been well explained why the emissions of NO_x are insensitive to ignition advance changes at low speeds.

The various tests have yielded enough information for the plotting of NO emission results against the degree of valve overlap (Fig. 148.9) and the amount of late closing of the exhaust valve (Fig. 148.10). It would appear that the delay in closing the exhaust valve is more significant than the actual valve overlap. Further reference to Figs 148.9 and 148.10 shows that a delay in closing the exhaust valve of 35° crankshaft, or a valve overlap of 70° with a symmetric camshaft, would give optimum results for reducing the NO_x emissions without modifying the engine response at low speeds.

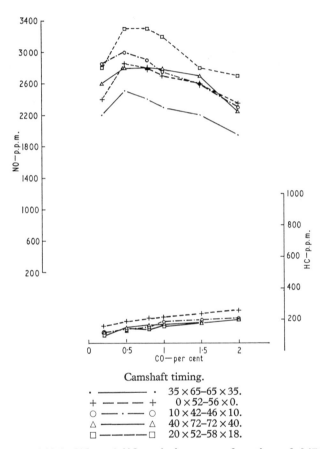

Fig. 148.6. NO and HC emissions as a function of A/F represented by exhaust CO content; 3500 rev/min; 24 hp; optimum ignition timing at 2% CO

Camshaft timing.
- · ——— · 35×65–65×35.
- + — — — + 0×52–56×0.
- ○ — · — ○ 10×42–46×10.
- △ ——— △ 40×72–72×40.
- □ — — — □ 20×52–58×18.

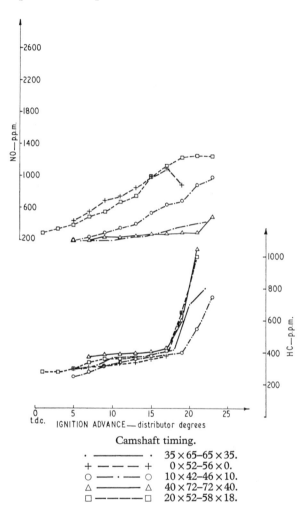

Fig. 148.7. NO and HC exhaust emissions as a function of ignition advance; 1000 rev/min; 3 hp; 2% CO; optimum ignition timing

Camshaft timing.
- · ——— · 35×65–65×35.
- + — — — + 0×52–56×0.
- ○ — · — ○ 10×42–46×10.
- △ ——— △ 40×72–72×40.
- □ — — — □ 20×52–58×18.

Table 148.2. Car trials results

Camshaft	Tappet clearance, mm		CO		HC		NO	
	Inlet	Exhaust	%	g/mile	p.p.m.	g/mile	p.p.m.	g/mile
16/52–52/22 . .	0·25	0·30	0·95	17·1	62	0·59	982	2·9
16/52–52/22 . .	0·15	0·20	0·83	14·9	56	1·54	855	2·52
16/52–52/36 . .	0·25	0·30	1·03	18·5	61	0·58	814	2·4
16/52–52/36 . .	0·15	0·20	1·12	20·1	66	0·63	732	2·16

THE INFLUENCE OF VALVE OVERLAP ON OXIDES OF NITROGEN EXHAUST EMISSIONS

VEHICLE RESULTS WITH VARIATIONS IN EXHAUST VALVE CLOSING TIMING

The car trials were carried out following the Californian 1972 procedure, i.e. 1972 U.S. Federal C.V.S. test plus two hot Californian cycles when the emissions of NO_x are measured with non-dispersive infrared analysers. As far as possible the air/fuel ratio was kept constant throughout the various modes of the test cycle. Table 148.2 shows the results obtained.

The figures given in Table 148.2 show that the efficacy of using late exhaust valve closure to reduce NO_x emissions has been proved. A decrease of emissions of about 170 p.p.m. for the whole cycle was obtained for an increase in exhaust valve closing of 14° crankshaft. The gain reached 250 p.p.m. with a reduction in valve clearance which increased simultaneously both the inlet valve opening angle and the exhaust valve closing angle. It must be pointed out that valve overlap increases very rapidly with reduction in valve clearance, but that the valve lift remains very small.

With the above two camshafts, the increase in valve overlap with modification to valve clearance is about 36° crankshaft, while increase in overlap with normal valve clearance is but 14°.

RESULTS IN NO_x EMISSION OF A VEHICLE FITTED WITH A LARGE OVERLAP CAMSHAFT

The necessity to produce an engine with a very large specific power has led to the use of camshaft timings of 40°–70° inlet, and 72°–40° exhaust, thus with a valve overlap of 80° crankshaft. This type of engine has yielded 450–500 p.p.m. NO_x following hot Californian cycles. These results should be compared with normal NO_x emissions of 1100–1200 p.p.m. when the engine is functioning with a 30° valve overlap. The above will not be

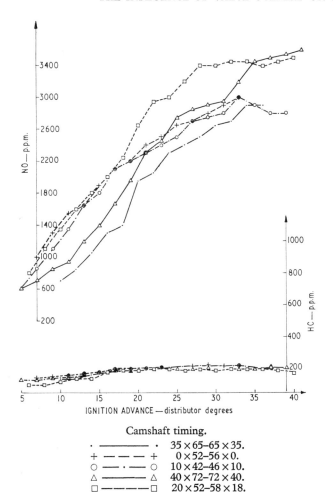

Fig. 148.8. NO and HC exhaust emissions as a function of ignition advance; 3500 rev/min; 24 hp; 2% CO; optimum ignition advance

Camshaft timing.
- · ———— · 35 × 65–65 × 35.
- + — — — + 0 × 52–56 × 0.
- ○ — · — ○ 10 × 42–46 × 10.
- △ ———— △ 40 × 72–72 × 40.
- □ — — — □ 20 × 52–58 × 18.

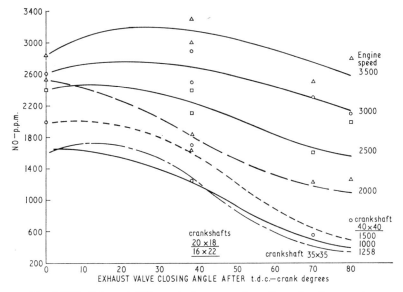

Fig. 148.9. Exhaust NO_x emissions as a function of valve overlap

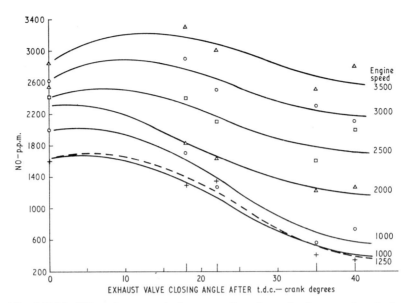

Fig. 148.10. NO_x exhaust emissions as a function of exhaust valve timing

discussed in detail as there are numerous problems with large overlap camshafts, such as high HC emissions and instability at low speeds.

CONCLUSIONS

Good correlation was found between the reduction of NO_x and valve overlap—or more accurately the delay in closing the exhaust valve—on a production engine. The road tests have confirmed the results, which in some cases enable low NO_x emissions to be obtained without having to use exhaust gas recirculation. Nevertheless, if the valve overlap is too great other problems appear, such as poor engine response and higher exhaust HC emissions at low speeds, which are not compatible with touring cars produced in large volumes.

ACKNOWLEDGEMENTS

The author is indebted to the Directors of the Régie Nationale des Usines Renault for permission to publish this paper.

C149/71 KINETICS OF NITRIC OXIDE FORMATION IN INTERNAL-COMBUSTION ENGINES

WINFRIED E. BERNHARDT*

The formation of nitric oxide in spark-ignition engines is examined in the light of elementary chemical reaction kinetics. A simple mathematical model describing the formation of nitric oxide inside the engine is developed, which consists essentially of a reaction mechanism suitable for reciprocating-engine combustion and of a thermodynamic analysis of the combustion process. On the basis of the conclusions obtained from this concept, a single-cylinder engine is employed in an attempt to influence the combustion process by means of pertinent combustion controls, such as stratified charge, in such a manner that the formation of nitric oxide will be restrained as the result of excessively low temperatures (<2600°K), even though considerable oxygen may be present. Furthermore, the temperatures are sufficiently high to permit the rapid oxidation of carbon monoxide and unburnt hydrocarbons. Thus, the results clearly demonstrate the advantage of a theoretical model based on a kinetic scheme in permitting simultaneous reduction of exhaust pollutants to low levels.

INTRODUCTION

THE INVESTIGATION of non-equilibrium processes in internal-combustion engines, including the kinetics of elementary chemical reactions, has been in the foreground of scientific interest for quite some time (1)–(11)†. In view of the drastic measures demanded by legislative authorities against air pollutants emanating from automotive exhaust systems, it is now an urgent necessity to find ways and means for reducing the pollutants carbon monoxide (CO), nitric oxide (NO), and unburnt hydrocarbons (HC) to obtain a more complete combustion, and as a result cleaner exhaust gases. (Although the term NO_x is commonly used to refer to the total oxides of nitrogen, the primary compound is NO (at least 98 per cent of the total oxides of nitrogen).)

In this paper the thermal formation of NO inside a combustion engine will be the subject of a theoretical investigation because of its special significance with regard to environmental control. A model of reciprocating engine NO formation has been developed and its transferability to an actual engine has been examined. On the basis of the resulting conclusions, attempts were made to control the engine combustion process in such a manner that NO concentrations would be rather low and that, in addition, the oxidation processes required for a reduction of HC and CO would still be almost complete.

The MS. of this paper was received at the Institution on 15th July 1971 and accepted for publication on 24th August 1971. 23
* Volkswagenwerk AG, Research and Development Department, Wolfsburg, Germany.
† References are given in Appendix 149.1.

No attempt will be made to discuss the possibilities of NO reduction by recycling exhaust gases or by using so-called 'diluents', for example, CO_2, H_2O, He, Ar, N_2, etc., by the use of reduction catalysts, by the selection of unconventional fuels, or by adding certain additives for promoting combustion. In this connection, refer to Meguerian (7).

MODEL OF NO FORMATION IN THE ENGINE CYLINDER

Reaction mechanisms for NO formation

The thermal formation of NO can be identified by the gross reaction

$$N_2 + O_2 \rightleftharpoons 2NO \qquad . \quad . \quad (149.1)$$

Such a reaction consists of a number of basic reactions which, seen individually, describe the collisions between molecules, radicals, or atoms, by means of which new substances will be formed. The basic reactions are therefore the actual chemical reaction while the gross reaction reflects only the total result of these basic reactions without considering the intermediate products.

It was determined from a review of the literature on basic chemical reactions in a N—O-system that 16 reactions may be important for the formation of NO_x [nitrous oxide (N_2O), nitrogen dioxide (NO_2), and NO].

These possible elementary reactions for the formation of NO_x follow:

$$N_2 + M \rightleftharpoons 2N + M \qquad . \quad . \quad (149.2)$$
$$N_2 + NO + O_2 \rightleftharpoons N_2O + NO_2 \qquad . \quad (149.3)$$

$$N_2 + NO_2 \rightleftharpoons NO + N_2O \quad . \quad (149.4)$$
$$N_2 + O \rightleftharpoons N + NO \quad . \quad (149.5)$$
$$N_2 + O_2 \rightleftharpoons NO + NO \quad . \quad (149.6)$$
$$N_2 + O_2 \rightleftharpoons O + N_2O \quad . \quad (149.7)$$
$$N_2 + O_2 \rightleftharpoons N + NO_2 \quad . \quad (149.8)$$
$$NO_2 + M \rightleftharpoons NO + O + M \quad . \quad (149.9)$$
$$N_2O + O \rightleftharpoons NO + NO \quad . \quad (149.10)$$
$$N_2 + O + M \rightleftharpoons N_2O + M \quad . \quad (149.11)$$
$$O + N_2O \rightleftharpoons N + NO_2 \quad . \quad (149.12)$$
$$N + O + M \rightleftharpoons NO + M \quad . \quad (149.13)$$
$$O + NO \rightleftharpoons N + O_2 \quad . \quad (149.14)$$
$$O_2 + M \rightleftharpoons 2O + M \quad . \quad (149.15)$$
$$O_2 + N + M \rightleftharpoons NO_2 + M \quad . \quad (149.16)$$
$$O + NO_2 \rightleftharpoons O_2 + NO \quad . \quad (149.17)$$

However, previous investigations indicate that only some of these basic reactions are primarily responsible for the formation of NO_x in engine-related combustion processes. Under conditions closely related to those occurring in engines, two mechanisms which are probably controlling the formation of NO_x are given below.

The reaction mechanism I, detailed below, has been reported recently by Newhall and Shahed (**1**); the reaction mechanism II by Caretto et al. (**2**). However, equations (149.9) and (149.19) to (149.21) were already shown in the reaction mechanism presented in 1945 by Vetter (**12**), while equations (149.24) to (149.29) were recently reported by Newhall and Starkman (**13**), as well as by Lavoie et al. (**3**). Since both computations and experimental tests have shown that NO is the only oxide of nitrogen of significance in an engine (**14**) (**15**), the basic reactions, which contain NO_2 or N_2O as reaction participants, may be omitted in these reaction mechanisms without causing major errors.

The following equations give the reaction mechanisms that probably control the formation of NO_x in internal-combustion engines.

Mechanism I

$$N + O + M \rightleftharpoons NO + M \quad . \quad (149.18)$$
$$O_2 + N \rightleftharpoons NO + O \quad . \quad (149.19)$$
$$N_2 + O \rightleftharpoons NO + N \quad . \quad (149.20)$$
$$NO_2 + O \rightleftharpoons NO + O_2 \quad . \quad (149.21)$$
$$NO_2 + M \rightleftharpoons NO + O + M \quad . \quad (149.9)$$
$$N_2O + O \rightleftharpoons NO + NO \quad . \quad (149.10)$$
$$N_2 + O + M \rightleftharpoons N_2O + M \quad . \quad (149.11)$$

Mechanism II

$$N_2 + O \rightleftharpoons NO + N \quad . \quad (149.22)$$
$$O_2 + N \rightleftharpoons NO + O \quad . \quad (149.23)$$
$$OH + N \rightleftharpoons NO + H \quad . \quad (149.24)$$
$$N_2 + O_2 \rightleftharpoons N_2O + O \quad . \quad (149.25)$$
$$N_2 + OH \rightleftharpoons N_2O + H \quad . \quad (149.26)$$
$$N_2O + O \rightleftharpoons NO + NO \quad . \quad (149.27)$$
$$N + O + M \rightleftharpoons NO + M \quad . \quad (149.28)$$
$$N_2 + O + M \rightleftharpoons N_2O + M \quad . \quad (149.29)$$

Table 149.1. Selected reaction rate constants

j	Reaction	Rate constant, cm³/mole s
1	$N_2 + O \rightarrow NO + N$	$7 \times 10^{13} \exp(-75\,000/RT)$
2	$NO + N \rightarrow N_2 + O$	1.55×10^{13}
3	$O_2 + N \rightarrow NO + O$	$13.3 \times 10^9 T \exp(-7080/RT)$
4	$NO + O \rightarrow O_2 + N$	$3.2 \times 10^9 T \exp(-39\,100/RT)$

Rate constants K_1 and K_4 from reference (**16**). Reverse rate constants K_2 and K_3 determined from equilibrium constants. $K_{eq12} = 2.5 T^{0.07} \exp(-75\,000/RT)$ and $K_{eq34} = 4.6 \exp(-31\,850/RT)$.

The third body M may be any particle from the reaction area, since it serves only for removing energy.

Recently Bernhardt (**4**) showed that with regard to engine NO formation the reactions

$$N + O + M \rightleftharpoons NO + M$$

and

$$OH + N \rightleftharpoons NO + H$$

are of subordinate significance; also compare (**5**). Therefore, two equations remain from each of the two reaction mechanisms shown above, and they are known as Zeldovich mechanisms (**6**). These are equations (149.19) and (149.20) of mechanism I, and equations (149.22) and (149.23) of mechanism II.

For computing the rate constants in the Zeldovich mechanism the reaction kinetic data of Wary and Teare (**16**) were used (see Table 149.1).

Model of flame propagation and thermodynamic analysis of combustion process

The model of flame propagation used in this paper is based on the assumption that there are regions of considerably varying temperatures in the combustion chamber during the combustion process (Fig. 149.1). The region already passed by the flame front contains a burnt mixture of high temperature; the region not yet reached by the flame contains an unburnt charge of relatively low temperature.

The assumption is that there is an intensive mixing in both regions, so that the temperature differences will be quickly reduced in each region and the release of energy

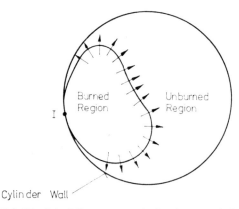

Fig. 149.1. Model of flame propagation in a spark-ignition engine cylinder. **I** denotes ignition point

between the unburnt and the burnt area of the charge will occur so quickly that the temperature of the unburnt charge will increase to the higher temperature of the exhaust gases quite suddenly.

Efforts to establish a generally valid analytical relationship for the propagation of the flame front inside a combustion engine have not been successful. But it is assumed that the model used here describes the processes adequately, particularly since the model is based on the actual cylinder pressure, which should be uniform throughout the combustion area. The volume V is given as a function of time by the kinematic equations for the piston motion. The total volume according to this model concept is composed of the volume of the unburnt and the volume of the burnt charge.

The temperature in the unburnt region is computed under the assumption of an isentropic change of state of a perfect gas. This is a coarse approximation, since the prerequisite for applying equilibrium thermodynamics is not provided. But since the temperature of the unburnt charge has but a slight influence, the computations can be made as described.

The temperature in the burnt region can be determined via the chronological progress of the conversion of chemical energy of the fuel into the internal energy of the combustion gases. But this requires a thermodynamic analysis of the engine combustion process first. This analysis must take into account the fact that the chemical components keep changing during the process as the result of chemical reactions. The result of such a combustion cycle analysis for a VW Type 3 engine (engine speed $n = 3000$ rev/min, 10 per cent fuel-rich mixture, full load operation) is shown in Fig. 149.2. The fraction mass burnt that is needed for the calculation of the temperature in the burnt region is obtained from Fig. 149.2. The result is shown in Fig. 149.3.

A detailed description of the method for determining the energy conversion in an internal-combustion engine is shown by Hinze (**17**) and Bernhardt (**18**). No detailed discussion is therefore provided here. Fig. 149.4 provides the temperature variation for the investigated combustion process in the unburnt and the burnt region of the combustion chamber. In addition, the mean temperature prevailing in the charge is recorded as it would occur in an adiabatic mixing process. Determination of the temperature has been based on the assumption that the behaviour of state of the reacting substances can be described by the equation of state of the perfect gas.

Calculation of kinetics of NO formation

The model of flame propagation described here is used for computing the kinetics of NO formation. The temperatures and volumes or densities computed with the assistance of this model are particularly needed. For the Zeldovich mechanism the chronological change of the NO concentration for isochoric processes, in which the density remains unchanged, is given by:

$$\frac{d\sigma_{NO}}{\rho \, dt} = k_3 \sigma_{O_2} \sigma_N - k_4 \sigma_{NO} \sigma_O + k_1 \sigma_{N_2} \sigma_O - k_2 \sigma_{NO} \sigma_N \quad (149.30)$$

In this equation ρ is the density, σ_i denotes the moles of species i per unit mass of mixture, t is the time, and k_j denotes the rate constant of the appropriate reaction j listed in Table 149.1.

Using the differential equation (149.30), there is some uncertainty due to the determination of the O and N concentrations. However, experimental results indicate that NO formation is slow relative to probable rates for post-flame relaxation of atoms and radicals to equilibrium concentrations. Therefore the temperature and pressure of the burnt gases (post-flame gases) were used to obtain the chemical equilibrium concentrations of O_2, N_2, O, and N.

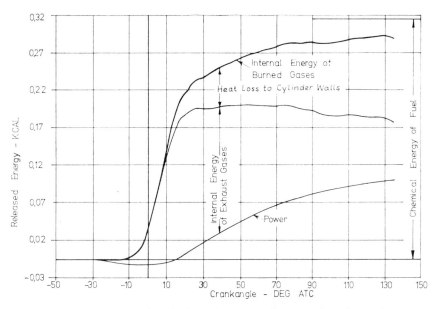

Fig. 149.2. Typical variation of energy release during the cycle

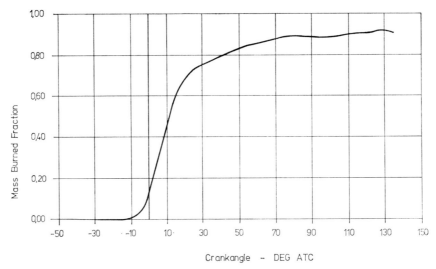

Fig. 149.3. Mass burnt fraction versus crankangle

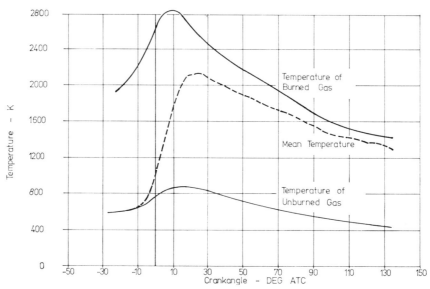

Fig. 149.4. Temperature of fresh and burnt gas during combustion process

For known concentrations σ_{O_2}, σ_{N_2}, σ_O, and σ_N, for known density and known temperature, the differential equation (149.30) can then be solved by means of the Runge–Kutta procedure. The NO formation both in the unburnt and in the burnt region is mathematically derived.

Fig. 149.5 shows the variation of the NO concentrations in the burnt region in a VW Type 3 engine for an engine speed of $n = 3000$ rev/min, ignition 28° b.t.d.c., and 10 per cent fuel-rich mixture (equivalence ratio 1·1). Starting from the ignition point, the NO formation during combustion and expansion is shown diagrammatically. In Fig. 149.5 NO does not appear between time of ignition and t.d.c. because NO concentration is very low—less than 30 p.p.m.

As expected on the basis of earlier investigations (19) made with the assistance of the theory of chemical equilibrium distribution, the NO formation in the unburnt region is neglected compared with the NO formation in the burnt charge because of the low temperatures (<1000°K) there. The NO concentration in the unburnt charge is so low that it cannot be shown in Fig. 149.5.

Characteristic of the thermal NO formation in an Otto engine is the fact that it is noticed only rather late, i.e. when about one-third of the entire combustion period is already gone. But this applies only to the average NO concentration of the burnt charge examined here. The concentration of NO formed in the first mass to burn has quite a different concentration history (2). The maximum of the average NO concentration will come about when two-thirds of the combustion period have passed. Upon reaching the maximum, a change in the NO concentration is hardly noticed during the continuing expansion. This result is in excellent agreement with the experimental

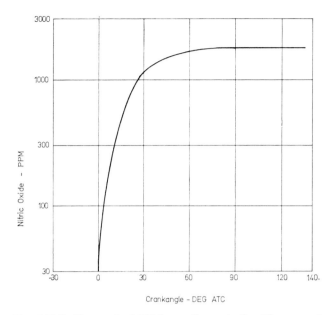

Fig. 149.5. Theoretical NO formation rate for 10 per cent fuel-rich mixture (equivalence ratio 1·1)

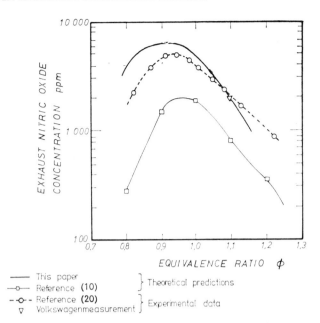

Fig. 149.6. Comparison of measured and predicted exhaust NO concentration

investigations of Newhall and Starkman (8) (13), who recorded the monochromatic emission of the NO by means of a direct spectroscopic determination during expansion, and with the reaction kinetics computations of Newhall (9).

Based on several mean cylinder pressures versus crankangle functions for equivalence ratios between 0·8 and 1·1 and for engine speed $n = 3000$ rev/min, the method described above was used to calculate exhaust NO concentrations as a function of the equivalence ratio. The predicted exhaust concentrations are shown in Fig. 149.6 along with experimental exhaust NO concentrations. Because of the absence of complete Volkswagen NO measurements [at present, only one VW measurement (for the equivalence ratio 1·1) is available], the results of the theoretical calculations may be compared with the measurements made by Huls and Nickol (20) for very similar engine operating conditions (engine speed 2500 rev/min, full load operation, spark timing 40° b.t.d.c.).

As shown by Fig. 149.6, theoretical calculation based on the Zeldovich mechanism and the model of flame propagation previously discussed predicts NO formation for stoichiometric mixtures as well as for fuel-rich mixtures with excellent accuracy. However, the results presented in Fig. 149.6 show also that predictions are only within about a factor of 2 of the experimental values for fuel-lean mixtures. In this region, predicted values lie above the measured values. The reason for this discrepancy may be:

(1) inaccuracy in the burnt gas temperature calculation,

(2) incomplete reaction mechanism for NO formation, and

(3) inaccuracies in the reaction rate constants.

The effects of these uncertainties on the model will be examined in a later paper.

Other models only recently proposed by Heywood et al. (10) and by Lavoie et al. (3) show, contrary to the model developed above, that for fuel-lean and stoichiometric mixtures predictions are within a factor of 2 below experimentally measured values. Furthermore, for fuel-rich mixtures the discrepancy is about an order of magnitude. Thus, the agreement between model predictions and experimental measurements shown in Fig. 149.6 is indeed better than previously presented results.

COMBUSTION CONTROL MEASURES FOR NO REDUCTION

The computation of the kinetics of NO formation shows that the formation of NO in the burnt region of the charge requires temperatures above 2600°K (referring to Figs 149.4 and 149.5). In comparison, the HC and CO oxidation processes occur already at considerably lower temperatures. Therefore, if the combustion process in the engine can be controlled in such a manner that the temperature of the combustion gases is slightly below 2600°K, the formation of NO will be restrained; which will most certainly result in lower NO concentrations in the exhaust gases.

Another important finding from the model of the formation of NO is that major concentrations of NO will be formed in the burnt gases after the passage of the flame front. Therefore the NO formation reactions are 'post-flame reactions'.

On the basis of the knowledge gained from the model of NO formation described above in the Research Department of the Volkswagenwerk AG, a new method for simultaneously reducing engine exhaust emissions was

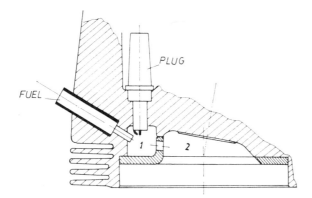

Fig. 149.7. Schematic drawing of combustion system with plug–nozzle configuration

tested in a Co-operative Lubricant Research (CLR) single-cylinder engine under actual combustion conditions. The experimental results indicate that both the HC and CO, as well as the NO concentrations, can be simultaneously reduced to low levels.

In principle, it is a prechamber arrangement in which an almost stoichiometric or fuel-rich mixture supplied by a fuel injector is ignited and burnt. The combustion system is designed in such a manner that, after rushing through the prechamber, the flame front passes through one or several orifices into the main combustion chamber, where it encounters the load-controlled main charge at relatively low temperature. (The main combustion chamber may even hold clean air, which is drawn into the cylinder through the intake valve during the inlet stroke of the operating cycle.)

As a result, the hot combustion gases are rapidly quenched to lower temperatures, so that in spite of excess O only little NO formation can occur. But the resulting temperature is still high enough to permit rapid oxidation for CO and unburnt HC. A prechamber arrangement similar in principle was recently presented by Newhall and El-Messiri (21).

Fig. 149.7 shows the arrangement of prechamber (1) and main combustion chamber (2) on a VW Type 3 cylinder head. The ratio of prechamber volume to the volume of the main combustion chamber was between 1:10 and 3:10 during the experiments. The exhaust gas emissions of this prechamber arrangement on a CLR single-cylinder engine at an engine speed of 2000 rev/min and under full load operation were measured. They were found to be very low, down to 40 p.p.m. NO, 40 p.p.m. HC (measured on hexane basis), and 0·1 vol. % CO at equivalence ratios of 0·3 to 0·5.

Finally, the combustion processes in the combustion chamber were filmed with the assistance of a high-speed camera (6000 frames/s) to check whether the combustion results obtained were in agreement with the knowledge gained from the model for engine NO formation. The pictures were recorded through a window in the piston, which allows an unobstructed view of the entire main combustion chamber.

Fig. 149.8 shows 64 frames of a single combustion process which were taken with a high-speed camera at 2000 frames per min, spark timing 30° b.t.d.c., a mean effective pressure of $p_{me} = 2·5$ kp/cm^2, an equivalence ratio of 0·4 in the main combustion chamber, and an overall equivalence ratio of 0·62. The experimental observations

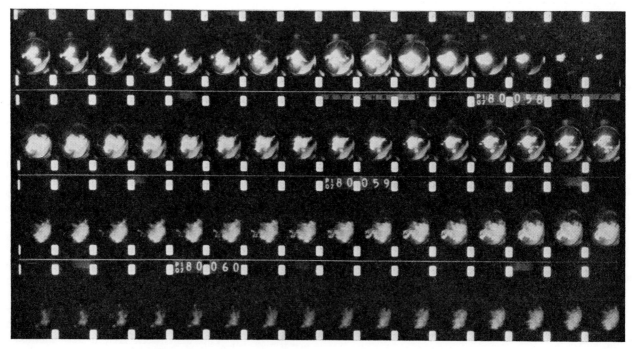

Fig. 149.8. Typical set of flame pictures of a single combustion process

consist of pictures showing successive positions of the flame at intervals of 2·0 crankshaft degrees during a single combustion process. Under these conditions 170 p.p.m. NO were measured, more than one order of magnitude less than for conventional combustion chambers.

CONCLUSIONS

Since the fuel burning is a slow process relative to the time scale and NO formation a fast process, the two processes can be virtually treated independently, as is the case in the model described in this paper.

For stoichiometric and fuel-rich mixtures, the exhaust NO concentrations predicted by the model of NO formation in engine-related combustion processes are in acceptable agreement with measured exhaust NO concentrations. It should be noted that the temperature of the post-flame gases has a main effect on the model predictions.

On the basis of the conclusions resulting from the model, basic changes of the conventional combustion system, such as stratified charge process, were made. The exhaust gas emissions from a CLR single-cylinder engine with a modified VW Type 3 cylinder head were measured and found to be for NO up to 10 per cent, and for HC up to 25 per cent of those levels observed with conventional combustion processes.

Prechamber-type engines, combined with a stratified charge concept, might therefore be a possible solution for a considerable reduction of pollutants emanating from automotive exhaust systems. But in spite of the good results obtained so far, it is nevertheless doubtful whether this method alone will be sufficient to meet the drastic emissions requirements for 1975 and 1976 while maintaining good fuel economy and high output performance.

ACKNOWLEDGEMENTS

The author wishes to acknowledge the assistance of Messrs E. Hoffmann, I. Geiger and G. Decker for their contributions to this paper. Mr E. Hoffmann has been extremely helpful in measuring engine data and developing the computer program for the thermodynamic analysis of the engine combustion process. Messrs I. Geiger and G. Decker tested the prototype combustion system and made the motion pictures of engine flames.

APPENDIX 149.1

REFERENCES

(1) NEWHALL, H. K. and SHAHED, S. M. 'Kinetics of nitric oxide formation in high-pressure flames', Paper presented at *Thirteenth Int. Symp. on Combustion*, Salt Lake City, Utah, 1970 (August).

(2) CARETTO, L. S., MUZIO, L. J., SAWYER, R. F. and STARKMAN, E. S. 'The role of kinetics in engine emission of nitric oxide', Paper presented at the A.I.Ch.E. Meeting, Denver, Colorado, 1970 (August).

(3) LAVOIE, G. A., HEYWOOD, J. B. and KECK, J. C. 'Experimental and theoretical study of nitric oxide formation in internal-combustion engines', *Combust. Sci. Technol.* 1970 **1**, 316.

(4) BERNHARDT, W. 'Untersuchungen der Nichtgleichgewichtsvorgänge der im Brennraum von Verbrennungsmotoren ablaufenden Stickoxid-Reaktionen', Paper presented on the Thermodynamik-Kolloquium of the VDI in Würzburg, 1970 (5th October); *Staub-Reinhaltung Luft* 1971 **31** (No. 7), 279. Also refer to *Brennst.-Wärme-Kraft* 1971 **23** (No. 2).

(5) CAMPBELL, I. M. 'Chemical mechanisms relevant to the production and emission of nitric oxide and carbon monoxide from combustion engines', Lecture 7 in *A short course on fundamentals of engine exhaust pollution*, 1970 (September) (Univ. of Leeds).

(6) ZELDOVICH, YA. B. 'The oxidation of nitrogen in combustion explosions', *Acta phys.-chim. URSS* 1946 **21**, 577.

(7) MEGUERIAN, G. H. 'Nitrogen oxide formation, suppression, and catalytic reduction', Paper PD 23, *World Oil Congress*, Moscow, 1971 (June).

(8) STARKMAN, E. S. 'Formation of exhaust emission in the combustion chamber', Paper No. 15.3.D, *Thirteenth Congress of FISITA*, Bruxelles, Belgium, 1970 (June).

(9) NEWHALL, H. K. 'Kinetics of engine-generated nitrogen oxides and carbon monoxide', *Twelfth Int. Symp. on Combustion* 1969, 603 (Combustion Inst., Pittsburgh).

(10) HEYWOOD, J. B., MATHEWS, S. M. and AVEN, B. 'Predictions of nitric oxide concentrations in a spark-ignition engine compared with exhaust measurements', S.A.E. Paper 71001, 1971.

(11) MUZIO, L. J., STARKMAN, E. S. and CARETTO, L. S. 'The effect of temperature variations in the engine combustion chamber on formation and emission of nitrogen oxides', S.A.E. Paper 710158, 1971.

(12) VETTER, K. 'Kinetik der thermischen Zersetzung und Bildung von Stickoxid,' *Z. Elektrochem.* 1949 **53**, 369.

(13) NEWHALL H. K. and STARKMAN, E. S. 'Direct spectroscopic determination of nitric oxide in reciprocating engine cylinders', S.A.E. Paper No. 670122 1967.

(14) WIMMER, D. B. and MCREYNOLDS, L. A. 'Nitrogen oxides and engine combustion', *S.A.E. Trans.* 1962 **70**.

(15) CAMPAU, R. M. and NEERMAN, J. C. 'Continuous mass spectrometric determinations of nitric oxide in automobile exhaust', *S.A.E. Trans.* 1967 **75**; Paper No. 660116, 1966.

(16) WARY, K. L. and TEARE, J. D. 'Shock-tube study of the kinetics of nitric oxide at high temperatures', *J. phys. Chem.* 1962 **36**, 2582.

(17) HINZE, W. 'Verfahren zur thermodynamischen Auswertung von Prüfstandsergebnissen untersuchter Verbrennungsmotoren', Dr Eng. Thesis, Technical University, Dresden, 1956.

(18) BERNHARDT, W. 'Thermodynamische Auswertung von Prüfstandsversuchen zur Ermittlung der Energieumwandlungsgesetze im Motor', VW Report T.327, 1969 (11th June) (unpublished).

(19) BERNHARDT, W. 'Bildung der Stickoxide in Verbrennungsmotoren', Abbreviated Report of Research 2, 1969 (18th December) (1942-drbe/st), unpublished; also VW Report V/70, 'Control of oxides of nitrogen', 5th Status Report of VWoA to California Air Resources Board, Los Angeles (unpublished).

(20) HULS, T. A. and NICKOL, H. A. 'Influence of engine variables on exhaust oxides of nitrogen concentrations from a multicylinder engine', S.A.E. Paper 670482, 1967.

(21) NEWHALL, H. K. and EL-MESSIRI, I. A. 'A combustion chamber concept for control of engine exhaust air pollutants emissions', *Combust. Flame* 1970 **14**, 155.

C150/71 EVAPORATIVE LOSSES FROM AUTOMOBILES: FUEL AND FUEL SYSTEM INFLUENCES

R. W. HURN* B. H. ECCLESTON*

The principal influences bearing upon evaporative hydrocarbon losses from automobiles have been studied in experimental work at the Bartlesville Energy Research Center of the U.S. Bureau of Mines at Bartlesville, Oklahoma, U.S.A. Thirteen fuels and thirty-one vehicles were used in the study. Hydrocarbon losses, both exhaust and evaporative, were measured during tests that simulated typical city driving with the vehicles. Results show that fuel composition markedly affects both the amount and chemical character of the losses; the amount of loss is also dependent upon ambient temperature and factors related to the vehicle. The manner and degree in which the specific fuel and vehicle factors influence the losses are discussed.

INTRODUCTION

HYDROCARBON VAPOURS lost from automobile fuel systems were suspect in the early 1960s as contributing appreciably to the total hydrocarbon emissions of automobiles. Subsequent experimental work provided good evidence that these so-called evaporative losses did, in fact, account for from 20 to 30 per cent of the total amount of hydrocarbon pollutants from automobiles. Among the proposals advanced as measures to reduce (and to control) this source of hydrocarbon were proposals for adjustments in fuel volatility and/or modifications to the type of hydrocarbon in that fraction of the fuel lost into the atmosphere via evaporation from fuel systems. Both proposals were supported by simplistic logic that, while superficially attractive, neglected to consider the possible effects of fuel change on emissions other than evaporative losses. Moreover, as late as 1966 neither the character nor the magnitude of fuel-change effects on evaporative losses *per se* had been adequately evaluated. Therefore, technical information was needed for use in evaluating both the benefits and the penalties that would be associated with each of the several fuel modifications that were suggested. Acting on this need, the Bureau of Mines of the U.S. Department of the Interior has completed a comprehensive experimental study of a number of fuel options. The work, extending over a five-year period, has been done in co-operation with U.S. industry technical groups. Member of these cooperative groups participated actively in designing the experiments and in interpreting the results.

The MS. of this paper was received at the Institution on 17th August 1971 and accepted for publication on 24th August 1971. 23
* *Bartlesville Energy Research Center, Bureau of Mines, U.S. Department of the Interior, P.O. Box 1398, Bartlesville, Okla. 74003.*

When the work was begun the data that were available (1)† (2) on this new and evolving problem provided only gross indication of the magnitude of fuel system losses. Moreover, techniques for the requisite measurements were only then being developed (3). Therefore, work first was done to refine and to verify the experimental procedures for evaporative loss measurement; this has been reported (4) and is not reviewed in this paper. Subsequently, the principal study was launched with the stated objective '... to determine the effect of fuel volatility and of front-end composition on quantity, composition, and photochemical reactivity of vehicle emissions including both tailpipe and fuel system losses'. In addition to fuel volatility and front-end composition, the test variables were to include engine fuel system features and ambient temperature. The objective was later extended to include mid-range volatility as a fuel variable, and the study was concluded upon completion of work with the mid-range test fuels.

THE EXPERIMENT

Test procedure

The experimental procedures have been described previously (4). Briefly summarized, data were acquired during chassis dynamometer tests that simulated city driving. Exhaust emissions were measured over the period of a typical average (7½ mile) city trip; evaporative emissions were measured during the trip and for a 1-hour 'hot soak' period immediately following. Tests were taken at four ambient temperatures‡, 20, 45, 70, and 95°F, and each

† *References are given in Appendix 150.1.*
‡ *The dynamometer room was equipped for both temperature and humidity control.*

fuel/vehicle combination was tested at all temperatures excepting certain small vehicles and clearly incompatible Reid vapour pressure (R.v.p.)/temperature combinations.

The exhaust emissions were sampled using a constant volume sampler: unburnt hydrocarbon, carbon monoxide, oxides of nitrogen, and aldehydes were measured. Tank evaporative emissions were measured using direct vapour recovery of fuel tank losses. Carburettor losses were calculated from data of physical measurements made on the fuel in carburettor bowls before and after the loss measurement period. Direct vapour recovery was also used for measuring evaporative losses from vehicles fitted with evaporative loss controls. These procedures and verification of their suitability are discussed in reference (5).

Photochemical reactivity of the emissions was experimentally observed through measurement of typical 'smog' manifestations that occurred during photo-irradiation of the emissions in a smog chamber (6).

Fuels

The terms 'front-end' and 'mid-range' volatility are used with reference to fuel boiling or distillation range. Front-end volatility refers to the volatility characteristics of the very light fractions as reflected in fuel vapour pressure. Mid-range volatility refers to the volatility characteristics of the approximately middle-volume fraction of fuel obtained in a distillation. Temperature for 50 per cent distilled (50 per cent point) and the percentage of fuel distilled at 160°F are used in this discussion as measures of mid-range volatility.

Three series of fuels were used in the experimental work:

(A) A set of four fuels, each closely similar except for having front-end volatility varied within the range of 5–12·5-lb R.v.p.

(B) A set of six fuels, of which three were 6-lb R.v.p. and three were 9-lb R.v.p. fuels. Within each of the 6- or 9-lb groups, mid-range volatility was varied within the nominal range of 190–230°F for 50 per cent distilled. A seventh fuel, studied with the sixth, was comparable to the base fuel of the set except that it contained no C_4 or C_5 olefin; to maintain the volatility balance of the base fuel, these olefins were replaced using the corresponding paraffin. Data from this fuel provided information on the effect of modifying hydrocarbon type and supplemented experimental data taken from the third set of fuels.

(C) A set of three fuels closely similar in physical properties but differing in the types of hydrocarbon in the fuel front-end. One of the fuels represented typical U.S. fuel composition; each of the other two represented an alternative fuel in which the olefin content of the light hydrocarbon fraction (C_{4-7}) was reduced as a

Table 150.1. Fuel properties

Gravity, °API	Reid vapour pressure, lb	50 per cent point, °F	Per cent evaporated at 160°F	Hydrocarbon composition (by gas chromatographic analysis)			
				Aromatic, mole %	Olefins		
					C_{4-5}, mole %	C_{4-6}, mole %	Total, mole %
A★ Series							
62·7	10·1	220	29·0	26·9	5·2	7·4	10·5
64·3	12·9	220	34·5	24·8	4·9	6·9	9·8
58·7	7·9	229	20·5	33·0	4·2	6·5	9·5
56·7	5·3	235	11·5	35·3	3·0	5·8	9·5
B★ Series							
63·3	8·9	207	29·0	22·1	4·7	7·0	10·2
65·2	9·4	208	29·5	22·6	Trace	2·1	5·0
68·2	9·5	185	36·6	17·9	1·5	2·9	4·8
62·3	9·3	231	20·9	24·3	1·6	2·9	4·9
65·4	6·4	194	27·3	18·5	1·5	2·9	4·8
61·4	6·7	210	22·0	25·0	3·0	5·6	8·6
59·2	6·4	226	13·5	29·1	2·5	4·8	7·8
C★ Series							
62·7	10·1	220	29·0	26·9	5·2	7·4	10·5
62·8	10·1	221	29·0	27·1	0·3	2·5	5·6
64·0	10·0	209	31·5	26·8	0·0	0·1	1·6

* 'A' series fuels correspond to fuels reported in reference (4); 'B' series correspond to fuels reported in reference (5); 'C' series correspond to fuels reported in reference (4).

means to reduce the photochemical reactivity of the evaporative losses*.

The 'typical U.S.' fuel of the third set was the same fuel as used for the base fuel of the 'A' set described above. Properties of the fuels are given in Table 150.1.

Vehicles

Thirty-one vehicles were used; all were conventional production-line passenger vehicles. Three were small cars (under 2500 lb vehicle weight), two were compacts (2500–3000 lb), and the remainder were a mix of light-to-heavy sedans. Engine sizes of the light-to-heavy group ranged from 232 to 455 in^3 displacement. Typically, each vehicle was equipped with such power accessories and air conditioning as would be representative of equipment sold with the particular vehicle.

The vehicles were chosen either as being representative of a high volume model or as having some feature (of the engine or fuel system) that held particular interest in connection with fuel volatility. Detailed information on the vehicles is available in references (4) and (5).

RESULTS

Role of fuel volatility

Results of the first study showed that, of the fuel parameters, fuel vapour pressure has dominant influence over the amount of the tank evaporative loss (Fig. 150.1). Subsequent work confirmed the first findings. This relationship follows from the fact that fuel vapour pressure reflects the driving force of the more volatile fraction of the fuel and this driving force remains relatively undisturbed† through repeated cycles of tank heating and cooling. Thus, the higher the fuel vapour pressure, the greater the percentage of hydrocarbon in the vapour/air equilibrium over the liquid fuel (and thus in the mixture displaced to the atmosphere during each heating period).

It follows that tank loss is reduced as fuel volatility is lowered. Based on the average of results with 16 vehicles tested at 70 and 95°F ambient temperature, tank loss was reduced by 55–70 per cent when R.v.p. was lowered by 3 lb from values typically found in practice. Representative data are given in Table 150.2.

The relative insensitivity of tank loss to mid-range volatility is indicated by experimental data showing tank loss reduced by less than 20 per cent in changing fuel 50 per cent point from 190 to 230°F.

Carburettor loss is found to be influenced both by fuel vapour pressure and by mid-range volatility, with the greater influence being exerted by mid-range volatility. The dependency upon the mid-range characteristic arises from the fact that fuel in the carburettor bowl reaches temperatures that correspond with temperatures that are

---

* *In this paper the term 'front-end olefin' means, loosely, C_6 and lower-boiling olefins. 'Olefin replacement', 'olefin elimination', or 'olefin reduction' should in all cases be interpreted to mean that olefin is replaced with the corresponding paraffin.*

† *The cumulative loss of fuel vapour during repeated trips and diurnal cycles typically amounts to less than 1 per cent of the fuel in place.*

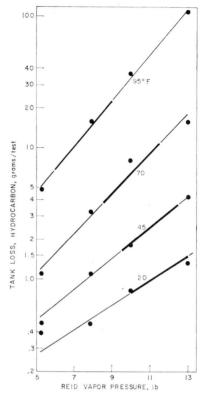

Note: Solid portion of curves indicates R.v.p. range of interest at given temperature.

Fig. 150.1. Effect of fuel volatility on tank loss (4)

Table 150.2

Ambient temperature, °F	Typical R.v.p., lb	Tank loss		
		With typical fuel, g/test	With R.v.p. lowered 3 lb, g/test	Percent change
20	13.0	1.5	0.7	55
45	12.5	3.7	1.6	57
70	11.5	10.8	3.9	64
95	9.0	23.0	6.9	70

well inside the boiling range of the fuel. Thus, for a cross-section of U.S. automobiles (1966–69 models), carburettor loss was found to correlate well with the percentage of fuel evaporated at 160°F (Fig. 150.2). The strong mid-range influence also is shown in the results of work with the 'B' series of fuels in which 50 per cent point was a fuel variable. Carburettor loss with a 230°F 50 per cent point fuel was found reduced to about one-half of the carburettor loss with a 190°F 50 per cent point fuel.

As indicated above, fuel vapour pressure also significantly affects carburettor loss; representative data companion to that given for tank loss are presented in Table 150.3.

Detailed information on the experimental work related to mid-range volatility is given in reference (5).

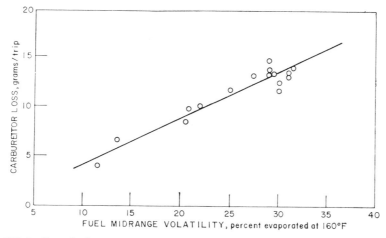

Fig. 150.2. Correlation of carburettor loss with percentage evaporated at 160°F

Table 150.3

Ambient temperature, °F	Typical R.v.p., lb	Carburettor loss		
		With typical fuel, g/test	With R.v.p. lowered 3 lb, g/test	Percentage change
20	13·0	5·6	3·0	45
45	12·5	10·2	5·3	48
70	11·5	14·0	7·7	45
95	9·0	13·8	6·2	55

Table 150.4

Emission source	Relative photochemical effect	
	Base fuel	Fuel free of olefin through C_5
Tank	1·00	0·34
Carburettor	1·00	0·41
Combined tank and carburettor losses	1·00	0·37

One objective carried throughout these studies was to determine the influence of fuel modifications on exhaust emissions. While the data are extensive, they can be quite succinctly summarized in saying that fuel volatility changes had minor, but real, effect on exhaust hydrocarbon. Directionally, exhaust hydrocarbon was found to increase with decreased fuel volatility. Lowering R.v.p. by from 2 to 4 lb from values typical of moderate ambient temperatures resulted in exhaust hydrocarbon being higher by from 3 to 7 per cent. The effect of increased mid-range volatility on exhaust hydrocarbon was less consistent but followed the trend observed for fuel vapour pressure.

Fuel front-end hydrocarbon type

Findings of the studies clearly established that with the amount of emissions remaining unchanged, the photochemical reactivity of evaporative emissions is markedly reduced by eliminating fuel front-end olefin. Data showing relative photochemical effect of emissions associated with a typical and a modified fuel are shown in Table 150.4.

Extending the fuel modification to exclude olefin through the C_7 fraction further reduced the photochemical effect of the emissions. However, the incremental improvement over that to be had with the less severe modification (olefin-free through C_5) was not impressive. This follows from the fact that the character of vapour lost from the tank is influenced primarily by the type of C_4 and C_5 hydrocarbons in the fuel. Therefore, modification of heavier components is only faintly reflected in modification of the evaporative loss.

Whereas volatility adjustment to reduce evaporative losses may have slight adverse effect on exhaust hydrocarbon, replacing front-end olefin has little—or slightly beneficial—effect on exhaust hydrocarbon. With respect to degree, the data are not conclusive; calculated reactivity values show exhaust hydrocarbon reactivity reduced by replacing front-end olefin, while data from direct experimental observation of the emissions reached in a smog chamber show no significant effect*. However, on balance, it appears that replacement of the low-boiling olefins has no significant influence on the reactivity of exhaust hydrocarbon. Without question, there is no *adverse* effect to mitigate the substantial advantage that this fuel modification shows towards reducing the photochemical reactivity of evaporative losses. In general, with respect to exhaust hydrocarbon reactivity, the effect of front-end olefin treatment does not appear to be significant.

Temperature and volatility interactions

Volatility and temperatures interact strongly as shown by the data of Fig. 150.3. These data (Fig. 150.3a) illustrate that vapour pressure dominates in controlling carburettor loss at low ambient temperatures, but, due to

---

* *Reasons for differences between 'calculated' and 'observed' reactivities are discussed in reference* (**6**).

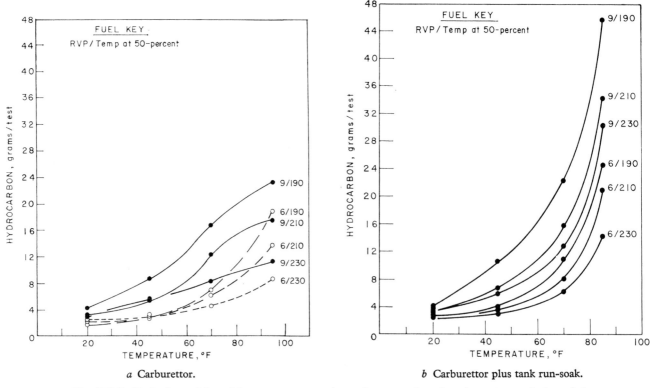

a Carburettor.
b Carburettor plus tank run-soak.

Fig. 150.3. Variation with ambient temperature in tank run-soak and carburettor emissions (5)

strong temperature interaction, 50 per cent point becomes the dominant factor at high ambients. Vapour pressure is, under all conditions, the dominant fuel factor influencing tank losses, and the relationship of tank to evaporative losses is quantitatively such that this dominance is characteristic of the combined losses as illustrated by the curves of Fig. 150.3*b*.

Another temperature/loss dependency is illustrated in Fig. 150.4. Three elements are prominent in the processes by which hydrocarbon is lost from a fuel tank: (1) fuel temperature; (2) fuel vapour pressure; and (3) the temperature differential that provides the force driving air/vapour mixture from the tank. The data of Fig. 150.4 show losses under one condition of vapour–air equilibrium —70°F ambient (Fig. 150.4*a*)—strongly contrasting much higher losses under another condition—95°F ambient (Fig. 150.4*b*). The principal factor that accounts for the higher losses at the higher ambient is the richness of the air–vapour mixture that is displaced to the atmosphere in tank-breathing action.

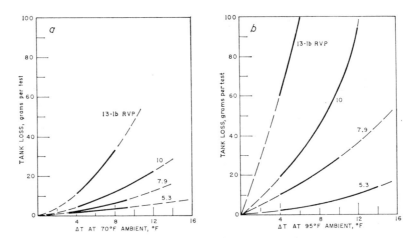

Note: Solid portions of curves are established by experimental data—remainder is extrapolated.

Fig. 150.4. Effect of fuel temperature increase on tank loss (4)

While the data of Fig. 150.4 do not relate to vapour volume, they do call attention to this factor which is basic in the loss problem. Inasmuch as thermal expansion is the primary mechanism by which tank vapour is discharged, it follows that loss is essentially proportional to vapour volume. This assumption is verified in the findings of the study. The data also clearly show the role of fuel tank heating and the sensitivity of tank loss to this variable; 1–2 degF in tank temperature rise may mean 25–50 per cent difference in the amount of hydrocarbon discharged. It therefore follows that heat absorption from exhaust components or from a hot air stream has marked influence on tank loss. Here it may be noted that only a few degrees difference in tank heating can have an effect equivalent to the effect of changing fuel volatility by several pounds. Similar observations are valid with respect to the effects of carburettor heating, wherein carburettor bowl temperature and the relevant fuel volatility characteristics are involved.

The vehicle factor

It is beyond the scope of this paper to discuss the large vehicle-to-vehicle differences that were found in measuring evaporative losses. However, large differences were found. In the latest tests that involved 15 vehicles, combined tank and carburettor loss (excluding diurnal loss) for the individual vehicles ranged between about 2 and 10 g/test for compacts; between about 10 and 25 g/test for 1968–69 light sedans; and between about 2 and 10 g/test for 1970 light sedans with evaporative loss control. Excepting the 1970 vehicles, the individual differences were explainable on the basis of system volumes and temperatures. It is therefore readily apparent that vehicle factors can easily out-weigh fuel factors in the influence they exert on the evaporative losses.

CONCLUSION

Automotive fuel evaporative losses are related to vehicle and to fuel factors in the following manner.

Fuel tank loss

The fuel characteristic that most prominently influences tank loss is front-end volatility; Reid vapour pressure or true vapour pressure correlate well with tank loss. Regarding vehicle factors, tank loss is strongly dependent upon fuel temperature, vapour-space volume, and the magnitude of temperature difference that provides a driving force for tank breathing.

Carburettor loss

Mid-range fuel volatility as well as front-end volatility has significant influence on carburettor loss. The percentage of fuel evaporated at 160°F correlates well with carburettor loss from a broad cross-section of U.S. automobiles. Ambient temperature has relatively less influence on carburettor than on tank loss. As a consequence, as ambient increases, carburettor loss becomes relatively less significant in the total of evaporative losses.

Fuel modification to reduce evaporative emissions

Tank evaporative losses at moderate to high (i.e. 70–95°F) ambient temperatures are reduced markedly when fuel R.v.p. is lowered by as much as 2–3 lb below traditional levels. Lowering mid-range fuel volatility has only a minor influence on tank loss but does reduce carburettor loss. The combined carburettor and tank losses, therefore, are most effectively reduced by alteration to both front-end and mid-range volatility. While lowering fuel volatility tends to reduce evaporative emissions, the adjustment has a small, but apparently real, adverse effect on the amount of exhaust hydrocarbon.

Replacing light olefins with the corresponding paraffin has no effect on the quantity of evaporative losses but does reduce the photochemical reactivity (i.e. the smog effect) of the losses. If olefin through C_5 is replaced, the magnitude of the reduction may range upwards of 60 per cent of the reactivity otherwise contributed by the evaporative losses.

Comparative effect—fuel and vehicle factors

The influences of vehicle or system factors (vapour volume, fuel system component temperatures, vapour recovery action) far outweigh the influences that fuel factors have on evaporative emissions. While this is true, it remains that fuel characteristics bear heavily upon the design and performance of fuel and vapour-handling systems. The principles that relate evaporative emissions and fuel thus become principles that relate fuel and vapour load in recovery and storage systems.

ACKNOWLEDGEMENT

The experimental work discussed in this paper was done in co-operation with and through financial support of the American Petroleum Institute and later the Coordinating Research Council acting through its Air Pollution Research Advisory Committee. The study was conducted at the Bartlesville Energy Research Center of the U.S. Bureau of Mines as a part of its fuel and air pollution programmes. The National Air Pollution Control Administration (now Office of Air Programs, Environmental Protection Agency) contributed materially to this work through financial support of related fuel studies in the Bureau of Mines programme.

APPENDIX 150.1

REFERENCES

(1) WENTWORTH, J. T. 'Carburetor evaporation losses', *S.A.E. Tech. Progr. Ser.* 1964 **6**, 146.
(2) 'Control of air pollution from new motor vehicles and new motor vehicle engines', *Federal Register* 1966 **31** (No. 5170, March).

(3) MULLER, H. L., KAY, R. E. and WAGNER, T. O. 'Determining the amount and composition of evaporation losses from automotive fuel systems', S.A.E. Paper No. 660407, 1966 (June).

(4) ECCLESTON, B. H., NOBLE, B. F. and HURN, R. W. 'Influence of volatile fuel components on vehicle emissions', Bureau of Mines Report of Investigations 7291, 1970 (February).

(5) ECCLESTON, B. H. and HURN, R. W. 'Effect of fuel front-end and midrange volatility on automobile emissions', Bureau of Mines Report of Investigations, 1971 (in press).

(6) DIMITRIADES, B., ECCLESTON, B. H. and HURN, R. W. 'An evaluation of the fuel factor through direct measurement of photochemical reactivity of emissions', *J. Air Pollut. Control Ass.* 1970 **20** (No. 3, March), 150.

FACTORS AFFECTING EMISSIONS OF SMOKE AND GASEOUS POLLUTANTS FROM DIRECT INJECTION DIESEL ENGINES

I. M. KHAN* C. H. T. WANG*

This paper studies the effect of injection timing, injection rate, nozzle configuration, and air swirl on the emission of smoke and gaseous pollutants from direct injection diesel engines. The paper also shows that fuel injection equipment and other variables affect the production and emission of pollutants through their effect on the fuel–air mixing process. The experiments presented show that an increase in the rate of injection or air swirl reduces the exhaust smoke, and that a retard in injection timing can cause a significant reduction in the concentration of nitric oxide in the engine exhaust. It is suggested that high rates of fuel–air mixing and retarded injection timings may obtain a substantial reduction of nitric oxide from direct injection engines.

INTRODUCTION

THERE IS A GROWING CONCERN on the emissions of pollutants from all types of combustion engines used in transport and elsewhere, including the diesel engine. Work has been and is being carried out in the C.A.V. Research Department to study the effect on smoke and gaseous pollutants of the various factors which are considered likely to affect emissions. Almost all this work has been carried out on direct injection type diesel engines as these are in wide use. This paper is confined to the presentation and discussion of results of the effect of fuel injection equipment (FIE) variables (injection timing, injection rate, and nozzle configuration) and air swirl levels on the emissions of smoke and gaseous pollutants from direct injection type diesel engines. It has to be emphasized, however, that the experiments and results presented in this paper are primarily exploratory in nature. Their implications as far as practical application is concerned require further study.

EXPERIMENTAL DETAILS

Three single-cylinder direct injection engines and a four-cylinder direct injection engine with swept volumes of 1–1.3 litre/cylinder (engines A, B, C, and D, see Table 151.1) and having top rated speeds of 2700–3000 rev/min were used.

Experiments were carried out for a range of injection timing, injection rate, fuelling, air swirl level, and engine speed. As the effect in qualitative terms of the FIE and other variables is not generally influenced by engine speed, most of the results presented are obtained at 2000 rev/min,

The MS. of this paper was received at the Institution on 14th May 1971 and accepted for publication on 24th August 1971. 23
* C.A.V. Ltd, Research Dept, Larden Road, Acton, London W.3.

which is close to 60 per cent of the rated speed for the engines used. Rates of fuel injection were varied by using various 'improvised', but conventional type, injection systems. Some experiments were also done with a number of fuel injection nozzle configurations (number and diameter of nozzle holes). Table 151.2 lists the injection rates and injection nozzles used during the experiments. In the text the lowest injection rates are termed as normal, and fast injection rate is the highest used on any given engine. Reference to the intermediate rates is not made without giving injection periods, etc.

During these experiments smoke measurements were made by using a Bosch–Dunedin smokemeter. Nitric oxide (NO) and carbon monoxide (CO) were monitored by non-dispersive infrared gas analysers. The unburnt hydrocarbons were measured by a heated hydrogen flame ionization detector with the sampling line and oven temperatures maintained at about 150°C.

DISCUSSION OF RESULTS

All the variables experimented with in this paper are expected to affect diesel emissions primarily through their

Table 151.1. Some details of the direct injection engines used in the experiments

Engine	No. of cylinders	Capacity per cylinder, litre	Nominal compression ratio	Piston bowl to engine bore ratio	Type of bowl
A	1	0.97	16:1	0.62	Disc
B	1	1.0	16:1	0.6	Disc
C	1	1.36	16:1	0.5	Toroidal
D	4	1.0	16:1	0.6	Disc

Table 151.2. Injection rates and injection nozzles used during the experiments

Engine	Injection nozzle, no. of holes × hole diameter, mm	Full load fuelling, mm³/st	Period, °c.a.	Actual rate of injection, mm³/°c.a. per litre at n.t.p. at 2000 rev/min
A	4 × 0·28	60	20·5	2·92
	4 × 0·28	60	17	3·53
	3 × 0·33	60	20·5	2·92
	5 × 0·25	60	20·5	2·92
	4 × 0·30	60	18	3·33
	4 × 0·33	60	17	3·53
B	4 × 0·28	60	22	2·73
C	4 × 0·3	80	24	2·45
	4 × 0·3	80	20	2·94
	4 × 0·33	80	21	2·80
	4 × 0·33	80	18	3·27
D	4 × 0·28	60	20	3·00
	4 × 0·28	60	18	3·33
	4 × 0·28	60	17	3·53
	4 × 0·28	60	15·5	3·87

effect on the details of the fuel–air mixing process, i.e. the history of the spatial distribution of fuel in the combustion chamber. Fuel–air mixing generally will be briefly discussed as an introduction to the discussion of results on the emissions of smoke and gaseous pollutants from diesel engines presented in this paper. When necessary, reference will also be made to other data published previously or at this Symposium.

The discussion in this paper is restricted to the conventional direct injection type diesel engine with a multihole nozzle more or less centrally located in the combustion chamber housed in the piston top. However, various aspects of this discussion also apply to other types of diesel engine.

Fuel–air mixing

Basically the rate of fuel–air mixing depends upon the rate of fuel entrainment within the fuel sprays or fuel jets ('macromixing') and the intimate mixing of fuel with air within the fuel jets ('micromixing'). The rate at which the fuel jets penetrate to the walls of the chamber housed in the piston, and the rate at which it subsequently develops on the wall, are important factors in determining the rate of air entrainment. The rate of development of the fuel jets is determined by the combined effect of injection pressure and diameter of the nozzle holes (**1**)* (**2**). High-speed photography of the direct injection combustion system done at the authors' laboratory (**3**) and elsewhere (**4**) indicates that the fuel jets reach the wall within a very short time (typically 3° crank angle (c.a.) or 0·5 ms at 1000 rev/min) and that the air entrainment primarily takes place after jet impingement on the wall, i.e. in the wall jet (or the jet in the vicinity of the chamber wall). The micromixing is expected to depend upon the rate of

* *References are given in Appendix 151.1.*

evaporation of fuel droplets and the rate of turbulent diffusion of fuel vapour in the whole volume of the jet, and it determines the equivalence ratio and temperature distribution within the jet. The temperatures are also affected by changes in ignition delay when these are present.

The work on fuel–air mixing done at the authors' laboratory has been reported previously (**1**) and is brought up to date in brief terms in an appendix to another paper presented at this Symposium (**2**). The prediction of the effect of FIE variables and air swirl on the fuel–air mixing process (heat release) based on the conceptions outlined above has been found to be in fairly general agreement with the experimentally observed effect of these variables on the rate of heat release. In developing the abovementioned heat release prediction model it was assumed that the fuel evaporation is 'instantaneous' and that the micromixing within the fuel jets is entirely controlled by the turbulent diffusion of the fuel vapour into the whole volume of the jets, and that the essential effect of an increase or decrease in air swirl is to scale the air entrainment in the jet up or down by a factor without affecting the rate of advancement of the jet. The work so far would seem to confirm that this assumption is justified for the high-speed diesel engine. No detailed account has yet been taken of temperature and fuel/air ratio within the jet.

Exhaust smoke

General considerations

The results presented in this paper generally confirm and substantiate the earlier observations regarding the effect of FIE variables on exhaust smoke (**5**).

Due to relatively large amounts of data involving a greater number of variables available on a number of engines and due to the development of soot prediction models, being dealt with in another paper at this Symposium (**2**), there has been a significant improvement in the understanding of the soot formation process in the direct injection diesel engine. It is now possible to explain satisfactorily the effect on smoke of almost all the major variables—injection timing, rate, fuelling, air swirl, etc. The fundamental factors affecting soot formation are the amount of incompletely mixed fuel within the fuel sprays at the time of ignition, its equivalence ratio, and temperature history during the burning period (see Fig. 140.7 of reference (**6**)). The soot burn-up during and after the end of heat release is controlled by the temperature and oxygen concentration variations during the cycle.

Soot combustion experiments and calculations (**7**) indicate that the soot consumption after the end of heat release is relatively unimportant, i.e. there is little difference between the soot release at the end of heat release and the soot in the engine exhaust. The soot burn-up during the main burning period may be more important, but even then, as discussed in detail previously (**5**), the variations in the exhaust smoke generally reflect the soot formation. This means that any effect of the variables on soot formation is reflected in the variations of exhaust smoke.

Effect of injection timing and rate

The effect of injection timing on smoke for a number of injection rates is shown in Figs 151.1*, 151.2, and 151.3, where results are given for engines A, C, and D at a speed of 2000 rev/min. A retard in injection timing increases the exhaust smoke to a peak value at a timing of about 5° b.t.d.c. at this engine speed. A retard in injection timing beyond this timing reduces exhaust smoke in all cases. The peak in the exhaust smoke versus timing curve corresponds with the trough in the ignition delay versus timing

* On this, and all other figures, the rate of injection is given in $mm^3/°$ c.a. per litre of air at n.t.p.

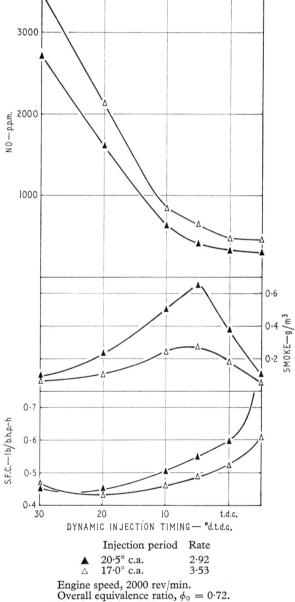

	Injection period	Rate
▲	20·5° c.a.	2·92
△	17·0° c.a.	3·53

Engine speed, 2000 rev/min.
Overall equivalence ratio, $\phi_0 = 0.72$.
Nozzle: 4 hole × 0·28 mm diameter.

Fig. 151.1. Effect of timing and rate of injection on emission and s.f.c. (Engine A)

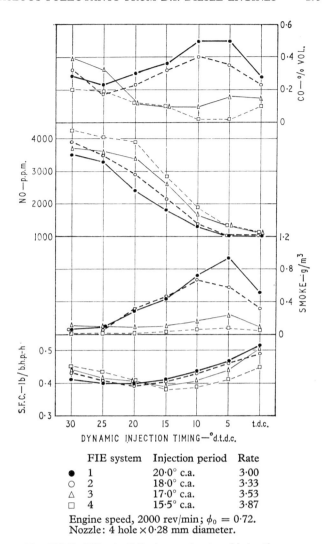

	FIE system	Injection period	Rate
●	1	20·0° c.a.	3·00
○	2	18·0° c.a.	3·33
△	3	17·0° c.a.	3·53
□	4	15·5° c.a.	3·87

Engine speed, 2000 rev/min; $\phi_0 = 0.72$.
Nozzle: 4 hole × 0·28 mm diameter.

Fig. 151.2. Effect of timing and rate of injection on emission and s.f.c. (Engine D)

curve. An increase in the rate of injection for a given injection timing reduces exhaust smoke. It also reduces the magnitude of the effect of injection timing on smoke without altering the shape of the exhaust smoke versus injection timing curve (Figs 151.1–151.3). The higher the injection rate the less important the timing variations become in controlling smoke (Figs 151.1 and 151.2).

The observed effects of injection timing and rate on exhaust smoke can be explained as follows. With a given injection nozzle (Fig. 151.4), an increase in the injection rate at a given timing or an advance in injection timing for a given rate of injection both decrease the amount of incompletely mixed fuel present in the sprays (and constituting what can be described as cores of rich premixed air–fuel mixture) as well as its equivalence ratio during the main burning period when the temperatures are expected to be high. These changes reduce soot formation and are, in the case of injection timing changes, due to an increase in ignition delay with an advance in timing.

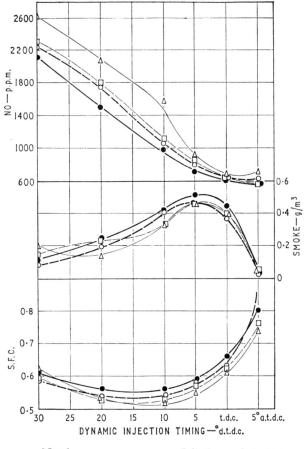

Fig. 151.3. Effect of timing and rate of injection on emission and s.f.c. (Engine C)

Nozzle		Injection period	Rate
△	4 hole × 0·33 mm diameter	18° c.a.	3·27
○	4 ,, × 0·33 ,, ,,	21° c.a.	2·80
□	4 ,, × 0·30 ,, ,,	20° c.a.	2·94
●	4 ,, × 0·30 ,, ,,	24° c.a.	2·45

Engine speed, 2000 rev/min; $\phi_0 = 0.72$.
Maximum air swirl level, $\alpha = 180°$ (see Fig. 151.6).

The long delay available leads to lower overall equivalence ratio within the sprays and also a lower equivalence ratio in the fuel-rich zone of the spray (see Fig. 142.11 of reference (2)). Similar effects in the case of an increase in the rate of injection are due to an increase in rate of air entrainment and in micromixing within the spray (see Fig. 142.11 of reference (2)). The reduction in exhaust smoke levels observed at timings more retarded than 5° b.t.d.c. (see Figs 151.1–151.3) is due to the combined effect of an increase in ignition delay (minimum delay occurs at about 5° b.t.d.c.—see Fig. 151.5) and a fall in cycle temperatures at these timings. A reduction in cycle temperature is known to decrease exhaust smoke (5).

Effect of air swirl levels

The effect of air swirl level was studied on engine C by using a 120° subtended angle shroud on the inlet valve. Previous experience on another engine fitted with a

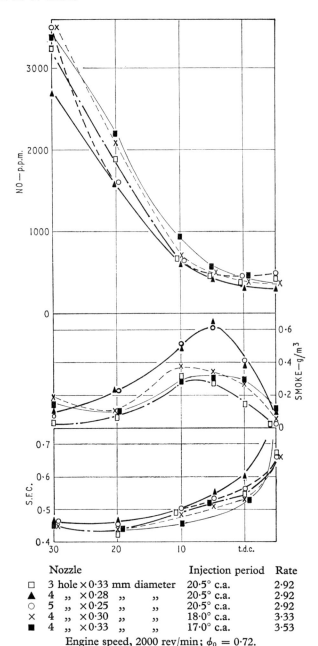

Nozzle		Injection period	Rate
□	3 hole × 0·33 mm diameter	20·5° c.a.	2·92
▲	4 ,, × 0·28 ,, ,,	20·5° c.a.	2·92
○	5 ,, × 0·25 ,, ,,	20·5° c.a.	2·92
×	4 ,, × 0·30 ,, ,,	18·0° c.a.	3·33
■	4 ,, × 0·33 ,, ,,	17·0° c.a.	3·53

Engine speed, 2000 rev/min; $\phi_0 = 0.72$.

Fig. 151.4. Effect of nozzle configuration on emissions and s.f.c. (Engine A)

shrouded valve and where air flow measurements in the motored engine were made by using a hot wire anemometer had shown that, all other things being equal, the NO concentration in the exhaust was proportional to the air swirl levels. On this basis three positions of the shrouded valve ($\alpha = 180°$, 0°, and 60°) were chosen to give 'maximum', 'intermediate', and 'low' swirl levels and it was verified that at 2000 rev/min—the engine speed used for the experiments—the volumetric efficiency was almost unaffected at the three shroud positions used. The experiments were done for a range of injection timings while

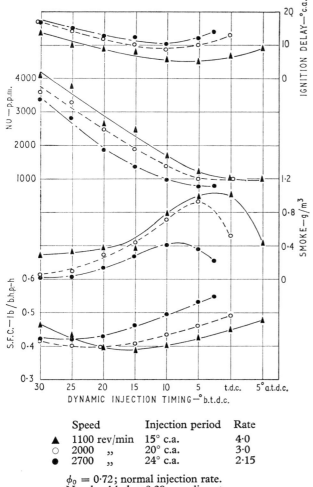

Fig. 151.5. Effect of engine speed on emissions, s.f.c. and ignition delay (Engine D)

	Speed	Injection period	Rate
▲	1100 rev/min	15° c.a.	4·0
○	2000 ,,	20° c.a.	3·0
●	2700 ,,	24° c.a.	2·15

$\phi_0 = 0.72$; normal injection rate.
Nozzle: 4 hole × 0·28 mm diameter.

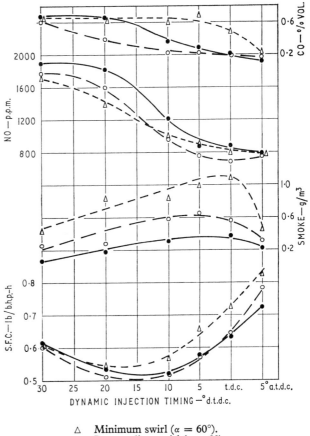

△ Minimum swirl ($\alpha = 60°$).
○ Intermediate swirl ($\alpha = 0°$).
● Maximum swirl ($\alpha = 180°$).

Engine speed, 2000 rev/min; $\phi_0 = 0.72$.
Nozzle: 4 hole × 0·3 mm diameter.
Injection period = 24° c.a.; rate = 2·45.

Fig. 151.6. Effect of injection timing and air swirl on emissions and s.f.c. (Engine C)

using an injection rate of 2·45 mm³/° c.a. per litre of air. The results given on Fig. 151.6 show that the smoke versus timing curves are essentially the same in character as those of Figs 151.1–151.4. It can also be noted on Fig. 151.6 that an increase in swirl reduces smoke and that the effect of injection timing on exhaust smoke becomes less and less pronounced as the air swirl levels are increased.

All other conditions (injection timing and rate, nozzle configuration, overall equivalence ratio) being the same, the effect of air swirl on smoke would seem to be explained by its effect on the rate of air entrainment and the rate of mixing with the fuel jets. An increase in air swirl increases the rate of air entrainment, leading to lower equivalence ratios in the richer zones of the jets (**2**). An increase in swirl will, in addition, be expected to increase the rate of intimate mixing (fuel vapour transport) in the fuel jets and will also reduce the equivalence ratios in the rich zone and increase the temperatures due to an increase in rate of heat release. The experimental observation on engine C (Fig. 151.6)—where, within the swirl range used, an increase in swirl always reduces exhaust smoke—would seem to indicate that the effect of air swirl variations on the equivalence ratio of the unmixed fuel within the fuel jets is the major factor involved.

An optimum swirl for exhaust smoke may exist and this will be expected to be the same as that giving the maximum rate of air entrainment. This is likely to be due to the bending of fuel jets which will occur at very high air swirl levels. This is expected to increase the time taken by the fuel jets to reach the chamber walls, thus affecting the relative part played by the wall jets. These are believed to be responsible for the bulk of the air entrainment within the jets (**2**). A small-diameter fuel jet is more prone to bending than a large-diameter fuel jet (**8**) and this would mean that the optimum swirl (that giving lowest exhaust smoke) would be expected to be lower for a nozzle with smaller diameter holes. As already mentioned these optimum swirls are expected to be rather high (the bending of jets in reference (**8**) was obtained with air swirl ratios of around 20:1) and are not normally likely to exist in a

Engine to engine variations of exhaust smoke

The engine to engine variations in exhaust smoke on engines of the same size when the overall equivalence ratio, injection timing, injection rate, and the nozzle are the same would be accounted for by the effects of air swirl levels (which have already been discussed), piston bowl to engine bore ratio, and actual compression ratio. A change in the piston bowl diameter to engine bore ratio will affect the time taken by the fuel jets to reach the bowl wall with the consequences on the rate of air entrainment pointed out in an earlier section. Variation in the engine compression ratio will affect the ignition delay and thus change the amount of unmixed fuel at ignition and the equivalence ratio history during the burning period. An increase in ignition delay will be expected to reduce both the amount of unmixed fuel and its equivalence ratio. Cycle temperatures will also be affected by changes in compression ratio. The preceding remarks would also be expected to be valid for engines of different sizes provided the systems are geometrically similar, i.e. have the same number of sprays

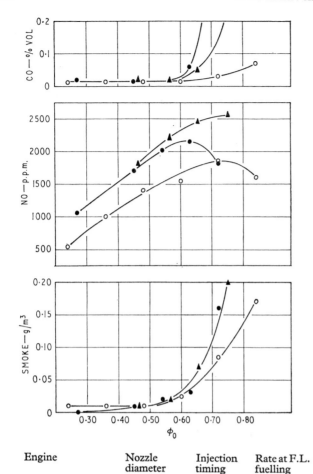

Engine	Nozzle diameter	Injection timing	Rate at F.L. fuelling
○ A	4 × 0·25 mm	20° b.t.d.c.	2·95
● C (high swirl)	4 × 0·30 mm	20° b.t.d.c.	2·45
▲ C (without shroud)	4 × 0·33 mm	25° b.t.d.c.	2·80

Fig. 151.7. Effect of overall equivalence ratio (ϕ_0) on emissions of smoke and gaseous pollutants

direct injection diesel engine. It may be that any observation (see, for example, reference (**9**)) on the existence of optimum swirl which is fairly close to the normal may be explained by either an appreciable reduction in the volumetric efficiency (which will increase the overall equivalence ratio (see Fig. 151.7) at a constant fuelling) or by the use of nozzles with hole diameters which are too small.

Effect of nozzle configuration

An increase in nozzle hole diameter of four-hole nozzles used on engines A and C increases the rate of injection, which reduces exhaust smoke (Figs 151.3 and 151.4). A reduction in the number of nozzle holes of the same total flow area in the case of engine A (Fig. 151.4) and engine B (see Fig. 4 of reference (**3**)) has, in general, resulted in a reduction in exhaust smoke. The changes in nozzle configuration are believed to affect exhaust smoke through their effect on the rate of fuel injection, rate of air entrainment, and other details already referred to in preceding sections.

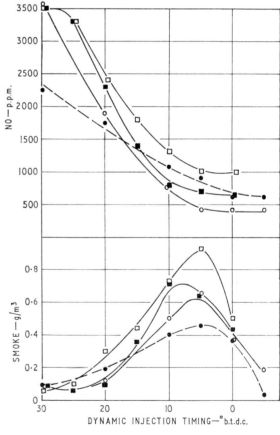

Engine	Nozzle		Rate
○ A	4 × 0·28 mm diameter		2·92
■ B	4 × 0·28 ,,	,,	2·73
● C (high swirl)	4 × 0·33 ,,	,,	2·80
□ D	4 × 0·28 ,,	,,	3·00

Engine speed, 2000 rev/min; normal injection rate, $\phi_0 = 0·72$.

Fig. 151.8. Engine to engine variations of smoke and nitric oxide emissions

with the same rate of spray penetration measured as a fraction of engine cylinder radius reached as a function of time.

Enough appropriate experimental data are not available as yet to check the validity of the above statements, particularly those concerning engines of different bores.

The results from engines A, B, C, and D under similar experimental conditions are given in Fig. 151.8. The actual and individual compression ratios of the engine were not known at the time of writing, although they all have the same nominal compression ratio. Engines A, B, and D have almost identical bores and piston bowl to bore ratio. Engines B and D are believed to have the same air swirl levels. The air swirl levels in engines A and C are not known. Hence, it is not possible to discuss the relative importance of the various factors mentioned above in producing the engine to engine variations in smoke shown in Fig. 151.8.

Nitric oxide

Effect of overall fuel/air ratio

The overall equivalence ratio (ϕ_0) affects NO formation through its effect on the equivalence ratio and temperature in the NO formation zone within the fuel sprays. The results presented in this paper seem to suggest that the equivalence ratio involved is substantially higher than the overall equivalence ratio. This will become clearer in the following discussion.

It is of interest to examine the effect of ϕ_0 at a number of timings, in the case of engine C (Fig. 151.9a). These results are fairly typical. Similar results have also been obtained on engine B. Fig. 151.9a is transformed into Fig. 151.9b by assuming that, throughout the ϕ_0 range, the equivalence ratio in the NO formation zone is equal to the maximum ϕ_0 (0.77) used during the experiments. Neither of these two assumptions is stictly true but they are helpful in illustrating the part played by the dilution of the combustion products with the excess air in determining the concentration of NO in the engine exhaust. As a matter of fact the equivalence ratio in the NO formation zone may not be the same as the maximum overall equivalence ratio used in the present case. Furthermore, a range of equivalence ratio rather than a single one may be involved. Nevertheless, the equivalence ratio in the NO formation zone will be expected roughly to follow the variations in ϕ_0. This means that the relative variations in the undiluted NO concentrations in Fig. 151.9, as a result of the assumption involved, are overestimated. The essential nature of these curves, however, is expected to be unchanged because the relative variations of equivalence ratio in the NO formation zones are much smaller than that of the overall equivalence ratio, as will be seen.

Fig. 151.9b shows that the effect of ϕ_0, or the undiluted NO concentration, depends upon the injection timing, i.e. the cycle temperature (in the present context injection timing mainly affects cycle temperature). At the most advanced timing (i.e. at the highest cycle temperature

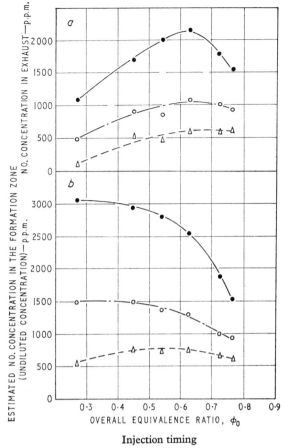

Injection timing
● 20° b.t.d.c.
○ 10° b.t.d.c.
△ 0° b.t.d.c.

Engine speed: 2000 rev/min; high swirl ($\alpha = 180°$).
Nozzle: 4 hole × 0.3 mm diameter; normal rate of injection, 2·45.

Fig. 151.9. Effect of ϕ_0 on formation and emission of nitric oxide at various injection timings (Engine C)

used) a decrease in ϕ_0 appreciably increases the 'undiluted' NO concentration, and this effect is particularly important towards the higher end of the ϕ_0 range. As the injection timings are retarded, or as the cycle temperatures are reduced, the effect of ϕ_0 (which characterizes the equivalence ratio in the NO formation zone) on undiluted NO concentration becomes progressively less important and almost disappears at an injection timing of 0° b.t.d.c. This is in agreement with the fact that at a constant temperature the relative changes in equilibrium concentrations of NO in a premixed fuel–air mixture with changes in its equivalence ratio become less and less important with a fall in temperature (see Fig. 142.2 of reference (**2**)). The concentration of NO is likely to be controlled by the kinetics of the mechanism rather than by chemical equilibrium considerations (**2**) (**10**). In the case of a diesel engine, however, there is an approximate correlation between the concentration of the rate of controlled NO and the equilibrium concentration which corresponds to the same equivalence ratio and temperature conditions (**2**).

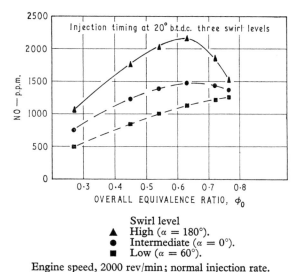

Swirl level
▲ High ($\alpha = 180°$).
● Intermediate ($\alpha = 0°$).
■ Low ($\alpha = 60°$).

Engine speed, 2000 rev/min; normal injection rate.
Nozzle: 4×0.3 mm diameter; rate at full load = 2·45.

Fig. 151.10. Effect of ϕ_0 on nitric oxide concentration at various air swirl levels and dynamic injection timings (Engine C)

The effect of ϕ_0 on NO concentration in exhaust at three levels of air swirl is given in Fig. 151.10. The curves on this figure are strikingly similar in nature to those obtained with a range of timings (Fig. 151.9a). An increase in swirl has basically the same effect as an advance in injection timing. In both cases the burning period is shortened and temperatures are increased. Hence, all the above discussion on the effect of ϕ_0 at various timings also applies in the present case (Fig. 151.10).

It would seem to follow also from the above discussion that the peaks in the exhaust NO concentration versus ϕ_0 curves are due to the combined effect of a reduction in the equivalence ratio in the NO formation zone and the 'dilution' of combustion products with the excess air which steadily increases in importance with a reduction in ϕ_0, rather than by the combined effect of average temperature and average oxygen concentration as has been previously suggested (**11**). The peak in the NO exhaust concentration curve becomes more and more pronounced as the cycle temperatures and, hence, NO concentrations in the exhaust increase (due to an advance in timing or an increase in swirl, Figs 151.9a and 151.10), and occur at relatively lower values of ϕ_0. This is explained by the changes in the magnitude of the effect of equivalence ratio (equivalence ratio in the NO formation zone) with variations in temperature levels, as has been already pointed out.

Furthermore, the changes in fuel–air mixing conditions which lead to high NO concentrations in the exhaust are also accompanied by a decrease in the equivalence ratio in the NO formation zone due to an increase in air entrainment in one case (increase in air swirl) and to an increase in the time for mixing in the other (increase in ignition delay due to an advance in injection timing). However, if the reduction in the equivalence ratio in the NO formation zone were large enough, the relative insensitivity of the undiluted concentration versus ϕ_0 curve for the lower half of the ϕ_0 at the injection timing of 20° b.t.d.c. (Fig. 151.9b) could not be explained. The same would apply to the high swirl in Fig. 151.10 if the corresponding NO exhaust concentration curve were transformed into the undiluted concentration curve. It follows that the bulk of the NO formation takes place in zones within the combustion chamber where the equivalence ratios are not very much lower than 1. It seems possible that the equivalence ratio range in question may not be very different from the inflammability limits referred to in the context of flame propagation in homogeneous fuel–air mixtures—e.g. the gas engine (0·7:1)—as the bulk of NO formation is expected to occur in close proximity to the high-temperature reaction zones. This also indicates that, in the diesel engine, NO formation at high values of ϕ_0 is controlled by equivalence ratio in the NO formation zones, but at low values of ϕ_0 the controlling factor is the temperature in these zones. This would also help to explain why the differences in NO concentration in exhaust with variation in timing or air swirl (i.e. in temperature) are smallest at the top end of the ϕ_0 range (Figs 151.9a and 151.10).

Effects of FIE variables and other factors

For a given overall equivalence ratio (ϕ_0) the observed effects of injection timing (Figs 151.1–151.4 and 151.6), injection rate (Figs 151.1–151.3), injection nozzle (Figs 151.3 and 151.4), and air swirl ratio (Fig. 151.6) would seem to be primarily determined by the combined influence of temperature and equivalence ratio in the NO formation zone or zones. An advance in injection timing, or an increase in injection rate, or a rise in air swirl level reduces the burning period and increases the temperature and probably also decreases the equivalence ratio in the zone within the combustion chamber where NO formation occurs. Any of these changes will lead to an increase in the formation of NO and explain why, in all the three cases mentioned, an increase in the emission of NO is observed. A reduction in heat release period or the time during which combustion occurs, for the reasons already given, is not expected to be of any great consequence in the present context.

Any change in the configuration of the injection nozzle is expected to intervene through an increase or a decrease in the rate of mixing. An increase may result from an increase in the diameter of holes with the same or greater number of nozzle holes, or it may be due to a decrease in the number of nozzle holes of a constant flow area nozzle, if the injection pressures are maintained fairly constant. As with other FIE variables an increase in the rate of mixing obtained with changes in nozzle configuration increases the NO emission. This is generally borne out by results given in this paper (Figs 151.3 and 151.4).

The variations in the slope of NO exhaust concentration versus timing curve for any given engine (Figs 151.1–151.4) are also to be explained by changes in the rate of mixing. It would seem that the relative increase in cycle temperature with an advance in injection timing increases with an increase in the rate of injection (or rate of fuel–air mixing).

The reduction in NO exhaust concentration with an increase in engine speed (Fig. 151.5), observed at a given injection timing, would primarily seem to be due to the combined influence of temperature and equivalence ratio in the NO formation zone and the times for the reaction of an element of fuel (which are expected to be controlled by fuel–air mixing). The latter two will be reduced with an increase in the engine speed. Analysis of a cylinder pressure diagram taken under relevant conditions seems to indicate that the third factor, i.e. temperature, is almost unaffected (sometimes somewhat reduced) with an increase in engine speed.

The engine to engine variations observed at a given speed (Figs 151.7 and 151.8) will normally be expected to be accounted for by the combined effects of various factors involved which have already been discussed.

Carbon monoxide

The experimental observations on the variation of CO with injection rate, injection timing, swirl, and fuelling (Figs 151.2, 151.6, and 151.7) may be explained by two formation mechanisms. One seems to operate under lean mixture conditions and the other under rich mixture conditions.

The first mechanism seems to be largely controlled by chemical kinetics; it has been observed that lean mixtures, incapable of sustaining flame propagation when heated, generate CO whose concentration is dependent upon the equivalence ratio of the mixture and the temperature to which it is subjected (12). Formation of CO by the second mechanism, involving rich mixtures, would seem to be primarily controlled in the case of the diesel engine by the rate of fuel–air mixing or by the rate at which the fuel rich zones within the fuel jets are fed with air.

CO formation at high rates of injection, advanced injection timings, and high air swirls within the normal range of engine fuellings seems to be controlled by the first mechanism. All the three factors increase the amount of fuel involved in the 'lean' zones within the diesel combustion chamber as well as the temperatures to which these zones are subjected. The CO concentration in exhaust in the case of the first mechanism is, however, controlled by the formation which occurs during the initial phase of burning as well as by its consumption during the rest of the burning period, the latter being the important one as we shall see. Generally, all the relevant experimental observations, some of which have been given in this paper (e.g. the effect of injection timing at high rates of injection, Fig. 151.2) and elsewhere (see Fig. 8 of reference (3)), are in agreement with the above statements. The fact that an increase in swirl generally decreases CO concentration in exhaust (an increase in CO formation with an increase in air swirl will be expected) indicates that CO consumption during the later phase of the heat release (which is likely to increase with an increase in air swirl levels) is an important factor (3).

The second mechanism of CO formation would appear to be operative under conditions of low injection rates, retarded timings, low air swirls, and high fuellings. Again this is generally substantiated by the experimental observations (see Figs 151.2 and 151.6).

Unburnt hydrocarbons

There would seem to be similarities between the origins of unburnt hydrocarbons in the diesel exhaust and the spark-ignition engine exhaust. In the case of the spark-ignition engine, hydrocarbons escape combustion due to flame quenching at the walls as well as 'blowout' in the over-lean zones within the chamber (see, for example, reference (13)). Flame 'blowouts' may also occur under conditions of overrich mixtures. Turbulence in the chamber plays an important part in determining these lean and rich inflammability limits (14). The experimental observations on diesel engines show that, at a given fuelling, an advance in injection timing (i.e. an increase in fuel impingement) and an increase in fuelling at a given injection timing generally increases emission of hydrocarbons (Fig. 151.11 and Fig. 8 of reference (3)). It would seem that under conditions of normal fuellings the unburnt hydrocarbons in the exhaust result from quenching at the chamber walls as well as airborne 'blowout' in the lean zones which may be present under these conditions. An increase in hydrocarbon emissions sometimes observed at

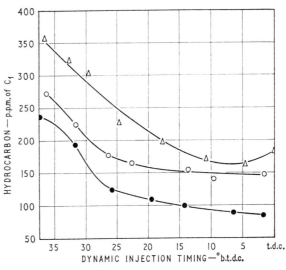

	Injection shape	Injection period	Rate
△	Steep front	17° c.a.	3·35
○	Pilot front	22° c.a.	2·73
●	Steep front	22° c.a.	2·73

Engine speed, 2000 rev/min; $\phi_0 = 0.72$.
Nozzle: 4 hole × 0·28 mm diameter.

Fig. 151.11. Effect of timing and rate of injection on hydrocarbon emissions (Engine B)

the lower end of the engine load range (15) will appear to be controlled by the latter mechanism. At high fuellings, i.e. under overfuelling conditions, the predominant mechanism is likely to be the 'quenching' or extinguishing of reaction in the fuel rich zones which are expected to exist under these conditions.

PRACTICAL IMPLICATIONS

There is increasing concern on meeting existing or future legislation on the emission of smoke and gaseous pollutants from diesel engines. The general legislative situation in the U.K., Europe, and the U.S.A. has recently been examined in detail (16).

It is of obvious interest to try to judge the extent to which the factors examined in this paper may be used to reduce emissions of smoke and NO which is the main gaseous pollutant from the diesel engine. There are indications that an increase in the rate of fuel-air mixing obtained by any of the methods employed during the present investigation which reduces exhaust smoke as well as its sensitivity to changes in injection timing may be used to reduce the emissions of NO by operating at retarded injection timings without appreciably affecting smoke emissions. It may be possible to retard the normally used injection timing by about 10° c.a. without any significant sacrifice in fuel economy. As we have seen, the 'slope' of the NO exhaust concentration versus injection timing curve varies from engine to engine. In the engines used in the present investigations a 10° c.a. retard in injection timing reduces NO concentration levels by a factor of 1·5–2·5. This also generally holds throughout the load and speed range of the engines.

The Californian legislation on the emissions of gaseous pollutants (CO, unburnt hydrocarbons and NO expressed as nitrogen dioxide or NO_2) specifies the levels of pollutants to be met, in terms of brake specific mass emissions (grammes of pollutants/b.h.p. h) worked out according to the 13-mode E.M.A. cycle along with the specified procedure (17). In this cycle, with the exception of the idling mode whose contribution is relatively very small, all the modes (range of loads at rated and 60 per cent rated speed) are equally weighted. It may, therefore, be said that in approximate terms the brake specific emissions of NO_2 from an average direct injection may possibly be halved by an increase in injection rate combined with 10° c.a. retard in injection timing. It must be emphasized, however, that the extent to which the above-mentioned possibilities are capable of practical application can only be ascertained by further study. Also, factors other than those dealt with in this paper may be used to reduce the emission of smoke and gaseous pollutants.

CONCLUSIONS

Though the investigation has been concerned with direct injection type diesel engines some of the fundamental aspects of the findings with respect to formation of soot and gaseous pollutants are expected to be applicable to other combustion systems where burning is controlled by fuel-air mixing.

It seems that the FIE variables (injection timing, injection rate, and nozzle configuration) and air swirl mainly affect the formation of pollutants through their effect on the details of the fuel-air mixing process, i.e. spatial distribution histories of fuel/air ratio and temperature in the combustion chamber. Variations in ignition delay, when present, will also affect temperatures. An increase in the rate of fuel injection and in air swirl, within the limits of variation employed in the present investigation, resulted in an increase in the rate of fuel-air mixing in the four engines used. The essential effect of the changes in nozzle configuration used also seems to be on the rate of fuel-air mixing. With no appreciable reduction in injection pressures or the number of nozzle holes, an increase in hole diameter of a nozzle with a constant number of holes maintains the same total flow area and injection pressures, but results in an increase in the rate of air entrainment in the fuel jets which is followed by an increase in the rate of fuel-air mixing.

These increases in the rate of mixing, by whatever method they are obtained, are generally confirmed by the fact that an increase in the rate of mixing is always accompanied by a retard in injection timing for minimum fuel consumption or maximum engine efficiency.

The exhaust smoke is reduced at a given injection timing and becomes less and less sensitive to the changes in injection timing as the rate of fuel-air mixing is increased. This applied throughout the range of injection rates and air swirl levels used during the experiments.

The formation of soot in the diesel engine, and its emission in the engine exhaust, is determined by the combined effect of incompletely mixed fuel at ignition (fuel injected after ignition plus unmixed fraction of the fuel injected during ignition delay), its equivalence ratio, and temperature history during the main combustion periods. The variables studied in this paper through their effect on the fuel-air mixing process affect these fundamental factors.

The same applies to NO formation in the diesel engine with the difference that the fundamental factors in this case are the equivalence ratios in or in the proximity of zones where intense reaction occurs and the associated temperatures. It would appear that the equivalence ratios in question, throughout the normal fuelling range for a diesel engine, are not greatly different from those normally associated with limits of inflammability in a premixed flame. In any case, the range of equivalence ratios involved is much smaller than the overall equivalence ratio (ϕ_0) range, and the equivalence ratios are substantially higher than ϕ_0 at the lower end of the ϕ_0 range. Under typical conditions, it would appear that when the cycle temperatures are high enough the concentration of NO formed in the NO formation zone or zones increases steadily with a decrease in ϕ_0 during the top half of the ϕ_0 range and is nearly constant during the remainder of the ϕ_0 range. The typical NO exhaust concentration versus ϕ_0 curve, with a peak at the higher values of ϕ_0, is due to the dilution of the zones involved in the NO formation in their reaction with the excess air.

An increase in the rate of fuel–air mixing at a given injection timing increases the concentration of NO in the exhaust. This increase is, however, relatively small compared to the reduction in NO concentration with a retard in injection timing which becomes possible with an increase in the rates of mixing. The implications are that an increase in the rate of fuel–air mixing, in combination with a retard in timing, may be used to reduce the emissions of NO from a direct injection diesel engine without affecting exhaust smoke levels and without any substantial increase in specific fuel consumptions. Further work is required to determine the extent to which these possibilities can be put into practice.

ACKNOWLEDGEMENTS

The authors wish to thank the Directors of C.A.V. Limited for permission to publish this paper. In particular they are indebted to Dr A. E. W. Austen, Technical Director, and Mr W. E. W. Nicolls, Chief Engineer (Special Projects), for their extremely helpful discussions.

APPENDIX 151.1

REFERENCES

(1) GRIGG, H. C. and SYED, M. H. 'The problem of predicting rate of heat release in diesel engines', *Symp. Diesel Engine Combustion, Proc. Instn mech. Engrs* 1969–70 **184** (Pt 3J), 192.

(2) KHAN, I. M., GREEVES, G. and PROBERT, D. M. 'Prediction of soot and nitric oxide concentrations in diesel engine exhaust', Paper 142 of this Symposium.

(3) KHAN, I. M. and GRIGG, H. C. 'Progress of diesel combustion research', *C.I.M.A.C.* 1971.

(4) ALCOCK, J. F. and SCOTT, W. M. 'Some more light on diesel combustion', S.A.E. Paper 872A, 1964 (June).

(5) KHAN, I. M. 'Formation and combustion of carbon in a diesel engine', *Symp. Diesel Engine Combustion, Proc. Instn mech. Engrs* 1969–70 **184** (Pt 3J), 36.

(6) BROOME, D. and KHAN, I. M. 'The mechanisms of soot release from combustion of hydrocarbon fuels with particular reference to the diesel engine', Paper 140 of this Symposium.

(7) KHAN, I. M., WANG, C. H. T. and LANGRIDGE, B. E. 'Coagulation and combustion of soot particles in diesel engines', to be published in *Combustion and Flame*.

(8) KNIGHT, B. E. 'Similarity considerations in assessing diesel engine fuel spray requirements', *Symp. Diesel Engines—Breathing and Combustion, Proc. Instn mech. Engrs* 1965–66 **180** (Pt 3N), 10.

(9) LYN, W-T. 'The spectrum of diesel combustion research', *Symp. Diesel Engine Combustion, Proc. Instn mech. Engrs* 1969–70 **184** (Pt 3J), 1.

(10) CARETTO, L. S., MUZIO, L. J., SAWYER, R. F. and STARKMAN, E. S. 'The role of kinetics in engine emission of nitric oxide', *Combust. Sci. Tech.* 1971 **3** (No. 2, April), 53.

(11) MCCONNELL, G. 'Oxides of nitrogen in diesel engine exhaust gas: their formation and control', *Proc. Instn mech. Engrs* 1963–64 **178** (Pt 1), 1001.

(12) KHAN, I. M. Unpublished work done at the Institut Français du Pétrole, Rueil-Malmais, France.

(13) JACKSON, M. W., WIESE, W. M. and WENTWORTH, J. T. 'The influence of air–fuel ratio, spark timing and combustion chamber deposits on exhaust hydrocarbon emissions', S.A.E. Paper 486A, 1962.

(14) SOKOLIK, A. S. 'Self-ignition, flame and detonation in gases', *Israel Program for Scientific Translations* 1963.

(15) YUMLU, V. S. and CAREY, A. W., jun. 'Exhaust emission characteristics of four-stroke, direct injection, compression ignition engines', S.A.E. Paper 680420, 1968 (May).

(16) SPIERS, J. and VULLIAMY, N. M. F. 'Diesel engine smoke and pollutants', *Proc. Diesel Engrs Users Ass.* 1971 (February).

(17) BASCOM, R. C. and HASS, G. C. 'A states report on the development of the 1973 California diesel emissions standards', S.A.E. Paper 700671, 1970 (August).

C152/71 EXHAUST EMISSION CONTROL SYSTEM FOR THE ROTARY ENGINE

T. MUROKI*

Efforts have been made towards reducing the hydrocarbon concentration in the exhaust emission of the rotary engine. The amounts of hydrocarbon, carbon monoxide, and oxides of nitrogen emitted from the rotary engine are basically affected by the air/fuel ratio, spark advance, load, etc. Various modifications have been made to the design of the rotary engine in an attempt to reduce exhaust emissions, but the researches have been directed towards an air injection system and a thermal reactor. This paper will describe the current development of these devices.

INTRODUCTION

SINCE TOYO KOGYO first put the rotary engined car on the market in 1967, we have been carrying out intensive studies on the exhaust emission of the rotary engine. As a result of the various tests, it became clear that the hydrocarbon (HC) emissions from the Wankel rotary engine were high while the oxides of nitrogen (NO_x) emissions were low. Our research and development studies were therefore directed mainly towards reducing the HC concentration. Our first studies were to understand the emission characteristics more completely, and our efforts were then concentrated on discovering how to decrease the emission level. Thus, in autumn 1969 we succeeded in passing the exhaust gas certification test laid down by the U.S. Government, and the R100 mounted with the 500 $cm^3 \times 2$ rotor rotary engine and the RX-2 with the 573 $cm^3 \times 2$ rotor rotary engine have been exported to the U.S.A. since 1970–71. The progress made in our development studies—particularly on HC emissions—and the results of these studies are presented in this paper.

EMISSION CHARACTERISTICS

The basic characteristics of the exhaust gas emissions from the rotary engine are qualitatively almost the same as those from the reciprocating engine. These characteristics—according to air/fuel ratio, ignition timing, and combustion chamber—are discussed in the following sections. The results were obtained on the original engine, which had no emission control device, and the concentrations of the pollutants were measured by NDIR analysis.

Wankel rotary engine

Prior to dealing with the emission characteristics of the Wankel rotary engine, I wish to refer to the structure and operation of the engine to help the audience understand the paper more fully.

As shown in Fig. 152.1, a rotor, which is formed by the inner envelope of the trochoid, makes a planetary motion inside the rotor housing having a trochoidal curve. The three

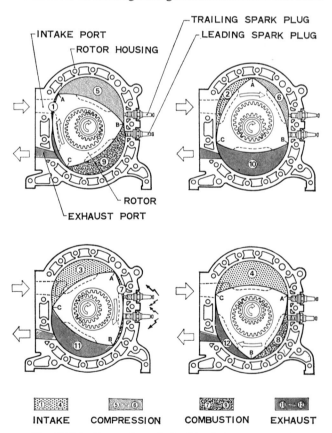

Fig. 152.1. Operation of the Wankel rotary engine

The MS. of this paper was received at the Institution on 4th May 1971 and accepted for publication on 24th August 1971. 43
* Toyo Kogyo Co. Ltd, 6047 Fuchu-machi, Aki-gun, Hiroshima, Japan.

chambers formed by the housing and the rotor repeat the combustion cycle as shown in the figure, changing their volume respectively.

On our engine, the side intake port and the peripheral exhaust port are adopted, and two spark plugs—leading and trailing—are installed.

In the original engine a bath-tub type recess is provided on the central part of the rotor flank and forms part of the combustion chamber.

Characteristics of air/fuel ratio

The concentrations of CO, HC, and NO_x respectively as a function of air/fuel ratio were measured with constant ignition timing. The results are shown on the left side of Fig. 152.2. The 'T' and 'L' in the figure show the spark advances of the trailing and leading spark plugs respectively.

First, a large deviation in the concentration of CO is seen from the theoretical value at an air/fuel ratio of approximately 14·8, i.e. excess air coefficient $\lambda = 1$. Even in the range of $\lambda > 1$, $CO\% \neq 0$, and in the rich mixture range of $\lambda < 1$ the CO has a tendency to gradually approach the theoretical line. The reason for this is that in the Wankel rotary engine the mixture flows strongly in the direction of the rotation of the rotor and thus the combustion flame does not extend to the trailing corner of the combustion chamber. Therefore, in this area there is a high concentration of residual oxygen. Considering the combustion chamber as a whole, there is no general mixing of the gases in this corner, and the residual oxygen is presumed to be emitted without being used.

Next, let us consider the HC concentration. Generally, on the internal combustion engine the HC concentration has a tendency to be high when the mixture becomes rich. This is because combustion of the mixture is impeded owing to a lack of oxygen, and the concentration of unburnt HC therefore increases.

On the other hand, when the mixture becomes extremely lean, misfiring occurs and the unburnt HC will again increase. In the Wankel rotary engine particular conditions have to be considered for the formation of HC. The flame does not extend to the trailing corner in the combustion space, and because of the quenching effect due to an excessively high S/V ratio on the trailing area, excessively rich unburnt HC is formed on the wall surface. In addition, most of this unburnt HC layer is presumed to

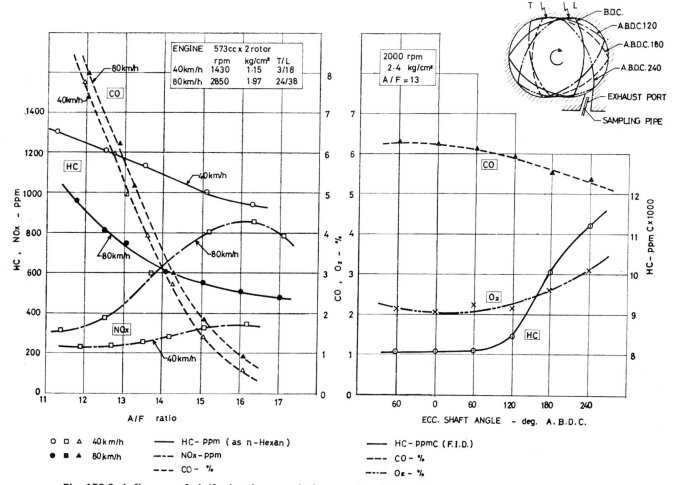

Fig. 152.2. Influence of air/fuel ratio on emissions, and emission distribution in combustion chamber

be scraped off by the apex seals and side seals, and emitted from the exhaust port. Therefore, the HC concentration at the exhaust stroke end becomes high. For the above-mentioned reasons the HC concentration of the Wankel rotary engine is inevitably at a higher level than that of the reciprocating engine.

The right side of Fig. 152.2 shows the results of our test, substantiating our theoretical reasons for the formation of CO and HC. The results are analysed from the exhaust gas taken at various positions of the rotor through a pipe installed at the exhaust port. In this test, measurements were made of the changes in the concentrations of pollutants at each position of the rotor (angle of the eccentric shaft). From Fig. 152.2 it can be seen that although the concentration of CO is low at the trailing side of the rotor, the concentration of oxygen rises, indicating that the flame does not propagate to the trailing corner.

Also, the concentration of HC suddenly increases from approximately 180° a.b.d.c., confirming that the HC is concentrated at the trailing corner.

Lastly, it is generally considered that NO_x concentration is the factor in the peak temperature of the combustion gas and the concentration of oxygen in the combustion gas. In our tests the maximum NO_x concentration appeared at an air/fuel ratio of 15 or 16. Contrary to the HC concentration, the NO_x value can be held at a considerably lower level with a rotary engine than it can with a reciprocating engine. This is because the S/V ratio is higher than that of the reciprocating engine, and the peak temperature is lower owing to the combustion delay due to the gas flow and the geometric shape of the combustion chamber.

Characteristics of ignition timing

CO concentration is not essentially affected by ignition timing. Concentrations of HC and NO_x have interesting characteristics brought about by the locations and number of the spark plugs and changes in the ignition timing. These characteristics are shown in Fig. 152.3. 'T' and 'L' in the figure indicate the trailing and leading spark plugs. In the case of both spark plugs the ignition timings of the spark plugs in the trailing side and leading side are set to be the same. First, the concentration of HC tends to be lower when the ignition timings of both spark plugs or a single spark plug are retarded. In the case of the trailing spark plug alone, or the leading spark plug alone, the HC concentration is lower than when both spark plugs are retarded. It is presumed that HC emissions are lower with a single spark plug than with two spark plugs because of the reduced quenching of the flame by the trailing area of the rotor. Similarly, quenching of the flame is reduced and the HC level is lowered when the ignition timing is retarded. Because the flame speed opposite the direction of rotation is slower with a single spark plug than it is with both spark plugs, the flame arrives at the trailing portion of the charge well after t.d.c. when S/V ratio at the trailing zone has already become small. As a result, the so-called 'after combustion' at the trailing corner occurs and a

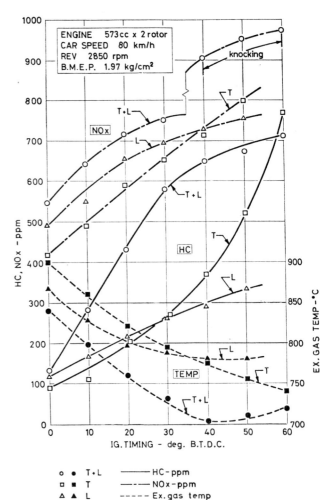

Fig. 152.3. Influence of ignition timing on HC and NO_x emissions

larger amount of HC is burnt, and the concentration of HC emitted from the exhaust port is therefore reduced.

In the case of a single spark plug the exhaust gas temperature rise is also caused by combustion delay, which is presumed to have a decreasing effect on the HC concentration.

With regard to NO_x, the same tendency occurs when the ignition timing is changed. By retarding the ignition timing, or by using one spark plug, the combustion is delayed and the peak temperature of the combustion gas is reduced.

Combustion chamber

The shape of the combustion chamber is closely related to a decrease in HC emission with respect to the quenching area of the local wall surface of the trailing corner and the combustion delay. That is (as shown in Fig. 152.4), if the main recess in the combustion chamber is moved to the trailing side, the S/V ratio of that area becomes low and the formation of the unburnt HC is presumed to decrease.

Next, when the shape of the combustion chamber is

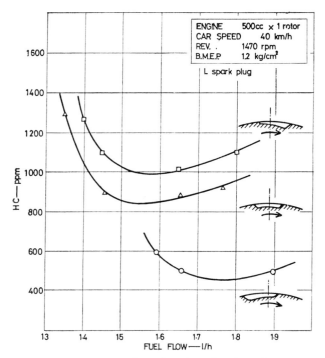

Fig. 152.4. Effect of combustion chamber shape on HC emission

in order to meet the 1970–71 standards demanded by the U.S.A., reduction rates of about 90 and 50 per cent are further required for HC and CO, respectively. Therefore, we must use the air injection with reactor (AIR) system where secondary air is injected into the exhaust gas and the pollutants are effectively oxidized in the thermal reactor.

Reduction characteristics by the AIR system

The reduction characteristics of CO, HC, and NO_x were examined at their steady condition by injecting secondary air from the nozzle installed inside the exhaust port and with the thermal reactor just behind the exhaust port. The results are given in Fig. 152.5. The amounts of CO and HC are decreased most effectively at an air/fuel ratio of 14/13; that is, when the AIR system is used, much lower levels of CO and HC can be obtained with a considerably richer mixture than can be obtained without the use of this system. The NO_x levels were in no way affected by the system.

As the result of engine modifications and the adoption of the AIR system, we were able to obtain the following values on the RX-2, as against the required values in our

thus changed, the mixture mass in the leading side becomes small and the combustion delay becomes comparatively great and, as mentioned in the previous section, 'after combustion' of the unburnt HC at the trailing corner is promoted and the HC level is reduced. Also, as shown in the figure, if the air/fuel ratio in each combustion chamber shape becomes high, misfiring occurs and the HC level tends to rise. Especially in the case of the combustion chamber recess in the trailing side, the throttle opening becomes large to obtain the same output and so the fuel flow increases and the HC level rises at a rate of 1·6 litre/h.

A lower level of HC can be obtained from the combustion chamber, but this has a tendency to reduce the engine performance.

Where the main recess is on the leading side of the combustion chamber, the HC level increases, with an improvement in engine performance.

Although it is not significant, other tests have shown that the NO_x level reacts in the same manner as the HC level with respect to the shape of the combustion chamber. For instance, if the main recess of the combustion chamber is moved to the trailing side, there is a tendency towards delayed combustion, the peak temperature is reduced, and the NO_x level drops.

AIR INJECTION WITH REACTOR SYSTEM

We were able to reduce the HC emission by approximately 40 per cent in, for instance, the U.S. Federal test procedure (7 mode × 7 cycle) by the above-mentioned engine modifications—that is, by changing the ignition timing, spark plug location and combustion chamber. However,

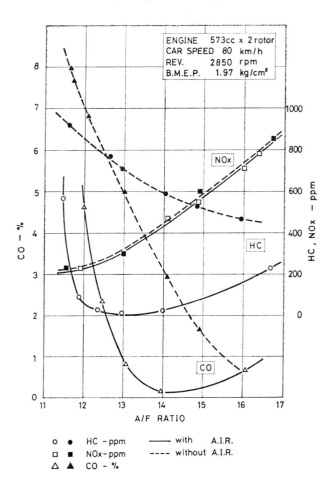

Fig. 152.5. Effect of the AIR system on HC, NO_x, and CO emissions

certification test at the Environmental Protection Agency (E.P.A.) in the U.S.A.: 0·9% CO, 177 p.p.m. HC, and 139 p.p.m. NO_x. The permissible emission standards in the U.S.A. for the RX-2 are: 1·55% CO, 277 p.p.m. HC, and 1650 p.p.m. NO_x.

Thermal reactor

As a result of the various tests, the so-called 'reflection type' thermal reactor (shown in Fig. 152.6) was initially produced to meet the reduction requirements necessary for the Wankel rotary engine. The reactor was mounted on the R100 and put on sale in the U.S.A. In our development of this reactor, durability was the most difficult problem encountered. Some of the reasons for this problem are listed below:

(1) Erosion, corrosion, and creep, by being exposed to high-temperature exhaust gas for lengthy periods.
(2) Cracks and breakage caused by thermal fatigue.
(3) Distortion of shells caused by thermal and mechanical restrictions and exhaust gas pressure.

To overcome these difficulties the development of proper materials and reactor design was carried out. During the development of the materials it was found that the fuel additives caused corrosion of the reactor material; halide forms of bromine and chlorine existing with the tetraethyl or tetramethyl lead increased the chemical corrosion on the material surface at high temperatures (approximately 800°C). Also, it was found that the presence of phosphorus caused further corrosion. As the result of various tests, the material finally selected was a 20% Cr–3% Al–Fe alloy.

Next, we had to overcome the difficulties stated in items (2) and (3) above. These were solved by improving the structure so that the reactor members were not mechanically or thermally restricted.

Finally, the temperature was controlled to prevent the wall temperature of the thermal reactor from becoming extremely high. This was achieved by (1) control of the ignition system, (2) control of the bypass system for the secondary air, and (3) development of a system and reactor structure by which the reactor inner shell is air-cooled by the bypassed secondary air. In particular, the solution to item (3) gave good results. The outline of the reactor structure and the reduction in the temperature of the reactor inner shell are shown in Figs 152.6 and 152.7. As these figures show, we were able to reduce the inner shell temperature by approximately 100–160 degC in the middle and high speed ranges. By using this type of reactor we were also able to lower the cost of the material and improve its productivity. This reactor was therefore installed in the 1971 model RX-2.

CONTROL SYSTEM

On the developed engine the ignition system, the amount of secondary air injected, and the flow of air required to cool the thermal reactor are controlled in order to obtain a low emission and to improve the durability of the thermal reactor.

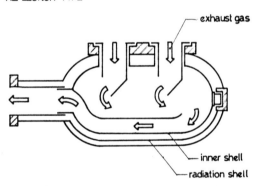
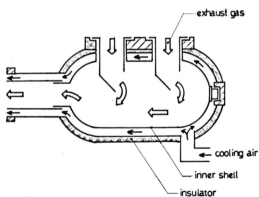

Fig. 152.6. Thermal reactors for the rotary engine

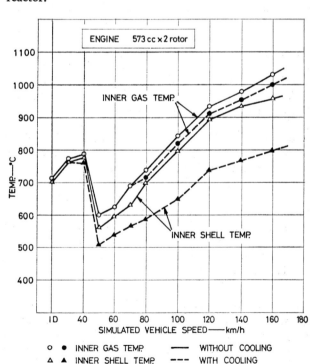

Fig. 152.7. Effect of cooling air on reactor temperature

Control of ignition system

As it is particularly necessary to decrease the HC level at low speeds and light loads, it is better to operate the engine with a single spark plug in these ranges. Judging from the relationship between the HC concentration and the fuel consumption, it was evident from previous tests that it was better to have a leading spark plug than a trailing spark plug, but the ignition of the trailing spark plug is cut in at low speeds and light load where the engine coolant has not yet been sufficiently warmed up, and the HC and CO emission levels are high. However, when starting the engine, and during idling, the trailing spark plug ignites in order to prevent difficult starting and fouling of the trailing spark plug. The trailing spark plug also ignites to avoid a reduction in engine performance at almost full throttle. When the engine and the thermal reactor have sufficiently warmed up, the HC and CO emission levels decrease even at low speeds and light loads and, therefore, in such conditions, both spark plugs ignite over the entire range of operation, decreasing the possibility of a reduction in performance.

The control systems are shown in Fig. 152.8. First, a system is adopted in which the number of engine revolutions is picked up from the ignition coil of the leading side, and the voltage pulse goes into the modulator where it is changed and amplified to the analogue voltages. The ignition circuit of the trailing spark plug is then cut by the operation of the relay when the number of revolutions drops below a certain value. Next, the intake manifold vacuum is picked up from upstream of the throttle valve of the carburettor, and the trailing spark plug is ignited at low vacuum by the vacuum switch operated by the intake manifold vacuum. At heavy load the vacuum switch is actuated and the trailing spark plug is ignited, thus ensuring good engine performance. When the engine has sufficiently warmed up, the trailing spark plug is ignited through the whole operational range by action of the thermosensor picking up the coolant temperature.

Secondary air flow control

As the HC emission level is especially low when the car is running at high speed, all the secondary air bypasses the injection passage, part of it being relieved to the air cleaner and the remainder being used for air-cooling of the thermal reactor to prevent a rise in temperature of the inner shell.

First, the number of engine revolutions is picked up and the relay is caused to actuate in the same way as described in the previous section, giving the on and off signals to the solenoid of the air control valve. That is, when the revolutions exceed a certain number, this solenoid opens the relief valve in the air control valve and the secondary air is not injected into the exhaust port, part

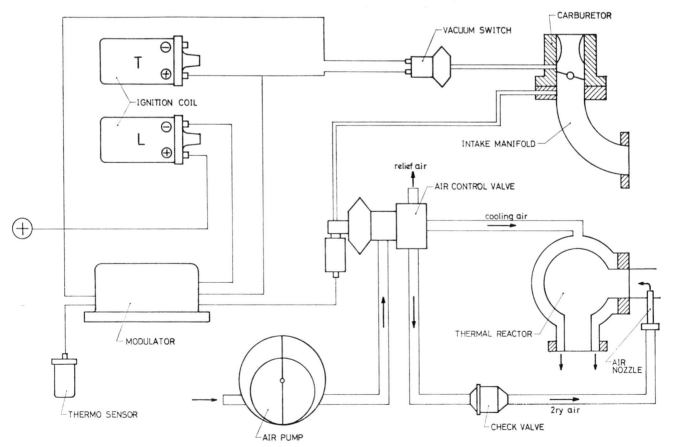

Fig. 152.8. Ignition and secondary air control system

Fig. 152.9. Emission controlled engine in RX-2

of it being bypassed to the air cleaner and the remainder to the air-cooling passage of the thermal reactor. Also secondary air is controlled according to the load condition by picking up the intake manifold vacuum directly as in the previous section and by directly opening and closing the relief valve in the air control valve.

In order to meet increasingly stringent emission regulations, work is now being carried out mainly to modify the engine and to improve the AIR system. The emission controlled engine mounted in the RX-2 is shown in Fig. 152.9.

CONCLUSIONS

Although the exhaust emission characteristics of the Wankel rotary engine are fundamentally the same as those of the reciprocating engine, the HC level of the Wankel rotary engine is considerably higher. On the other hand, its NO_x level is low because of the combustion mechanism peculiar to this type of engine. As these characteristics are closely related to the spark plugs and the combustion chamber, we made modifications in various combinations of these which were considered to be most appropriate for our requirements. However, we were able to achieve lower levels of emission by developing the AIR system further.

The thermal reactor that we have developed should fulfil an important role in meeting emission regulations which are becomingly increasingly stringent.

ACKNOWLEDGEMENTS

The author wishes to express his deep appreciation to the Manager, Mr Yamamoto, for his competent guidance, and to Messrs Shigetake Yoshimura, Motoyuki Hayashida, Terutoshi Tsumura, and several others who have helped with this work.

C153/71 AUTOMOTIVE EMISSION RESEARCH—A REVIEW OF THE C.R.C. PROGRAMME

C. E. MOSER*

The Coordinating Research Council (U.S.A.) automotive emissions research programme, involving over 40 projects, was initiated in 1967 and funded at approximately $4 million annually. This programme, expected to be active for over six years, consists of engineering, atmospheric, and medical projects largely carried out by contract with universities, research institutes, and government and industrial laboratories. Results of completed projects are now becoming available in the public domain for use by the government and industry to aid in abating the automotive emissions problems. This programme has been supported by the U.S.A. Environmental Protection Agency, the American Petroleum Institute, and the Automobile Manufacturers' Association.

INTRODUCTION

FOR ALMOST 50 YEARS, the Coordinating Research Council (U.S.A.), sponsored by the Society of Automotive Engineers (U.S.A.) and the American Petroleum Institute, has endeavoured through scientific co-operative research to improve the adaptation of automotive vehicle equipment and the fuels and lubricants used in their operations. Not only has the C.R.C. served as a vehicle for the joint efforts of the automotive and petroleum industries, but also as a means of co-operation with the government on matters of national interest in the transport field.

In 1967 the Air Pollution Research Advisory Committee (A.P.R.A.C.) was added as an entity to the Aviation, Diesel, and Motor Committees then existing in the C.R.C. organizational structure. Further details concerning the C.R.C. organization and the initiation of the C.R.C.-A.P.R.A.C. programme are contained in reference (1)†. C.R.C.-A.P.R.A.C. acts in an advisory capacity to the C.R.C. project manager who directs the automotive emissions research programme as shown in Fig. 153.1. The C.R.C.-A.P.R.A.C. comprises seven members from the petroleum industry, seven members from the automotive industry, and four members from the U.S.A. Environmental Protection Agency (E.P.A.).

The primary objective of the C.R.C.-A.P.R.A.C. programme is to develop basic information concerning the nature and effects of vehicular air pollution needed by industry to further reduce emissions and needed by government to formulate air quality criteria and control techniques. The programme is not to develop automotive hardware or petroleum products.

The research administered by C.R.C.-A.P.R.A.C. is carried out principally by contract with industry, university, government, and independent research organizations. Each contract is under the direction of a project group of technical specialists drawn from the automobile and petroleum industries and the E.P.A.

Financial support for the programme is shared equally by the Automobile Manufacturers' Association and the American Petroleum Institute. The E.P.A. provides financial support on a project-by-project basis depending upon the degree of public benefit and the programme's established priorities, and is currently supporting a majority of the projects. For those projects supported by the government, the cost is normally distributed equally among the three sponsors. During the first three years of the programme, which began in January 1968, over $12 million have been committed towards contractual work, and the sponsors have approved in principle an additional three-year unit estimated to require an additional $10 million.

In addition to the contract work, some research projects are being carried out co-operatively in the laboratories of individual companies and the government. The cost of these 'in-house' programmes, as they are known, along with the considerable costs associated with the manpower made available by industry and government for supervising and co-ordinating all research efforts, is absorbed by the individual organizations and represents a substantial addition to the direct cost of contractual studies.

The research programme consists of three groupings of

The MS. of this paper was received at the Institution on 26th July 1971 and accepted for publication on 24th August 1971. 33
* *Texaco Inc., P.O. Box 509, Beacon, New York 12508.*
† *References are given in Appendix 153.1.*

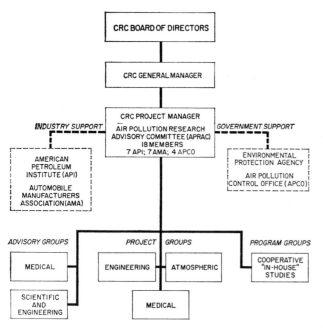

Fig. 153.1. Functional chart for C.R.C. air pollution research activities

projects: engineering, atmospheric, and medical. Subsequent discussion will deal with each group separately.

ENGINEERING RESEARCH

The purpose of the engineering projects is to study the interaction of fuel and engine design upon exhaust emissions. These projects may be broken into the three categories shown in Table 153.1 with a listing of the individual projects under each category.

Development and evaluation of testing techniques

The first project in the testing technique category has as

Table 153.1. Engineering research

Development and evaluation of testing techniques
    Exhaust odour measurements
    Characterization of exhaust particulates
    Improved methods for determining polynuclear aromatics in engine exhaust
    Exhaust gas oxygenate content

Fuel, lubricant, and engine interrelationships
    Gasoline additive effects on carburettor and PCV system performance and relationship to exhaust emissions
    Effects of leaded and unleaded gasolines on exhaust emissions caused by combustion chamber deposits
    Fuel volatility and automotive design as related to vehicle emissions
    Vehicle refuelling emissions
    Polynuclear aromatics and phenols in vehicle exhaust
    Kinetics of combustion in gasoline engine exhaust systems

Consumer use and vehicle emissions
    Urban vehicle operating patterns
    Fuel system temperature during automobile use
    Vehicle inspection and maintenance studies

its objective the development of information needed for the identification and measurement of engine exhaust odour. After a number of years of co-operatively sponsored work by several contractors on diesel exhaust, the present contract project in the Arthur D. Little laboratories is showing good progress. The sample preparation scheme involves collection from the exhaust by condensation at 0°C, extraction with pentane and chloroform, and separation of the odorous organic extract into three basic working fractions by liquid chromatography. Utilizing primarily high-resolution mass spectrometry, the oily-kerosine complex odour is due principally to indans and tetraline with contributions from the alkyl benzenes as shown in Table 153.2. Present data suggest the most important odour contributions from the smoky-burnt fraction are made by the dienones, phenols, benzaldehydes, indanones, tetralones, and naphthols.

While further contract work is proceeding at Arthur D. Little, studies are being conducted in the laboratories of the sponsors to determine the variations encountered in selected odour measurement techniques.

The second testing technique project concerns determining the physical and chemical characteristics of particulate matter from the internal combustion engine as a function of sampling procedure, engine operating conditions including emission control systems, fuel composition, and residence time in the atmosphere. In the exploratory programme carried out by Battelle Memorial Institute, particulate matter generated in a small single-cylinder engine using leaded gasoline was collected in a cascade impactor and characterized by electron probe and electron microscopy. Calcium, phosphorus and iron were found in larger particles; in turn, these were found to be aggregates. The most prevalent particles were ≤1·0 μm where lead was the major component and associated with bromine and sulphur. Future studies will be made with 1970 model autos with leaded and non-leaded fuels using an urban driving cycle on a chassis dynamometer.

Under a contract with the Esso Research and Engineering Company, the third testing technique has resulted in an improved method for determining the polynuclear aromatics benzo(a)pyrene and benz(a)anthracene in auto exhaust. The method, which employs alumina column chromatography followed by a combined chromatographic–ultraviolet technique, results in a substantial saving in time and cost over the past chromatographic–radioactive tracer method. Research is continuing to

Table 153.2. Diesel exhaust condensate extract

Odour free fraction	Oily-kerosine fraction	Smoky-burnt fraction
(Paraffins)	Indenes Indans Tetraline Alkyl benzenes Naphthalenes	Dienones Naphthols Phenols Benzaldehydes Indanones Tetralones

reduce instrumental complexity and broaden the range of analysis.

The fourth project on exhaust gas oxygenate analyses involves two contracts. The first being performed at the U.S. Bureau of Mines laboratories seeks to obtain data on the relative levels of carbonyl (e.g. aldehydes and ketones) and noncarbonyl oxygenates in exhaust samples from a current model vehicle under urban driving conditions. This task will be followed presumably by work to develop an analytical method for noncarbonyls.

The second contract with the Scientific Research Instruments Corporation provides for the application of chemical ionization mass spectrometry in an instrument to continuously monitor the major carbonyls in auto exhaust. Tests with the instrument for a vehicle on a dynamometer indicated generally satisfactory results for the simultaneous determination for nine compounds ranging from formaldehyde through benzaldehyde. Further work to improve the instrument is under way.

Fuel, lubricant, and engine interrelationships

The first project under the 'fuel, lubricant, and engine interrelationships' category shown in Table 153.1 was conducted by the Scott Research Laboratories and consisted of a 48-vehicle test programme to evaluate the effects of gasoline detergent and dispersant additives on carburettor and positive crankcase ventilating valve performance as they relate to exhaust emissions. The programme was terminated after 12 000 miles of operation with evidence that more severe operation and increased additive concentrations would give more definitive results.

Before undertaking a contract programme on the effect of leaded and unleaded gasoline on exhaust emissions, currently available data from 18 industry laboratories were analysed by a committee of representatives of the C.R.C.–A.P.R.A.C. sponsors. This study concludes that for vehicles equipped with first generation (1966 and 1967) exhaust emission control systems, cars operated with leaded gasolines have higher equilibrium hydrocarbon emission levels than those operated with unleaded fuel. This combustion chamber deposit effect under consumer-type driving conditions is about 7 per cent, and under rapid mileage accumulation conditions is about 20 per cent. No effect was noted for either carbon monoxide or oxides of nitrogen.

The next project in the 'fuel–engine' category has as its objective the study of the interaction of fuel volatility and automotive design as they relate to driveability, evaporation losses, and exhaust emissions. Twelve late model (1968–69–70) vehicles are being tested for emissions by the U.S. Bureau of Mines with eight test gasolines varying in front-end and mid-range volatility at ambient temperatures from 20 to 95°F. The Ethyl Corporation has performed the driveability testing, and a final report should be available soon.

The project to determine the magnitude and significance of fuel spillage and vapour evolution involved with vehicle refuelling is being conducted by the Scott Research Laboratories. A pilot test programme carried out in 1969 indicates that refuelling losses increase with fuel volatility, dispensed fuel temperature, quantity of fuel dispensed, careless attendant technique, and filler pipe design. Losses were not affected by dispensing rate or tank shape. A mathematical model for prediction of losses is being constructed after a recent five-city survey and laboratory studies of typical dispensing pumps and vapour loss measuring techniques.

The polynuclear aromatic (PNA) and phenol project has as its objective the determination of emission rates of these materials in vehicle exhaust as influenced by fuel composition, engine operating variables, and exhaust emission control systems. Esso Research and Engineering in contract work to date in this complex field using typical emission controlled and uncontrolled vehicles on a 7-mode cycle conclude that recent production low-emission vehicles contributed much less to atmospheric PNA than do earlier uncontrolled vehicles. Composition of the fuel and combustion chamber deposits have complex effects on PNA production, with increasing aromatics and benzo(a)pyrene in the fuel generally increasing PNA.

The presence of lead in engine deposits appears to be important with respect to PNA. Phenol emission is largely controlled by fuel aromaticity, though emission controls reduce phenols slightly at low fuel-aromatic levels. This programme is being continued.

To those involved in the design of automotive components, a very interesting project is the study of combustion kinetics in gasoline engine exhaust systems. The University of Michigan has undertaken this contract to develop an understanding of the oxidative phenomena of combustion under temperatures, pressures, compositions, and flow conditions typical of the exhaust gas from gasoline engines. Baseline emission data have been obtained for a Chevrolet 350 CID V-8 engine including hydrocarbon type (paraffins, olefins, and aromatics) data over a range of air/fuel ratios and spark advance.

A single-cylinder two-tank exhaust reactor has been fabricated and installed on the engine for further work, and a computer model will be constructed to simulate full-scale reactor behaviour. The information from this project should be helpful to those engaged in the design of practical reactors.

Consumer use and vehicle emissions

The first project under the 'Consumer use and vehicle emissions' category has as its objective the collection and evaluation of urban traffic patterns and vehicle modal operating characteristics to permit the development of an average driving cycle. Early cycles were developed exclusively upon Los Angeles information. In this study, data were obtained for Los Angeles, Houston, Cincinnati, Chicago, Minneapolis/St Paul, and New York by the System Development Corporation using tacographs installed in owners' cars.

The modal operating parameter information (speed, acceleration, deceleration, engine intake vacuum, etc.) on test vehicles operating on typical consumer patterns was obtained by Scott Research Laboratories. Traffic pattern reports are available, and the modal report for Los Angeles should be available shortly. The results should supply the information necessary to simulate accurate field conditions in the laboratory.

Fuel system component time–temperature histories representing car use patterns at high temperatures were determined in Los Angeles using the traffic pattern survey information. Data obtained by the Scott Research Laboratories are available for use by all investigators. As would be expected, the variance in fuel tank liquid temperature is dependent primarily upon the ambient temperature, and the carburettor bowl temperature varies from vehicle to vehicle.

An important and extensive project under 'Consumer use' is that involving the question of the economic effectiveness of various approaches to maintaining vehicles in the best possible condition from the point of view of air pollution control. A TRW Inc. study with Scott Research Laboratories as a subcontractor has been under way since 1969 to develop acceptable methods for inspecting and maintaining cars with emission control devices in the best possible condition for air pollution control. To date, over 250 vehicles have been inspected and the degree of maladjustment evaluated. Based on laboratory information, the degree of improvement which could be obtained by various types of inspection followed by the maintenance indicated has been estimated. Some preliminary conclusions of the study follow.

- Most effective engine parameters are: fuel/air ratio, idle rev/min, idle timing, ignition system, PCV valve, air cleaner.
- Maximum emission reductions by inspection of engine and control system, and maintenance above six items.
- Net cost effective procedure: idle emission measurement to diagnose basic timing, fuel/air ratio, and engine speed.
- State inspection lanes more effective than franchised garages.
- Infrared instruments preferred for State lane applications.

Maximum emission reductions are achieved by direct inspection of engine and control system components followed by maintenance of the six parameters defined. The current programme entails a strengthening of the economic-effectiveness model for accurately predicting the effect of various inspection/maintenance strategies.

ATMOSPHERIC RESEARCH

The A.P.R.A.C. atmospheric studies are devoted to the development of fundamental information on chemical and physical atmospheric processes in order that pollutant behaviour in the atmosphere may be explained and perhaps related to vehicle emissions. Some seven project areas are listed below, with more than one contract investigation active in some areas:

Irradiation chamber research.
Hydrocarbons in the atmosphere.
Measurement of atmospheric pollutants in urban areas.
Carbon monoxide in the atmosphere: carbon monoxide sinks in the biosphere; carbon monoxide sinks in the atmosphere; sources of carbon monoxide.
Plant damage by air pollutants.
Haze formation.
Atmospheric modelling.

Irradiation chamber research

Co-operative studies have been conducted for several years by some eleven laboratories to obtain data on reactions of vehicular emissions in irradiation chambers, simulating the reactions of the atmosphere. Owing to the variability of results, a contract has been taken with the Lockheed Missiles and Space Company to study the design and operating variables such as surface material, surface to volume ratio, and ultraviolet light intensity and spectral distribution which affect the reactions occurring in environmental chambers. Simple mixtures of hydrocarbons and nitric oxide will be used for the first phase of the photochemical reaction programme.

Hydrocarbons in the atmosphere

An exploratory programme has been supported at the Riverside Campus of the University of California for a number of years where ambient air samples were analysed for the light hydrocarbons and sources of these compounds hypothesized from their relative amounts before and after further irradiation (2). The current programme is emphasizing the extension of such work to aromatics and oxygenates. An improved chromatograph for the work has been obtained and results should be available before the end of 1971.

Measurement of atmospheric pollutants in urban areas

In this study, thousands of air samples were collected and analysed by the Scott Research Laboratories from the Los Angeles basin at ground level and at altitudes up to the inversion layer during the 1968 and 1969 smog seasons. The purpose of this work is to obtain a comprehensive picture of the atmospheric reactions of varying concentrations of the oxides of nitrogen.

Air contaminants determined in the samples include carbon monoxide, total hydrocarbons, individual hydrocarbons (C_1–C_{10}), aliphatic aldehydes, formaldehyde, acrolein, nitric oxide, nitrogen dioxide, total nitrogen oxides, total oxidants, ozone, and peroxyacetyl nitrate (p.a.n.). Pertinent meteorological data are also taken. A preliminary analysis of the data indicates that mathematical modelling of the atmospheric processes appears possible.

A general data analysis of the Los Angeles measurements has been carried out. In 1970 data were obtained in the New York City–New Jersey area during the summer season, including free-flying kite meteorological patterns. Helicopter sampling was done by following the kites.

A subsequent but similar project will be carried out over San Francisco Bay by the Stanford Research Institute. From all of these measurements it is expected that a more complete understanding of the photochemical processes over urban areas will result.

Carbon monoxide in the atmosphere

In the past the production and occurrence of carbon monoxide in the earth's atmosphere was attributed largely to the incomplete combustion of fossil fuels, and since no known sinks existed it might accumulate and become a fatal poison to all mankind. On closer examination it was apparent that the average global concentration was static and that unidentified processes were at work to remove carbon monoxide from the atmosphere at a rate equal to its production (3). For these reasons, projects were established in the A.P.R.A.C. programme to establish whether sinks existed in the biosphere and the atmosphere. More recently, studies indicated that marine (4) (5) and land sources existed beyond the fossil fuel combustion, especially in the automobile, and a third project was established to evaluate the natural production of carbon monoxide.

In the biosphere, Stanford Research Institute workers have found that unsterilized soil, apparently due to the activity of soil micro-organisms, removes carbon monoxide from the atmosphere at a rapid rate (6). As shown in Fig. 153.2, a concentration of 120 p.p.m. of carbon monoxide can be depleted to near zero within 3 h at 30°C. The future programme will involve the isolation of the soil micro-organisms involved and the role of higher order plants as sinks.

While the stratosphere offers a number of possibilities for removing carbon monoxide, studies by GCA Inc. indicate that transport of carbon monoxide from the troposphere to the stratosphere and subsequent conversion to carbon dioxide is a significant removal process but not sufficient to account for the static concentration.

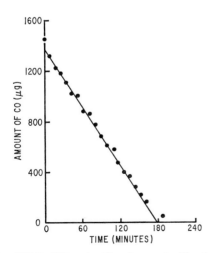

Fig. 153.2. CO reduction by unsterilized soil

The third area of investigation for sources of carbon monoxide has been undertaken by the Argonne National Laboratories and should provide information on the relative importance of land, marine, and fossil fuel sources in the atmosphere. Early results using ratios of the stable isotopes of carbon and oxygen in atmospheric carbon monoxide appear promising.

Plant damage by air pollutants

The true economic damage to plants by air pollution has not been well documented and studies are under way to determine the economic impact on crop production where significant air pollution exists. To date, estimates are being developed by use of a newly developed computer program by the Stanford Research Institute utilizing existing literature on the prevalence of pollution and plant sensitivities. Field verification of the estimates remains to be performed.

Haze formation

Where haze is one of the earliest documented effects of air pollution, it is difficult to distinguish between the haze from man-made sources and from natural phenomena. Two studies are under way in the atmospheric haze programme: one to determine the chemical composition of hazes, especially over urban areas, and one to determine the feasibility of using an optical instrument to distinguish between those from natural and man-made sources. Battelle Memorial Institute is carrying out the first study and Science Spectrum Inc. the second.

Atmospheric modelling

The atmospheric modelling project under way has as its objective the development of a mathematical model which will predict the manner in which automotive exhaust emissions diffuse through a given area under specified meteorological conditions. Initial work has been limited to the relatively inert gas carbon monoxide. The Stanford Research Institute has developed a working model largely from existing monitoring and earlier modelling programmes. Initial results from validation of the model performed in a small downtown area of San Jose, California, are very encouraging. It is planned to study validation in an area in St Louis, Missouri, to obtain city canyons of greater depth.

MEDICAL RESEARCH PROGRAMME

The overall objective of the medical research programme is to determine the deleterious effects of automotive emissions or their atmospheric-derived products upon human health. Such work, for the most part, has been undertaken under contract at university medical schools.

Following the recommendations of its Medical Advisory Group, C.R.C.–A.P.R.A.C. has had work under way during the past year upon 11 projects. Owing to its known toxicity and presence due to combustion sources, seven have been concerned with the health effects of carbon monoxide, as indicated in the following list:

Exposure to carbon monoxide: effects of low levels of carbon monoxide on humans; effects of carbon monoxide on human performance; effects of carbon monoxide on cardiovascular system; determination of human carboxyhaemaglobin levels; effects of carbon monoxide on humans performing driving tasks; influence of carbon monoxide on motor vehicle accidents; effects of carbon monoxide on myocardial infarctions.

Exposure to nitrogen oxides and oxidants: effects of nitrogen oxides and oxidants on morbidity and mortality; effects of nitrogen oxides and oxidants on humans.

Toxicity of polynuclear aromatic hydrocarbons.

Synergistic effects of air pollutants.

Exposure to carbon monoxide

The initial step in undertaking work upon the human health effects of carbon monoxide was to have the National Academy of Sciences undertake a study of available information, and have a committee of knowledgeable physicians recommend those areas where additional data were needed (7). The problem areas of highest priority were then incorporated in the C.R.C.-A.P.R.A.C. medical programme.

The effects of low levels of exposure to carbon monoxide upon human response or performance was one of the first areas to require expanded work. An experimental study was carried out at the Medical College of Wisconsin upon healthy male volunteers wherein they were exposed to concentrations of carbon monoxide from less than 1 up to 1000 p.p.m. at established intervals for periods of 30 min to 24 h. From a battery of laboratory performance and physiological tests, significant conclusions were: (1) no untoward effects were observed in sedentary males exposed up to 100 p.p.m. for 8 h which produced maximum carboxyhaemoglobin saturations in the range of 11–13 per cent; and (2) exposures producing carboxyhaemoglobin saturations greater than 15–20 per cent resulted in delayed headaches, changes in visual evoked responses and impairment of manual co-ordination (8). This programme is being expanded to include additional variables such as age, sex, cigarette smoking, exertion, and certain drugs.

No final results have become available from the remaining projects listed above utilizing carbon monoxide. Studies are under way in Jefferson Medical College of Philadelphia, Harvard University, Ohio State University, Johns Hopkins University, and Columbia Medical School. While in one or two cases contracts have not yet been awarded, it is hoped that results from a number of these programmes will become available in the near future.

Exposure to nitrogen oxides and oxidants

Very little reliable information is available on the effects of low-level exposure of humans to nitrogen oxides and oxidants. Accordingly, contract work at Tulane University Medical School has resulted in a research design for a five-year morbidity and mortality study, integrating meteorologic, aerometric, and health observations. Based on the recent U.S. Government report (9) on the health effects of nitrogen oxides and oxidants and the Tulane research design, an epidemiological study will be formulated and contracted.

Toxicity of polynuclear aromatic hydrocarbons

An active project at the University of Nebraska College of Medicine has shown that a combination of haematite and benzo(a)-pyrene injected through the trachea into the lung of the hamster will produce lung cancer. Since the haematite may play more than a passive role in this system, other carriers are being examined and work extended also to auto exhaust particulates.

Synergistic effects of air pollutants

An extensive project is being completed at the TRW Inc. Hazleton Laboratories, where the physiological action of five common air pollutants is being investigated by the ventilation of rodents and primates. High concentrations and normal atmospheric concentrations of pollutants taken singly and in combination include gases (CO, NO_2, SO_2) and particulates ($PbClBr$ and $CaSO_4$). In general, those animals exposed to nitrogen dioxide were the only ones to indicate significant physiological effects, and additional work in this field will be considered. A series of papers on the Hazleton work is being released for publication in 1971.

CONCLUSIONS

The C.R.C. air pollution research programme has contributed significantly in five areas:

(1) It has served to define the many unknown or only partially explored areas dealing with automotive emissions effects and their control.

(2) It has helped to co-ordinate a number of research efforts in these areas.

(3) It has acted as a stimulant for additional studies by other researchers.

(4) It has contributed significant information about engineering, atmospheric, and medical problems.

(5) It has demonstrated a mechanism by which government and industry can co-operate in scientific studies.

A summary status report of recent work sponsored by A.P.R.A.C. may be obtained free of charge from the Coordinating Research Council (10).

APPENDIX 153.1

REFERENCES

(1) BARTH, D. S., HEINEN, C. M. and McREYNOLDS, L. A. 'The air pollution research program administered by the Coordinating Research Council', *Eighth World Petroleum Congress*, Moscow, 1971 (June).

(2) STEPHENS, E. R. and BURLASON, F. R. *J. Air Pollut. Control Ass.* 1969 **19**, 929.

(3) ROBBINS, R. C., BORG, K. M. and ROBINSON, E. *J. Air Pollut. Control Ass.* 1968 **18**, 106.

(4) SWINNERTON, J. W., LINNENBOM, V. J. and LAMONTAGNE, R. A. *Science* 1970 **167**, 984.

(5) WILSON, D. F., SWINNERTON, J. W. and LAMONTAGNE, R. A. *Science* 1970 **168**, 1577.

(6) INMAN, R. E., INGERSOLL, R. B. and LEVY, E. A. *Science* 1971 **172**, 1229.

(7) *Effects of chronic exposure to low levels of carbon monoxide on human health, behavior, and performance* Book No. 309-01735-1, 1969 (National Academy of Sciences, Washington, D.C.).

(8) STEWART, R. D., PETERSON, J. E., BARETTA, E. D., BACHAND, R. T., HOSKO, M. J. and HERRMANN, A. A. *Archs envir. Hlth* 1970 **21**, 154.

(9) 'Air quality criteria for nitrogen oxides', *AP-84*, Environmental Protection Agency, Air Pollution Control Office, Washington, D.C., U.S.A., 1971 (January).

(10) *APRAC Status Report* 1971 (January) (Coordinating Research Council Inc., New York).

Discussion during Session 1

R. G. Gippet Montlhery, France

After reading Paper C143 I should like to ask the authors two questions.

(1) In the Japanese test procedure, the brake load values can be calculated from a given equation. Is the chassis dynamometer accurately programmed for brake load value, or does the equation allow for differences between the true brake load value in current circulation of the car and the effective brake load value on the chassis dynamometer during the test cycle?

(2) With regard to diesel-engine vehicles, the method of chemi-luminescence has proved valuable for mixtures of NO and NO_2. Since the limit will in the coming year only be 5 g/hp, would this method be more convenient than non-dispersive infrared techniques?

With regard to the 1972 Federal procedure, it can be demonstrated that a large number of errors in the measurement are caused by basic inadequancies in the testing procedure as a result of air impurities. Consequently, accuracy cannot be obtained with the analyser sets available at the present time unless considerable and costly care is taken in the practical operating procedure.

With regard to both petrol and diesel engines, it would be possible, and interesting, to define standards of mass emissions in the light of new criteria such as useful load, useful speed or combinations of power and such parameters, to render pollution a function of service qualities.

I would like clarification on one point. There are two types of chassis dynamometer. For most of those now in use, the correct correspondence between brake values on the chassis is correctly supplied for two points, corresponding to two given figures of speed. A dynamometer such as the Siemens type, issued recently, gives, I believe, a quite good relation of agreement between effective and actual brake values for every speed all over the cycle.

W. E. Bernhardt Germany

I should like to ask the author of Paper C141 a question concerning the thermal efficiency of the three systems which he discussed. Comparing the thermal efficiency in these systems, is there no chance for the Rankine cycle?

M. S. Bolton Sheffield

The author of Paper C141 presented the stratified charge engine as an alternative system for meeting the 1967 requirements but added that there was usually an exhaust treatment device and recycling. Has he an alternative power unit, if he is still having to use the same clean-up devices as at the moment, and what are the emission characteristics of this power unit without exhaust clean up?

My next point concerns both Papers C135 and C145 but more particularly the fuel injection paper. Because of changes in ambient conditions, the paper discusses all sorts of monitoring, for instance, of atmospheric pressure, temperature and so forth, to compensate for these changes. The reason why all this compensation must be made is because of the effect on the air–fuel ratio, and this is because all carburation systems are feed-forward control

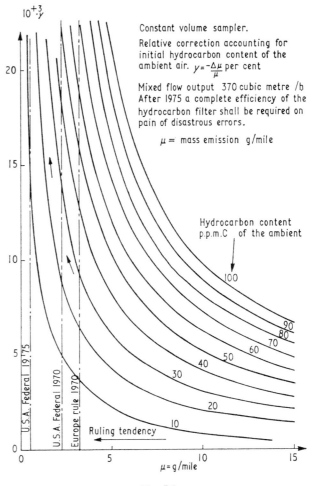

Fig. D1

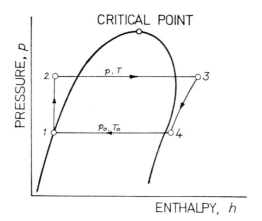

Fig. D2. Analysis of a Rankine cycle

systems. The same goes for the Zenith carburetter: the compensation for fuel, temperature, etc., must be applied because it is a feed-forward system. What this means is that the carburetter is putting into the engine what it thinks the engine wants; however accurately you meter everything, you have no idea of what is coming out at the other end.

At Sheffield we have developed a feedback system to compensate for all these factors. We have developed a very cheap and simple exhaust detector which is to be mounted permanently in the exhaust of the car. A signal from this detector, which is sensitive to carbon monoxide and hydrocarbons, controls the fuel flow into the engine. So this is the closed loop feedback system. There is no compensation for anything like atmospheric pressure, fuel temperature, etc., because these are all compensated for in the exhaust detection unit. Rather than making a system more complicated, this would make it simpler, especially in the case of the Lucas injection system. In the extreme, the only monitoring you would need to do would be of throttle angle: you would need this to maintain feed-forward control to keep a fast response. But everything else could be done away with by having a feedback loop which could be applied very cheaply. I would like some comment on this.

Discussion during Session 2

W. J. D. Annand Manchester

What I have to say refers to Papers C129 and C132. The authors have said that the effects on the pollutants which they find are really indirectly related to the fuel properties in a sense, in that fuel properties really change the mixture strength and the mixture distribution which I would call 'primary variables'. It is annoying when someone says to the author of a paper, 'Why didn't you do something quite different?', but it would be helpful in interpretation if one knew how big the effect of the changes in fuel properties had been on the primary variables, so that we could get a clearer interpretation. I am not certain how important this is, since Paper C129 seemed to suggest that fuel variability did not matter very much and Paper C132 seemed to suggest that perhaps it does.

On a more positive point, as regards the emissions of nitric oxide in particular, the second paper shows that, in regard to fuel volatility, some cars do show a quite firm and positive correlation and others a negative one. It is remarked also, in the first paper, that there is no consistency of direction of effects. I would like to point out that this is to be expected at least in some cases. If you look at the plot of nitrogen oxide emissions against fuel–air ratio for a single cylinder engine in perfect conditions, you find that you get a maximum at 5 or 10 per cent weak. If you now take, say, a four-cylinder engine with two rich and two weak cylinders centred upon that maximum, you find it gives less emissions because each cylinder is working to either side of the peak. On the other hand, if your fuel ratio is centred far down the rich branch of the nitric oxide–fuel ratio curve, the curve becomes concave up and maldistribution will increase the average emission. So it all depends where you are on the fuel ratio plot as to what effect a deterioration in distribution will have. This is very typical for the NO_x case where there are concave up parts and peaks in the range.

Finally, it is just worth mentioning that the effects of such things as the hydrogen–carbon ratio can perhaps be got at more directly by calculation since we can now do cycle calculations including NO kinetics, thanks to several of the contributors here. If one does this for the hydrogen–carbon ratio, which can be varied in the sums without changing anything else in the fuel, one finds it has quite a strong effect. One case, which I hastily pushed through the computer after receiving the papers—though perhaps it is not really representative—where the hydrogen–carbon ratio varied from 6·1 to 7·4 gave a 60 per cent increase in NO purely from the change in temperature arising from the fact that the stoichiometric ratio of air to fuel drops.

A Speaker

I hate to labour the point of effective aromatic content in hydrocarbon emissions, but I wonder if the authors of Paper 129 could clarify two slides shown and also two figures in their paper. In Fig. 129.8 they show the effect of increasing fuel aromaticity on non-dispersive infrared (NDIR) hydrocarbons: this is a negative slope showing a decrease. In Fig. 129.9 they show the ratio of flame ionization detector (FID) to NDIR, but here they have plotted increased paraffins. So presumably for increased aromatic content one has to read this right to left. If one takes, say, a 60 per cent aromatic, one finds that one now has a 3·5 ratio of FID to NDIR. Would this not turn one of those slopes given in the previous figure to a positive slope?

C. L. Goodacre London

In Fig. 129.8 of the paper by B. J. Kraus and B. C. Richard there is an increase in aromatics up to apparently 60 per cent. Did the authors observe that when running on high aromatics there was increased particulate or free carbon content in the exhaust?

P. J. Moss Birmingham

I should like to make a few comments on Paper C129. Some time ago I was involved with work to evaluate both the quantity and quality of motor vehicle engine exhaust emissions. Early on, while using a proprietary 5-star fuel, I noticed that the chromatograms of the exhaust gas were very similar to those of the parent fuel, the only difference being at the front end (C_1 to C_4 hydrocarbons). This characteristic was confirmed using pure octane (224 TMP) as a fuel and at a later date a similar indication was obtained with propane as fuel. Exhaust gas chromatograms showed octane and propane respectively to be the predominant exhaust hdyrocarbon accompanied by small proportions of C_1 to C_4 hydrocarbons.

The fact that exhaust hydrocarbons are substantially the same as the parent fuel and that most commercial fuels are similar suggest that non-dispersive infrared (NDIR) analysis would be reasonably repeatable and therefore a suitable measure of exhaust hydrocarbons. This is certainly borne out in practice and I have found that when using the 5-star petrol as fuel the response of the 'Hexane' NDIR analyser to exhaust hydrocarbons was 0·6 of the flame ionization detector response.

M. A. I. Jacobson London

I would like to ask the authors of Paper C129 whether, in their view, as far as pollution is concerned, they have also taken into account the main objections on the part of the public, that is, 'smelliness'. It is a difficult thing to define. The public at large appear to be concerned with three problems: the first is that it should cost them as little as possible to run their cars, second, that the cars should not smell, and third, that the cars should start under all conditions of operation. Could the authors say what effect variables have had on these three factors?

P. Draper Dorset

I was interested to read Paper C133 since I was involved in very similar work 32 years ago and came to rather similar conclusions; but we were then using the motor method CFR test so we had a different rubber yardstick from the one used by the authors. When we hear of the various preferences for different cars and different fuels, all it means is that the CFR test method is in error in not following what the cars require. Even those many years ago, we did suggest a method somewhere in between the motor and research methods, which gave a pretty good correlation; had this been adopted we would not have had this paper today.

The paper has shown that lead could be reduced or even omitted, if necessary, for use in some cars. The car I am running now, for instance, would be quite happy on unleaded fuel from the point of view of anti-knock value because it is not one which requires a very high anti-knock level. However, there is more to the matter than this, which brings me to my question. I wonder whether the authors could say anything about the resistance of valve and valve seat materials to lead-free fuels; because, if it were not for possible trouble there, I would like to see many more lead-free fuels for a number of reasons into which I have no time to go now.

To those of us who are interested in pollution control Papers C135 and C145 are of most practical interest. I will confine my comments mainly to Paper C135. The adoption of such carburetters, in conjunction with, one presumes, some slight engine modification, could well be the solution to the problem of air pollution from car exhausts by bringing it to acceptable levels in Europe, but this course might not necessarily bring it down to the very severe U.S. levels. It might be of interest to say that the National Society for Clean Air arranged to test a number of cars off the roads; we arranged with the police to pull in a few cars quite at random so that we would do some quick CO tests. Having tested over 140 cars, we found that on average the CO at idling was about 6·5 per cent. What is even more interesting is that all makes of cars showed that at least some of the models could achieve CO emissions of under 4 per cent. This shows that you have only to 'titivate' those engines or fit good carburetters to achieve results. A minor but interesting point is that 6 sports-type cars in the test, which one would expect to have very rich mixture under idling because of the need to 'jump away' quickly, fell in line with most of the other cars.

A. G. Bell Chester

I should like to make a few comments on Paper C133. My first point is perhaps one of clarification; the point came out quite clearly in the written paper, but perhaps has not been made so clear in this afternoon's presentation. It is that in the octane requirement experiment all the cars ran on leaded gasoline during the mileage build-up period, and that there may be either a plus or a minus in terms of octane requirement if the cars had run on unleaded gasoline up to the point they were rated. The size of this plus or minus is at present being studied in a large scale experiment by CORC and CEC.

To come back to the paper, the low speed knock results surprised me a little in that they showed rather a large benefit for the leaded fuel and also suggested that front-end quality, Delta ON (100°C), was not important for unleaded gasolines. We ourselves did not find this in some work we have done in this area; and I wonder whether those results, if re-plotted taking into account motor octane number as well as research octane number, would show the same effect.

I am interested to see that, at high speed, the authors' results confirm the results of our work, i.e. that there is a bigger lead bonus with tetramethyl than with tetraethyl lead; but I wonder whether they could give us an indication of how many octane numbers that bonus is worth. In the paper it is plotted in terms of knock limited spark advance. Could the authors give us please an approximate conversion from knock limited spark advance into octane numbers?

G. L. Lawrence Stanmore, Middx.

I would like to return to Paper C128 and ask a question about the method in which the idle level was set, because the authors mentioned a 7 per cent idle for vehicles of 1962 vintage. This is entirely normal: vehicles set up on a sensory perception of the best idle would be somewhere in this area. But the paper goes on to show figures deteriorating from idles set at levels of 2 to 3 per cent, which indicates that they were set to some other parameter than sensory perception.

I would also like to ask whether there were some accelerated means used to invoke the contamination because in 20 years of experience I have never seen anything approaching this degree of contamination or the associated problems described.

Perhaps the authors could also tell me whether, with

the SU carburetter that was shown to have progressive contamination, there was also a fall-off in idle performance as apart from some possible reduction of idle speed this would also be contrary to experience?

In reply to M. S. Bolton concerning exhaust detection units, everyone is attracted by this principle, but I do not think we ourselves see it as a solution which looks like being commercially viable.

J. G. Anderson Sunbury-on-Thames

We were extremely interested in Paper C135 because it shows that some positive moves are being taken to iron out the emission variations caused by metering in the carburetters. Clearly, the paper is intimately related to the two fuel effects papers which were discussed earlier.

BP have been actively engaged in an examination of the CO-at-idle problem with compensation methods, and our work has led us to two main conclusions which I think agree with what the author has said. Firstly, fixed jet carburetters are subject to interference to flow by vapour bubbles in the idle and progression tubes, quite apart from the effects of vapour entrainment from internal vent systems and evaporative loss systems as well as possible percolation problems. Secondly, the variable jet or air valve carburetter operates over a wide range of temperature without disturbance of the fuel flow through the jet. Flow rates are, therefore, primarily influenced by the viscosity of the fuel. Again, mixture strength would also be influenced by vapour entrainment from evaporative loss systems and also by percolation.

To illustrate my point, Fig. D3 shows some typical results obtained. We constructed the same carburetter in three different forms. The left-hand curve shows discharge at the throttle blade edge as it would be in a fixed jet carburetter (showing the same relationship as the author's). The next curve shows the relationship with the fuel discharging at the Venturi or bridge area of the variable jet carburetter but with an orifice suitable for idling only and without a needle in it. Thirdly, and the most important curve, shows the normal relationship we expect with a variable jet carburetter discharging at the needle in the hole: increasing CO content with increasing fuel temperature. The point to observe is that at very high temperatures peculiar things happen. A very remarkable fall in CO level occurred round about 70°C in this case. The temperature at which the fall occurred could be varied according to the type of fuel being used. This would also depend to some extent on whether the engine would keep running at 7 per cent or 8 per cent CO; it could stall there or the CO could fall off to a very low level and it would stall at that point.

Looking at the air valve carburetter in more detail, we constructed a glass replica shown in Fig. D4. This enabled us to study the flow of fuel through the glass jet. We discovered that at very high carburetter bowl temperatures this carburetter behaved in much the same way as the fixed jet carburetter giving a very rapid drop in CO level. We observed that the fall in CO was the direct result of vapour blockage of the jet. This effect has been observed on the CDSE carburetters and has been commented on

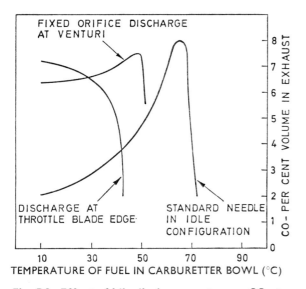

Fig. D3. Effect of idle discharge system on CO at engine idle

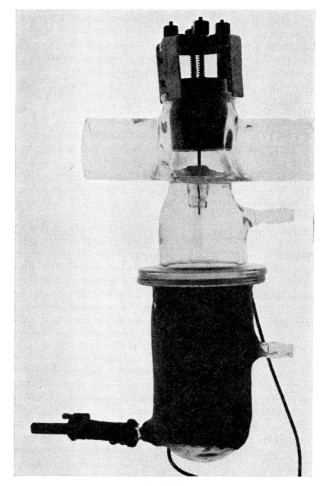

Fig. D4

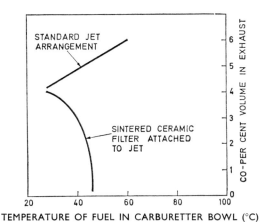

Fig. D5. Effect of ceramic filter on CO at engine idle using glass curburetter

before by Zenith engineers who have helped us. We feel that this aspect is very important because we are aware, from vapour lock testing work, that some cars have carburetters which operate at temperatures in the 70°C to 90°C region in hot climates. These are, of course, expected to idle satisfactorily. I think many of you will understand this point and perhaps the author would care to comment on it.

Secondly, we too have been actively trying to compensate for temperature and gasoline quality by various carburetter design changes and have hit upon the idea of purposely forming vapour (not air) bubbles in the jet of an air valve carburetter. To do this, we attached a sintered ceramic filter to the bottom of the jet tube through which the fuel supply to the orifice was passing and we obtained a dramatic change in the response of the CO to temperature.

With the standard jet arrangement, we observed the normal, expected increase in CO, with increase in fuel temperature, Fig. D5. With the filter at the base of the jet, the trend was completely reversed. We are progressing with this in an attempt to strike a compromise on filter pore size which will allow satisfactory, all-round vehicle performance and yet not overload the whole system with vapour bubbles. We should be able to take care of both temperature effects (such as thinning of the fuel due to increased temperature) and gasoline volatility effects by creating vapour restrictions proportional to gasoline volatility.

Both very high and very low temperatures can be experienced in practice, and we should take a much closer look at emission levels under such conditions. Blind allegiance to test cycles could lead us into major marketing problems because customer demands on our cars and fuels are often far more stringent than the requirements of the emission test procedures.

P. P. Swatman Birmingham

I think that G. L. Lawrence, in referring to his carburetter as an air valve type, does the SU an injustice. It is also confusing because there are many air valve carburetters which are not variable choke carburetters and do not inject their fuel into a high velocity region.

He starts his paper by extolling the virtues of this kind of carburetter with its single jet system and then goes on to describe a carburetter which has no less than eight additional metering orifices, many more than are found in the fixed choke carburetter which he infers gives so much trouble in service!

One is left wondering whether the reason for all these holes is not only to cure a technical problem but to redress loss of service sales previously mentioned!

In his attempt to eliminate viscosity effects, has he not introduced a greater volatility hazard? He excites us with a reference to the depressingly large number of arrangements tried for all these holes, but then leaves us high and dry and in a state of anticlimax, because he has omitted to define the critical control features. Why, for instance, has the bias of the needle been reversed in direction from Fig. 135.3 to Fig. 135.5? Which of the holes meter the air? How do you consistently remove burrs at the intersection with a precision orifice of about 0·100 in diameter? Which part of the jet does the actual fuel metering? Is the position of the cross-holes critical within the length of the jet, and, if so, how critical? Has admission of air reduced the depression signal on the metering portion of the jet and is the carburetter now more sensitive to fuel level changes and gravitational effects?

Have tests been carried out on various reference fuels to show the effect of changing from summer to winter grade at inappropriate ambients? If an inlet temperature control is necessary for reliable results, what happens when the ambient is outside the range of dictated temperature control? Does the admission of air really create turbulent flow in the jet, or does it just cause separation of fuel from the wall in the manner of a ventilated weir?

To many people, it will seem retrogressive to complicate the essential simplicity of the variable choke carburetter and its basic one jet concept in this manner. My company feels that effective control over viscosity can be achieved by moving the jet axially to the needle as a function of fuel temperature.

With regard to the section headed 'Idle "dipping"', whereas I agree that the problem is aggravated by an entrapment of fuel anterior to the throttle disc, I feel I must disagree with the author as to its cause. I submit that this dipping is not due to the inertia of the piston, as he postulates, and therefore fundamental to the variable choke carburetter, but is due to loss of metering signal due to friction between the piston rod and its bearing. This friction can vary jet depressions as much as 40 per cent during throttle blipping. My company has developed an anti-friction guide which substantially eliminates this pressure drop variation: this not only prevents the dipping problem occurring but essentially prevents peaks and valleys in CO traces during other transient modes further up the air flow range. I consider that the lack of reference to piston rod friction is a serious omission from the paper. Perhaps the author would comment.

Towards the end of the paper, reference is made to a fixed needle-jet relation. Since the needle is moved relative to the jet, could we please have an explanation of this apparent paradox? Even with the somewhat familiar throttle disc by-pass emulsion arrangement I understand that the throttle has to be moved to trim the speed, which means that even at idling there is not a fixed relation between the needle and the jet.

Discussion during Session 3

W-T. Lyn Columbus, Indiana

I shall confine my remarks to Papers C140, C142 and C151.

Paper C140 provides a general qualitative background of the mechanism of carbon formation. It will be recalled that it was for this very purpose that a special symposium (*1*) was organized in 1961, and I, therefore, suggest that this particular proceeding might be usefully included in the reference. While I do not deny the usefulness of such an exercise, the difficulty in applying to a specific case is well known, due essentially to the lack of detailed knowledge of the local physical conditions, such as pressure, temperature concentration and how these change with time. For this reason, little specific conclusion can be drawn.

Khan and Wang, Paper C151, report experimental results on the effect of injection timing, injection rate, some nozzle configuration and swirl on emission and generally confirm the results reported in (*2*) (*3*) (*4*) and Paper C124 in this Symposium. Of particular interest is the effect of injection rate and swirl. In the swirl study, the authors had kept all other variables constant and found, quite reasonably, that the NO concentration went up with swirl. My question is that if the nozzle configuration was optimized at each swirl level for efficiency, would the authors still arrive at the same conclusions? Have the authors done such experiments? If not, what are their views? Obviously, this is quite an important issue.

Paper C142 is the first paper on smoke model and third paper on NO model (*5*) (*6*) that I am aware of. My own view is that our understanding of the physical basis of smoke formation and depletion is far less certain than that of NO and until that is made clearer, a satisfactory working model would be difficult. On the NO model under discussion, my main objection is the way in which the O_2 concentration in equation (142.2) is defined, i.e. $[O_2] = [O_2]_0 - [NO]/2$ and $[O_2]_0 = 0.21(1-\phi_p)$; that is to say, the $[O_2]$ available for NO formation is limited to the excess oxygen from the mixture under consideration. The consequence of such assumption is that there would be no NO formation until the mixture is leaner than stoichiometric, as is shown by the phase lag of some 20° C.A. of the NO and ϕ_p curves in Fig. 142.16. This, of course, is not true because firstly we do have NO at rich mixture and secondly there is experimental evidence, both in the case of petrol as well as diesel engines, that the maximum NO formation rate concurs with the beginning of combustion and pressure rise.

From the data presented by the authors of the three papers, and from what one of the delegates has told us about the stratified charged engine (after all a stratified charge engine is essentially a diesel engine with spark ignition!) and some of the results regarding pollution control, I, personally, feel that diesel engines are still a very attractive power plant in this day of ever-increasingly stringent anti-pollution requirements.

REFERENCES

(*1*) Five papers on the Mechanics of formation and adhesion of deposits arising from the combustion of liquid fuel, I. Mech. E. Conference, 1961.

(*2*) MARSHALL, W. F. and FLEMING, R. D. 'Diesel emissions as related to engine variables and fuel characteristics', S.A.E. Paper 710836, *S.A.E. National Meeting*, St. Louis, October, 1971.

(*3*) BASCOM, R. C., BROERING, L. C. and WULFHORST, D. E. 'Design factors that affect diesel emissions,' S.A.E. Paper 710484.

(*4*) ABTHOFF, J. 'Measurement and variation of emission of oxides of nitrogen from i.c. engines,' *Research Report* No. 2-216/1, 1968, **79**, Combustion Engine Research Association, Frankfurt Main.

(*5*) BASTRESS, E. K., CHNG, K. M. and DIX, D. M. 'Models of combustion and nitrogen oxide formation in direct and indirect injection compression-ignition engines,' Paper 719053, *Intersociety Energy Conversion Engineering Conference*, Boston, 1971.

(*6*) BRACCO, F. V. 'A model for the diesel engine combustion and NO formation,' Combustion Institute Central States Spring Meeting, March 1971, Ann Arbor.

F. J. Wallace Bath, Somerset

Papers C140, C142 and C151 have been remarkably informative and provide a veritable feast of information.

Papers C151 and C142 are complementary, the one being based entirely on experimental results and the other seeking to give us a working model.

The authors of Paper C151 have explored those parameters, such as injection rate, injection timing, engine speed and so on, which suggest themselves as the natural ones in the case of the normally or naturally aspirated engine. Can they give a brief preview of what we might expect from high boost engines? We know that increased charge density has a fairly profound effect on the combustion pattern in general. It reduces rates of pressure rise and it reduces delay periods, so there is a beneficial effect for a given quantity of fuel injected, or even for quantities which go up proportionate to charge density. I would like to know about the general trends regarding nitric oxide and soot formation under otherwise similar conditions, e.g. similar engine ratings, swirl speeds, etc., as the boost is increased. One often hears that the development of the very high output engine—the very high boost engine with very high rates of fuel injection—has to some extent been arrested by the claims of impending legislation. Where would boost levels of the order of three to four atmospheres lead us?

With regard to Paper C142, W-T. Lyn has just questioned the complexity of the two crucial equations for soot formation and NO formation. The authors have indicated that they do regard these equations as forming a useful working basis. With what degree of confidence would they apply these two equations to the special case of the high boost engine?

W. Jellings Cambridge

I wish to question the authors of Paper C142 on their assumption that 'The time for droplet evaporation is negligible'. The reason, I suggest, for this assumption is that it makes the problem easier to solve and not that droplet evaporation times are in fact negligible. This, to me, is not a convincing basis for even the simplest model in such an important area of research.

I am convinced that the part played by fuel droplet size, in the mechanism of emission formation within a diesel engine, is of the utmost importance. By using a device which generates a spray of uniform size droplets and known mass distribution across the spray, I intend to demonstrate the aforementioned statement in order that the fallacy of negligible droplet evaporation time will cease to perpetuate.

M. S. Janota London

There is evidence from steady flow rig experiments that the fuel droplet size has only a secondary effect on the carbon formation rates in burning fuel sprays. Have the authors of Paper C142 observed a similar relation between the fuel droplet size and carbon formation rates in their non-steady diesel engine experimental work?

W. Tipler Peterborough

Papers C151 and C142 contain a great deal of useful experimental data, but their value is limited by omission of one piece of information, namely the properties of the fuel used in the engine tests.

Fuel properties may be varied in any location in order to meet changed requirements imposed by climatic changes with the seasons; similarly these properties vary between geographic locations; in addition fuel properties can vary as a result of changes in the crude oil with which a refinery is fed, or of changes in the product demand pattern which the refinery must meet.

To assume that the fuel is the same simply because successive batches comply with B.S. 2869:Class A is supremely optimistic in view of the wide limits set by that standard. For example, if one refers to Paper C137, three of the five fuels quoted, namely A, E and F, would appear to comply with the above-mentioned standard, yet the same paper shows that these fuels give appreciably different levels of emissions.

I would, therefore, recommend most strongly that research workers should take samples of their test fuels for detailed chemical analyses at regular intervals, to ensure that uncontrolled variables are eliminated from their work. This information is also needed in order to make valid comparisons between research conducted in different locations.

T. J. Williams London

I have one short question to ask relating to Paper C151. In figures such as Fig. 151.10, the authors give results relating to high, intermediate and low levels of swirl using different valve mask positions, but they have not given any information relating to the use of an unmasked valve. Shrouding can produce both swirl and turbulence, which can be quite different in their influence on combustion. It may well be that the results should have been related to turbulence rather than swirl and that the function of the mask was to produce turbulence in this instance in an otherwise swirling flow.

D. K. Roberts Southall, Middlesex

I would like to make some comments on Paper C151. The suggestions put forward for exhaust opacity improvement incur significant engine design changes and these could have an adverse effect on the noise level of the engine. There would be little point in improving a vehicle from one aspect of its effect on the environment whilst causing a deterioration from another standpoint. Can the authors advise whether the higher injection and swirl rates proposed affect the noise level of the engine, also what the effect is on maximum cylinder pressures?

Although four engines were used by the authors, they were all of the same order of size (one litre per cylinder), had a constant compression ratio of 16:1, and the combustion chambers were similar, as indicated by the relatively constant piston bowl to cylinder bore ratio. Can the authors state whether they would expect similar emission improvements to apply to other cylinder capacities, higher and lower compression ratios and to differing configurations of combustion chamber?

A. E. W. Austen London

My comments are not on the practical application of this emissions work, on which we have said about as much as we know at present. Paper C140, however, has received rather faint praise and for it I cannot escape some measure of responsibility, since, as they have disclosed, I was concerned in setting it going.

Some three years ago we recognized that there was a great deal of information in the literature of combustion, from academic sources and research institutions on carbon formation and combustion. We felt that some of this might have a bearing on carbon in diesel engines. So we set going on the subject people we were sure were competent to extract any information relevant to the diesel problem.

It is true that, having reviewed all the information very carefully, they came to the conclusion that although the basic mechanisms were established, no great light was shed on the solution to the carbon in diesel engines problem. Nevertheless, having made this review it was in my view well worth putting a brief account on record, if only to save others from having to do it all over again.

G. Kershaw Shrewsbury

We have followed a similar line to that of the authors of Paper C151 with respect to smoke and fuel consumption, but not NO_x on an automotive engine. Using a directed port, we have increased the swirl level by using a mask trapped behind the valve seat insert, and varied the level by varying the size of the mask. Optimizing the swirl ratio has enabled us to burn an 18:1 air–fuel ratio for a smoke level of 1 Bosch unit at one speed, but with a variable speed engine, the swirl level has to be a compromise between high and low speed operation. The high swirl level suits the low speed and vice versa. Too high a swirl level produces smoke at all loads but especially at light load where it becomes a mixture of blue and black.

K. Lea Leyland, Lancs.

The authors of Paper C137 have raised a point upon which I would like some clarification. Has any work been done yet on the effect of additives on wear rates, e.g. with regard to exhaust valves?

K. Alcock Leeds

My company makes diesel powered machinery for underground working in mines and all the problems which now affect automobile engineers have been concerning us for about 30 years. We have done a considerable amount of research on fuel additives, including isoamylnitrate, which we have now dropped. Two problems led us to discontinuing this research. First, we measured a big difference of our results when the additive had been mixed with the fuel just before the experiment and when the additive had been mixed with the fuel and left to mature for several days before carrying out the experiment. Has the author noted this phenomenon? Second, although the fuel additive tested by us reduced considerably the measurable levels of carbon monoxide and NO_x the exhaust actually smelt far worse than when running on conventional fuels. Have the authors noted this fact and, if so, has the source of the smell been identified?

I understand that the NO_x tubes for the Drager system are affected by moisture. Has the author or any of the audience any quantitative knowledge of this fact?

We have carried out some preliminary work on water injection which is well known to be a very good way of reducing pollutants. Have any of the authors any experience regarding engine life when the water injection system is used?

Do any of the authors have any quantitative information regarding hydrocarbons, which are present in diesel exhaust, adhering to the smoke particles? In the Hunslet system we wash out most of the solid particles in the diesel exhaust by passing the gases through a water scrubber. The large quantity of sludge formed in the scrubber even during relatively short periods of running is a measure of how much solid particles are removed.

N. F. Crouch Sheffield

I have been in this field for the past 4 years, but I have only started research a month ago. In paper C142, the calculation of the rate of formation for NO is based upon the Zeldovitch chain mechanism. There are a few recent papers which demonstrate that this mechanism could be unrepresentative. Most computational methods to predict the amounts of oxides of nitrogen which will be formed in any system consistently underestimate the amount formed in practical systems, and it is thought that a modified mechanism will explain why this is happening. The Zeldovitch mechanism is thought to be incorrect with regard to the first equation at the top of page 207; the mechanism assumes that the oxygen radicals are equilibrated with the oxygen molecules, but it is now thought that this is not so, and a modified mechanism (7) can account for this non-equilibration.

REFERENCE
(7) THOMPSON, D., BROWN, T. D. and BEER, J. M. 'The formation of oxides of nitrogen in a combustion system.'

I. M. Khan London

I would like to point out that the explanation of the effect of cetane number on exhaust smoke given by the authors of of Paper C137 is not at all in keeping with the known facts. It is clear that for a given fuel the formation of soot in a diesel combustion chamber depends upon the amount of completely mixed fuel at ignition (or the amount of fuel burning as a diffusion flame), and its equivalence ratio and temperature history throughout the burning period, (8) and Papers C151 and C142. When a fuel of a different chemical structure is used or when an additive is mixed with a fuel the chemical reactions leading to soot formation and their reactions are altered, Paper C140. These changes for a given set of mixing and temperature conditions will be expected to lead to different rates and hence amounts of soot formation. This is clearly shown by the results obtained by the authors while using cyclohexane as an additive (see Fig. 137.2). It can be seen that there is no change in ignition delay (and hence in mixing conditions) but there are appreciable changes in smoke emission. The other results obtained with fuel of different cetane numbers (changes obtained either using different fuels or by mixing additives with fuel) shown on Fig. 137.2 are explained as follows: A reduction in ignition delay with an increase in cetane number increases the amount of fuel involved in diffusion burning. This is expected to increase smoke formation and emission. The fact that with an increase in cetane number of up to 55 or 60 the exhaust smoke decreases with an increase in cetane number can only be explained by the fact that the reduction in smoke formation due to changes in the chemical nature of fuel overshadows the increase due to the reduction in ignition delay. An increase in cetane number beyond 60 causes an increase in smoke emission and this is explained by the fact that under these conditions the increase in smoke emission due to the increase in the amount of diffusion burning is preponderant. My own experiments on the use of isopropyl nitrate and their analysis presented elsewhere (9) is in line with the above explanation.

The experiments of Burt and Troth (10) with a number of fuels each containing the same class of hydrocarbons giving different cetane numbers are also in agreement with my remarks on the effect of ignition delay and chemical structure of fuel on smoke formation and emission. Burt and Troth have shown that when fuels containing the same class of hydrocarbons are used an increase in cetane number increases smoke.

REFERENCES

(8) KHAN, I. M. 'Formation and combustion of carbon in a diesel engine', *Proc. Instn mech. Engrs* 1969–70 **184** (Pt 3J), 57.
(9) KHAN, I. M. Reply to discussion, *Proc. Instn mech. Engrs* 1969–70 **184** (Pt 3J), 284.
(10) BURT, R. and TROTH, K. A. 'Influence of fuel properties on diesel exhaust emissions', *Proc. Instn mech. Engrs* 1968–69 **183** (Pt 3E), 171.

Discussion during Session 4

D. Downs Shoreham-by-Sea, Sussex

My contribution is not so much a question to the author of Paper C153, but a comment. We have long admired on this side of the Atlantic the work of the Co-ordinating Research Council (C.R.C.) in the United States, and its manager is a familiar figure in this country and in Europe. Some 16 years ago, in the aftermath of the World Petroleum Congress in Rome, we tried, with his help, to get something similar to the C.R.C. started in this country. At that time we failed, but some years later the British Technical Council of the Motor and Petroleum Industries was set up and we now have, through C.E.C., similar organizations in Europe. However, when we compare the activities of the British Technical Council (B.T.C.) even now, seven years after its formation, with those of the C.R.C., we cannot help feeling that what we are doing is rather inadequate. What I am asking is whether we in this country and our friends in Europe, through C.E.C., ought perhaps to take more account of the work going on in the C.R.C. We could see for instance whether we could get some more meaningful programmes going in Europe through the co-operation of the motor and petroleum industries and—an important third partner—the governments. I should like to suggest that the B.T.C. and the C.E.C. take this very much to heart and try to see if we can get some co-operative research programmes going in this country and in Europe along the lines of the C.R.C. programmes.

D. G. Burton London

I should like to comment on Fig. 134.8 of Paper C134 which was also shown as a slide. I think it has been incorrectly marked because it is not consistent with the words used in the text. The diagram indicates that the 8-holed nozzle is less sensitive to swirl than the 6-hole, but the text says the opposite. Could there be something wrong with the key or the indications on this diagram?

T. J. Williams London

One of the most important points made in Paper C134 is that very similar conditions exist during compression, whether the engine is firing or not. For that reason, the results obtained using hot-wire anemometry with motored engines gain greater validity. The spark probe itself has an advantage in that it measures velocities directly, whereas the anemometer measures several quantities, velocity, density variations and temperature. However, looking at a picture of the device in Fig. 134.2b, it does appear liable to interfere with the flow to quite a degree and to distort the flow as it comes over the spark gap and moves towards the probe. This distortion appears to be shown in the calibrations in Fig. 134.3, where the results tend to be unsymmetrical with regard to the zero position. In addition, I find it very difficult to understand why one gets a constant reading as probe A is rotated in the diagram on the left. One would have expected some sort of variation with the angle θ.

Turning to Fig. 134.4, there are variations shown between the fired engine and the motored engine during the induction period. One explanation for this is that the probe itself is not in a very typical position as regards the induction process; judging from the diagram, the probe is placed a little way from the stream of air coming from the inlet valve into the cylinder; and one could expect a great deal of eddy motion during induction in this particular region. If readings had been obtained lower down in the engine cylinder, these might have shown better agreement between motoring and firing.

The results shown in Fig. 134.5 with a cavity in the piston exhibit an increase in cylinder velocities during compression. I was hoping that these results would have told us whether squish existed or not. On the whole, the results are not very informative since the squish velocity as calculated is generally quite negligible in comparison with the measured velocities. It would have been helpful if the authors had attempted to calculate the increase in velocity expected as the mass of air transfers from the main cylinder body into the cavity, assuming the law of conservation of angular momentum. Are these measured increases comparable to what one would expect from such a calculation?

Discussion during Session 5

W. J. D. Annand Manchester

My comments are on Paper C149. All of us who try to do calculations concerning NO are really just trailing along in the wake of E. Starkman and his absent collaborators making use of what they and other American teams have produced at Berkeley, Wisconsin and M.I.T.

I should like to ask the author of Paper C149 if he would explain something about Fig. 149.3, which shows the results of an analysis from, I take it, measured pressures in an engine. This seems to show 10 per cent of unburnt material persisting right to the end of the expansion. Are we to read this as really meaning 10 per cent of unaltered material in the chamber or rather that 90 per cent of the expected energy release has been attained? I find it hard to believe that there is 10 per cent of unburnt mixture there.

Concerning the combustion pictures (Fig. 149.8) I should like to know more about the photographic technique used. We have tried to take such photographs and have found that we could not achieve sufficient luminosity to get a reasonable photograph. I wonder if perhaps a special film was used for the purpose.

We have made calculations very similar to those of the authors but the difference is that our programme is written to compute the combustion process more or less from scratch; it does not use a measured pressure diagram to start from. We have embodied the type of reaction scheme which Lavoie set up and use rather more equations than does the author. Our output figures for nitric oxide use single-zone models as does the author. For multi-zone models very large computers and a lot of spare time is needed. Unfortunately, we ran out of both space and time. However, with single-zone models we get quite sensible numbers, perhaps because, as E. Starkman's Fig. 138.5 showed, the mean value of NO is very near to that found in the '50 per cent burned' slice of the charge. Our output numbers are very like those of the authors for the variation of NO with the fuel–air ratio in similar cases. The problem of getting matching figures on the weak side is that the quantities one gets seem to be more sensitive to the flame travel time (affecting the peak temperature) and to the amount of residuals one thinks one has got to start with, than on the rich side, so that better agreement might be obtained merely by adjusting the assumed values of these quantities, within the limits of the uncertainty with which they are actually known.

Referring to Paper C122, the measurements presented are very interesting; I think there is no doubt about what they show, that the addition of water does reduce the NO. I would not like to comment on the practicability of this line of attack; I am more interested in the author's analysis and I find it very hard to understand as something about it seems inside out, or wrong way up. I should have thought that the only thing water does is to reduce the cycle temperatures. As far as I know, it has no direct influence on either flame propagation speed or the rate of reactions that control nitrogen oxide. There are some calculations which I think support this idea; whether you use exhaust recirculation or water, what you are doing in both cases is to reduce the temperature. The analysis presented does not seem to show this; we see higher temperatures with an increase of water, even if we look at points taken at the same fuel–air ratio, which is a parameter as well as temperature. Yet we see that the flame speed is reduced when water is added. I cannot understand why this happens unless the temperature is lowered; there is no other way I know of in which flame speed could be reduced. I would be interested to learn what the author thinks is the mechanism here.

This leads me to wonder whether the analysis method should be re-examined, and possibly also the measurement of pressure. We have had some difficulty in getting reliable pressure measurements in spark ignition engines, even with the recent sophisticated and accurate pressure transducers; they are still subject to thermal shock, particularly at high speed, in spark ignition machinery. I would like to know whether any particular precaution was taken to protect the transducers from the transient temperature of the flame.

I am rather troubled by the argument that water can be neglected in the analysis since, to my mind, the only thing water does (if it does anything) is to reduce the temperature. Synthesis calculations which I have made show that the amount of water vapour used in the present case lowers the temperature rise during compression perceptibly, and reduces the temperature of the burned gas appreciably, even if the reduction of intake temperature due to evaporation of the water is neglected. The evaporation alone should produce a measurable reduction also. The calculations indicate, too, that the general level of the peak burned-gas temperatures quoted in the paper is improbably low. These results support the view that the analysis method should be critically reviewed.

W-T. Lyn Columbus, Indiana

I would like to make a comment on modelling in general. About ten to fifteen years ago, cycle simulation, that is, thermodynamic analysis of the events in the cylinder, was quite fashionable. Unfortunately, it did not produce many applicable results so the industry in general has rather neglected it. Then came the problem of emissions and everyone started rushing into cycle simulation again because it is an essential part of the emission prediction. The important point is that whatever the emission model, one needs a very accurate thermodynamic input to that model. Even now, we do not know how accurate our thermodynamic model is and I was surprised that W. J. D. Annand did not pick up the particular point of uncertainty regarding heat transfer. We know how sensitive NO is in relation to temperature, where 100°C can make all the difference. In the gasoline engine, the cycle analysis is perhaps easier, but even then I doubt if the model is general enough. If one looks at tomorrow's MIRA paper and tries to apply the gasoline engine model to all the different chambers, I think one will get a surprise, because of the uncertainty in the flame temperature, particularly near the wall. It is worse still with diesel engines and it is very difficult indeed to make a correct prediction of the local flame temperature. I hope W. J. D. Annand could put a lot more effort into this area. I feel most strongly that until we are more sure of our cycle analysis, which should include local flame temperature, we cannot push very far ahead with emission prediction.

W. Tipler Peterborough

At first consideration, the concept of the importance of the location of the exhaust valve in reducing emissions looks extremely promising, but I have one query. If one retains a higher NO concentration in the cylinder gases, then surely one starts the next cycle of combustion at a higher base level of NO than would otherwise be the case? Could the authors of Paper C138 please explain what is the effect of the base level of NO on the overall emission of the engine?

E. S. Starkman dealt with this point rather quickly but it appears to me that his suggestion can only be valid if the NO retained in the cylinder is entirely dissociated in the next cycle. Is there any evidence that this is in fact the case, please?

D. R. Blackmore Chester

Fig. 138.5 of Paper C138 interests me, particularly with respect to the first burned curve—I should like to know more about the shape and position of this. What experimental evidence is there for the *decrease* in NO, which I presume arises out of his mechanism? The position of this whole curve is no doubt a function of the oxygen atom concentration, which appears to be calculated on an equilibrium basis. The practical implication is that local high concentrations may be the hub of the problem and these high concentrations in the first-burned fraction are very dependent on any decrease in NO. If one could cope with these high concentrations, one could make substantial practical reductions in NO_x levels.

C. L. Goodacre London

In Fig. 138.9 of Paper C138 it suggests putting the exhaust valve into the 'end gas zone', and the authors wonder why Chrysler took no notice of this suggestion. Surely if one does this one is breaking one of the golden rules of engine design, in that you must initiate combustion in the engine from the region of the exhaust valve, or as near as possible, otherwise if you have the exhaust valve in the end gas zone you run into high speed detonation, which very soon turns into pre-ignition leading to piston failure. Did the authors, with this arrangement of exhaust valve in the end gas zone, run the engine at high power, under lean mixture conditions?

With regard to Paper C122, did the author notice, with water injection, an increased rate of engine wear? We have made several experiments in the past with water injection in race engines and aero-engines and whilst we agree that water is a very good anti-knock and pre-ignition suppressant we have found that the rate of engine wear, even with 10 per cent water, can be catastrophic.

P. Draper Dorset

I should like to address my questions to the author of Paper C122. First of all, did the author employ pure distilled water or did he have any effect from deposits in ordinary water? The answer to this makes the point one of academic or practical use. Secondly, has he any information on the effect of the water injection on the octane requirements of the engine, because this could prove a bonus?

S. Papez Wolfsburg, Germany

I should like to comment on Paper C122. The phase position of the gas oscillation in the exhaust pipe at the moment when the exhaust valve is closing governs the amount of residual gas in the cylinder.

By optimizing this gas oscillation the amount of residual gas can be increased for a desired range of rotational speed, so that one can influence the NO_x emissions in the exhaust gas (Fig. D6).

Did the author measure this influence?

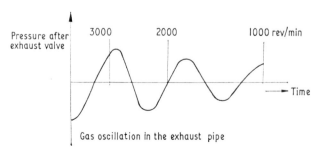

Fig. D6. Gas oscillation in the exhaust pipe

P. K. Doerfler Baden, Switzerland

Factors that lower the heat release rate, such as making the mixture more lean or increasing the residual gas fraction, generally increase the cyclic variations of cylinder peak pressure and the indicate mean effective pressure. Did the author observe such an increase when he applied water injection?

R. Lindsay London

I would like to ask the author of Paper C148 whether the changes in exhaust valve timing caused variations in pulse effects, etc., which in turn altered the pattern of the mixture going into the different cylinders; also whether the air-fuel ratio of the different cylinders was changed as a result of changing the exhaust valve timing and whether this in itself would cause some quantifiable effect on oxides of nitrogen?

Discussion during Session 6

S. E. Voltz Paulsboro, U.S.A.

I have a question for the authors of Paper C144. Have the authors tried any monolith catalysts, or do they plan to do so in the future?

In the I.I.E.C. programme, we have developed mathematical models for thermal reactors and CO–hydrocarbon converters. Some work on CO–hydrocarbon catalysts and converters was presented at the S.A.E. meeting in January, 1971; we recently determined the kinetics of CO–hydrocarbon oxidation on platinum catalysts and will publish the results soon. A considerable amount of research has been started in American universities on monolith catalysts which is of a fundamental nature and includes reaction kinetics, heat transfer processes and fluid dynamics.

J. H. Alden Luton, Beds

With regard to Paper C144, I am not entirely clear from the remarks which have been passed of the nature of the pellets used. Are they pellets of metal or do they have some medicinal properties, and were all the tests carried out on unleaded fuels?

C. L. Goodacre London

I would like to ask the author of Paper C146 if he noticed any increase in the rate of engine wear with the reactors, with exhaust gas recirculation, and whether he has done any realistic runs with thermal reactors of the kind a commercial traveller might do when travelling from, say, London to Edinburgh or Glasgow at an average speed of, for instance, 50 mile/h, running normally, perhaps cruising at 80 mile/h, in a car with a 1·2 l engine.

With regard to Paper C144, have the authors subjected their catalytic afterburner to sub-zero temperature conditions and, if so, have they found a loss of catalyst?

Going from the sublime to the ridiculous, had there been no pollution of the environment in the last hundred million years, there would have been no crude oil which, as you know, is derived from decaying bodies and vegetation, marine and land-borne.

M. A. I. Jacobson London

I have a practical point which I would like to address to the authors of Papers C130, C146 and C144. Has any of the hardware been subjected to a M.I.R.A.-type test of a thousand miles of Belgian pavé to check it for durability under all conditions of rough road vibration?

At the risk of being considered even more provocative than I intend to be, I want to throw out a straightforward challenge. I contend that it is not in the interests of motorists in Europe to support the extravagant demands of the American market. I would go as far as to say that it may well be that European and British manufacturers cannot meet the market requirement after 1976 in a number of other respects either. It is not only a question of the air pollution problem; there is the matter of crash survivability, of small to medium sized cars. The question surely must be asked whether there does not come a point at which European manufacturers should agree amongst themselves that Europe must set standards of its own and that the effort and the considerable cost involvement of man hours should be considered on the European problem. Otherwise, I foresee that the taxpayers of Europe, more particularly the motorists of Europe, will ultimately have to support, to the tune of something like £50 per car, the cost of development work for perhaps what may turn out to be a rapidly shrinking and marginal American export market for European cars.

There surely should be some understanding of the range of real problems involved. If Europe is large enough as a total market, it can have its own sensible set of regulations which will meet the requirements for the next five, ten or fifteen years. By that time one hopes that some sense of realism will have taken hold of the American legislators and that it will not merely be a case of 'We can do it in the laboratory; therefore let us attempt to do it in mass production, regardless of cost'.

J. S. Campbell Teesside

I wish to comment on the question raised as to whether or not the catalysts we are working with at the moment had been exposed to sub-zero temperature. This has not been done on a car, as J. H. Weaving pointed out, but this is one of the tests which we apply to catalysts in the laboratory. We cycle the catalyst between ambient and sub-zero temperature in the presence of a certain amount of water vapour in order to see whether the catalyst breaks up or suffers other damage. I have not got the figures with me, but the results of this kind of test are very promising indeed; the catalysts in which we are most interested at the moment show no signs of attrition or break-up.

G. L. Lawrence Middlesex

I should like to ask a general question of the authors of Papers C130, C146 and C155. In view of the 1975 legislation, meaning 1975 model year production very early in 1974, and in view of the amount of work which stands between the present state of the art and the actual production of the system, is it considered that it will be possible to reach a developed system in time for production, or will 1975 need to be met with less exotic treatment?

E. S. Starkman Michigan, U.S.A.

I should like to make some comments from my own experience. In 1970, one of the University of California 'clean air cars' went from M.I.T. to Caltech equipped with an exhaust manifold reactor that had an insulation of ceramic material. When it arrived in Detroit, it could no longer be started; on being disassembled, the cylinder bore wear was found to be excessive. Apparently the ceramic liner of the thermal reactor had been swallowed back through the exhaust valves and into the combustion chambers. The ceramic was left out; Chrysler Corporation very kindly provided a new short block and the engine back was put together and very satisfactorily finished the race.

But beyond that, the question of thermal reactors should be subjected to analytical treatment along the lines that were described this morning in a context of what might be called 'implicit' solutions. We have been discussing external or 'explicit' solutions to the problems this afternoon. I would like to see more attention given to the way in which the products of combustion are treated chemically. We heard earlier of the A.P.R.A.C. programme. A part of this has to do with obtaining some good data on the chemistry of the reactions that take place between hydrocarbons and oxygen in the exhaust system.

I would be interested in knowing to what extent the chemistry and thermodynamics and fluid mechanics have been utilized in optimizing exhaust thermal reactors. I am sure that some application has already been made, and I look forward to more work in this direction because whether we cure the problem explicitly or implicitly, optimization can be effected with application of available knowledge.

The situation in Los Angeles in the early 1950s was not very much different from what I have been seeing and hearing with respect to Europe within the last two weeks: the contention that there was no problem, that really this whole business was imagination. I can assure you it is not.

Whereas people cannot be proved to be dying from air pollution, I think the population in general is willing to pay for a better environment (at least this is true for the United States). As far as Los Angeles is concerned, the discoloration of the atmosphere had more to do with the promulgation of legislation than did a list of deaths or even illnesses published in the papers, because the latter could not be proved.

I had the occasion last week of flying into Paris through layers of air which had a familiar appearance. They were very much like the layers of air over Los Angeles and San Francisco and Madrid and Mexico City, and other places.

We have also heard that in Fleet Street the oxides of nitrogen, nitric oxide specifically, on some occasions get up to very high levels. Not all of this can be coming from automobiles.

It may be that the value of trade considerations for European automobile manufacturers may mitigate against attempting to satisfy the present host of requirements for American automobiles. If the rest of the United States wishes to follow the State of California, this is for their own politicians to decide. But, in any event, I think that what will come out of this, if nothing else, is a means for decision on how much Europeans are going to be willing to pay to decrease pollution, against a background of increasing vehicle populations in Europe. I do not think that the effects on this side of the Atlantic are wasted. I believe a point will come where it will be decided that in order to get rid of the whisky-brown colour in the sky, and perhaps to reduce pollutants, carbon monoxide and so forth, some controlling of vehicles in specialized locations, if not on the whole, will be necessary.

I do not want to leave the impression that I am in favour of the present Federal pollution standards. I want to make that clear. The standards which existed before the most recent were ones in which I had a hand in promulgating. These previous standards were based on the discoloration of the air, on the formation of oxidants, and we thought that this was good enough for California for the interim period. Those people still on the committees on which I served will agree with me that the stringency specified for 1975 by California was sufficient.

I can only speculate on the reason for the differences in ratio of NO against NO_2 in London and California. Los Angeles is blessed with sunshine, which attracts people and which also provides high temperatures and for the manufacturing of ozone in the atmosphere. The combination of high temperature and ozone enhances the conversion of NO to NO_2. I think it is basically a question of different climates and environments, when one compares Los Angeles and London.

H. Connor London

We have examined some of the chemistry of the catalytic oxidation mechanisms of hydrocarbons and carbon monoxide in a catalytic system which was intended for nitrogen oxide reduction. We found that the ammonia produced from nitrogen oxide reduction would appear to stem from the pressure of water vapour. We synthesized the exhaust gas and in some instances added water vapour; when it was not added, we found no formation of ammonia and concluded that it was the reforming reaction of hydrocarbons with water vapour on the catalyst leading to the formation of hydrogen which reacted with the nitrogen oxides to produce ammonia. Our development work has led us to the formulation of a catalyst which intrinsically suppresses this primary reforming reaction, successfully enabling us to avoid the formation of ammonia. Maybe this is part of the examination process which E. S. Starkman has in mind.

J. J. Mikita Wilmington, U.S.A.

In connection with the question which has been raised on engine wear we have had some experience which I would like to recount for you, because it might be of value to those people seriously considering E.G.R. systems. In 1970, we supplied to the California Air Resources Board twelve Chevrolet cars equipped with 350 in^3 displacement, V-8 engines. Half these cars were standard and half were equipped with a thermal reactor system, including exhaust gas recirculation. The system-equipped cars had CO emissions of 8 to 10 g/mile by the seven-node California cycle and NO levels of about 1 g/mile. The cars were operated in normal service, and after nine months—roughly 20 000 to 25 000 miles—all the cars, including three other cars each of which had operated for about 30 000 miles on normal-type driving, showed excessive wear. Excessive wear occurred almost exclusively in the timing chain, on the cam followers and on the tips of the rocker arms; these are the points of high lubricating stress. We traced the difficulty down to the fact that the E.G.R. system was inhaling sub-micron metal oxide particles from the exhaust system into the engine, and these particles were of such a fine nature that they got into the lubricating oil and from there down into the crankcase. It now seems apparent to us that if one wishes to use E.G.R. taken from the exhaust system, then the complete exhaust system should be made of oxidation resistant metal. The other alternative is to take the exhaust gas from the exhaust cross-over passage as is done in the 1972 Buick. I should like to point out that the interior cores of the thermal reactors used in the California field test were 310 stainless steel which is far less resistant to oxidation than Inconel 601. This latter steel was not available when the reactor cores were constructed for the California cars.

In connection with thermal reactors, I should like to make the observation that all the work done to date, our work and that of others to the best of our knowledge, has really been with tack-on devices. As E. S. Starkman said this morning, we are now in a new ball game, and we have to get away from the idea that this problem can be solved by tacking something on to the engine. We must go further. On thermal reactor systems, we believe it is essential to place the thermal reactor as close to the cylinder head as possible, which means that cylinder heads will have to be re-designed. It is very important to tailor the air–fuel mixture carefully to get best results in each of the operating modes of the emission test cycle and to be able to do this we are becoming convinced it will be necessary to use an electronic fuel injection system and, in addition, a fully electronic ignition system. My point is that we must get away from the idea that we can simply tack something on cars to meet the standards. I think we shall have to go much further.

In our laboratory with our latest thermal reactor systems on V-8 engines, we are able to run the 1975 C.V.S. procedure and get hydrocarbons down to ridiculously low levels with no problems; we get CO down to about 5 g and NO to 0·5 g/mile. We can do this repeatedly, but at a large sacrifice in fuel economy, about 25 per cent. We think we can do better with close-coupled reactors (new cylinder heads), insulated exhaust ports, electronic fuel injection and fully electronic programmed spark advance and longer duration spark.

D. Downs Shoreham-by-Sea, Sussex

What I have to say is not so much in the nature of a question, but rather a comment which I think is of importance. It is clear that some of the more promising catalytic systems, for both reduction and oxidation, require the use of unleaded fuel. The big question is: how unleaded is 'unleaded'? We know the importance of trace amounts of lead in fuel from the point of view of the performance and in particular of durability of some of these catalytic systems. We have not only lead but also phosphorus and sulphur in the fuel and the various metal additives barium, calcium, zinc, etc., in the lubricating oil. There is very little sign of work being done, on this side of the Atlantic, to study this very important question which is of interest to both motor manufacturers and oil companies.

One of the difficulties is that the motor manufacturers are exporting motor cars to the United States and, therefore, have an interest in the American situation, but the oil companies on the other hand are not exporting their petroleum to the States and, therefore, have only a limited interest in the American situation. I mentioned earlier that we should make better use of the B.T.C. and it seems to me that this is an ideal case to be referred to it for a joint study by the motor and petroleum industries, i.e. the specification of the fuel and of the lubricating oil to permit catalytic devices to give optimum performance.

J. P. Soltau Birmingham

I would like to thank the authors of Papers C121, C147 and C139 for their interesting papers. Nevertheless, throughout the discussion I had the impression that I had heard and seen it all before.

We know thermal reactors work, we know flame afterburners work, it was shown ten years ago. We know that oxidizing and reducing catalysts work, we know that lean carburetter and fuel injections work. Is it not time that we pooled our knowledge and developed reliable commercial products, instead of proving over and over again basic accepted principles?

Perhaps there is a breakdown in communications, or are the new developments vastly superior to the work done a decade ago?

P. Draper Dorset

One is generally asked to state one's affiliation to a commercial interest; fortunately, I have none, as you will see from what I am going to say.

Papers C121, C147 and C139 show the absurd nature of the 1976 specifications, which are not based on any real requirement. I think it is a great pity to waste the efforts of these eminent research engineers present when they might be doing much more useful work. The authors of Paper C147 have created a whole lot of complicated apparatus and I shudder to think what will happen if and when that gets into the hands of garage mechanics.

I am a great believer in clean air, and I think we should prevent any forms of pollution, in so far as methods of doing so are reasonably practical. I do not think the 1976 specification is practical. We have a number of organizations working on this. Could they not get some sense into the heads of the legislators? I would press for that very hard indeed. The whole subject of air pollution has now got rather akin to trying to improve the serious water pollution problem, which does exist, by specifying that by 1980 all rivers should attain the level of distilled water.

R. Lindsay London

I should like to ask the authors of Paper C121 about the lean-reactor car. I have always been impressed by the work at Ethyl Corporation on mixture preparation and distribution, and this has been an area of interest for me. I wonder, in the state that has now been achieved, of getting to about 18:1 air–fuel ratio and then applying recycle, whether the authors are hopeful of achieving complete control of nitric oxide by running leaner still. My company have done some work using a dynamic apparatus which allowed us to run a car under transient conditions in the laboratory, which showed that this seemed possible. I would be grateful for comments in the light of practical road experience.

I would also like to ask the authors of Paper C139 a question about oxygen measurement in exhaust. It is an increasingly important property to measure, and we are bedevilled by insufficient instruments giving an inaccurate and slow response. Do they or anyone else know of something which gives response times comparable with those which are used for measuring nitric oxide and the like?

Looking at the problem of emissions control, it seems it would be much simplified if we could have a catalyst that would decompose nitric oxide. I wonder if any of the catalytic experts present would care to speculate on the possibility of being able to find something that will decompose nitric oxide, while at the same time allowing the engine to run under lean mixture conditions.

D. R. Blackmore Chester

This is a slightly more mundane, practical question, which could represent quite a lot in terms of development cost and time. The authors of Paper C139 mentioned four different tests for catalysts. I would be interested to hear more details of these, particularly with regard to durability testing, if the authors got that far.

W. E. Bernhardt Germany

I would like to add one or two points to this discussion.

It seems to me that the situation in Germany is somewhat different to that in the United States, especially in California. There may, of course, come a time when there is some smog in Frankfurt or Stuttgart, but at the present time it is not the sort of smog which is seen over Los Angeles, it is more a problem of particulate matter. It could be that our government is looking at NO_x standards for Germany and for Europe, but it is impossible to measure NO_x emission using the European driving cycle. Furthermore, it is doubtful whether NO_x standards are necessary for Europe because of the very different climatic and geographic conditions.

Y. Tatsutomi Hiroshima, Japan

The air pollution problem in Japan is gradually getting serious. The emissions regulations to relieve the air pollution have been in force in Japan since 1966, and the government is now considering introducing more stringent regulations than those being considered in the United States. In the regulations now proposed, however, cold starting is not included.

A. H. Ball Birmingham

I would like to return for a moment to M. A. I. Jacobson's comments and put him in the picture and bring him up to date on the question of different standards for Europe. I would say, first of all, that they already exist and that there is a body, associated with the United Nations, under the Economic Commission for Europe, meeting in Geneva, which is called Working Party 29; this deals with these matters and all topics of safety which are of international concern. A body with headquarters in Brussels called the 'Bureau Permanent' (B.P.I.C.A.), on which European motor manufacturers are represented, also deals with this subject, along with others. I can assure M. A. I. Jacobson that all the engineers and the administrations concerned in this type of work are very well aware of the different conditions in Europe compared with those in the U.S.A. They are also aware of the technical and economic penalties which would be incurred if European manufacturers adopted the Federal standards.

I should like to ask a question arising from Paper C144, in which the authors mentioned an exhaust bypass. Fig. 144.3 showed a bypass pipe, presumably to protect the reactor in the event of plug or partial ignition failure. Has a satisfactory protective system been developed for catalytic reactors? If so, could the authors enlarge on this?

M. S. Bolton Sheffield

Following on the description by the authors of Paper C144 of their valve in 'fairly exotic materials' which I suppose means 'a bit expensive', I would have thought this would be an ideal application for a fluidic switch, which would be a lot simpler.

Discussion during Session 7

D. K. Roberts Southall, Middlesex

As I understand it, in Fig. 125.2 the line P represents the overall pressure ratio achieved by both the turbocharger and the Roots blower. Looking at the conditions pertaining to 1000 rev/min, the turbocharger gives a pressure ratio of about 1·2:1·5 which would be augmented by the Roots, giving a pressure ratio of 1·6:1·7, and therefore I would have expected the overall ratio to have been something of the order of 2 or even more. Is my interpretation of the diagram wrong?

F. J. Wallace Bath, Somerset

The authors of Paper C125 raise a few points upon which I would like some clarification. The first is concerned with Fig. 125.2.

Taking the $\frac{1}{4}$, $\frac{1}{2}$, $\frac{3}{4}$ and the 1/1 in the lower part do these refer to the load?

My other question concerns the full curve. Did the authors arrange matters so that the maximum pressure ratio of 2 occurs at an engine speed of 1000 rev/min?

Finally, was the adoption of this fairly complicated mechanism dictated equally by emission considerations and by torque back-up considerations? I have not read the paper in detail and cannot readily extract the torque back-up, but Fig. 125.4 seems to provide an indication. Was the motivation roughly the same on both counts?

I. M. Khan London

Use of gaseous (or finely atomized) fuels by manifold introduction at the top load end to increase smoke limited power output is not new but its use for controlling gaseous emissions is. The points I am going to make are somewhat academic as the authors have mainly devoted themselves to developing a practical system for using gaseous fuels in diesel engines.

I found the second devoted to 'knock' somewhat confusing. The authors of Paper C126 are dealing with diesel knock and the gasoline knock in the same section at the same time. Diesel knock is due to rapid rates of pressure rise which follow ignition of injected fuel and obviously presence of gaseous fuel will increase this rate of pressure rise. The gasoline knock encountered in gasoline engines, gas engines or dual fuel engines is due to autoignition of the end gas, i.e. of the fuel change which is not at that time consumed by normal combustion. The phenomenon of autoignition of end gas, i.e. knocking, is very well documented (**11**) (**12**). The important factor is the ignition delay of the end gas. This ignition delay (for a given temperature history) is a function of equivalence ratio of the mixture which constitutes the end gas and of the octane number of the fuel involved. The minimum ignition delay is usually observed with lean mixtures (equivalence ratio of about 0·6).

The important point to make here is that contrary to the statement made by the authors there is no evidence that the limits of inflammability are in any way involved in knocking. Knocking of very weak mixtures well outside their limits of inflammability has been observed (**13**). Hence the authors are not justified in using the mixture strength of the gaseous fuel alone to identify the type of knock that may occur.

This contribution relates to the work done by the authors of Paper C124.

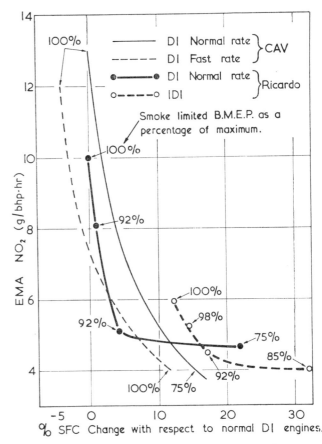

Fig. D7. Mass emission of NO_2 versus b.s.f.c.

I would like to say how pleased I am to have a paper where a lot of directly comparable results obtained on direct injection (d.i.) and indirect injection (i.d.i.) engines of the same size are presented.

It is of interest to compare the performance of d.i. and i.d.i. engines in terms of: mass emissions of nitrogen dioxide (NO_2) calculated with the 13 mode-EMA cycle versus changes in the brake specific fuel consumption (b.s.f.c.). This is done in Fig. D7, where results given by the authors in their Fig. 124.3 are represented as NO_2 emission versus b.s.f.c. The nitric oxide, NO, concentrations given by the authors in Fig. 124.3 are transformed into NO_2 mass emissions by using empirical data obtained at my laboratory.

The results given in Fig. 151.2 of Paper C151 obtained with injection periods of 20°CA (normal rate) and 15·5°CA (fast rate) are also replotted in Fig. D7 on the same basis. All the b.s.f.c. variations shown on Fig. D7 are with respect to the minimum b.s.f.c. obtained on the d.i. engines with normal rate injection. As is obvious from statements made above the various curves on Fig. D7 are generated by retarding injection timing from optimum. The variation in smoke limited b.m.e.ps are also given. It would seem from Fig. D7 that the d.i. engine would maintain its advantage in efficiency without loss in smoke limited output while giving NO_2 emissions similar to those of the i.d.i. engine if high enough rates of injection are used.

REFERENCES
(11) TAYLOR, C. F., TAYLOR, E. S., LIVENGOOD, J. C., RUSSEL, W. A. and LEARY, W. A. 'Ignition of fuels by rapid compression', *S.A.E. q. Trans.* 1950 **4** (No. 2).
(12) KARIM, G. A., KLAT, S. R. and MOORE, N. P. W. 'Knock in dual-fuel engines', *Proc. Instn mech. Engrs* 1966–67, **181** (Pt 1), 453.
(13) KHAN, I. M. Contribution to 'Knock in dual-fuel engines', by A. Karim *et al.*, *Proc. Instn mech. Engrs* **181** (Pt 1), 453.

D. G. Burton London

I would like to ask a question referring to Fig. 126.15. It is well known that when the injection timing of a d.i. engine is retarded, the NO falls rapidly, but there is an increase of smoke and reduction in power. The surprising result with partial use of LPG, as shown in the figure, is that the NO falls similarly with retard; but even when this is as much as 9° the power is almost maintained and smoke is no worse than at normal timing without LPG. I wonder if the authors could offer a theory to explain this interesting difference in behaviour, which might be useful for general understanding of NO formation.

H. O. Hardenberg Stuttgart, Germany

I would like to thank the authors of Paper C126 for presenting such an interesting paper. The Saurer engine which was used in the tests seems to me to be not quite the typical example of a direct injection engine, as can be seen from the paper.

A b.m.e.p. of less than 7 kp/cm² giving a smoke density of 50 Hartridge units seems to me a poor example. It is the same with the exhaust emissions given in Fig. 126.16, giving about 16 g/b.h.p. h NO_2+HC, which is considerable. A modern standard production engine of that dimension and the same hp should not have been more than 20 Hartridge units and not more than 5 g/b.h.p. h CO and 8 g/b.h.p. h NO_2+HC at normal injection timing. Most of today's direct injection engines are below these values, as we find in tests with makes of different companies.

I think that the engine used by the authors and which gives twice the exhaust emission in terms of NO_2+HC and in terms of Hartridge units would lead to the wrong conclusions.

We carried out similar tests as the authors using a standard production but fully developed modern direct injection engine; we found that the use of neither LPG nor PNG gave advantages in terms of smoke and exhaust emissions.

The reason for these contrasting results could be that the authors' results were possible because the injection period was too long. With more modern engines, with much shorter injection periods, one can retard injection a lot more, and the advantages gained by the authors would then disappear.

As can be seen from Fig. 126.16, retarding injection gives a decrease in NO_2+HC of about 1 g/b.h.p. h per one degree crank angle. But when one uses an engine of lower values, as we did, one does not get the same slope of the curve—we found that the decrease is only 0·5 g/b.h.p. h NO_2+HC per one degree crank angle. That means that at retarded injection, both the curves—the authors' and ours—approach each other, so that one gets the same exhaust emissions from diesel fuel as from dual fuel. Why then complicate the engine by dual fuelling?

There is another point which the authors did not take into account. That is when dual fuelling the engine, the rate of pressure rise and the peak pressures were considerably higher than they had been before. Can the authors comment on these points?

W. Tipler Peterborough

I should like to make a few comments on Paper C126 and C124.

When the Saurer engine was derated by 25 per cent on diesel fuel, was any rematching of the fuel injection equipment carried out? Some attention to this matter might have yielded better results than those reported.

The investigations were carried out with LPG which was either 95 per cent pure propane or 95 per cent pure butane. These figures cannot be regarded as typical of normal commercial supplies of LPG and the paper itself shows that the composition of the gas is a significant variable influencing engine performance.

If the technique described in this paper were to be widely accepted, would LPG be available in sufficient quantities to meet the demand? Or is this merely an academic exercise unlikely to be repeated on a wider scale?

The authors of Paper C124 have attempted to make deductions concerning the effects of combustion chamber geometry and compression ratio (Figs 124.6 and 124.7). Reference to Table 124.2 shows that five variables have been changed in this test series, and it is difficult to see how any firm deductions can be made concerning any individual characteristic. Could the authors please explain the deductive processes used to separate the variables?

W-T. Lyn Indiana, U.S.A.

I have two comments to make, one in connection with Paper C124 and the other in connection with my earlier comments.

I said earlier that the NO model is something about which we still have a lot to learn. We are perhaps more conscious of it, because we have the opportunity to examine the model at different situations. For example, curves produced by S. Timoney show that NO concentration increases with compression ratio at a given speed, but not at another. I do not think variation of ignition delay is the reason, because I have never seen any experimental data on ignition delay which reverses the direction as compression ratio increases, and so it must be due to something else. We should know a lot more about the diesel side, particularly in relation to local temperatures.

In Fig. 124.2 and Fig. 124.3 the HC concentration for direct injection (d.i.) is higher than for indirect injection (i.d.i.) but the general experience is the other way round: the d.i. is typically around 0·7/0·9 mg on the 13 mode cycle while the Comet type i.d.i., depending on the engine size, sometimes gets over 3 mg. Can the authors comment on this?

I think it is interesting that no one has talked about HC in this symposium, but it is very important. We are tightening down every 0·1 g of emission and the contribution of HC is very significant. It needs to be much more carefully understood.

Discussion during Session 8

W. D. Holt Coventry

This contribution relates to the work done by the author of Paper C152.

With the thermal reactor in such a position and running with a skin temperature of 800°C, has the author had any under-bonnet problems and problems of carburation due to temperature?

P. S. Wilson Coventry

This question relates to Paper C152. If one has two spark plugs per reactor with different timings and one of them fails, will one not get unacceptable conditions in the reactor?

C. L. Goodacre London

Fig. 152.1 of Paper C152 shows the engine with peripheral ports on the perimeter of the epitrochoid. The author's earlier engines had side entry ports. Has the author abandoned the side entry port in favour of the peripheral port as used on the NSU engine?

J. P. Soltau Birmingham

Could the author of Paper C152 please say whether he has tried spark plugs in his reactors, and if so, did it improve the performance? I would also like to know if the reactor skin-cooling air has been used as a means of providing heated air to the exhaust port nozzles.

P. Draper Dorset

I presume that the important parts of the engine are the seals on the periphery and on the side walls. Could the author of Paper C152 say if these are the limiting factors as regards maintenance of the engine?

Paper C150 has shown the necessity for preventing the hydrocarbons from escaping. I do not think the oil companies can really do anything other than accept that point.

There are a couple of points I would like clarified. I do not understand what is meant by 'compacts' or 'light sedans'.

With regard to filtration of exhaust, Paper C123, it would seem to me that this is a practical piece of ironmongery that could be incorporated in the exhaust with very little trouble and expense; it might well counter the complaints about lead emission. It would at least reduce such emission quite appreciably.

I have watched combustion chamber shape development for many years and it seems that, as a result of this symposium, there are several aspects that should be included. For instance, can a reciprocating engine combustion chamber achieve the same results as the Wankel engine by having a better squish section and therefore better cooling of the end gases? Several points have been raised where combustion chamber designers might be able to improve matters.

J. A. Hoare Shoreham-by-Sea, Sussex

My contribution is really to show results obtained at Ricardo's which are complementary to those of the author of Paper C131.

Using a variable compression engine with a disc combustion chamber, Fig. D8, we have had somewhat better luck with NO measurements and Fig. 131.3 shows the effect of compression ratio on NO emissions at two speeds and two mixture strengths. The emissions were plotted on a mass basis and the b.m.e.p. was kept constant under all conditions. The effect of compression ratio is negligible at the rich mixture but the emissions decrease with ratio at the weaker mixture. The effect of induced swirl is shown by the red curves where, at both speeds and mixture

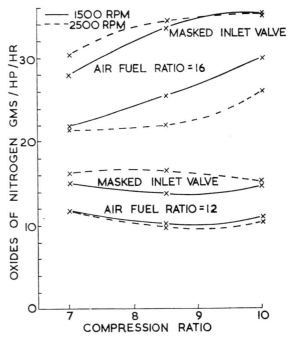

Fig. D8. E-6 single cylinder engine constant load

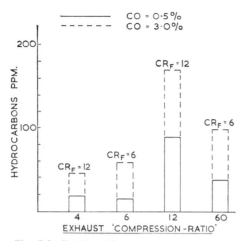

Fig. D9. 'Varicom' engine constant load

strengths, the NO emissions are increased. Like the author we have also shown a reduction in hydrocarbons with reduced compression ratio and, in an attempt to examine this more closely, we have constructed a rather peculiar engine which, by means of eccentrics, can be operated with different and variable clearance volumes on the exhaust and compression strokes.

In Fig. D9, we have along the horizontal axis what I have chosen to call 'exhaust compression ratio' and CR_F is the firing compression ratio. The figure shows that the general effect of reducing the exhaust compression ratio is to reduce the hydrocarbon emissions at both CO levels used. This effect is believed to be due to the retention in the cylinder of unburnt hydrocarbons on the walls and piston crown, due to the low scavenge efficiency. There is, however, an effect on hydrocarbons from the firing compression ratio which is probably caused by the lower exhaust temperatures at the higher ratio. Fig. D10 shows this effect on NO emissions, where they are shown to be directly proportional to exhaust compression ratio and independent of firing compression ratio. This is undoubtedly due to the retained residuals giving the same effect as exhaust recycle.

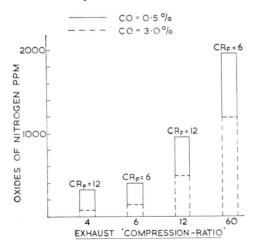

Fig. D10. 'Varicom' engine constant load

J. H. Alden Luton, Beds

I would like to thank the author of Paper C131 for his interesting paper; we should encourage M.I.R.A. and others to pursue this question. This could get down to the very fundamentals of the problem where, instead of hanging on all this hardware, could we fully oxidize the combustibles inside the combustion chamber?

There are two other observations to be made in a symposium of this kind. The first is that we should emphasize the difficulty we have in Europe with the smaller types of engine and the kind of rotational speed we use. We have heard references to 350 in^3 and 450 in^3 engines, V-8s and so on, but the problem is much more difficult as the engine size decreases. This may lead to a world-wide tendency to pull two ways: on the one hand, to decrease the size of the engine for economic reasons and, on the other hand, to increase its size due to the performance standard decrease to conform to the Federal cycle. It might be of interest to observe that our own company, to comply with severer exhaust emission standards, has dropped the 1200 cm^3 and 1300 cm^3 engines for certain markets and are fitting only the 2 litre size.

The other point I would like to make is that we really need to unite on harmonizing international regulations. Those who work on the vehicle production side are fearing that the days of mass production are passing because of the complexity of the different types of equipment required by different countries. We are discussing the future with a number of other countries working to the same standards. It would be a good thing if we could somehow induce the rest of the world to follow those standards; we talk about the severity of the haze in Tokyo and other places but really there are so many other human beings with cars living in other parts of the world and we are at least harmonizing our standards to an economic level for the majority in Europe. Even if we cannot harmonize the actual limits to which we should work, we might go some way to harmonizing the test cycle. We get more and more sophisticated machinery to measure these different limits; the equipment sometimes works and sometimes does not. Therefore, if we could have one set of equipment and one set of standards, this would help enormously.

A. W. E. Henham Guildford, Surrey

I would like to ask the author of Paper C131 in view of the importance of part-load conditions in urban driving, whether he has a breakdown of Figs 131.5 and 131.6 showing the order of merit for part-load and full-load conditions separately. In other words, is there a good cylinder head for part-load driving and a different cylinder head for full-load driving? It is not suggested that one could change them in between, of course!

Returning to compression ratio, it would be interesting to know whether this really does affect road efficiency since most of the time one is driving one is deliberately destroying peak pressure by throttling the intake. It was pointed out in the 19th century that it is the maximum expansion ratio which is required for efficiency, not maximum compression ratio. By deliberately retarding spark timing, we are reducing the expansion ratio relative to the compression ratio. Is it absolutely necessary to keep the high compression ratio?

This leads me to wonder whether anyone has done any tests on the road with two exactly similar cars, one with a low compression engine available for certain markets and one with a high compression engine; and similarly whether anyone has done tests on fuel consumption for two identical vehicles with different sizes of engines.

R. W. Mellor Basildon, Essex

I have some familiarity with one of the chambers shown in Paper C131. At the recent Motor Show lecture, a gentleman from Jaguars made some comparisons between chambers of the kind described by the author as Type I and Type J. In those remarks, he attached a lot of importance to the presence of valve cut-outs in the flat head of the conventional Heron bowl-in-piston chamber such as Type I. In Paper C131, it is not entirely clear whether this chamber did have deep valve cut-outs in it, and whether any experiments were made to determine if they had an effect.

Referring to Paper C123, I find a problem in the way in which leaded and unleaded fuels were quoted as if everyone knew what they were. I would find it helpful if the fuel composition on which results in Fig. 123.8 were based were included in the discussion, as the lead content varies between different leaded fuels of the same octane number. Therefore it would be useful to know the grammes per litre of lead in the fuel.

D. A. Browne Norwich

I was intrigued by Paper C136 and by the way that the air–fuel ratio is varied. Presumably at light load the air–fuel ratio at the spark plug must be between normal ignitable limits, roughly, 10:1–17:1. Yet at full load it must also be between those limits at the plug, whilst the amount of fuel injected must be considerably greater. I should like to know how the ignitable air–fuel ratio is not exceeded at the electrode if a too weak mixture is not provided at light running. In addition, this air–fuel ratio must be richer for cold starting. I would like to know how this is arranged with a fixed injector position.

W. E. Bernhardt Germany

My comment relates to Paper C136. Under full load conditions, at about 2000 rev/min, the NO concentration was about 1500 p.p.m. in the Southwest Research Institute's stratified charge engine. It may be that the stratified charge concept combined with pre-chamber arrangements is a better method for reducing the NO concentration. In my paper I pointed out that we measured only about 170 p.p.m. nitric oxide in the exhaust gas under very similar engine conditions. However, it is interesting to note that the hydrocarbon emissions are high in the investigated region in both cases (in the swirl stratified charge engine as well as in the pre-chamber stratified charge engine).

It would be useful if we could discuss, in addition to the discussion on the lead trap, the Ethyl particulate trap presented in a previous paper.

D. Broome Shoreham-by-Sea, Sussex

In many ways stratified charge combustion systems such as that described in Paper C136 are closer to diesel operation than the conventional gasoline engine, and it is of interest to compare the results of Table 136.2 with typical diesel emission levels, such as those presented in Paper C124.

Taking NO first, concentrations are markedly a function of air–fuel ratio, as with the diesel, although the result of run 4 (high load at 1000 rev/min) appears anomalous. While the overall mixture strength may have approached the value for minimum NO at this point, this cannot be confirmed from the data given, and even then the result is surprising. The absolute levels recorded are broadly equivalent to a moderately retarded direct injection diesel and rather worse than those of a swirl chamber engine, and the variations with speed may be due to changes in the relative spark timing, since the stratified-charge engine would be expected to be sensitive to this feature just as the diesel is to the start of combustion.

While CO levels are, as the author says, good by untreated gasoline standards, they are higher than normal for a diesel operating with a clean exhaust. It is to be noted from the results at 3000 rev/min that if over-fuelled the stratified-charge engine gives rapidly increasing CO levels, again just like the diesel. Just why the stratified-charge engine is some 2–3 times worse than the diesel under normal operating conditions is not clear, but it may be due to the mixing pattern, which being the reverse of the diesel with the fuel concentrated at the centre of the chamber to give the necessary rich mixture at the spark plug, cannot bring oxygen from the outer spaces to this fuel.

The HCs of the stratified-charge engine are very high, as noted in the paper, although it is not clear from the published text if the figures quoted are p.p.m. C. However, assuming this to be so, then measurements on the stratified-charge engine using a volatile fuel and unheated F.I.A. give results some 30–100 times higher than a good diesel operating on a heavier fuel but tested with F.I.A. at 180°C. The author has outlined his views on why this occurs and I believe him to be right, although I think that cause (a) stemming from the different mixing pattern as noted earlier may be the major problem. Cause (c) would be expected to be influenced by fuel volatility, and tests with different fuels might be of interest here.

While catalytic treatment of the exhaust will be effective against HC and CO emissions, there are likely to be difficulties at part load where the exhaust temperature is low, exactly as we have found at Shoreham on diesel engines. It also seemed to me that if such devices are deemed necessary then this is in effect an admission that the increased complexity of the system has not enabled the goal of combustion control within the cylinder to be achieved.

I cannot, however, agree with the author's closing statement that catalytic devices could be used to control NO, since the excess oxygen always present in the exhaust precludes any reducing chemical reactions known today. This is an area of great practical importance, where the conventional gasoline engine has an inherent advantage over stratified-charge, diesel or gas turbine power plants.

T. J. Williams London

I should like to make some comments on Paper C136. I refer particularly to the mechanism by which a stratified charge is produced and the explanation given in the paper. Referring to Fig. 136.2, one can see from the second diagram, which relates to the compression process, that there will be an inward acting drag on the fuel spray which should tend to concentrate the charge in the region of the cylinder centre line. Assuming a solid body swirl with a rather turbulent central zone, then the drag directed towards the cylinder centre will become less as the jet continues its penetration.

Having accepted this explanation of the production of the stratified charge, I was a little puzzled to see from Fig. 136.5, that in the actual engine configuration the nozzle had been moved out of the region of high cylinder swirl to approximately the half radius position. In this position there is a danger of the drag force dispersing the spray outwards rather than stratifying inwards. Without further information on the penetration distance of the nozzle spray and the swirl level, it is difficult to form any conclusion relating to the pre-combustion location of the charge.

Presumably the new position of the nozzle is connected with the use of a bowl in the piston crown which will intensify the swirl velocity level in the central region of the cylinder as the piston approaches top dead centre. Does the author have any information concerning the behaviour of the stratified charge as the air moves inward from the periphery of the cylinder into the bowl? Further, is there any danger of the fuel being thrown outwards as a result of the increasing rotational velocity in the bowl? Both these factors could result in increased charge mixing. Finally, since injection seems to be directed almost parallel with the cylinder head, it is a little difficult to imagine where the stratified charge will be located in the piston bowl. Is it contended that the charge will form a central 'pipe' surrounding the centre line? Conditions in the actual engine could well be very much more complicated than Fig. 136.2 would have us believe, since the flow pattern in the combustion chamber of the engine shown in Fig. 136.5 is very much three-dimensional as compared with the two-dimensional model used to explain the stratification process.

M. T. Hall Swindon, Wiltshire

I am interested in Table 123.1 of Paper C123, showing the amounts of particulates emitted from so-called 'minor' sources. One wonders whether the footwear emission in the United Kingdom would be much higher, since one has a mental picture of the average American having four wheels rather than two legs!

The paper is entitled 'Vehicle particulate emissions' but appears to be devoted to engine particulate emissions. What is the particulate emission from braking systems, for example? I do not wish to appear biased, but, working for a public transport concern, one feels that a lot of our problems in the automotive field, which are certainly pronounced in city environments, could be greatly reduced by spending a lot more money not necessarily on research but on investment in public transport, rather than attempting to prolong the existing situation.

W-T. Lyn Indiana, U.S.A.

I should like to make a few comments on Paper C123. I was interested to note that the amount of particulates is equal in leaded and unleaded fuels. I think the presence of metals in fuel has a lot to do with the formation of carbon and its consumption. Earlier we were discussing diesel engine emissions and the importance of the local flame temperature. Here again the presence of metals in the flame could have a significant influence on local flame temperature due to radiation, hence the formation of NO. Diesel engineers have complained for a long time that the petroleum industry never does anything useful for diesel; this might be a good line of attack. I would like to hear the petroleum industry's view on this and also on the basic mechanism of carbon formation.

M. A. I. Jacobson London

One of the factors which seems to be of paramount importance, according to this morning's papers and discussion, is the problem of under-bonnet temperature. If one could only reverse the current tendency of this temperature to increase, surely some of our problems would be minimized? Does the industry think that there is any reasonable chance of reducing such under-bonnet temperature problems? Could the temperature be reduced to a level which we used to find acceptable some three or four years ago?

R. W. Wheeler Shoreham-by-Sea, Sussex

I would like to thank the authors of Paper C123 for their paper and say that this sort of work should have been repeated by a lot of other people. I know the extent of this research and it is relatively small compared with other researches into the emission problem. It goes a long way towards answering the exaggerated criticisms of the anti-lead brigade.

The authors have made the point that this work is very difficult because there is no definition of particulates. This seems to be inhibiting us from having a reasonable particulate control which is badly needed in Europe; we all know of this problem from diesels and even from oil burning petrol engines. Perhaps this problem of the definition of a particulate should really be left until later, but let us have a simple particulate emissions regulation.

In any event, it should not be too hard to find a reasonable control based on the authors' filtration experiments, and I think we should really make a start on this.

W. J. D. Annand Manchester

In the table at the beginning of the Paper C123, we do not actually see the vehicle particulates on the same basis of tons per day. We have a statement underneath that in California they come out as 2 per cent, but what are they in tons?

D. R. Blackmore Chester

Have the authors of Paper C123 any suggestions as to the mode of action of their lead trap? It seems at first sight a little extraordinary how the trap tends to be more efficient for the smaller particles.

Communications

B. Caldicott Derby

I should like to make some comments on the road test. Following on the work described in Paper C128, we have completed further trials to establish the repeatability of the test method and its sensitivity to changes in the test conditions. In the original trials, the results of which were illustrated in Fig. 128.2, there were only four variations from normal operation:

(1) The cars were run on a controlled circuit consisting of a mixture of rural, urban and congested conditions.

(2) No carburetter adjustments were made apart from the idle speed which was reset to 600 rev/min during each emission check. This adjustment was considered necessary in order to derive comparative curves of change of CO level due to carburetter fouling. The constant idle speed was needed to avoid apparent variations in CO which would be due not to changes in carburation but changes in speed.

Had the test been run with no adjustments, stalling would have occurred at an early stage on the untreated fuel cars, preventing further measurement of idle CO and making these cars undriveable.

The adjustment of throttle position was also considered valid as being the one most likely to be made by an owner driver or mechanic. Our own checks have shown that drivers are completely unaware if their vehicles are idle at 1 per cent CO or 12 per cent CO but are extremely critical of either too high or too low an idling speed.

(3) The crankcase ventilation system was modified from that used by the manufacturer of the particular test vehicle, to one in which all the blow-by was fed into the engine via the air cleaner, this being typical of a large percentage of the vehicles on the road today.

(4) The crankcase oil was bought from a local garage and represented one of the poorer quality lubricants commercially available.

Since these initial runs, a test has been carried out to check the effect of changes in crankcase oil quality and crankcase ventilation system.

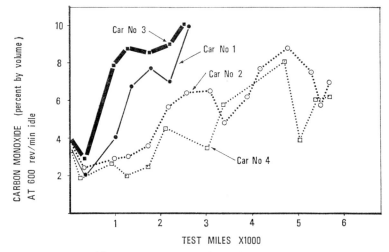

All cars run on untreated premium gasoline

	Crankcase ventilation		Crankcase oil	
	Simple	Federal	Low quality	MIL-L-2104B
Car 1	★		★	
Car 2		★	★	
Car 3	★			★
Car 4		★		★

Fig. D11. European test—variation of idle carbon monoxide emission

The four cars were again run on the same circuit but this time, two with oils of MIL-L-2104B quality and two with the lower grade oil. Two cars were equipped with a Federal ventilation system—in which all the blowby was fed into the inlet manifold via a valve—and two with the blowby piped to the air cleaner. The results are shown in Fig. D11. All cars were run on commercial gasoline that did not contain dispersant additives.

From this it can be seen that the change of lubrication oil quality had no measurable effect on the rate of increase of carbon monoxide emission, and that the change in engine ventilation system, while slowing the rate down, did not prevent the steady increase in emission.

B. H. Croft Coventry

I should like to make a few comments on Paper C129. It is questionable whether the inclusion in the sample of vehicles fitted with exhaust post-treatment devices was wise on the grounds that:

(i) It introduced a further very significant variable.

(ii) The response of post-treatment devices could well be in opposing directions depending upon the particular air–fuel ratio pertaining during the test, i.e. the result could be very specific and may be in fact misleading in predicting trends.

Since the sample contained 21 per cent of vehicles with post-treatment devices, it would be interesting to know if the results show similar trends with these vehicles removed from the sample.

With regard to the fuel injection vehicles, the classification under mechanical or electronic systems appears irrelevant unless the control system actually corrects for the variations in fuel properties under consideration. The significant feature is the metering arrangement employed, and while it would appear from the paper that the mechanically controlled systems used volume metering, and the electrically controlled system had fixed orifice metering, this is not entirely clear, and more specific detail on this point would add to the value of the paper.

The increase in CO with increase in specific gravity for the volume metering fuel injection systems agrees with the anticipated result, but the reason for the insensitivity of the fixed orifice metering systems to specific gravity is not obvious. From the following simple relations

Mass = f(volume, specific gravity)–Volume metering

Mass = f(pressure across orifice, specific gravity)$^{1/2}$– Orifice metering

it would be expected that the latter would respond at about half the rate of the former for small changes in specific gravity. The authors' explanation of the response of the pressure regulator to specific gravity and associated changes does not seem very plausible, since in practice the pressure regulator has to cope with different operating conditions producing much greater flow rate changes while maintaining a sensibly constant fuel pressure. Consequently the metering orifice should be subject to constant

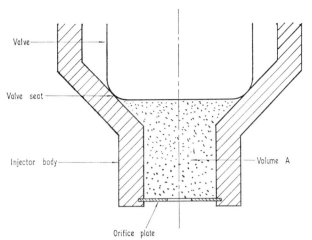

Fig. D12. Inlet manifold

pressure conditions and the above relation would apply. It would be interesting to know if the authors have given further consideration to the explanation of some of the more unexpected trends.

The CO response of the fuel injection vehicles to R.v.p. is at first sight surprising. However, recent experience of the performance of the fixed orifice metering injector shown diagrammatically in Fig. D12 may illustrate the effect of detail mechanical design deficiencies in overall system response. The injector delivers fuel when the valve is open, the delivery being proportional to the open period. The engine tests were conducted with a constant open period, with one opening every engine cycle, at various manifold pressures, and with the fuel pressure drop across the orifice being held constant.

Results are shown in Fig. D13 with fresh and conditioned fuel which had the lighter fractions removed. The increased delivery at low absolute pressures is explained by evaporation from space 'A' during the valve closed period. Reducing the volume 'A' similarly reduced the additional delivery.

It is suggested that most practical fuel handling systems have analogous deficiencies, and it is partly due to these effects that correlations such as the authors have attempted should be interpreted with due regard for the complexity of the problem, and the results applied with caution.

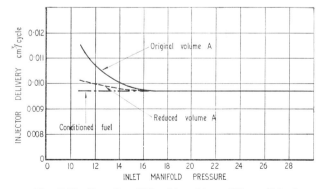

Fig. D13. Results of 'fresh' and 'conditioned' fuel

H. Daneshyar and J. R. Orme Cambridge

Referring to Papers C130, C144, and C146 on reduction of emissions by using an afterburner in the exhaust pipe of an engine, little attention has been paid to the effect of this device on engine performance. We have recently carried out some tests on the effect of an Engelhard P.T.X.3 afterburner on the performance of a Ricardo E6 (single cylinder) variable compression-ratio engine. This afterburner consists of a venturi suction device for air addition and a combustion chamber incorporating a grid of platinum catalyst. Tests at a fixed compression ratio and engine speed were carried out, over a range of mixture settings, with and without the afterburner in the exhaust pipe, for several engine speeds. At the same fixed compression ratio, maximum power tests over a range of engine speeds were carried out with and without the afterburner in the exhaust pipe; motoring tests were similarly carried out.

It was found that whilst the effect of the afterburner at engine speeds of less than 1500 rev/min was negligible (Fig. D14), at 2500 rev/min addition of the afterburner to the engine caused a 16 per cent loss in maximum power and a 20 per cent increase in minimum specific fuel consumption (Fig. D15). The motoring tests revealed that this magnitude of the power loss could not be entirely

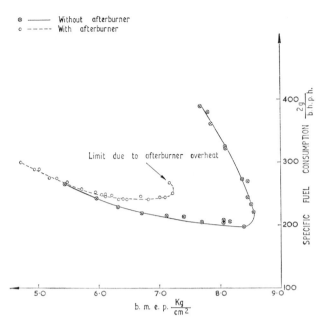

Fig. D15. The performance of a Ricardo E6 spark ignition engine over a range of fuel air ratios, at 2500 rev/min, with and without an Engelhard P.T.X.3 afterburner in the exhaust pipework

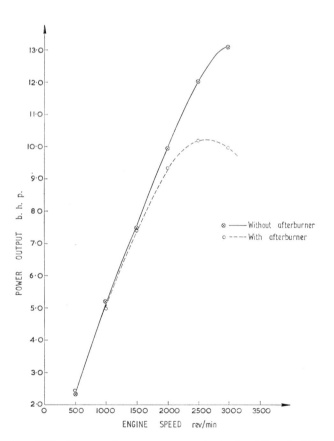

Fig. D14. The maximum power output of a Ricardo E6 spark ignition engine, over a range of engine speeds, with and without an Engelhard P.T.X.3 afterburner in the exhaust pipework

attributed to the venturi air suction device, and that the pressure drop across the reactor was a major cause of the decrease in maximum power output of the engine.

It is possible from the results to obtain a qualitative understanding of the effects of the afterburner. The change in engine performance must be due to a change in the exhaust back pressure during the exhaust valve open period brought about by the venturi suction device and the reactor. It is known that heat addition to flow of a perfect gas (in a duct of constant cross-sectional area) produces a loss of static pressure (Rayleigh flow). Therefore heat addition to the exhaust gas flow (combustion of the exhaust gases in the reactor) is an important factor in the pressure loss in the afterburner and thus the power loss at higher engine speeds.

To summarize, the afterburner produces appreciable loss of power at high engine speeds under full load. Further work is being carried out to obtain a better understanding of the nature of this loss. However, from this preliminary examination it would seem that the static pressure drops across the venturi suction device and the reactor are major contributors to the loss.

P. J. Ivens Basildon, Essex

Referring to Paper C144 on catalysts, my company have been pursuing the same goals along somewhat different paths.

Owing to the shifting sands of Federal legislation we have aimed at the preparation of various emission packages which can subsequently be put together in combinations according to the needs of each model year. Typical packages, under universal consideration, are: Imco, Thermactor, E.G.R., CO–HC Catalyst, Thermal Reactor, NO_x Catalyst.

Here we give a brief account of our work on CO–HC catalysts. With the aim of preparing such a catalyst for possible introduction in model year 1974, it was desirable to avoid major problems such as provision of an adequate by-pass valve, or slow warm-up in the C.V.S. cold test. Thus we sought a catalyst which had to be small (that is, very active), which could be fitted close to the engine, with safe burn-out, and with established durability. These considerations pointed to the fixed support noble metal catalyst.

Dynamometer test procedures were set up to distinguish between makes and models of catalysts, and to arrive at first sight calibrations of air pump and carburetter. NO_x was to be controlled by carburation sufficiently rich to meet 1974 California development level (about 1·5 g/mile). Typical curves of catalyst efficiency and resulting tail pipe emissions are shown in the accompanying graphs, Figs D16 and D17.

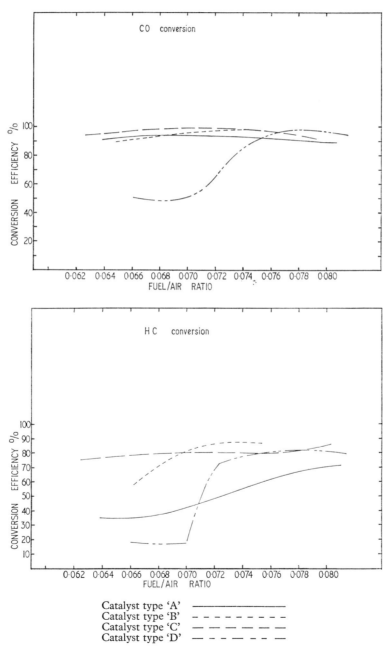

Fig. D16. Different types of oxidation catalyst compared at 2000 rev/min, 65 lb/in² b.m.e.p. Optimum secondary air. Results corrected for air dilution

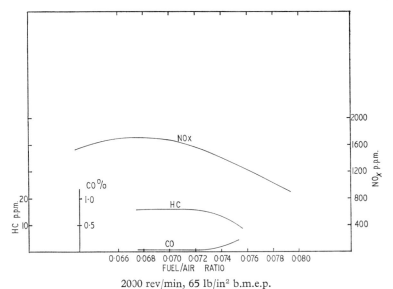

2000 rev/min, 65 lb/in² b.m.e.p.

Fig. D17. Tailpipe emission levels from catalyst type B

We have a fairly reliable method of forecasting vehicle results from dynamometer tests, which shortens total development time considerably. Thus in a matter of weeks we were able to set up a vehicle, which with minor adjustments gave about 10 g/mile CO, 0·5 g/mile HC and 1·5 g/mile NO_x in C.V.S. hot test, with a 'used' catalyst, the car being otherwise of 1971/2 type. Its driveability was acceptable, and the fuel consumption penalty was small.

Subsequent attention to the choke yielded CVS-C levels on the development targets for 1974 California. We are chary of quoting vehicle test results in detail since there are so many variables, not least in the tests themselves; we can only be satisfied when the average of a number of tests is somewhat below the target.

Another difficulty in reporting is due to the change of catalyst efficiency which may be expected to take place as mileage builds up; presently we do not quote results from 'new' catalysts. Thus far in the development we had a satisfactory catalyst from the point of view of emissions, installation and performance; it remained to show that it would not overheat in the worst 'customer usage' simulation. Tests of trailer towing, and of gradient—using test track and wind tunnel—showed that the makers' intermittent temperature rating of 1000°C would probably not be exceeded. A valve to divert air from the catalyst is helpful.

It then remained to demonstrate the durability, and to verify that the result of an engine malfunction, such as plug failure, is not unsafe; the fixed support catalyst burns out very rapidly, so that although the metal becomes quite hot it may not last long enough to endanger the vehicle. We believe that one of the major objections to the pelleted catalyst is that it takes much longer to burn out and therefore requires a by-pass valve.

S. Timoney Dublin

I would like to add one or two points to the discussion on Paper C124. All engine experimenters must feel continually in debt to Ricardo and Company for their liberality in showing the results of their research on so many aspects of engine development down through the years.

Our work on emissions at University College, Dublin, has been confined until recently to the investigation of parameters affecting NO emissions from an open chamber engine. While agreeing substantially with the Ricardo re-

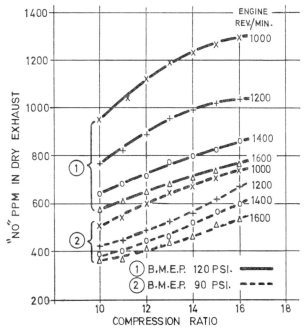

Fig. D18. NO emissions relative to compression ratio

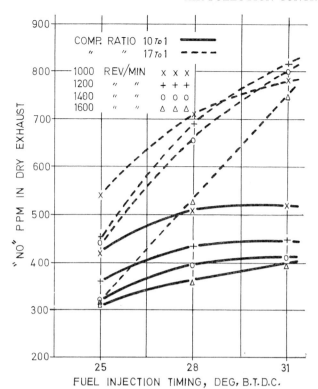

Fig. D19. NO emissions at constant b.m.e.p. (90 lb/in²) for varying injection timing

sults in most respects, we have found variations in compression ratio to be of much greater significance.

The engine used was a 3-cylinder, turbocharged two-stroke opposed piston engine of 199 in³ displacement having a facility for variation of compression ratio from 18 to 1 down to 9 to 1. The results plotted in Figs D18 and D19 show the effect of compression ratio and injection timing on the quantity of NO in the exhaust gases. The variation of NO quantity with injection timing is comparable with that shown in Fig. 124.7, but the variation with compression ratio is considerably greater. It is noticeable that reduction of compression ratio is significantly more effective in reducing NO emissions at advanced injection timing, when the emissions are high, than it is with retarded timing. This may explain the lack of response recorded by the Ricardo engineers with the indirect injection engine which has relatively low emission levels at all injection timings.

The opposed piston, open-chamber engine shows significantly lower values of NO at comparable load conditions than does the open-chamber engine used in the Ricardo experiments. This may be due to the presence of more residual exhaust gases in the trapped charge of the two-stroke engine as compared with the four-stroke engine.

It would be interesting to examine the possibility of maintaining high b.m.e.p. values in a two-stroke engine by using a high exhaust back pressure to keep trapped charge pressure high, while reducing the volumetric delivery ratio to reduce scavenge efficiency and therefore increase the trapped residual gas. The higher residual content may reduce the formation of NO, as is suggested by the Ricardo hot recirculation tests of Fig. 124.4.

Combinations of various techniques will probably have to be used to get NO levels in diesel engines down to acceptable levels without unacceptable derating of their power output.

Authors' Replies

J. W. E. Adams, H. J. Gibson, D. A. Hirchler and **J. S. Wintringham**

R. Lindsay has asked whether we are hopeful we can accomplish the needed control of nitrogen oxides by running the engines at air/fuel ratios leaner than those we now use (16:1–18:1) and using exhaust gas recycle. We do not believe this to be likely in a vehicle with reasonable driveability. It might be possible in an engine operating under conditions of constant load and speed.

R. Lindsay asked about the possibility of a catalyst to decompose nitric oxide under lean mixture conditions. There are many catalysts which will bring about some measure of decomposition in the presence of some oxygen but the best we know of lose the greater part of their effectiveness when 1 per cent or more oxygen is present in the exhaust. Research in this area should continue even though the chances of success are quite small.

P. Draper has stated his views on the lack of need for emissions controls as severe as those specified by United States law. I can only concur with his views and hope that our Congress will see fit to modify the 1970 law on the basis of a realistic study of the cost versus the benefits to be realized from the last incremental reductions in allowable emissions. I hope that the Japanese government will not slavishly copy the 'Muskie bill', as Y. Tatsutomi has indicated they propose to do. Instead, I hope they will determine what is needed in Japan and do that instead. M. A. I. Jacobson and W. E. Bernhardt plead for doing what is necessary in each country rather than copy the United States. I hope their respective governments will do just that. Unfortunately, it now appears they are more likely to follow us. Again, I can only plead for a realistic study of the need–cost–benefit relation in each country.

In reply to a question by D. Downs concerning unleaded gasoline, platinum catalysts are sensitive to small concentrations of a number of contaminants. For them unleaded probably should contain no lead at all, i.e. 0·00 g. Since this is almost impossible, a realistic maximum might be 0·05 g/gal (U.S.). These same catalysts are likely to be poisoned by phosphorus, sulphur, and zinc, so a really sterile fuel is needed and metals from the oil should be excluded as well. There are a number of base metal catalysts which can tolerate reasonable concentrations of such contaminants and they may well prove to be the catalysts of choice.

The remarks by E. S. Starkman serve to illustrate the many problems which must be solved to be sure the highly sophisticated, highly stressed control systems that are being studied will work satisfactorily in the hands of the public. Relative to his question as to the use of fundamental principles in the study of thermal reactors I can answer only that our lean reactors can be treated with the same principles as any other chemical reactor.

G. L. Lawrence asked whether the systems now being developed can be made to meet 1975 requirements and be ready for production in 1974. Based on our experience I can say that it is unlikely they can meet these deadlines but I should like to know what 'less exotic' treatments are available which can?

The following corrections to Paper C121 should be noted.

Page 3, Fig. 121.3 should be as given on facing page.

Page 4, Fig. 121.4. For INCREASE IN RECYCLE (vertical axis) read MILES PER GALLON.

Page 5, Table 121.3, PnO should read PbO and PaSO$_4$ should read PbSO$_4$.

Page 5, second column, fifth line. For LPG read pentanes.

J. L. Addicott and **D. Barker**

Replying to A. G. Bell's first point, the miles on all the cars were accumulated on leaded gasolines before the octane requirements were determined. Fuel rating work with other cars referred to in our paper indicated that there may be a further increase in octane requirements where mileage is accumulated entirely on unleaded gasolines. It is expected that a programme currently in hand by C.O.R.C. will more fully establish whether or not we can expect an increase in octane requirement as a result of mileage accumulated with unleaded fuels.

Our results showed a decided anti-knock benefit for lead when the vehicles were accelerated from low speeds but most gain was obtained at high constant speeds. While we chose to compare between the anti-knock values of unleaded, low and highly leaded gasolines either by research octane number, for the octane requirement work, or by knock limited spark advance, for fuel gasoline ratings, we have reason to believe that in terms of motor octane number the benefits of lead are still apparent. For example, with the octane requirement work we showed (Table 133.4) when accelerating and at constant speed on the fuel type B, 51 per cent and 83 per cent of the cars had a lower octane requirement with the high lead fuel series

Fig. 121.3. Exhaust reactor

compared with 13 per cent and 6 per cent which had their lower octane requirement with unleaded; and 36 per cent and 11 per cent of the cars showed no preference. These results were assessed in terms of research octane number because the octane requirement reference fuels were blended in increments of one research octane number. From the Table D1 it may be seen that the sensitivity, and hence the motor octane number, of fuel type B was very close, at any research octane number level, for the leaded and unleaded fuel series; thus showing the octane requirements of the cars were generally lower with the leaded gasolines than the unleaded when assessed by motor octane number.

In answer to A. G. Bell's last question we estimate that for the cars in our experiments one crankshaft degree of knock limited spark advance was equivalent to a mean change of 0·6 research octane number.

We cannot agree with P. Draper that the differences between the anti-knock behaviour of leaded and unleaded fuels are solely because there is an imprecise correlation between results on the C.R.F. engine and a multitude of car engines. We believe that our work has shown that, as lead contents are reduced so there will be a need for gasoline refiners to make up for a loss of antiknock previously given by lead and, for car manufacturers to reduce the anti-knock requirements of their engines.

It was not the intention of our paper to comment on the effect of unleaded gasolines upon engine wear. However, in reply to P. Draper's question we would comment that there is plenty of evidence to show that valve seats ground

Table D1. High lead (0·7 g Pb/l) versus nil lead
Octane requirement data.
Effect of sensitivity (Motor ON).

	ORRF. 2 0·7 g Pb/l Tel	ORRF. 4 Nil lead
RON	SENSITIVITY	
91	8·1	8·6
93	9·7	10·2
95	11·2	11·4
97	12·5	12·6
99	13·5	13·3
Accelerating conditions Percentage cars preferring (100 per cent = 223 cars) . . .	51	13
Constant speed Percentage cars preferring (100 per cent = 103 cars) . . .	83	6

directly into cast-iron cylinder heads will suffer abnormally high rates of wear if the engine is run entirely on unleaded fuels. Our own investigations, and others, have shown that the wear occurs at high engine speeds.

W. E. Bernhardt

In reply to W. J. Annand, with regard to Fig. 149.3 approximately 10 per cent of the system's combustible mass is unburnt. With an engine speed of 3000 rev/min, under full load, with spark timing 28° b.t.d.c., a mean effective pressure of 8·32 kp/cm², and an equivalence ratio of 1:1 the calculated mass burnt fraction was 0·8946. This result is in excellent agreement with the calculations of mass burning rates (**14**). Contrary to that conventional methods for calculating the mass burnt fraction mostly indicate a 100 per cent combustion (**15**) and (**16**) though there are combustible compounds such as hydrocarbons, carbon monoxide, hydrogen etc. It was found that errors in measured pressure data affect the magnitudes of calculated mass burnt fraction curves.

Therefore, for this study it was decided to base the thermodynamic analysis of the engine combustion process on an average pressure trace which relies upon more than 1000 individual cycles.

Furthermore heat transfer to the combustion chamber walls is calculated using the Woschni heat transfer coefficient (**17**). The wall temperatures were estimated with the sleeve, piston, head, and valves, each at a different assigned temperature. The perfect gas law was assumed to be valid in this case.

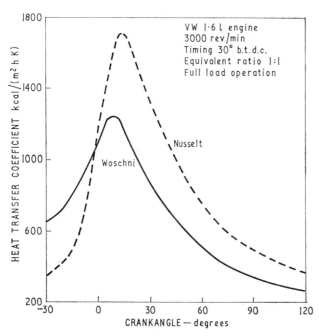

Fig. D20. Comparison between the Nusselt and the Woschni heat transfer coefficient for a typical thermodynamic analysis of an engine combustion process

With respect to the comprehensive procedure used here it is believed that the computed fraction mass burnt is quite correct.

Concerning the combustion photographic technique used here, working with Kodak Ektachrome Film EF 7242 (Tungsten 3200 K; ASA 125) which is developed especially for high speed cameras we got reasonable flame pictures as shown in Fig. 149.8. The luminosity of the combustion process itself was sufficient when using our recording technique.

With regard to W-T. Lyn's comments, during our studies it was found that an inaccuracy in the heat transfer to the cylinder walls affects the results of a thermodynamic analysis of an engine combustion process, it is, therefore, very difficult indeed to make a correct prediction of the local temperature of the burnt gas.

Fig. D20 shows, for example, heat transfer coefficients calculated from different equations given (**17**) and (**18**). It can be seen that the engine heat transfer problem is still to be solved because of this discrepancy of the calculated data. When such inaccuracies are involved the agreement between model emission prediction and experimental measurements cannot be very satisfactory.

The following correction to Paper C149 should be noted.

Page 284, second column, fourth paragraph, third line, 2000 frames per min should read 2000 rev/min.

REFERENCES

(**14**) PETERS, B. D. and BORMAN, G. L. 'Cyclic variations and average burning rates in a S.I. engine', S.A.E. Paper No. 70 00 64 1970.
(**15**) MILLAR, G. H., UYEHARA, O. A. and MYERS, P. S. 'Practical application of engine flame temperature measurements', *S.A.E. Transactions* 1954 **62**, 514–524.
(**16**) RASSWEILER, G. M. and WITHROW, L. 'Motion pictures of engine flames correlated with pressure cards', *S.A.E. Journal* 1938 **42**, 185–204.
(**17**) WOSCHNI, G. 'Beitrag zum Problem des Wärmeübergangs im Verbrennungsmotor', *Motortechn. Zeitschr.* 1965 **26**, 128–133.
(**18**) NUSSELT, W. 'Der Wärmeübergang in der Verbrennungskraftmaschine', *Forsch.-Arb. Ing.-Wes. H.* 264 1923.

D. R. Blackmore

In reply to W. J. D. Annand the fuel effects we measure are small and variable. We can only guess at what the mechanisms are. Obviously two variables are air/fuel (A/F) (or more properly equivalence) ratio and mixture distribution. But other effects may be present as well. A completely different and very extensive type of testing would be necessary to unravel all the reasons. Our objective was to quantify the overall fuel effects.

The reason for the variability in NO_x as mentioned by the contributor may be correct: on the other hand, it may not. I rather doubt if any of the cars we tested operated at equivalence ratios 5–10 per cent weak, and therefore near the NO_x peak. Since NO_x emission is so dependent on throttle position and spark timing, small changes in engine conditions, brought about by fuel property changes, can also bring about large changes in NO_x.

I very much doubt if engine cycle calculations are yet far enough advanced to predict with accuracy experimental effects of, for instance, hydrogen/carbon (H/C) ratio. Uncertainties in rate constants, gas flow patterns and temperature distributions probably make even simple predictions subject to large error. However, the results quoted are interesting in that they point to large potential changes and these relate to changes in equivalence ratio under conditions of perfect mixture distribution, but apparently at constant A/F ratio. In our experiments, the associated physical properties of the fuel (e.g. specific gravity in this case) will affect metered A/F ratio and mixture distribution, and may help to explain why we actually measure much smaller changes. In Table 132.8, the result of a simple calculation of the turbulent flow through a carburetter jet indicates no change at all in equivalence ratio as specific gravity of the fuel changes.

The following corrections to Paper C132 should be noted.

Page 115, second column, third paragraph, fifth and sixth lines. The sentence 'Both HC and NO emissions are very insensitive to volatility changes' should read 'HC emission decreases whereas NO emission is insensitive to increases in volatility.'

Seventh line, (cars 4 and 5) should read (cars 3, 4 and 5), ninth line, (cars 3, 6 and 7) should read (cars 6 and 7).

Page 116, first column, first line, (cars 4, 5 and 8) should read (cars 3, 4, 5 and 8).

Second paragraph, fourth line, (except car 7) should read (except car 6), tenth and eleventh lines should read, 'NO (strongly for cars 4 and 10, both with variable jet carburetters). Increase of viscosity'.

Line fourteen should begin 'notably, for cars 3 and 5)'.

Page 116, second column, second paragraph, sixth and seventh lines should read, 'jet emission controlled carburetted cars were also insensitive to the influence of specific gravity on HC. Few generalizations'.

Page 117, Table 132.8, third column, third line, volatility section, for 'Insens' read 'Dec.' Fourth column, second line, insert 'Dec.', the volatility section should then read:

$$\begin{cases} \text{Inc.} \\ \text{Dec.} \end{cases}$$
$$\text{Dec.}$$
$$\text{Dec.}$$

M. G. Brille and Y. E. Baguelin

In reply to D. K. Roberts, in Fig. 125.2, the a curve represents the turbocharger pressure ratios, and the b curves the Roots speed ratios.

Knowing that the 4-stroke engine has a 5·28 dm³ displacement for two revolutions and that the Roots has a 2·4 dm³ swept volume for one revolution, an overall ratio of the order of 2 or more should be obtained. However, from 1000 rev/min and below, the relative leakage of the Roots increases rapidly and the p curve in this area falls to under 2.

In reply to F. J. Wallace's comments, the $\frac{1}{4}$, $\frac{1}{2}$, $\frac{3}{4}$ and 1/1 refer to load and the curves represent the speed ratios of both the Roots and the engine against the engine speed and the load obtained by the regulator 15.

In reply to the question relating to emissions and torque back-up considerations, the accepted complexity of the whole is justified by the simultaneous search for good road traction qualities (torque, weight) and a reduced pollution level. The increased prices due to additional devices are offset by obtaining a higher specific power from a basic engine with low displacement.

In connection with the complication of the mechanism which was mentioned several times, we would like to point out that this complication is rather insignificant when compared to that of steam engines aiming at the same results.

In fact, it is more apparent than real, if the following points are examined:

(1) The lay-out in Fig. 125.1 has not been drawn to scale, which makes the space occupied by the whole out of proportion with that occupied by the engine alone.

(2) A complex design is not inconvenient; on the contrary, it meets the requirements; it can be criticized if it increases the weight, space occupied and cost and reduces viability. As far as the 'moteur compensé' is concerned, this complexity is examined in the following paragraphs.

(3) The turbocharger is already in use in current turbocharged engines.

In Fig. 125.1 the air–water exchanger (1) is commonly used in the the same engines, but only for cooling purposes. Here, the device is similar but better used as the heat exchange takes place in both ways.

The oil–water heat exchanger (2) is utilized in almost all the heavy-duty engines, even in naturally aspirated engines in which pistons are cooled by oil. In this case, the water comes from the normal circuit, which requires a bigger radiator.

The additional radiator (4) of the same size as a Renault R16 involves an increase in price but also implies, to a certain extent, that the main radiator size is reduced. It is difficult to set the necessary radiator between the truck side-members and for this reason, the breaking down in several units, as it is carried out here, constitutes a satisfactory solution.

The extractor (5) is set at the same position as the usual muffler and occupies about the same space.

(4) In fact, the true additional complexity is created by the Roots blower and its variable pulley drive together with the electric motor (14) (only when the generator cannot be used) and the longer additional pipes of the secondary water circuit.

(5) The Roots blower is a simple device whose weight is relatively low and which is not very expensive. The space it occupies does not increase the dimensions of the parallelepiped circumscribed to the engine, Fig. 125.7.

(6) The variable pulley drive of the blower constitutes the most critical complexity, especially from the point of view of the belt (13) and controller (15) viability. If the belt (13) breaks or the controller (15) does not work properly, this does not imply a complete breakdown as the vehicle can still continue its journey but at reduced power and producing worse pollution conditions.

On the other hand, with the second generation engine now being made, the belt can be easily and rapidly removed. The work in progress on this unit enables us to expect a viability similar to that of the other engine components.

7. An accident of the secondary circuit pipes does not stop the vehicle which can proceed at reduced power. In this case too, the quality specifications will avoid these incidents.

8. Finally, it is very important to note that the increase in price, which will constitute the only additional complex characteristic accepted must be offset by the increase in power.

The aim is to have the same manufacturing cost per horsepower as in current diesel engines.

The following corrections to Paper C125 should be noted.

Fig. 125.2, page 36. The three curves right at the top should be marked, in descending order, $\frac{1}{4}$, $\frac{1}{2}$ and $\frac{3}{4}$.

Page 38, second column, tenth and eleventh line should read, 'm.e.p. more than 13 kg/cm² which gives a P max to m.e.p. ratio less than 10, when operated with petrol as well as gas oil'.

Page 40, fifth paragraph, sixteenth line should read, 'At overrun, the air temperature gains 15 degC in some 200 m of downhill run, which is further'.

J. J. Brogan

In reply to questions by M. S. Bolton, the present generation of stratified charge engines, when operated without exhaust after-treatment, are inherently cleaner than their carburetted counterpart.

Exhaust emissions from the standard L-141 engine with the engine adjusted to military specifications, which include a rich air/fuel ratio, are as follows:

HC 6 g/mile*
CO 120 g/mile
NO_x 2 g/mile

When adjusted to a more conventional air/fuel ratio, the standard engine exhibits the following exhaust levels:

HC 5 g/mile
CO 60 g/mile
NO_x 4 g/mile

In contrast, typical stratified-charge engine emissions, without after-treatment, are:

HC 3 g/mile
CO 12 g/mile
NO_x 2 g/mile

* *Emission levels per 1975 Federal test procedure; the values given are typical, but as much as 30 per cent variance from these levels might be encountered in any given test.*

It is worth noting that the above levels represent untreated emissions from a stratified-charge engine optimized to deliver lowest emissions when operated with after-treatment provisions. They are not the lowest raw exhaust levels achievable. If the engine is adjusted for minimum untreated emissions, the following levels would be expected:

HC 1·5 g/mile
CO 3 g/mile
NO_x 2 g/mile

The solution to this is the planned expenditures on the Rankine cycle programme of $26·3 million from fiscal year 1970 to fiscal year 1975. The Federal Government fiscal year runs from 1st July, thus fiscal year 1970 ran from 1st July, 1969 to 30th June 1970. Planned expenditures on gas turbine will total $14·2 million and the total planned for the stratified-charge engine is $3·7 million over the same fiscal years.

K. Campbell and P. L. Dartnell

In reply to W-T. Lyn, the catalytic action of PbO on the oxidation of carbon is well known and documented in the literature. Undoubtedly this reaction cotributes to the reduction of carbon emissions observed with leaded gasoline.

For R. W. Mellor's information, the lead content of the leaded fuel was 0·80 g Pb/l and that of the unleaded fuel was less than 0·01 g Pb/l as referred to in Fig. 123.8. The lead content of the fuel used for testing the catchment efficiency of the filters was 0·8 g Pb/l. We selected this high lead content because it would produce the most arduous test conditions for these filters. We have run other tests which have shown these filters to be equally effective over a very wide range of lead content of the fuels.

In reply to W. J. D. Annand, the estimates given in the paper of the percentage particulate matter from vehicles (2 per cent) relate to the United States of America and not just to California. The estimates based on 1968 data indicate that the weight of particulates in the U.S.A. arising from motor vehicles is approximately 20 000 tons per day.

In reply to D. R. Blackmore, it has been shown that above 200°C alumina begins to react with lead salts to form non-volatile lead complexes. The lead halides react at lower temperatures and at a faster rate than the other lead salts. Data on the composition of the fine lead particulates emitted by motor vehicles indicate that these are predominantly lead halides. Consequently, the efficiency of alumina-coated filters in reducing the fine particulate content may be due to the preferential reaction of the alumina with the halides which are the source of the fine material. At the moment we cannot say which mechanism predominates.

J. W. Dable

In reply to G. L. Lawrence's comments, I believe his first point refers to data which were not published in the paper,

Table D2

Registration letter	Approximate year of manufacture	Carbon monoxide emission at idle of sample cars—percentage volume		
		Minimum	Maximum	Average
K	1971	3.9	8.9	6.8
J	1970	1.0	10+	6.3
H	1969	3.1	10+	6.7
G	1968	7.1	10+	8.7
F	1967	0.4	9.4	6.6
E	1966	5.2	8.9	7.6
D	1965	0.3	10+	5.6
C	1964	4.0	9.0	6.5
B	1963	1.5	10+	6.0
—	Before 1961	2.0	10+	8.4

but were presented as illustrating the level of exhaust emissions in cars currently running on the road. We stated that the average level of carbon monoxide emission during idle for 62 cars taken at random in a roadside check was 7.1 per cent.

We obviously did not present these figures very coherently, since G. L. Lawrence interpreted them as meaning that the cars were 1962 models. In fact, as it happens, most of the cars were quite recent models. Due to the nature of this random sampling, we did not determine precise details of the cars involved, but simply noted the make, model and final registration letter (indicating the approximate year of manufacture). A summary of the emission levels obtained is given in Table D2, which illustrates the apparent independence of the idle carbon monoxide level upon the year of manufacture. Although this is an extremely small sample, the figures indicate the serious nature of idle levels of carbon monoxide emissions which may be issuing from the majority of cars on the road today.

In reply to the second point I would confirm that accelerated means were adopted in order to allow the test to be carried out within a reasonable period of time. The deviations from normal conditions of operation were as follows:

(1) All cars were run on a controlled circuit consisting of mixed rural, urban and congested city traffic conditions.

(2) The crankcase ventilation system was modified from that fitted originally to one in which all the blowby was directed through the carburetter via the air cleaner, which, although non-standard, is typical of a large percentage of cars on the road today.

(3) The crankcase oil, although bought from a local garage forecourt, represented one of the lower quality lubricants currently available commercially.

Finally, the early engine tests using S.U. carburetters, covered in Table 128.3, were designed only to compare deposit levels in carburetters of a fairly light nature which had no measurable effect on engine operation. However, two of the cars used in the 'old car' fleet reported in Table 128.5 had constant vacuum carburetters, and they both exhibited a considerable 'clean-up' when switched to the additive treated gasoline.

A. E. Dodd

The data presented by J. A. Hoare are very interesting, particularly the results obtained on the E6 engine. Use of such an engine gives an ideal means of identifying effects due to compression ratio which can be varied between limits with minimum interference from other factors such as valve timing, chamber deposits and the indeterminate effects that can arise from stripping and rebuilding an engine. Results obtained from the 'Varicom' engine are also of great interest although the conclusions drawn from Fig. D10 do not appear to be supported by the data presented. However, we await with some eagerness further emission results from this intriguing engine.

I have little comment to make on J. H. Alden's contribution except to thank him for his very kind remarks. As he states, it is more difficult to achieve reductions in emissions on engines with small cylinder volumes and this fact resulted in the initiation of the work undertaken by M.I.R.A.

In reply to A. W. E. Henham, I would like to refer to Fig. D21 which presents mean emission results from the different combustion chambers at two levels of load; overall means for the chambers are also indicated. As would be expected, peak NO emissions at half load are lower than at full load but the relations between hydrocarbon emissions at the two load levels are dependent on chamber configuration. With chambers D, G and H, similar hydrocarbon concentrations were produced at both load levels and, while the data are insufficient to give complete certainty that the concentrations emitted are completely independent of load, the indications are there. Thus, the best chamber 'H' produced almost equal emissions at both full and half load. It is of interest to note that the chambers exhibiting this characteristic are of a more quiescent type.

On the question of compression ratio, road efficiency is affected because, although peak pressures are deliberately reduced by throttling, the effective expansion ratio is basically dependent on compression ratio and valve timing while peak pressure can have an influence only at the lighter loads. In improving emissions one has a choice of means for achieving reductions in both hydrocarbon and nitric oxide levels but unfortunately both lower compression ratio and retarded ignition result in reduced efficiency. Possibly retarded timing at certain loads and speeds is preferable because it may allow the benefits of high compression to be obtained under other operating conditions.

Within the motor industry, comparative tests between similar cars with engines of differing compression ratios must be fairly common but M.I.R.A. have no data on the subject. A comparison between similar cars fitted with

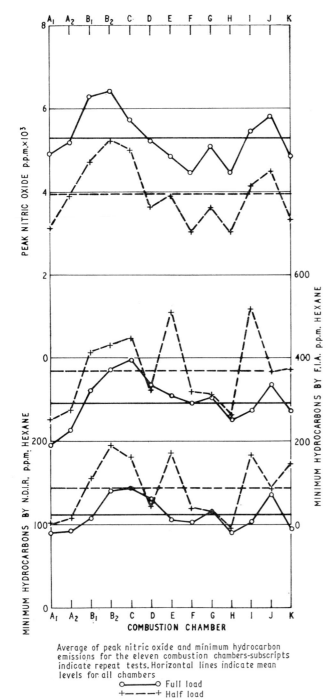

Average of peak nitric oxide and minimum hydrocarbon emissions for the eleven combustion chambers—subscripts indicate repeat tests. Horizontal lines indicate mean levels for all chambers
○———○ Full load
+ — — — + Half load

Fig. D21

to avoid cut-outs would result in a less turbulent chamber and, from the data previously referred to, this might be expected to result in lower overall hydrocarbon emissions. This would tend to confirm H. Mundy's comments on Jaguar findings reported at a Motor Show lecture.

D. R. Fussell

D. Downs asks the question 'How unleaded is "unleaded"?' The United States oil industry has, of course, considered the specification level it could meet for lead in unleaded gasoline. It has submitted evidence to the United States Government that average lead levels could be held to about 50 p.p.m. (0·05 g/l of lead) for gasoline distributed at the pumps. An average level substantially below this could not be guaranteed while oil companies are marketing both leaded and unleaded fuel, a situation likely to exist for some years.

H. Daneshyar and J. R. Orme are correct in saying that relatively little attention has been paid to the effect of thermal and catalytic reactors on engine performance. It is a question of priorities. The principal requirement is to find which methods meet the stringent 1975/76 U.S. emissions requirements. Having identified the most promising techniques we can then turn our attention to minimizing the adverse effects on engine performance afterwards. With the reactor-equipped M.G.B. we have reported a power loss sufficient to drop the maximum speed about 5 per cent. This we believe to be equivalent to about 10 b.h.p., or just over 10 per cent of the net power output from this engine.

C. D. Haynes and J. H. Weaving

In reply to C. L. Goodacre, I.C.I., with whom we have been working, do, in the development of catalysts for catalytic reactors, subject the material to water sprays followed by freezing as part of the development.

We agree with M. S. Bolton that a fluidic switch might be a possible solution for by-passing catalytic reactors when temperatures become excessively high.

In reply to H. Daneshyar and J. R. Orme we do not use a venturi for aspirating air in our catalytic system because inexpensive air pumps which have a low power consumption are available. Catalytic reactor boxes are designed to have a similar pressure drop to that of the normal silencer system which they replace. Due to the complexity of the flow, back pressures are measured experimentally with a typical system and modifications made as required to produce acceptable values.

In reply to G. Lawrence, it is our view that it would be difficult to get into production with any device to meet the regulations for 1975 but the catalytic approach shows more promise than the alternatives. In addition, as very low nitric oxide figures are required for 1976, an approach for 1975 only would appear to be undesirable.

J. Soltau comments that the time has arrived to pool our knowledge and develop reliable commercial products. This point is well taken but we do not feel that communi-

engines of 1200 cm³ and 1500 cm³ of which I have experience was complicated by the fact that the smaller-engined car was fitted with a higher axle ratio to maintain performance; the larger-engined car was the more economical.

In reply to R. W. Mellor, I can confirm that the piston of chamber I had cut-outs 0·16 in deep to give clearances to the valves but no tests were carried out to determine the effects of these. However, a reduction in piston height

cations are as bad as he suggests. Research and Development (R & D) must precede commercial exploitation and due to the continually shifting target in the guise of the American regulations, R & D has had to continue with progressive intensity to meet these stringent requirements which now bear no relation to what they were ten years ago. The situation now seems ripe for the development of a reliable commercial product but it should be appreciated that it is the limited time available that precipitates this stage rather than the feeling that all necessary research has been done. It certainly has not!

We support the comment of D. Downs with regard to specifying a suitable fuel and lubricating oil to permit catalytic devices to give their optimum performance for the very long duration that Federal regulations require, namely, 50 000 miles.

With reference to H. Connor's observations on the production of ammonia when exhaust gases are passed over reduction catalysts, as stated in our paper, we feel that the water–gas shift reaction produces hydrogen rather than a reforming reaction of hydrocarbons although at this juncture this could not be proved. I.C.I., however, have developed catalysts that form practically no ammonia.

A. H. Ball asks whether a satisfactory by-pass arrangement has been devised. We think the answer to this is 'no' and the by-pass system shown in our paper is more of a thermostatic system to avoid the catalyst overheating in any normal driving. We do not think its time response would be quick enough to deal with an abnormal condition, namely: the failure of a sparking plug while the engine is under a high load condition.

In reply to S. E. Voltz, we have tried many forms of catalyst including monolithic ones. Whilst the monolithic form is currently the favoured version, development of granulated and extruded forms continues.

D. Lyon, A. H. Howland and W. L. Lom

We agree with D. G. Burton that the effect of retarding injection timing in combination with dual fuelling yielded an encouraging result. Our explanation in the paper was that the time required to introduce the total fuel into the engine was less in the dual fuel mode and hence more time was available for combustion. As injection timing is retarded, less time is available for completion of combustion and hence the duration of the injection period becomes more critical. Any method of reducing this time should show advantages and it was interesting to note that the authors of Paper C151 obtained similar effects by increasing the fuel pump delivery rate.

Although decreasing the time required to introduce the fuel may in part explain the effect of dual fuelling under retarded conditions it does not explain the differences obtained using propane and butane. In our experiments on the Petter AV-1 engine we noted that butane had a less detrimental effect on the ignition delay period and also tended to give more rapid combustion than propane (see Table 126.4 of Paper C126), and this may be the reason for the better results obtained under retarded conditions with this fuel.

In reply to H. O. Hardenberg, our choice of the Saurer engine to initiate our studies in the field of dual fuelling was dictated by the fact that it was the only engine equipped for that purpose available on the market. Certainly, if a Daimler-Benz engine similarly equipped was available at the time we would have considered using it.

We are extremely interested in the observation that the Daimler-Benz engine failed to respond to dual fuelling and regret that no detailed results of the tests were presented at the symposium. Since our work on the Saurer engine, we have progressed to studies on other more modern engines and these studies confirm our original observations.

H. O. Hardenberg claims most direct injection engines emit less than 8 g/b.h.p. h for $NO_2 + HC$. This is certainly contrary to our findings and also other workers in the field, and we wonder whether these figures refer to an indirect injection engine which is known to emit approximately half the emissions of a direct injection engine.

Higher rates of pressure rise and peak pressures were recorded by us only with high butane concentrations in conditions of obvious rough running (Fig. 126.2). In normal practice with concentrations well below the limits of inflammability these were not observed.

In reply to W. Tipler's comments, the lower limit of smoke emission from L.P.G. dual fuelling is dictated by the smoke level at 75 per cent diesel fuelling. The L.P.G. premixed with the air merely restores the original power with only a minor addition to the 75 per cent diesel fuelling smoke level. We, therefore, agree that any optimization at 75 per cent diesel fuelling should show benefits, but we feel they would be of second order magnitude. The Saurer engine as supplied to us was designed to operate in either the dual fuel or the 100 per cent diesel mode and therefore the fuel injection equipment was optimized for the higher fuelling level.

The experiments were carried out with commercial grade propane and butane which are typically of 95 per cent purity in the component specified.

Obviously, it would be difficult to meet immediate demands for L.P.G. if there was a massive shift to dual fuelling. However, we envisage this approach as best suited to municipal bus fleets and large road haulier firms where bulk delivery could be made as a normal extension of industrial L.P.G. business. Concentration in this way on, for example, municipal bus fleets, would have the added advantage of maximizing the benefits in areas where air pollution is at its highest, i.e. city centres.

T. Muroki

In reply to W. D. Holt, the space between the inner and outer shell in our thermal reactor is filled with insulation material and therefore the temperature of its outer surface is considerably lower than the skin temperature of the inner shell. Besides, the thermal reactor mounting position is lower on the rotary engine than on the reciprocating engine. For these reasons, problems due to temperature scarcely occur.

This reply relates to a question by P. S. Wilson concerning different timings of reactor spark plugs. Our rotary engines have two rotors, each of which is provided with two spark plugs. Therefore, even if one of these spark plugs fails, it will be the fail-safe point. Very seldom will two spark plugs fail at the same time. But should both spark plugs fail to work, the engine output will be reduced to half which will be sensed by the driver who then will stop his car for checking. In the case of the 4 to 8-cylinder reciprocating engines, it will be relatively difficult to discover the misfiring which has occurred on one of the spark plugs.

In reply to C. L. Goodacre's question concerning Fig. 152.1, though the inlet port type may not be clearly identified in the figure, our rotary engines are provided with side entry ports. We have never abandoned the application of the side entry port system to passenger car engines.

Replying to P. Draper's question, in our experience, it is seldom necessary for the apex seals to be replaced before the car covers 100 000 km. There is no problem with the side seal. However, if overhauling is found to be necessary then gas sealing is one of the important factors.

In reply to J. P. Soltau, from my experience I have found that it is very difficult to burn the pollutants in the thermal reactor using the spark plug.

With regard to the question concerning the use of reactor skin cooling air, this is not used as a means for providing heated air to the exhaust port nozzles.

Under the operation with light load and/or low speed of the engine after cold starting, the cooling air is not passed through the air jacket controlled by the air control valve so that the cooling air will not cool the reactor.

S. Ohigashi, Y. Hamamoto and S. Tanabe

As pointed out by D. G. Burton, Fig. 134.8 was incorrectly marked and corrections have been made at the end of this reply.

In reply to T. J. Williams, the unsymmetrical configuration shown in Fig. 134.3b and c would suggest, as he pointed out, that the flow is interfered with and distorted by electrodes, a probe and insulators. In this study, however, it is most essential that the direction of flow can be determined by the angles θ and φ in which the value of $1/\tau$ attains its maximum. The thickness of the discharge path is expanded by the diffusion of ions. Thus, in a plug of type A, as shown in Fig. 134.3a, the value of $1/\tau$ does not change as the probe is located in the range where the discharge path is expected to pass, that is $\pm 30°$ of θ in this case.

The probe cannot be placed lower down in the engine cylinder than that shown in Fig. 134.2a, because it would come into contact with the piston. It is presumed that the different values of the flow velocities are obtained by varying the measuring points in the cylinder during the intake stroke. It is considered, however, that the flow velocity coming into the cylinder in a firing operation may be higher than when motoring, due to the expansion of air heated by the residual gas in the cylinder or by the intake valve, the chamber wall, etc.

Assuming the conservation of angular momentum of the swirling air in the cylinder, the swirl velocity increases owing to the increase in the mass fraction of the gas in the piston cavity to the total cylinder charge during the compression stroke. The increase in swirl velocity is estimated as follows:

$$\frac{\omega_1}{\omega_2} = \frac{V_1 D^2 + v d^2}{\left(\dfrac{V_1+v}{V_2+v}\right)(V_2 D^2 + v d^2)}$$

where ω is the angular velocity,
  D the diameter of cylinder,
  d the diameter of piston cavity,
  V the volume of cylinder, except the piston cavity,
  and v the volume of piston cavity.
Suffix 1 is for the initial stage and 2 for the final.

The result shows that the velocity at 32° crank angle before t.d.c. increases by 18 per cent of that at 80° crank angle before t.d.c. It is noticed that this increase in velocity is much less than that caused by the piston squish. Therefore, the authors understand that the increase in velocity at the latter stage of compression stroke may be mainly caused by the squish action.

The following corrections to Paper C134 should be noted.

Page 133, Fig. 134.3. From left to right the diagrams are a, b and c respectively.

Page 135, first column, second paragraph, sixth line: 'fuel to air' should read 'air to fuel'.

Fig. 134.8: ○ 6-holed nozzle should read ○ 8-holed nozzle. ● 8-holed nozzle should read ● 6-holed nozzle.

B. J. Kraus and G. P. Richard

Regarding W. J. Annand's first point, I must remind him that detailed investigations of the effects of fuel properties on mixture strength and distribution, which must then be related to emissions under the conditions of a standard test including several driving modes, would prove to be very time consuming. Our object was to test a reasonable cross-section of the vehicle population, in several different laboratories, and bearing in mind the fact that the vehicles were not available to us for more than a few days, we felt that such a comprehensive investigation would be impractical. In consequence, we did not carry out individual measurements of cylinder to cylinder fuel distribution. In the case of mixture strength, however, we obtained where possible a measure of the mean mixture strength over the whole of the test cycle. These measurements indicated that fuel quality variations over a range of typical commercial fuels might produce a variation of 0·05 to 0·1 in the mean equivalence ratio at which a vehicle operates.

Although the scope of the individual tests was limited to the measurement of emissions and mean mixture strength under standard cycle conditions we feel that we have been able to demonstrate quite clearly the very important point that various combinations of fuel properties react with different fuelling systems in different ways, so that any given alteration in fuel quality can have a beneficial effect on emissions from one vehicle, and an adverse effect on another. Moreover, we have been able to show that for vehicles with effective emissions controls the magnitude of the effect is small compared with that which could be obtained by minor mechanical adjustments to the fuelling system.

I would also agree with W. J. Annand that variations in fuel distribution between cylinders in different cars is one of the factors which prevent us from observing consistent fuel quality effects on the concentration of nitric oxide to be found in the exhaust. It is true that one can carry out calculations which indicate quantitatively the effect that a change in fuel carbon–hydrogen ratio will have on nitric oxide under certain fixed conditions. Since, however, an alteration to carbon–hydrogen ratio also implies changes in fuel physical properties, which can affect mixture strength and distribution in various ways, it seems to me that such a calculation would be of limited use in the prediction of nitric oxide concentration in the exhaust of a typical vehicle.

Our finding that the ratio of F.I.D.–N.D.I.R. hydrocarbon varies according to fuel composition is in line with P. J. Moss's contention that exhaust hydrocarbon composition can reflect to a large extent the constitution of the parent fuel. We would also expect the observed F.I.D.–N.D.I.R. ratio to vary considerably from vehicle to vehicle depending on combustion chamber temperature and dimensions and on the effectiveness of the various emission controls. The figure quoted by P. J. Moss to quantify the relation between F.I.D. and N.D.I.R. measurements is equivalent to an F.I.D.–N.D.I.R. ratio of 1·67 and this is quite close to the generally accepted mean value. In these experiments we tended to get higher values, even for fuels comparatively low in aromatics content. We are at present unable to offer an explanation for this, but it does seem possible that a systematic investigation of the parameters controlling the ratio of the two measurements could yield useful information on the factors affecting the control of exhaust hydrocarbon.

In reply to a speaker in Session 2, to convert the N.D.I.R. reading as shown in Fig. 129.8 to a mean F.I.D. equivalent, the figures given on the left-hand side of the graph need to be multiplied by about 2·5, while those on the right-hand side must be multiplied by about 3·5. This considerably reduces the slopes of the lines, but in neither case do they become positive. In fact, the points plotted in Fig. 129.9 are the mean values we obtained for the ratio of F.I.D.–N.D.I.R. readings, the actual ratios obtained varied from car to car. Examination of the raw data showed that while there was a variable tendency for N.D.I.R. hydrocarbon to decrease as fuel aromatic content increased, the F.I.D. measurements were essentially independent of this variable.

In reply to C. L. Goodacre, it is really quite difficult to obtain reliable and reproducible measurements of particulates in vehicle exhausts, and in this series of tests, where the vehicles were not available to us for more than a day or two, we made no attempt to measure particulate emissions.

As M. A. I. Jacobson observes, 'smelliness' is difficult to define, but in general it does appear that the odour of vehicle exhausts increases as the quantity of partially burned material in the exhaust increases. Most people who stand anywhere near to an over-choked vehicle can vouch for this. While we did not carry out any systematic odour tests on our test vehicles, we believe that the effect of controlling emissions to the standards used in these cars would have a definite beneficial effect on exhaust odour. Whether any significant further improvement could be obtained by even more stringent emissions control is, however, quite doubtful.

In reply to comments by B. H. Croft, it is true that the inclusion of eight vehicles having exhaust post treatment devices introduced a further variable into the test. Since our object was to find the response of a reasonable cross-section of the vehicle population to changes in fuel quality we felt that it was necessary to accept this. The vehicles having post treatment devices are identified in the paper, and examination of the emissions results from these vehicles shows that their response to fuel quality was very similar to that of corresponding vehicles whose emissions were controlled by engine modification.

Regarding the difference between mechanically and electronically controlled injection systems, I would like to confirm that the former used volume metering, while the latter had fixed orifice metering. Our essential point was that those which employed fixed orifice metering were remarkably insensitive to any fuel parameter other than R.V.P.

Remembering that the discharge coefficient of the orifice is a function of Reynolds number and thus depends on viscosity, it seems to us quite possible that the response of the pressure regulator to simultaneous changes in gravity and viscosity would very possibly be as we have postulated. However, we have not subjected these systems to any detailed investigation, and we agree with B. H. Croft that some other, as yet undefined factor, could well be responsible for at least a part of the observed response.

Finally, B. H. Croft's results illustrating the response of a fixed orifice metering system to 'conditioned' and 'fresh' fuel, presumably having low and high R.V.P. respectively, are most interesting. His explanation of the fact that the 'fresh' fuel gives a high delivery at low manifold pressure seems very reasonable, but unfortunately does not help us to explain our observations of the effect of R.V.P., since in our case all three vehicles having orifice metering showed increased exhaust CO as fuel R.V.P. was reduced. It is apparent that considerably more work must be done before we really understand the complexities of such fuel metering systems.

J. P. Soltau and K. B. Senior

With reference to M. S. Bolton's question, we have used a self-adaptive form of control with a hot wire CO detector as the control device. Response was adequate for near steady-state work, e.g. idle warm-up, but it is felt that it is inadequate to satisfy the normal transient changes. It would seem that a memory type of control is required in order to give rapid response together with a CO measuring device which 'inches' the reference points of the control whenever the conditions are steady.

J. P. Soltau and R. J. Larbey

In reply to R. Gippet's questions, the Japanese regulations allow adjustment of the dynamometer according to the following formula:

$$F = \mu r W + \mu e A V^2$$

where F is the running resistance (kg),
W the gross vehicle weight (kg),
μr the rolling resistance coefficient,
μe the air resistance coefficient,
A the frontal projection area (m^2),
V the speed (km/h).

The regulations were attempting to ensure that the power absorption characteristics of the dynamometer matched the power absorption characteristics of the vehicle operating on a level road.

The first part of the formula $\mu r W$ gave the total mechanical drag of the vehicle; W was easily established but μr varied with vehicle condition, i.e. new or run-in, temperature, etc.

The second part of the formula $\mu e A V^2$ gave the aerodynamic drag of the vehicle. The frontal projection A could be relatively easily determined, but a lengthy procedure, either practical or empirical, was required to establish μe. F was calculated for various values of V.

At the design stage of a new vehicle this formula would indicate the capacity or range of the dynamometer required to conduct the test. However, the Japanese regulations allow matching of the dynamometer and car driving force values by setting to known manifold pressures after previously conducting a series of level road steady speed test recordings.

Most dynamometers currently used for emission test procedures tend to be of the hydraulic absorption type as for a fixed hydraulic rotor–stator arrangement and true propeller law force–speed characteristic was available. The popular method of matching the dynamometer to the vehicle, using only one speed and intake pressure datum point, could lead to some degree of mis-match and a better practice would be to use a minimum of four datum points to establish a higher degree of accuracy.

It was felt that many hydraulic absorption dynamometers currently in use did not have the adjustment capability to rectify load errors over particularly limited speed ranges without affecting the entire dynamometer characteristics. In this light, programmable electric absorption dynamometers such as the Siemens, Labeco and some Heenan and Froude types could offer advantages.

The real effect of dynamometer adjustment accuracy on exhaust emission testing is unknown and some beneficial work could be conducted to establish this point. It may be pointed out that early developments in clean engines employing both programmed dynamometers and water brakes did not show any significant differences in results. A more definite form of dynamometer setting procedure could well be beneficial as emission vehicles and test procedure become more discriminating.

A chemiluminescent NO_x analyser would be better than N.D.I.R. for the measurement of diesel NO_x emissions because the latter only senses NO and not NO_2, thus underestimating NO_x emissions. However, as the main outflow from a diesel engine is NO, N.D.I.R. analysers currently give valid results.

The last point is interesting and we would agree with Mr Gippet that improved gas analysis techniques are yet to come if we are to keep pace with the proposed future exhaust quality regulations.

P. M. Torpey, M. J. Whitehead and M. J. Wright

In reply to S. Timoney, it is most gratifying to see the universities conducting experimental work on this topic of a practical nature. Comparison of emissions between 2- and 4-stroke diesel engines is of particular interest since we have not covered this aspect in our paper.

In Fig. D18 shown by S. Timoney it appears that this engine has no automatic speed advance on the injection system. This means that the performance was not optimized over the speed range. Likewise, at reduced compression ratios the injection timing characteristic required for optimum performance over the speed range will be different. Thus the fall in NO concentration with compression ratio may be due in part to a non-optimized F.I.E. for minimum fuel consumption.

We agree with S. Timoney's observation that the effect of variation in engine parameters on NO reduction becomes less at lower levels, this also accounting for the greater effect of injection timing on open chamber engines compared to swirl chamber engines.

Although our experiments with hot exhaust gas recirculation were not all that encouraging as a means of NO reduction, experiments on a reduced scavenged 2-stroke engine would prove useful, particularly as this method involves the minimum of cost and effort.

We agree with W-T. Lyn that the control of HC emissions should be more fully investigated and understood. It has now been fairly well established that the specific HC mass emissions reduce with increasing engine size and its control in small engines is difficult. W-T. Lyn would agree that the value he quotes of 3 g/hp h must have come from a small engine of around 3 to 4 in bore size. In our experiments using $4\frac{3}{4}$ in bore engines a true comparison of emissions can be made and the results show that the indirect injection can have very low HC emissions in the order that W-T. Lyn expects to meet the legislative limits. In fact, the levels recorded for the indirect injection engine produce a 13-mode result of around 0·4 g/hp h while the direct injection engine result was 1 g/hp h.

The question by W. Tipler summarizes the problems existing for all engine investigators, namely, that of isolation of a single variable and its effect on engine performance.

A variation in engine compression ratio is a particularly difficult parameter to isolate due to its inherent effect in changing the engine geometry and surface–volume ratio. In the indirect injection engine this is made more difficult by the dividing of the two combustion chambers. However, a number of parameters were optimized for each engine build so that each gave as good a performance as possible. These were the area of the throat between chambers and the F.I.E. specification. We feel that the most important reason leading to our conclusions was the fact that although five other parameters were varied in order to obtain a reasonable range of compression ratio, each build still maintained similar peak NO concentrations. It is difficult to see how all five parameters could cancel each other out in every case.

We still believe that a reduction in compression ratio combined with sufficient injection retard to reduce the NO to acceptable levels will lead to difficult starting and large increases in light load HC levels due to engine misfire. On the other hand, a high ratio allows adequate retard to achieve 1975 C.A.R.B. levels without misfire, any increase in NO due to the high ratio being insignificant in comparison.

The following correction to Paper C124 should be noted.

Page 24, Fig. 124.2, second illustration on the right-hand side: 400 contour should be 200.

Page 31, Table 124.3, fourth column, sixth line, 4.2 should read 42; fifth columc second line, 11·5 should read 8·5.

S. E. Voltz

In reply to comments on our paper many of the basic principles of emission control systems have undoubtedly been known for many years and much of the current research may appear to be repetitious. However, considerable progress is being made in the development of improved thermal reactors, catalysts, and converters. Further technological breakthroughs will be needed to meet the 1975–1976 U.S. emission goals.

In reply to G. L. Lawrence, although significant progress has been made in the development of emission control systems during the past few years, many problems must be solved to provide practical and effective systems for 1975–76 production vehicles in the United States (U.S.). The National Academy of Sciences and other agencies are conducting studies of the entire situation and will provide recommendations to the Environmental Protection Agency on the feasibility of meeting the 1975–76 U.S. emission goals.

In reply to E. S. Starkman, within the I.I.E.C. programme, fundamental mathematical models of thermal reactors and CO–hydrocarbon catalytic converters have been developed. These models are based on reaction kinetics, thermodynamics, and fluid mechanics. They have been used to accurately predict the performance of thermal reactors and converters under many different operating conditions and to optimize emission control systems. I described some applications of the converter model in my presentation.

With regard to P. Draper's comment, the hardware development which I reported in my presentation is part of the I.I.E.C. programme and involves many companies in addition to Ford.

In reply to D. R. Blackmore, during the past year we developed four different screening tests for CO–hydrocarbon oxidation catalysts. They have been used to determine the oxidation activities of many widely different catalysts. Each test represented an improvement over the preceding one, particularly with respect to using more realistic reactant gases. The first two tests involved synthetic gas mixtures; we currently utilized doped engine exhaust gases.

Most of our initial ageing studies are conducted in a converter mounted on an engine dynamometer. We can simultaneously age about 100 catalyst samples in this converter. The oxidation activities are usually determined at increments of 5000 equivalent miles of ageing.

J. W. Wisdom

It has been freely acknowledged by all responsible engineers that whilst significant progress in the design of thermal reactors has been made over the last few years, much remains to be done. Improvements will be made in the packaging requirements, both from the optimization of the mixing processes and combustion performance and also from the use of improved heat insulation methods. However, in general, the long term performance index of the systems available at the moment and production lead times are probably the major technological factors controlling the use of these arrangements to meet the 1975 levels.

Whilst I am in no position to speak for the motor industry's capabilities, I personally feel that a decision to produce a particular design for 1975 would have to be made in the immediate future. Since it will not be possible to provide sufficient data on the long term operation of a representative number of units to give an adequate basis for a decision to be made, I therefore feel that, unless my view of lead times is completely wrong, it is unlikely that the 1975 levels will be met by the use of thermal reactors. With regard to the use of less exotic treatments to attain the 1975 levels, I know that significant advances have been made in lean burning air valve carburetter equipped engines. Some of the data available are contained within G. L. Lawrence's own paper for this symposium. At this stage it is too early to be positive, but I believe that there is sufficient basis for some optimism with regard to the possibility of these systems attaining remarkably low emissions coupled with good vehicle drivability.

I think J. P. Soltau was speaking with his tongue in his cheek to some degree. Nevertheless, his remarks are true in principle. A lot has been said before about thermal reactors and afterburners. They do work well in terms of one-off development systems. In the past the legislative controls did not have sufficient bite in them to produce the necessary engineering and commercial interest needed to design practical exhaust treatment systems. Rightly or wrongly, the bite is now on and the work intensity and coverage has increased accordingly.

In both this paper and our earlier one in 1970, we gave details of the effect of the reactor upon the engine power, torque and fuel consumption characteristics. In all cases, some loss in performance was recorded but the penalty was only of the order of 3–4 per cent of the standard levels. In other work we have experienced greater losses in performance. This was particularly marked in engines where the exhaust pulse was utilized as a tuning feature. In these cases an alternative design of reactor was used to restore the performance to near the original levels.

I find it difficult to believe that loss due to heat addition within the reactor plays a significant part in the increase in back pressure. Although the burning efficiency and thus temperature would be higher, I do not think that the hot losses would be much greater than those within a standard manifold and would in any case be small compared to the aerodynamic losses.

With regard to J. J. Mikita's discussion I would agree that theoretically a fuel injection system offers advantages in the control of fuel distribution. However, I would question if this potential advantage has in fact been realized in practice in sufficient magnitude to significantly affect the exhaust emission levels.

J. E. Witzky

The stratified-charge (s/c) engine, we believe, has some of the characteristics of the Otto and diesel engines. Therefore, some engineers call it a hybrid engine. Whether it is closer to the diesel cycle is a matter of opinion.

I have to agree with D. Broome's statement that NO concentration is a function of air/fuel ratio. According to well-known curves (Fig. D22) NO_x should reach zero at a ratio of 20:1. A diesel or the s/c engine works seldom richer than this ratio but the NO concentration of these two engines is high, indicating that the combustion temperature determines the NO concentration and the above mentioned curves apply only to gasoline engine.

The CO level of the s/c engine is low and in fact lower than the United States' specification for gasoline engines for 1975. Therefore, a CO emission control should not be necessary.

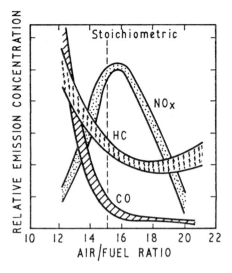

Fig. D22. Emission concentration in engine exhaust

The HCs are high on the s/c engine; it should be low, however, when operated with a 20:1 air/fuel ratio. I believe the reason is wall quenching, which cannot be avoided with the stratified charge engine.

I have to agree that at the present state-of-the-art the combustion control within the cylinder could not be achieved. But we think, considering the little work invested in this system, that the swirl stratified charge engine has a considerable potential

In reply to W. E. Bernhardt, it is a tragic fact that if we want to build a highly efficient engine the combustion temperature has to be high and consequently the NO concentration. The NO can be reduced by going to a less efficient combustion system like a precombustion chamber or a Wankel arrangement.

In reply to D. A. Browne, the particular characteristic of the swirl stratified charge principle provides an ignitable air–fuel mixture at the spark plug independent of load. We never experienced any difficulty operating the engine at idle with an overall air/fuel ratio of 100:1 and at full load where the air/fuel ratio was stoichiometric.

In this stratified charge principle, the fuel has to be injected against the air swirl; by injecting with the air swirl the engine did not perform at all.

In reply to T. J. Williams, Fig. 136.2 shows the theoretically perfect stratification of the fuel at the central location of the spark plug. As the high speed photograph of injection and combustion (Fig. 136.7) shows in the sequence photograph 6-7-8-9 that if the penetration is excessive the tip of the injection spray was carried with the air to the periphery of the combustion chamber. It is assumed that the fuel in the periphery is too weak to be consumed, a condition which should be avoided under all circumstances.

As mentioned in the paper, the nozzle first was located at the periphery of the combustion chamber (Fig. 136.4). The performance and overall behaviour of the engine, however, could be markedly improved by moving the nozzle into the cylinder head. The engine usually tells us what it wants without much consideration to the theory established.

We, of course, were trying to photograph the air movement using our picture window engine (Fig. 136.3), but air is difficult to see and, therefore, I am not able to give information of the air movement from the periphery into the piston bowl.

I have to agree that the actual engine condition and the actual stratification of the fuel–air mixture is certainly more complicated as the sequences shown in Fig. 136.2 indicate.

There are two ways an invention is born: one starts with an idea; the other is accidentally discovered.

K. Zeilinger

There is no evidence that water or recirculated gas only reduces the temperatures. Even the highest quantity of water, 5·5 mg per cycle, is a very small percentage of the charge and only consumes 4 per cent of the delivered energy. This causes very little decrease in peak temperatures. Water and recirculated gas as well must influence (Fig. 122.7 and Fig. 122.9 prove it) the rate of heat release by influencing the reactions.

The pressure was measured by means of a Kistler quartz pick-up situated 8 mm from the wall of the combustion chamber, connected with it by a hole of 4 mm in diameter. The quartz was cooled by water and there were no difficulties with regard to thermal shock.

The cyclic variations of the pressure did not change with water injection when the equivalence ratio was kept constant; it increased at higher equivalence ratios.

In reply to P. Draper and C. L. Goodacre, for all tests ordinary water was used; in order to avoid unnecessary wear of the engine, pure fuel was used during warm-up and for some five minutes after the end of the tests. Under these conditions neither deposits nor wear could be noticed. The influence of water injection on octane requirements was not investigated, but it is known that water is a very good means to prevent knocking.

In reply to W-T. Lyn, as can be seen from my conclusions there is full agreement with W-T. Lyn's ideas. The author, too, doubts whether a prediction of exhaust emissions on the basis of present day calculating models is exact.

List of Delegates

ACRES, G. S. K.	Johnson Matthey and Co. Ltd, London
ADDICOTT, J. L.	Shell Mex and BP Ltd, London
ALCOCK, K.	Hunslet Holdings Ltd, Leeds, Yorks.
ALDEN, J. H.	Vauxhall Motors Ltd, Luton, Beds.
ALEXANDER, A. L.	Department of the Environment, London
ANDERSON, J. G.	BP Research Centre, Sunbury-on-Thames, Middx.
ANNAND, W. J. D.	University of Manchester, Manchester.
APPLEBY, W. A.	Engineering, Research and Application Ltd, Dunstable, Beds.
ASTBURY, J. B.	Rolls-Royce Motors Ltd, Crewe, Cheshire.
ATKINSON, E.	Shell Mex and BP Ltd, London.
ATKINSON, T. R.	Chrysler U.K. Ltd, Coventry, Warwickshire.
AUSTEN, A. E. W.	C.A.V. Ltd, London.
AYRES, R. H.	Massey Ferguson U.K. Ltd, Coventry, Warwickshire.
BAGNALL-OAKELEY, J. M.	R.E.M.E., Bordon, Hants.
BAQUELIN, Y. E.	Saviem, Suresnes, France.
BAKER, R. C.	Imperial College, London.
BALL, A. H.	Joseph Lucas Ltd, Birmingham.
BANDUCCI, G.	Fiat-Laboratorio Gasdinatica, Turin, Italy.
BARDRAM, J.	Ford Motor Company Ltd, Basildon, Essex.
BARKER, D.	BP Research Centre, Sunbury-on-Thames, Middx.
BARUAH, P. C.	University of Manchester, Manchester.
BATH, M.	National Engineering Laboratory, Glasgow.
BEALE, N. R.	Cranfield Institute of Technology, Cranfield.
BEDDOWS, L. M.	Midlands Research Station, Solihull, Warwickshire.
BEGGS, M. C.	Agip (U.K.) Ltd, London.
BELCHER, L. J.	The Associated Octel Co. Ltd, Bletchley, Bucks.
BENNETT, M. C.	Lanchester Polytechnic, Rugby.
BERNHARDT, W. E.	Volkswagenwerk AG, Wolfsburg, Germany.
BERTIOLI, M. M.	Lucas Group Research Centre, Solihull, Warwickshire.
BETTS, T. W.	Royal Automobile Club, London.
BLACKMORE, D. R.	Shell Research Ltd, Chester.
BODDY, J. H.	Mobil Oil Co. Ltd, London.
BOLTON, M. S.	University of Sheffield, Sheffield.
BOOKEY, J. B.	I.C.I. Agricultural Division, Teesside.
BOOT, A. E.	Engineering Research and Application, Dunstable, Beds.
BOUCHER, G. S.	Rover Co., Solihull, Warwickshire.
BOYLE, J.	Cummins Engine Co. Ltd, Darlington, Co. Durham.
BRADLEY, D.	University of Leeds, Leeds, Yorks.
BRETT, A.	Luton College of Technology, Luton, Beds.
BRIGGS, G. C.	University of Sheffield, Sheffield.
BRIGHT, P. E.	Shell International Petroleum Co. Ltd, London.
BRILLE, M. G.	Saviem, Suresnes, France.
BROGAN, J. J.	Environmental Protection Agency, Michigan, U.S.A.
BROOME, D.	Ricardo and Co. Engineers (1927) Ltd, Shoreham-by-Sea, Sussex.
BROWN, C. D.	Cummins Engine Co. Ltd, Darlington, Co. Durham.
BROWNE, D. A.	Lotus Cars Ltd, Norwich, Norfolk.
BRUFFELL, W. K.	Holset Engineering Co. Ltd, Huddersfield, Yorks.
BRYAN, P.	Vauxhall Motors Ltd, Luton, Beds.
BUCHHEIM, R.	Volkswagenwerk AG, Wolfsburg, Germany.
BURNS, I. D. M.	Mobil Oil Co. Ltd, Coryton.
BURNSIDE, B. M.	Heriot-Watt University, Edinburgh.
BUSCAGLIONE, S.	Fiat S.p.A., Corso Apuelli, Turin, Italy.
BUTCHER, L. F.	Admiralty Oil Laboratory, Cobham, Surrey.
CACKETT, B. H.	B.L.M.C., Coventry, Warwickshire.
CADMAN, C. R.	Coopers Filters Ltd, Abergavenny.
CALDICOTT, B.	Lubrizol International Laboratories, Derby.
CAMPBELL, D.	B.S.A. Motor Cycles Research, Umberslade, Warks.
CAMPBELL, J. S.	I.C.I. Agricultural Division, Teesside.
CAMPBELL, K.	Associated Octel Co. Ltd, Bletchley, Bucks.
CARNUS, J. C.	Automobiles Peugeot, Paris, France.
CARVEY, D. M.	Loughborough University, Loughborough, Leics.
CAULTON, J. E.	B.L.M.C. Ltd, Coventry, Warwickshire.
CAYLESS, G. E.	Ford Motor Company, Basildon, Essex.
CHITTY, D. B.	BP Trading Ltd, Sunbury-on-Thames, Middx.
CHOPIN, J.	Automobiles Peugeot, Paris, France.
CHOVIN, P.	Central Laboratory of Police Prefecture, Paris, France.
CLARKE, B. E.	Burgess Products Co. Ltd, Hinckley, Leics.
COLWILL, D. M.	Road Research Laboratory, Crowthorne, Berks.
CONNOR, H.	Johnson Matthey Chemicals Ltd, London.
CONNOR, W. A.	University College, London.
CRISP, T. N.	B.L.M.C., Coventry, Warwickshire.
CROFT, B. H.	British Racing Motors, Bourne, Lincs.
CROUCH, N. F.	Sheffield University, Sheffield.
CYCLIAX, R.	Cummins Eng. Co. Tech. Centre, Essen, Germany.
DABLE, J. W.	Lubrizol Ltd, London.
DANESHYAR, H.	University of Cambridge, Cambridge.
DANIEL, N. H.	S.U. Carburetter Co., Erdington, Birmingham.
DARTNELL, P. L.	Associated Octel Co. Ltd, Bletchley, Bucks.
DAVIS, B. T.	Edwin Cooper & Co. Ltd, Bracknell, Berks.
DAWSON, J. G.	Zenith Carburetter Co. Ltd, London.
DE FORGE DEDMAN, A. S.	Ford Motor Co. Ltd, Basildon, Essex.
DERMOTT, H. J. R.	B.L.M.C., Coventry, Warwickshire.

LIST OF DELEGATES

DIXON, J. P.	Cummins Engine Co., Shotts, Lanarks.	HARDENBERG, H. O.	Daimler-Benz AG, Stuttgart, Germany.
DODD, A. E.	M.I.R.A., Nuneaton, Warwickshire.	HARDIMAN, M.	Micropore Insulation Ltd, Lipton.
DOERFLER, P. K.	Brown, Boveri & Cie, Baden, Switzerland.	HARMAN, T. W.	Garrett-Airesearch, Skelmersdale, Lancs.
DONNELLAN, G. L.	Holset Engineering Co. Ltd, Huddersfield, Yorks.	HARRISON, G. F.	Associated Octel Co. Ltd, London.
DOWNS, D.	Ricardo & Co., Shoreham-by-Sea, Sussex.	HARTLEY, J. R.	*Automobile Engineer*, London.
DRAPER, P.	Shipstal Cottage, Arne, Dorset.	HARVEY, D. C.	Lucas Group Research Centre, Solihull, Warwickshire.
DUGGAL, V. K.	I.S.V.R., University of Southampton.	HAWKINS, S. W.	I.C.I. Ltd, Agricultural Div. Billingham, Teesside.
DUNNE, J. M.	Perkins Engine Company, Peterborough, Northants.	HAWLEY, D.	Lubrizol International Laboratories, Hazelwood, Derby.
EDMONDSON, A. J.	SU Carburetter Company, Erdington, Birmingham.	HAY, N.	University of Nottingham, Nottingham.
EGEBACK, K. E.	AB Atomenergi, Studsvik, Nykoping, Sweden.	HAYNES, C. D.	B.L.M.C., Coventry, Warwickshire.
EGLII, H.	Airesearch Industrial Division, Los Angeles, U.S.A.	HEIRICH, H.	Garrett-Airesearch, Skelmersdale, Lancs.
EKLUND, J. K.	Saab Scania, Trollhattan, Sweden.	HEMMINGS, K.	Joseph Lucas Ltd, Birmingham.
ELLIS, D.	BP Research Centre, Sunbury-on-Thames, Middx.	HENAULT, C. H.	Henault Claude, Paris, France.
ELLBJAR, J. E. M.	AB Volvo, Gothenburg, Sweden.	HENHAM, A. W. E.	University of Surrey, Guildford, Surrey.
EYRE, K. G.	B.L.M.C., Birmingham.	HESLOP, D. A. G.	Ford Motor Co. Ltd, Basildon, Essex.
FARRIER, C. D.	Zenith Carburetter Co. Ltd, Stanmore, Middx.	HIERETH, H.	Daimler-Benz AG, Stuttgart, Germany.
FARROW, I. K.	Ford Motor Co., Basildon, Essex.	HOARE, J. A.	Ricardo & Co. (1927) Ltd, Shoreham-by-Sea, Sussex.
FERGUSON, J. N.	Ford Motor Co., Basildon, Essex.	HODGETTS, D.	Cranfield Institute of Technology, Bedford.
FIELDING, J. A.	B.L.M.C., Jaguar Cars Ltd, Coventry.	HOLT, W. D.	Triumph Motor Co. Ltd, Coventry, Warwickshire.
FIRTH, R. J.	Zenith Carburetter Co. Ltd, Stanmore, Middx.	HOPPER, W. C.	Stichting Concawe, The Hague, The Netherlands.
FIUSTERWALDER, G. S.	Kloeckner-Humboldt-Deutz, Porz, Germany.	HOUSTON, D. J.	Automobile Association, London.
FLETCHER, R. S.	Northern Research and Engineering Corporation International, Radlett, Herts.	HOWARD, J. A.	Ford Motor Co., Laindon, Basildon, Essex.
FLINDERS, W.	Tate and Lyle Transport Ltd, Croydon, Surrey.	HOWLETT, F.	Zenith Carburetter Co., Stanmore, Middx.
FOORD, D.	Associated Octel Co. Ltd, London.	HUCHO, W. H.	Volkswagenwerk AG, Wolfsburg, Germany
FORRESTER, R. J.	Associated Octel Co. Ltd, Bletchley, Bucks.	HUNT, G. A.	Chrysler United Kingdom Ltd, London.
FOWLER, J. H.	Rover Motor Co. Ltd, Solihull, Warwickshire.	HURN, R. W.	Department of Interior, Bureau of Mines, Bartlesville, Oklahoma.
FOWLER, J. G.	Ford Tractor Division, Basildon, Essex.	IRWIN, J.	S.I.G.M.A., Lyon, France
FRANCIS, P. G.	Petters Ltd, Staines, Middx.	IVENS, P. J.	Ford Motor Co., Warley, Essex.
FULLER, D. E.	University of Cambridge.	IVES, A. P.	Lucas Group Research Centre, Solihull, Warwickshire.
FUSSELL, D. R.	Esso Research Centre, Abingdon, Beds.	JACKSON, H. E.	Petrol Injection Ltd, Plymouth
GAY, E. J.	Associated Octel Co. Ltd, Detroit, Michigan, U.S.A.	JACOBSON, M. A. I.	Automobile Association, London.
GIBSON, H. J.	Ethyl Corporation, Ferndale, Michigan, U.S.A.	JAMES, E. H.	B.L.M.C., Coventry, Warwickshire.
GIESLER, D.	Adam Opel AG, Russelsheim, Germany.	JANOTA, M. S.	Queen Mary College, London.
GILES, J. G.	12, Whittington Road, Worcester.	JELLINGS, W.	Cambridge University, Cambridge.
GILKS, A. J. E.	Perkins Engine Company, Peterborough, Northants.	JOEL, S.	I.C.I. Ltd, Runcorn, Ches.
GIPPET, R. G.	U.T.A.C., Montlhery, France.	JOHNSON, E. W.	Triumph Motor Co. Ltd, Coventry, Warwickshire.
GIRGIS, N. S.	Lanchester Polytechnic, Rugby, Warwickshire.	JOHNSON, P. W.	University of Warwick, Coventry, Warwickshire.
GISSANE, W. J. M.	I.C.I. Mond Division, Runcorn, Cheshire.	JONES, B. E.	Royal Military College of Science, Shrivenham, Wilts.
GOLOTHAN, D. W.	Shell International Petroleum Co. Ltd, London.	JONES, D. L.	Dorman Diesels Ltd, Stafford, Staffs.
GOODACRE, C. L.	Norfolk Rd, London.	KENDALL, N.	Shell Research Limited, Chester, Cheshire.
GREEVES, G.	C.A.V. Ltd, London.	KERSHAW, G.	Rolls-Royce Motors Ltd, Shrewsbury, Wilts.
GREGORY, D. F.	AC Delco, Division of General Motors Ltd, Dunstable, Beds.	KERSLEY, P. H. J.	Ford Motor Co. Ltd, Basildon, Essex.
GRIFFITH, R. R.	Associated Octel Co. Ltd, London.	KHAN, I. M.	C.A.V. Ltd, London.
GRIFFITHS, S. T.	Associated Octel Co. Ltd, London.	KILLMAN, I. G.	Kloeckner-Humboldt Deutz AG, Port, Germany.
HAEFNER, G. W.	Daimler-Benz AG, Stuttgart, Germany.	KIMBERLEE, M. C.	B.L.M.C., Coventry, Warwickshire.
HAILSTONE, V. L.	Ford Motor Co., Basildon, Essex.	KING, B. D.	Engelhard Industries Ltd, Cinderford, Glos.
HALL, D. R.	Mullard Mitcham, Mullard Ltd, Mitcham, Surrey.	KLÖBER, W. E.	Cummins Eng. Co. Techn. Centre, Essen, Germany.
HALL, M. T.	British Railways, Swindon, Wilts.	KNIGHT, P. G. G.	S.U. Carburetter Co. Erdington, Birmingham.
HANSRANI, S. P.	University of Technology, Loughborough, Leics.	KUZMICKI, L.	Chrysler United Kingdom Ltd, Coventry, Warwickshire.

LIST OF DELEGATES

LAKE, A. D.	Cummins Ltd, Darlington, Co. Durham.
LAMBERT, A. J.	Burmah Castrol Ltd, Cheshire.
LARBEY, R. J.	The Associated Octel Co. Ltd, Bletchley, Bucks.
LATHAM, E. G.	Henry Wiggin & Co. Ltd, Hereford.
LATTER, R. W.	London Transport, Chiswick, London.
LAUGHTON, P.	Chrysler United Kingdom Ltd, Coventry, Warwickshire.
LAWRENCE, G. L.	Zenith Carburetter Co. Ltd, Stanmore, Middx.
LEA, K. E.	British Leyland, Leyland, Lancs.
LEAN, J. P. M.	B.L.M.C., Jaguar Cars Ltd, Coventry, Warwickshire.
LEWIS, R.	Zenith Carburetter Co. Ltd, Stanmore, Middx.
LINDSAY, R.	Shell International Petroleum Co. Ltd, London.
LITTLEHALES, H. J.	J. Lucas Electrical Ltd, Birmingham.
LUCAS, T.	Queen Mary College, London.
LYN, W-T.	Cummins Engine Co., Columbus, Indiana, U.S.A.
LYON, D.	Esso Research Centre, Abingdon, Berkshire.
MA, A. S. C.	Imperial College, London.
MACMILLAN, R. H.	M.I.R.A., Nuneaton, Warwickshire.
MACPHERSON, I.	Ford Motor Company Ltd, Basildon, Essex.
MADDOX, R. A.	Sandvik U.K. Ltd, Halesowen, Worcs.
MAGUIRE, D. G.	M.E.E.P., U.W.I.S.T., Cardiff, Wales.
MALINS, A. R.	The Associated Octel Co. Ltd, Bletchley, Bucks.
MANDERSCHEILD, P. R.	Volkswagenwerk AG, Wolfsburg, Forschungz, Germany.
MARCHESI, G. F.	Snam Progetti S.p.A., Milan, Italy.
MARSHALL, A.	Cummins Engine Co. Ltd, Darlington, Co. Durham.
MARTIN, J. B.	Coras Iompair Eireann, Dublin, Eire.
MAY, L. C.	Plessey Co. Ltd, Romford, Essex.
McEWAN, I. A.	Vauxhall Motors Limited, Luton, Beds.
McWHANNELL, D. C.	University of Southampton, Southampton.
McWILLIAMS, J. A.	Micropore Insulation Ltd, Kidderminster, Worcs.
MELLOR, R. W.	Ford Motor Co. Ltd, Basildon, Essex.
MIANES, J. M.	Shell Française, Paris, France.
MICHAELIS, W.	Zenith Carburetter Co., Stanmore, Middx.
MIKITA, J. J.	E. I. Du Pont de Nemours and Co., Wilmington, U.S.A.
MILES, C. O.	Calor Gas Ltd, Surrey.
MILLINGTON, B. W.	Ricardo and Co., Shoreham-by-Sea, Sussex.
MOAKES, P. J.	Calor Gas Ltd, Surrey.
MOBSBY, J.	Rolls-Royce (1971) Ltd, Derby.
MOLYNEUX, P. H.	Lubrizol Limited, London.
MOHTADI, M. F.	University of Calgary, Alberta, Canada.
MONTGOMERIE, G. A.	Serck Limited, Birmingham.
MOSER, G. F.	Texaco Inc., Beacon, New York, U.S.A.
MOSS, M. D.	Ford Motor Co., Laindon, Essex.
MOSS, P. J.	S.U. Carburetter Co. Ltd, Erdington, Birmingham.
MUROKI, T.	Toyo Kogyo Co., Ltd, Hiroshima, Japan.
MURHAGHAN, H.	Texaco Ltd, London.
NAILON, T. E.	Hobourn Eaton Manufacturing Co. Ltd, Rochester, Kent.
NARAIN, C.	British Oxygen Co., London.
NEALE, R. J.	Tate and Lyle Transport Ltd, Croydon, Surrey.
NICHOLS, P. N. R.	Mullard Mitcham, Mullard Ltd, Mitcham, Surrey.
NICOLLS, W. E. W.	C.A.V. Limited, London.
OCCELLA, S.	Fiat S.p.A., Turin, Italy.
OLDFIELD, T. A.	The Plessey Co. Ltd, Fareham, Hants.
ORME, A. P.	BP Research Centre, Sunbury-upon-Thames.
PACK, R. C.	Zenith Carburetter Co. Ltd, Stanmore, Middx.
PALIN, A.	Q.A.D. (F.V.E.) M.O.D., Royal Arsenal, West Woolwich.
PALMER, J. M.	Midlands Research Station Gas Council, Solihull, Warwickshire.
PAPEZ, S.	Volkswagenwerk AG, Wolfsburg, Germany.
PARKER, D. G.	Ford Motor Co., South Ockendon, Essex.
PARKER, K. G.	Garrett-Airesearch, Skelmersdale, Lancs.
PARKINSON, G. S.	Shell Mex and BP Ltd, London.
PARNAM, A. K.	Q.A.D. (F.V.E.) M.O.D., Royal Arsenal, West Woolwich.
PASK, W. J.	Ford Motor Co. Ltd, Laindon, Basildon, Essex.
PATRICK, W. A.	Lubrizol Great Britain Ltd, London.
PECASTAING, J. P.	Shell Française Research Centre, France.
PHILLIPPS, R. A.	Esso Petroleum Co. Ltd, Abingdon, Berks.
POTTS, M.	Joseph Lucas (Electrical) Ltd, Birmingham.
PRATT, N. H.	University of Southampton, Southampton.
PROBERT, D. M.	C.A.V. Ltd, London.
PULLEN, J.	G.K.N. Ltd, Smethwick, Worcs.
PURNELL, N.	S.T.D. Services Ltd, Walsall, Staffs.
REDDAWAY, M. A.	Consumers' Association, Halstead, Essex.
REYNOLDS, F.	Department of Air Pollution and Noise Abatement, Birmingham.
RICHARD, G. P.	Esso Research Centre, Abingdon, Berks.
RIJKEBOER, R. C.	T.N.O. Organization for Industrial Research, Delft, The Netherlands.
ROBERTS D. D.	Shell Research Ltd, Chester.
ROBERTS, D. K.	A.E.C. Limited, Southall, Middlesex.
ROBERTS, M. W.	Massey-Ferguson Manufacturing Co. Ltd, Coventry, Warwickshire.
ROBINSON, E. S.	Shell Research Ltd, Chester.
ROBSON, J. V. B. R.	Champion Sparking Plug Co. Ltd, Feltham, Middx.
ROSS, G. E. D.	Zenith Carburetter Co. Ltd, Stanmore, Middx.
ROSSINI, G.	Ente Nazionale Idrocarburi, Rome, Italy.
SALT, J.	Zenith Carburetter Co. Ltd, Stanmore, Middx.
SANDHAGEN, J.	Volkswagenwerk AG, Wolfsburg, Germany.
SAVAGE, J. D.	British Petroleum Co. Ltd, London.
SATCUNANATHAN, S.	University of the West Indies, Trinidad
SCHELTUS, P. I. T. H.	T.N.O. Organization for Industrial Research, Delft, The Netherlands.
SCHUSTER, H. D.	Daimler-Benz AG, Stuttgart, Germany.
SCOTT, R. G.	British Leyland (A.M.) Ltd, Birmingham.
SEARLES, R. A.	Johnson Matthey Chemicals Ltd, London.
SENIOR, K. B.	Joseph Lucas Ltd, Birmingham.
SHADBOLT, C. F.	Engineering Research and Applications, Dunstable, Beds.
SHIBATSUJI, M.	Atsugi Laboratory, Japan.
SJOSTROM, T.	AB Volvo, Gothenburg, Sweden.
SMALLEY, D.	B.L.M.C., Leyland, Lancs.
SMITH, C. G.	Engelhard Sales Ltd, Sutton, Surrey.
SMITH, E.	B.L.M.C., Leyland, Lancs.
SOLTAU, J. P.	Joseph Lucas Ltd, Birmingham.

LIST OF DELEGATES

SPIERS, J.	Perkins Engines Ltd, Peterborough, Northants.
SQUIRE, P. C.	R.A.F.C. Cranwell, Sleaford, Lincs.
STARKMAN, E. S.	General Motors Corporation, Michigan, U.S.A.
STEELE, R. J.	I.C.I. Agricultural Division, Teesside.
STEWART, H. N. M.	Department of Trade and Industry, Stevenage, Herts.
STRANSKY, H.	Chrysler U.K. Ltd, Coventry.
STURROCK, J. H.	Cheswick and Wright Ltd, Blackpool, Lancs.
SUMMERAUER, I.	Adolph Saurer Ltd, Arbon, Switzerland.
SUTTON, D. L.	The Rover Co. Ltd, Solihull, Warwickshire.
SVENSSON, N. G.	Saab-Scania, Trollhattan, Sweden.
SWATMAN, P. P.	S.U. Carburetter Company, Birmingham.
SWINSON, P. J.	Orobis Ltd, London.
TAKETOMI, A.	Toyo Kogyo Co. Ltd, Hiroshima, Japan.
TANABE, S.	Kyoto University, Japan.
TATSUTOMI, Y.	Toyo Kogyo Co. Ltd, Hiroshima, Japan.
TAUZIN, S. J.	S.A. Automobiles Citroen, Paris, France.
TAYLOR, A.	Amoco (U.K.) Ltd, Wembley, Middx.
TEMPLE, R. G.	University of Aston, Birmingham.
TEMPLE-PEDIANI, R. W.	South Bank Polytechnic, London.
THOBURN, A.	I.C.I. Mond Division, Runcorn, Cheshire.
THOMAS, J. R.	Rolls-Royce Motors Ltd, Shrewsbury, Wilts.
THRING, R. H.	Reading University, Reading, Berks.
TIMONEY, S. G.	University College, Dublin.
TIPLER, W.	Perkins Engines Ltd, Peterborough, Northants
TODD, J. W.	Texaco, Ware, Herts.
TUBB, R. J.	Petrol Injection Ltd, Plymouth, Devon.
TULLETT, J. E.	S.U. Carburetter Co., Birmingham.
TURNER, W. T.	Chrysler Technical Centre, Coventry, Warwickshire.
TWIGGER, T. R.	Brico Engineering Ltd, Coventry, Warwickshire.
ULLMAN, T. J.	Engineering Research and Application, Dunstable, Beds.
VAN DER ROS, T.	Philips, Eindhoven, The Netherlands.
VAN KEULEN, G. F. M.	Ketjen NV., Amsterdam, The Netherlands.
VARDE, K. S.	Loughborough University, Loughborough
VARGA, V.	I.C.I., Billingham, Co. Durham.
VON DER BECKE, R. P.	S.U. Carburetter Co., Birmingham.
VOLTZ, S. E.	Mobil Research and Development Co., Paulsboro, U.S.A.
VULLIAMY, N. M. F.	Perkins Engines Co., Peterborough, Northants.
WADSWORTH, A.	Associated Octel Co. Ltd, London.
WAITE, W.	Rubery Owen & Co. Ltd, Wednesbury, Staffs.
WALDE, N.	Saab-Scania, Sodertalje, Sweden.
WALL, D. J.	The Rover Co. Ltd, Solihull, Warwickshire.
WALLACE, F. J.	University of Bath, Bath, Somerset.
WALLMAN, S.	AB Volvo, Gothenburg, Sweden.
WALMSLEY, R. E.	Vandervell Products Ltd, Maidenhead, Berks.
WALZER, P.	Volkswagenwerk AG, Wolfsburg, Germany.
WANG, C. H. T.	C.A.V. Ltd, London.
WARD, R. J.	British Rail Research Dept., Glasgow.
WARE, P. G.	Dunlop Limited, Coventry, Warwickshire.
WEAVING, J. H.	B.L.M.C., Coventry, Warwickshire.
WEEDON, T. M. W.	B.L.M.C., Coventry, Warwickshire.
WEIGHELL, H. J. C.	Chrysler International S.A., Coventry, Warwickshire.
WEITZEL, H.	Adam Opel AG, Russelsheim, Germany.
WHEELER, R. W.	Ricardo and Co., Shoreham-by-Sea.
WHITE, A.	Gas Council, Midlands Research Station, Solihull, Warwickshire.
WHITE, T. H.	H. Wiggin & Co. Ltd, Hereford.
WHITEHEAD, M. J.	Ricardo & Co., Shoreham-by-Sea.
WHITEHEAD, P.	Associated Octel Co. Ltd, Bletchley, Bucks.
WILD, M. D.	Henry Wiggin and Co., Birmingham.
WILLIAMS, C.	Rolls-Royce Motors Ltd, Crewe, Cheshire.
WILLIAMS, C.	M.I.R.A., Nuneaton, Warwickshire.
WILLIAMS, M.	J. Lucas Ltd, Solihull, Warwickshire.
WILLIAMS, W. N.	Western Welsh Omnibus Co. Ltd, Cardiff.
WILLIAMS, S. G.	Lotus Cars Ltd, Norwich.
WILLIAMS, T. J.	King's College, London.
WILSDON, R. E.	Dept. of Trade and Industry, London.
WILSON, P. D.	Cummins Ltd, Darlington, Co. Durham.
WILSON, P. S.	Chrysler (U.K.) Ltd, Coventry, Warwickshire.
WINSHIP, A. T.	B.S.A. Motor Cycles Ltd, Umberslade, Warwickshire.
WINTERNITZ, F. A. L. W.	Associated Engineering Ltd, Leamington Spa.
WISDOM, J. W.	Engineering Research and Application Ltd, Dunstable, Beds.
WITHEY, D. S.	The British Oxygen Co. Ltd, London.
WOODS, W. A.	University of Liverpool, Liverpool.
WRIGHT, M. J.	Ricardo and Co. Ltd, Shoreham-by-Sea.
WRIGHT, N. C.	Ford Motor Co. Ltd, Basildon, Essex.
YATES, D. A.	Lanchester Polytechnic, Coventry, Warwickshire.
ZACZEK, B. J.	Enfield College of Technology, Enfield, Middx.
ZANONI, G. F.	Snam Progetti S.p.A., Milan, Italy.
ZEILINGER, K.	Technical University, Munich, Germany.
ZUBER, R. K.	Ford Motor Co. Ltd, Basildon, Essex.

Index to Authors and Participants

Names of authors and numbers of pages on which papers begin are in bold type.

Adams, W. E., 1, 352
Addicott, J. L., 120, 352
Alcock, K., 328
Alden, J. H., 334, 342
Anderson, J. G., 323
Annand, W. J. D., 321, 331, 345
Atkinson, T. R., viii
Austen, A. E. W., 327

Ball, A. H., 337
Baguelin, Y. E., 34, 355
Barker, D., 120, 352
Bell, A. G., 322
Bernhardt, W. E., 279, 319, 337, 343, 354
Blackmore, D. R., 110, 332, 337, 345, 354
Boddy, J. H., vii, viii
Bolton, M. S., 319, 337
Brille, M. G., 355
Brogan, J. J., 198, 356
Broome, D., 185, 343
Browne, D. A., 343
Burton, D. G., 330, 339

Caldicott, B., 346
Campbell, J. S., 334
Campbell, K., 14, 356
Chovin, P., 57
Connor, H., 335
Croft, B. H., 347
Crouch, N. F., 328

Dable, J. W., 62, 356
Daneshyar, H., 348
Dartnell, P. L., 14, 356
de Forge-Dedman, A. S., 264
Dodd, A. E., 101, 357
Doerfler, P. K., 333
Downs, D., viii, 330, 336
Draper, P., 322, 332, 336

El Nesr, M. S., 156
Eccleston, B. H., 286

Fogg, A., vii
Fosberry, R. A. C., viii
Fussell, D. R., 89, 358

Gibson, H. J., 1, 352
Gippet, R. G., 319
Goodacre, C. L., 321, 332, 334, 341
Greeves, G., 205

Hall, M. T., 344
Hamamoto, Y., 129, 360
Hardenberg, H. O., 339
Haynes, C. D., 232, 358
Henault, C., 273
Henham, A. W. E., 342
Hirschler, D. A., 1, 352
Howard, J. A., 264
Howland, A. H., 42, 359
Hoare, J. A., 341
Holt, W. D., 341
Hunt, G. A., vii
Hurn, R. W., 286

Ivens, P. J., 348

Jacobson, M. A. I., 322, 334, 344
Janota, M. S., 327
Jellings, W., 327

Kershaw, G., 328
Khan, I. M., 185, 205, 293, 328, 338
Koehl, W. J., 172, 363
Kraus, B. J., 77, 360

Larbey, R. J., 218
Lawrence, G. L., 137, 322, 335
Lawther, F. J., vii
Lea, K., 328
Lindsay, R., 333, 337
Lom, W. L., 42, 359
Lyon, D., 42, 359
Lyn, W-T., 326, 332, 340, 344

Mellor, R. W., 343
Mikita, J. J., 336
Millington, B. W., viii
Moser, C. E., 311
Moss, P. J., 321

Muroki, T., 304, 359
Muzio, L., 163

Newhall, H., 163

Ohigashi, S., 129, 360
Orme, J. R., 348
Osterhout, D. P., 172, 363

Papez, S., 332
Probert, D. M., 205

Richard, G. P., 77, 89, 358, 360
Roberts, D. K., 327, 338

Satcunanathan, S., 156
Senior, K. B., 241, 362
Sheahan, T. J., 62
Soltau, J. P., vii, viii, 218, **241,** 336, 341, 362
Starkman, E. S., 163, 335
Swatman, P. P., 324

Tanabe, S., 129, 360
Tatsutomi, Y., 337
Timoney, S., 350
Tipler, W., 327, 332, 339
Torpey, P. M., 21, 362

Voltz, S. E., 172, 334, 363
Vulliamy, N. M., vii, viii

Wade, D., 89, 358
Wallace, F. J., vii, viii, 326, 338
Wang, C. H. T., 293
Weaving, J. H., vii, viii, **232,** 358
Wheeler, R. W., 344
Whitehead, M. J., 21, 362
Wilson, P. S., 341
Williams, T. J., 327, 330, 334
Wintringham, J. S., 1, 352
Wisdom, J. W., 253, 363
Witzky, J. E., 147, 364
Wright, M., 21, 362

Zaczek, B. J., 156
Zeilinger, K., 7, 365

Subject Index

Titles of papers are in capital letters.

ACCOMPLISHMENTS OF THE I.I.E.C. PROGRAMME FOR CONTROL OF AUTOMOTIVE EMISSIONS, 172
ADDITIVES, THEIR CONTRIBUTION TO THE ELIMINATION OF AIR POLLUTION, 62
Additives, see Fuel additives
Air bush restriction, 67, 322, 355
Air cleaners, 269
Air flow control, rotary engine, 309, 341, 359
Air/fuel ratio; characteristics, rotary engine, 305
  effect, 2, 337, 352
Air injection with reactor, rotary engine, 307, 341, 359
Air sampling methods, 57
Air valve carburettors, see Carburettors, air valve
Air–water heat exchanger, 'moteur compensé', 35, 338, 355
Alkaline earths content, particulates, 15
Altitude, effect on fuel control system, 249
Alumina-coated filters, 18, 341, 344, 345, 356
America, United States of, Advanced Automotive Power Systems research;
  Clean Car Incentive programme, 201
  electric power systems, 200, 203, 319, 356
  gas turbine, 200, 319, 356
  hybrid power systems, 200, 202, 319, 356
  Rankine cycle engine, 200, 319, 356
  stratified charge engine, 200, 319, 356
  worldwide participation, 201
America, United States of, Coordinating Research Council programme; atmospheric research, 314, 315
  combustion study, 313
  consumer use, effect, 312, 313
  exhaust gas oxygenate analyses, 313
  fuel, lubricant, and engine interrelationships, 312
  fuel volatility and automotive design, 313
  functional chart, 312
  haze formation study, 315
  hydrocarbons and carbon monoxide in the atmosphere, study, 314, 315
  irradiation chamber research, 314
  leaded and unleaded fuel, effect, 313
  medical research, 315
  odour, measurement and identification, 312
  particulates, physical and chemical characteristics, 312
  petrol engine tests, 225, 319
  plant damage by pollutants, 315
  polynuclear aromatics, determination, 312, 313
  refuelling emissions, 313
  testing techniques, development and evaluation, 312
  vehicle maintenance study, 314
America, United States of, Inter-Industry Emission Control programme; catalysts and converter systems, 173, 174, 175, 334, 335, 336, 337, 363
  concept emission systems, 180, 181, 182, 183
  exhaust gas recirculation, 179
  mathematical model for catalytic converter systems, 175
  participants, 172
  thermal reactor, 178, 334, 335, 336, 337, 363
America, United States of; California test cycle, 51, 225, 230
  Clean Air Act, 5, 198, 319, 356
  Ethyl lean-reactor car, 1, 6, 335, 336, 337, 352
  Federal standards, penalties of adoption by European car manufacturers, 334, 337, 352
  Federal test cycle, 221, 222, 225, 243, 256, 259, 319
  fuel additive tests, 63, 69, 70, 74, 75, 322, 346, 355
  particulates, sources and composition, 14
  standards, 1, 5, 14, 23, 42, 51, 89, 137, 163, 172, 198, 226, 227, 232, 265, 319, 335, 336, 337, 342, 352, 356

stratified-charge engine development, 147, 200, 201, 319, 340, 343, 344, 356, 364
total vehicular emissions, 31
Ammonia content, particulates, 15
Ammonia levels, measurement, 234
Ammonia production, catalysts, 238, 335, 359
Aniline additive, 157, 327, 328
Annular combustion chamber, 101, 341, 342, 343, 357
Armstrong Siddeley car, fuel additive tests, 68, 322, 355
Aromatic content of fuel, effect on hydrocarbon emissions, 86, 321, 361
Aromatics, polynuclear; determining, 312, 313
  toxicity, 1, 316
Atmospheric research, 314
Austin cars; catalyst tests, 233, 234, 237, 334, 337, 348, 358
  fuel additive tests, 68, 322, 355
Australia, legislation, 219
AUTOMOTIVE AIR POLLUTION CONTROL, RECENT LEGISLATION IN THE U.S. AND A NEW APPROACH TO ACHIEVE CONTROL: ALTERNATIVE ENGINE SYSTEMS, 198
AUTOMOTIVE EMISSION RESEARCH—A REVIEW OF THE C.R.C. PROGRAMME, 311

Bath-tub combustion chamber, 101, 341, 342, 343, 357
Belgium, legislation, 219
Benzene vapour, inflammability limit, 44
B.L.M.C. dual catalyst system, 232, 334, 337, 348, 358
Bowl-in-piston combustion chamber, 101, 341, 342, 343, 357
Brake mean effective pressure, 9, 104, 250
Brake specific fuel consumption, 9, 52, 53
British Standard AU 141, diesel engines, 42
British Technical Council for the Petroleum and Motor Industries Analytical Methods Committee; aims and activities, 228, 231
  gas and car exhaust correlation programmes, 229
  instrument specifications, 229
Bromine content, particulates, 15
Buses, LPG dual fuel, 43
Butane; combustion pressure study, 46
  highest useful compression ratio, 44
  inflammability limit, 44
  influence on thermal efficiency, 45

Camshafts, overlap, 277
Canada, legislation, 219
Car fitted with dual catalyst and catalyst bypass systems, 235, 337, 348, 358
Carbon content, particulates, 15
Carbon, forms produced in combustion, 185
Carbon monoxide emission control; air valve carburettors, 144
  catalysts, 90, 173, 232, 334, 335, 336, 337, 348, 358, 363
  recycling, 3, 337, 352
  thermal reactor, 253
  thermal reactor and exhaust recycle, 89, 348, 358
Carbon monoxide emissions; American limits, 198
  at idling, 62, 64, 70, 138, 144, 319, 322, 323, 324, 346, 355
  engine variables, 101, 341, 342, 343, 357
  fuel additives, 68, 158, 327, 328
  fuel/air ratio, 2, 3, 301, 326, 327, 337, 352
  fuel composition, 80, 82, 110, 321, 322, 347, 354, 360
  injection rate and timing, 243, 301, 326, 327
Carbon monoxide emissions; effect on human health, 316
  estimated, American 1969, 199
  European standards, 64, 79, 220, 226, 322, 334, 337, 346
  in the atmosphere, study, 315

SUBJECT INDEX

measurement, 21, 23, 32, 58, 79, 220, 234, 319, 338, 339, 340, 341, 350, 362
monitoring network, Paris, 57
particulate, 271
predicted legal limits, 43, 338, 339, 359
production, 265
removal from the atmosphere by unsterilized soil, 315
rotary engines, 304, 341, 359
sampling points in cities, 59
Saurer engine, 49, 338, 339, 359
standards obtained to date, 89, 348, 358
stratified charge engine, 152, 319, 343, 344, 356, 364
Carburettors, air valve; adjustment means, idle/off-idle, 141
air density effect, 141, 319, 322, 323, 324
CD-4 and CDSE, 138, 141, 143, 144, 319, 322, 323, 324
emulsion air jet system, 139
idle 'dipping' correction, 142, 324
metering system development, 138, 319, 322, 323, 324
selective venting, 141
stabilized flow, 140
Stromberg, 138, 319, 322, 323, 324
Carburettors; 3-venturi, 2
adjustment, 62, 322, 346, 355
ceramic filter, 324
cleanliness and deposit removal, 64, 65, 68, 322, 346, 357
fixed choke (plain tube), 137, 144, 320, 323, 324
single and double venturi, 268
SU, 74, 138, 322, 346, 355
Weber, 74, 268, 322, 346, 355
Zenith 36VN, 145, 320, 322, 323, 324
Catalyst system; dual, 232, 334, 337, 348, 358
exhaust pipework and bypass system, 234, 337, 348, 358
in sub-zero temperatures, 334, 358
mathematical model, 175
underfloor box, 234, 348, 358
Catalysts; ageing, 173
ammonia production, 238, 335, 359
copper, 174, 237, 337, 348, 358
development, and desirable characteristics, 173
endurance test box, 234
engine valve seat wear, 239, 240
iron, 174, 237, 337, 348, 358
metallic alloys, 174, 237, 337, 348, 358
monolith, 334, 359
nickel, 237, 337, 348, 358
oxidation, 173, 232, 236, 237, 334, 335, 336, 337, 348, 349, 358, 363
oxides of nitrogen reduction, 174, 237, 334, 335, 336, 337, 348, 358, 363
'packaged', 348
pelleted, 234, 334
platinum, 237, 337, 348, 358
poisoning, 90, 239, 337, 348, 358
precious metal, 174
reduction, 232, 334, 337, 348, 358
secondary air flow, 238, 240
specifying lubricant and fuel for optimum performance, 336, 359
temperature effects, 238, 240
tests—furnace, engine testbed, vehicle, and testbed endurance, 233, 234, 337, 348, 349, 350, 358
CATALYTIC REDUCTION OF ATMOSPHERIC POLLUTION FROM THE EXHAUST OF PETROL ENGINES, 232
CD-4, and CDSE carburettor development, 138, 141, 144, 319, 322, 323, 324
Ceramic liner, thermal reactor, 335
Charcoal canister control system, effect of additives, 71, 322, 346, 355
Chemical kinetics, NO formation, 165, 167, 168, 279, 331, 332, 354
Chemiluminescent technique, NO measurement, 220, 225, 319, 362
Chevrolet car, exhaust recycle and reactor manifold, 90, 92, 99
Chlorine content, particulates, 15
Choke, automatic, 269
Clean Air Act, America, 5, 198, 319, 356
Clean air package system, effect of fuel composition, 79, 321, 322, 347, 360
Clean Car Incentive programme, 199, 201, 319, 356
Climate, effect, 335

Cold start and warm-up, fuel system, 252
Combustion chamber; effect of design, 101, 264, 341, 342, 343, 357
rotary engine, 306, 341, 359
Combustion pressure studies, LPG fuelling, 46, 338, 339, 359
Combustion process; 189, 313, 326, 327, 328
and nitrogen oxides, 163, 283, 331, 332, 354
effects of air swirl, 134, 330, 360
forms of carbon produced, 185
photograph of, 153, 284, 331, 354
stratified charge engine, 150, 343, 344, 364
thermodynamic analysis, 280, 331, 332, 354
Compression ratio; effect, 28, 29, 107, 269, 340, 341, 342, 343, 350, 357, 363
highest useful, various fuels, 44
Computer, fuel injection equipment control, 251
Condition of engine, effect on fuel injection equipment, 250
Constant Volume Sampling, particulate emissions; 16
Olson rig, 223
U.S. Federal system, 222, 319
Consumer use effect, 15, 312, 313
Control, fuel flow, closed loop feed-back system, 320, 362
Control parameters, fuel injection equipment, 241, 243, 251, 319, 322, 362
CONTROLLING EXHAUST EMISSIONS FROM A DIESEL ENGINE BY LPG DUAL FUELLING, 42
Coolant temperature, 108, 341, 342, 343, 357
Copper, as an NO reduction catalyst, 237, 238, 337, 348, 358
Copper oxide catalysts, 174
Corrosion, thermal reactors, 179
Crankcase emission control, 271, 357
C.R.C. programme, see America, United States of, Coordinating Research Council
Cyclohexane, additive, 158, 327, 328
Cyclone technique, 17, 341, 343, 344, 345, 356

Denmark, legislation, 219
DESIGN, DEVELOPMENT AND APPLICATION OF EXHAUST EMISSIONS CONTROL DEVICES, 264
Design of vehicle and fuel volatility, 313
Design, recycle system, 97
Designing vehicles for emission control systems, 183
Diaphragm distributor, double, 266
DIESEL ENGINE EXHAUST EMISSIONS AND EFFECT OF ADDITIVES, 156
Diesel engines; analysis of exhaust gases, 21, 22, 338, 339, 340, 341, 350, 362
anti-smoke additives, 42, 47, 338, 339, 359
British Standard AU 141, smoke, 42
Californian test cycle studies, 51, 227, 338, 339, 359
carbon monoxide measurement, 21, 23, 32, 338, 339, 340, 341, 350, 362
combustion characteristics, 189
compression ratio and combustion chamber design, 28, 29, 340, 350, 363
exhaust recirculation, 25, 27, 29, 338, 339, 340, 341, 350, 362
fuel additives, 156, 327, 328
fuel consumption, effect of additives, 161, 327, 328
high-speed direct-injection systems, operation, 21, 190
hill climbing test, 47, 338, 339, 359
hydrocarbons, detection by flame ionization method, 22, 338, 339, 340, 341, 350, 362
hydrocarbons, effect of injection rate and timing, 301, 326, 327
indirect-injection systems, operation, 21, 191
injection timing, 24, 29, 46, 50, 338, 339, 351, 359, 362
knock, 44, 338, 359
legislation, and smoke reducing techniques, 42, 43, 338, 339, 359
LPG dual fuelling, 44, 338, 339, 359
M-system, operation, 191, 326, 327, 328
Diesel engines, 'moteur compensé'; installation in truck, 39, 338, 355
multi-fuel, 37, 338, 355
Roots blower, 36, 338, 355
starting, 37, 338, 355
tests in vehicle, 38, 338, 355
Diesel engines, nitrogen oxides; chemiluminescent analysis, 319, 362
effect of fuel injection variables, 298, 299, 300, 326, 327
measurement, 22, 23, 29, 338, 339, 340, 341, 350, 362
prediction models, 208, 212, 214, 215, 326, 327, 328

SUBJECT INDEX

Diesel engines; noise reduction, 55
    quiescent direct-injection systems, operation, 190
    Saurer tests, 47, 49, 50, 338, 339, 359
    swirl effects, 129, 134, 293, 326, 327, 330, 360
    swirl-chamber and direct-injection tests and comparisons, 21, 338, 339, 340, 341, 350, 362
    temperature and pressure compensation, 34, 35, 338, 355
    thermal efficiency, 44, 338, 339, 359
    turbocharging, 42
    variables affecting smoke, 195, 293, 295, 297, 298, 326, 327, 328
    water and water–methanol injection, 27, 29, 338, 339, 340, 341, 350, 362
Diesel engines, *see also* Soot formation in diesel engines
Drager gas detector, 157, 234, 328
Driveability with emission control systems, 183, 232
Dual fuel operation; anti-smoke additives, 42, 47, 338, 339, 359
    experiments, Petter AV-1 engine and Morris diesel van, 44, 338, 339, 359
    hill climbing test, 47, 338, 339, 359
    knock, 44, 338, 359
    obtaining California test cycle, 51, 338, 339, 359
    thermal efficiency, 43, 338, 339, 359
Duplex manifold system, tests, 79, 321, 322, 347, 360
Dynamometer measurement, particulate emissions, 19
Dynamometer tests, 218, 319, 349, 350, 362

Electric discharge method, gas flow velocity measurement, 129, 330, 360
Electric power systems, American development, 200, 203, 319, 356
Electronic control, fuel injection equipment, 251, 319, 322, 362
Emulsion air jet system, carburettors, 139, 140, 319, 322, 323, 324
Engelhard P.T.X. 3 afterburner, performance, 348, 363
Engine, lubricant, and fuel interrelationships, 312, 313
Engineering research, 312
Environment, effect, 249, 335
Equivalence ratio, 79, 186, 207, 298, 300, 321, 326, 327, 328
Ethyl lean-reactor car, 1, 6, 335, 336, 337, 352
Europe; Economic Commission for, tests and standards, 64, 79, 220, 226, 322, 334, 337, 346
    legislation, 219
    penalties of adoption of American standards, 334, 337, 352
    total vehicular emissions, 31
European engine, fuel additive tests, 64, 66, 67, 72, 74, 75, 322, 346, 355
European road test, 64, 322, 346, 355
Evaporative emission control system, 71
EVAPORATIVE LOSSES FROM AUTOMOBILES: FUEL AND FUEL SYSTEM INFLUENCES, 286
EXHAUST EMISSION CONTROL: FURTHER EXPERIENCE IN THE APPLICATION OF EXHAUST THERMAL REACTORS TO SMALL ENGINES, 253
EXHAUST EMISSION CONTROL SYSTEM FOR THE ROTARY ENGINE, 304
EXHAUST EMISSION STUDIES USING A SINGLE-CYLINDER ENGINE, 101
Exhaust port liner, 3
Exhaust system design effect, 15
EXPERIMENTS IN THE CONTROL OF DIESEL EMISSIONS, 21
Exported vehicles, emissions control packages, 268
Extraction, exhaust gas samples, 8, 331, 332, 333, 365
Extractor, 'moteur compensé', 35, 338, 355

FACTORS AFFECTING EMISSIONS OF SMOKE AND GASEOUS POLLUTANTS FROM DIRECT INJECTION DIESEL ENGINES, 293
Fiat cars; lead emission tests, 18
    present emission standards, 89
Filters; alumina-coated, 18, 341, 344, 345, 356
    ceramic, 324
Filtration method, determination of particulate emissions, 16, 341, 343, 344, 345, 356
FIRST RESULTS OBTAINED WITH THE CARBON MONOXIDE MONITORING NETWORK INSTALLED IN PARIS, 57
Fixed choke carburettors, 137, 144, 319, 322, 323, 324
Flame ionization detectors, 22, 79, 101, 218, 224, 228, 321, 322
Flame propagation model, 280, 331, 332, 354
Flames, formation and combustion of soot, 185, 326, 327, 328
Flow visualization studies, thermal reactor, 260
Flywheels, hybrid power systems, 203
Ford cars; automatic choke, 269
    crankcase emission control, 271
    export, 264
    fuel additive tests, 68, 322, 355
    IMCO system, 264, 266, 269
    lead emission tests, 18
    particulate emissions, 271
    spark modulation system, 267
    thermactor package, 264
    Weber carburettors, 268
Formaldehyde emissions, effect of fuel additives, 158, 327, 328
France, legislation, 219
FUEL COMPOSITION, INFLUENCE ON AUTOMOTIVE EXHAUST EMISSIONS, 77
Fuel, *see also* Diesel engines *and* Dual fuel
Fuel additives; anti-smoke, 42, 47, 157, 338, 339, 359
    compatibility with emission control systems, 71, 322, 346, 355
    dynamometer, laboratory, and road tests, 68, 69, 322, 346, 355
    isoamylnitrate, aniline, and cyclohexane, 157
Fuel additives, effect on; carburettor cleanliness, 63, 64, 65, 68, 69, 71, 322, 346, 355
    cetane number, 156, 327, 328
    charcoal canister control systems, 71, 322, 346, 355
    consumption, 161, 327, 328
    crankcase ventilating valve performance, 313
    PCV valves, 71, 322, 346, 355
    soot formation, 186, 326, 327, 328
Fuel/air mixing, 2, 294, 326, 327, 337, 352
Fuel and lubricating oil specification for optimum performance of catalytic systems, 336, 359
Fuel causing corrosion of thermal reactors, 179
Fuel composition; 79, 321, 322, 347, 360
    effect on quantity of particulates, 15, 344, 356
Fuel consumption, effect of; additives, 161, 327, 328
    combustion chamber configuration, 101
    nitric oxide control with recycle, 4, 337, 352
Fuel consumption; I.I.E.C. concept emission system, 183
    with thermal reactor and exhaust recycle, 91, 99, 348, 364
Fuel evaporation; carburettor loss, 286, 288, 291
    modification of composition, 286, 288, 291
    tank loss, 286, 288, 291
    vehicle factors, 286, 288, 291
Fuel flow, control, closed loop feedback system, 320, 362
Fuel injection systems; 241
    air density effects, 249
    cold start and warm-up, 252
    control parameters, 243, 319, 322, 362
    effect of altitude, 249
    effect of engine condition, 250
    effect of fuel composition, 77, 347, 361
    ram effect, 250
    retarding, 24, 29, 53, 338, 339, 340, 341, 351, 362
    soot and nitric oxide formation, 205, 326, 327, 328
    timing, 46, 50, 243, 293, 326, 327, 338, 339, 359
    transient conditions, 250
Fuel, lead content, specification, 336, 358
Fuel, lead content reduction, effect on anti-knock performance of engines; fuel-rating tests, 123, 322, 352
    octane requirement surveys, 121, 322, 352
    road tests, 121, 322, 352
Fuel, leaded, advantages, 1
Fuel, leaded, emission reduction; cyclone/inertia-type devices, 17, 341, 343, 344, 345, 356
    Ethyl lean-reactor car, 1, 335, 336, 337, 352
    filter, 18
Fuel, leaded/unleaded; emission level tests, 313
    particulate emissions, 4, 20, 343, 344, 356
Fuel, lubricant, and engine interrelationships, 312, 313
Fuel, 'moteur compensé', 37, 338, 355
Fuel, octane requirements of vehicles; effect of water injection, 332, 365
    surveys, 121, 322, 352
Fuel rating, Uniontown technique, 123
Fuel, soot formation, 186, 192, 205, 326, 327, 328
Fuel, unleaded, disadvantages, 1, 322, 336, 352, 353
Fuel, various; highest useful compression ratio, 44
    inflammability limit, 44
Fuel volatility and vehicle design, 313
Fuel, volatility, viscosity, and specific gravity, 110, 321, 354
Furnace test, catalysts, 233, 337, 348, 358

Gas chromatograph data, unmodified and lean-reactor car, 6
Gas oscillation in an exhaust pipe, 332, 365
Gas recycling, 3, 25, 27, 29, 90, 92, 97, 273, 337, 338, 339, 340, 341, 348, 350, 352, 358, 362

374

SUBJECT INDEX

Gas sampling and analysis, 8, 21, 22, 152, 157, 164, 218, 222, 226, 319, 331, 332, 333, 338, 339, 340, 341
Gas turbine, American development, 201, 319, 356
GASOLINE BULK FUEL PROPERTIES, INFLUENCE ON EXHAUST EMISSIONS UNDER IDLE AND DYNAMIC RUNNING CONDITIONS, 110
Germany, West, legislation, 219
Grubb Parsons gas analysis trolley with Beckman FID, 224

Haze formation, 315
Heat engine/electric, heat engine/flywheel, American development, 200, 202, 319, 356
Heat exchangers, 'moteur compensé', 35, 338, 355
Hemispherical combustion chamber, 101, 341, 342, 343, 357
Hill climbing test, dual fuel operation, 47, 338, 339, 359
Hybrid power systems, American development, 200, 202, 319, 356
Hydrocarbon catalytic converter system, mathematical model, 175
Hydrocarbon emission; aromatic content, 86, 321, 361
    estimated, U.S.A. 1969, 199
    European standards, 220
    in the atmosphere, study, 314
    measurement, 22, 23, 29, 32, 79, 220, 234, 319, 338, 339, 340, 341, 350, 362
    'moteur compensé', 38, 338, 355
    particulate, 271
    pollution controlled vehicles, 89, 348, 358
    predicted legal limits, 43, 338, 339, 359
    production, 265
    rotary engines, 304, 341, 359
    Saurer engine, 49, 338, 339, 359
    stratified charge engine, 152, 343, 344, 364
    toxicity, 316
Hydrocarbon emission control; catalysts, 90, 173, 232, 334, 335, 336, 337, 348, 358, 363
    thermal reactor and exhaust recycle, 89, 253, 348, 358
    recycling, 3, 25, 27, 29, 90, 97, 273, 337, 338, 339, 340, 341, 348, 350, 352, 358, 362
Hydrocarbon emissions, effect of; engine variables, 101, 341, 342, 343, 357
    fuel/air ratio, 2, 3, 337, 352
    fuel composition, 80, 83, 321, 322, 347, 360
    fuel injection timing, 243, 301, 326, 327
    fuel properties, 110, 321, 354
    valve overlap, 273
Hydrocarbon vapour losses, fuel and vehicle tests, 286
Hydrogen content, particulates, 15

Idle 'dipping' correction, carburettors, 142, 324
Idling, carburettor design, 62, 64, 70, 138, 144, 319, 322, 323, 324, 346, 355
Ignition system control, rotary engine, 309, 341, 359
Ignition timing and; knock, 106, 341, 342, 343, 357
    recycle, effect on fuel consumption, 91
    mixture strength, 94
Ignition timing characteristics, rotary engine, 306
I.I.E.C. programme, see America, United States of, Inter-Industry Emission Control
IMCO system, Ford engines, 264, 266, 269
Inertia separators, 17, 341, 343, 344, 345, 356
Inflammability limit, various fuels, 44
Infrared analysis, 21, 79, 101, 218, 220, 221, 224, 225, 227, 228, 321, 362
Instrumentation for emission testing, 220
Intake manifold deceleration control valve, 268
Iron, as an NO reduction catalyst, 237, 337, 348, 358
Iron content, particulates, 15
Iron oxide catalysts, 174
Irradiation chamber research, 314
Isoamylnitrate additive, 157, 327, 328
Iso-octane, highest useful compression ratio, 44
Iso-octane vapour, inflammability limit, 44

Japan; regulations, 219, 337, 352, 362
    rotary engined cars, 304, 341, 359
    test method, 227, 319

Kent engine, thermactor package, 264

KINETICS OF NITRIC OXIDE FORMATION IN INTERNAL-COMBUSTION ENGINES, 279
Knock; dual-fuel operation, 44, 338
    effect of gasoline lead content, 120, 322, 352, 353
    ignition timing, 106

Lead catchment filter, 18, 341, 344, 345, 356
Lead compounds, melting points, 5
Lead content, particulates, 15
Lead, see Fuel
Lean-reactor car, Ethyl development, 1, 6, 335, 336, 337, 352
Legislation, see Standards
Liners, exhaust ports, 3
LPG dual fuelling, see Dual fuel operation
Lubricant, fuel, and engine interrelationships, 312, 313
Lubricant, specification for optimum performance of catalytic devices, 336, 359

M combustion process improvement, 34, 40, 338, 355
Maintenance of vehicle study, 314
MANIFOLD REACTION WITH EXHAUST RECIRCULATION FOR CONTROL OF AUTOMOTIVE EMISSIONS, 89
Materials; flywheels, hybrid power systems, 203
    thermal reactors, 178, 308, 336, 375
    valves, effect of lead-free fuels, 322, 353
    vapour control canister, 71
Mechanism of the problem, 63
MECHANISMS OF SOOT RELEASE FROM COMBUSTION OF HYDROCARBON FUELS WITH PARTICULAR REFERENCE TO THE DIESEL ENGINE, 185
Medical research, 315
MEETING FUTURE EMISSION STANDARDS WITH LEADED FUELS, 1
METABOLISM AND DIETETICS OF 'MOTEUR DIESEL COMPENSÉ B' AND ITS RESULT ON AIR POLLUTION, 34
Metallic alloy catalysts, 174
Metering fuel, 3
Methane; highest useful compression ratio, 44
    inflammability limit, 44
MGB car, reactor plus recycle, 93
Monel metal, as an NO reduction catalyst, 239, 337, 348, 358
Morris car, fuel additive test, 68, 322, 355
Morris diesel van, experimental dual-fuelling, 44, 338, 339, 359
Moteur diesel compensé B, see Diesel engines, moteur compensé
Multi-fuel 'moteur compensé', 37, 338, 355

Netherlands, legislation, 219
Nickel, as an NO reduction catalyst, 237, 337, 348, 358
Nitrate ion content, particulates, 15
Nitrogen oxide emissions, effect of; combustion chamber design, 101, 341, 342, 343, 357
    engine to engine variation, 298, 326, 327
    engine variables, 101, 341, 342, 343, 357
    fuel additives, 160, 327, 328
    fuel/air ratio, 299, 300, 326, 327, 364
    fuel composition, 80, 84, 321, 322, 347, 360
    fuel properties, 110, 321, 354
    overlap camshaft, 277
    valve overlap, 273
Nitrogen oxide emissions; effect on human health, 316
    estimated, U.S.A. 1969, 199
    leaded fuels, 1, 335, 336, 337, 352
    measurement, 22, 23, 29, 79, 220, 225, 319, 338, 339, 340, 341, 350, 362
    'moteur compensé', 38, 338, 355
    particulates, 271
    predicted legal limits, 43, 338, 339, 359
    relation with fuel injection timing, 243
    rotary engines, 304, 341, 359
    Saurer engine, 49, 338, 339, 359
    stratified charge engine, 152, 319, 343, 344, 356, 364
Nitrogen oxide emissions, reduction; catalysts, 90, 174, 232, 334, 335, 336, 337, 348, 358, 363
    CD-4 carburettor, 144
    combustion control, 283, 331, 332, 354
    exhaust gas recycling, 3, 90, 179, 334, 335, 336, 337, 348, 352, 358, 363
    lean-reactor car, 1, 335, 336, 337, 352
    standards obtained to date, 89, 348, 358
    thermal reactor, 253, 261
    thermal reactor and exhaust recycle, 89, 348, 358

SUBJECT INDEX

Nitrogen oxide formation; 10, 163, 265, 269, 332
   and emission, early studies, 164
   chemical equilibrium considerations and effects, 165, 332
   chemical kinetics, 167, 168, 332
   model, 212, 215, 279, 326, 327, 328, 331, 332, 354
   petrol engines, influence of water injection, 7, 331, 332, 333, 365
Noise reduction, 55, 327
Nozzle configuration effect, 298, 326, 327

Odour, identification and measurement, 312, 361
Oil–water heat exchanger, 'moteur compensé', 35, 338, 355
Olson constant volume sampling rig, 223
Operating conditions, effect on particulates, 15
Orsat gas analyser, 157
Oxidation of carbon monoxide and hydrocarbons, catalytic converters, 173, 334, 335, 336, 337, 363
Oxidizing catalysts, 232, 334, 337, 348, 358
Oxygenate content, exhaust gas, 312, 313

Paris; carbon monoxide monitoring network, 57
   hourly concentration of pollution, 59, 60
Particulate emissions; determination by filtration, 16, 341, 343, 344, 345, 356
   effect of exhaust system design, 15
   estimated, U.S.A., 1969, 199
   fuel and engine operation effect, 15, 341, 343, 344, 345, 356
   leaded/unleaded fuels, 3, 4, 20, 341, 343, 344, 345, 356
   measurement on dynamometer, 19
   mileage accumulation condition effect, 15
   physical and chemical characteristics, 312
   sampling systems, 16
   sources and composition, 14, 15, 345, 356
   trap, 5
   using the IMCO system, 271
PCV valves, effect of additives, 71, 322, 346, 355
Petrol, *see also* Fuel
Petrol engine tests, U.S.A., 225, 319
Petrol engines, catalyst systems, *see* Catalysts
Petrol engines, influence of water injection on NO formation, 7, 331, 332, 333, 365
Petrol injection, *see* Fuel injection
PETROL INJECTION CONTROL FOR LOW EXHAUST EMISSIONS, 241
Petter AV-1 engine, dual-fuelling, 44, 338, 339, 359
Photography, injection and combustion process, 153, 284, 331, 354
Pinto car, carburettor, 268
Plain tube carburettor, performance, 137, 144, 319, 322, 323, 324
Plant damage by pollutants, 315
Platinum catalyst, HC and CO oxidation, 237, 337, 348, 358
Plymouth Fury car, exhaust gas recycle, 90
Pneumatic integration, air sampling, 57
Pontiac car; exhaust recycle test, 3, 337, 352
   lean-reactor performance, 6, 337, 352
Power and torque, engine, effect of thermal reactors, 348, 364
Precious metal catalysts, 174
Pressure compensation, diesel engine, 34, 35, 338, 355
Pressure measurement, cylinders, 8, 331, 332, 365
Propane and propylene; combustion study, 46
   highest useful compression ratio, 44
   inflammability limit, 44
   influence on thermal efficiency, 45

Radiator, 'moteur compensé', 35, 338, 355
RAM thermal reactor, 99
Rankine cycle engine, American development, 200, 319, 356
Recycle and thermal reactor, 92, 348, 358
Recycling, exhaust gas, 3, 25, 27, 29, 90, 97, 273, 337, 338, 339, 340, 341, 348, 350, 352, 358, 362
REDUCTION OF GASOLINE LEAD CONTENT, EFFECT ON ROAD ANTIKNOCK PERFORMANCE, 120
Refuelling emissions, 313
Research, previously published, 264
Ricardo; E6 engine tests, 342, 348, 357, 363
   heated flame ionization detector, 22
   trolley and FID, 228
Roots blower, 'moteur compensé', 36, 338, 355
Rotary engines; air/fuel ratio characteristics, 305
   air injection with reactor system, 307, 341, 359
   combustion chamber, 306, 341, 359
   design, 304, 341, 360
   ignition system control, 309, 341, 359
   ignition timing characteristics, 306
   Japanese cars, 304, 341, 359
   operation, 304
   seals, 341, 360
   secondary air flow control, 309, 341, 359
   thermal reactor, 308, 341, 359

Saab 99, thermal reactor performance, 257
SAMPLING AND MEASUREMENT OF EXHAUST EMISSIONS FROM MOTOR VEHICLES, 218
Sampling; air, 57
   constant volume, 16, 222, 319
   exhaust gas, 8, 152, 164, 218, 226, 319, 331, 332, 333, 365
   exhaust particulate, 16
   instrument specifications, 229
   Olson rig, 223
   variable dilution, 223
Saurer engine; diesel fuel operation, 49, 338, 339, 359
   dual fuelling, 43, 47, 50, 338, 339, 359
Seals, rotary engines, 341, 360
SECOND GENERATION CARBURETTORS FOR EMISSION CONTROL: FIRST STAGE, 137
Silencers, incorporation of lead emission filter, 18
Simca car, fuel additive tests, 68, 322, 355
Single-cylinder engine, emission studies, 101, 341, 342, 343, 357
Smoke control; British Standard AU 141, 42
   by LPG dual fuelling, 42, 338, 339, 359
   fuel additives, 42, 47, 157, 327, 328, 338, 339, 359
   hill climbing test, 47, 338, 339, 359
Smoke, effect of; air swirl, 129, 296, 326, 327, 330, 360
   engine to engine variations, 298, 326, 327
   engine variables, 195, 295, 326, 327, 328
   nozzle configuration, 298, 326, 327
Smoke; legislation and reduction techniques, 42, 338, 339, 359
   Saurer diesel engine, 49, 338, 339, 359
SO_2 emissions, estimated, U.S.A. 1969, 199
Soil, unsterilized, removal of carbon monoxide from the atmosphere, 315
SOOT AND NITRIC OXIDE CONCENTRATIONS, PREDICTION IN DIESEL ENGINE EXHAUST, 205
Soot; composition and physical properties, 186, 326, 327, 328
   effect of type of fuel, 192, 326, 327, 328
   equivalence ratio, 186, 326, 327, 328
   laboratory studies of combustion, 185
   photomicrographs, 190
   pressure and turbulence effects, 187, 326, 327, 328
   temperature effects, 186, 326, 327, 328
Soot formation in diesel engines; chemical and physical factors, 186, 326, 327, 328
   coagulation, 188, 326, 327, 328
   mechanisms, 187, 189, 195, 206, 215, 294, 326, 327, 328
   model for prediction, and its validity, 208, 214, 326, 327, 328
   nucleation, 187, 326, 327, 328
   particle growth, 188, 189, 326, 327, 328
   pyrolytic reactions, 187, 326, 327, 328
Sources of air pollutants, 14, 15, 345, 356
Southwest Research Institute stratified charge engine, 149, 343, 344, 364
Spain, legislation, 219
Spark ignition, predicted emission limits, 43, 338, 339, 359
Spark modulation systems, 267
Spark plugs, timing, thermal reactors, 309, 341, 360
SPATIAL AND TEMPORAL HISTORY OF NITROGEN OXIDES IN THE SPARK-IGNITION COMBUSTION CHAMBER, 163
Spectrometry, 164, 313
Speed, effect on; car octane requirements, 124, 322, 352
   soot and NO formation, 205, 297, 326, 327, 328
Standards; American, 1, 5, 14, 23, 42, 51, 89, 137, 163, 172, 198, 226, 227, 232, 265, 319, 335, 336, 337, 342, 352, 356
   American, penalties of adoption by European car manufacturers, 334, 337, 352
   British, AU 141, 42
   Japan, 219, 337, 352
   obtained to date, 89, 335
   worldwide, 219, 342
Starting, temperature/pressure compensated diesel engine, 37, 338, 355
Steel, in the exhaust system, 3, 336

Steel wool/alumina filters, 20, 341, 344, 345, 356
STRATIFICATION AND POLLUTION, 147
Stratified charge engine; American development, 200, 201, 319, 356
  combustion analysis, 150, 343, 344, 364
  definition and characteristics, 148, 343, 344, 364
  exhaust sampling, 152, 343, 344, 364
  nitric oxide reduction, 279, 331, 332, 354
  performance and pressure–time diagrams, 151, 343, 344, 364
  Southwest Research Institute, 149, 343, 344, 364
Stromberg carburettor, 138, 319, 322, 323, 324
SU carburettors, 74, 138, 322, 346, 355
Sulphur poisoning of catalysts, 239, 337, 348, 358
Sweden, legislation, 219
SWIRL—ITS MEASUREMENT AND EFFECT ON COMBUSTION IN A DIESEL ENGINE, 129
Swirl, measurement; 129, 330, 360
  and effect on combustion, 134, 205, 296, 326, 327, 328, 330, 360
  gas velocity in cylinder, 131, 330, 360
Swirl stratified-charge engine, see Stratified-charge engine
Switzerland, legislation, 219

Taxis, carburettor deposits, 63, 322, 355
Temperature compensation, diesel engine, 34, 35, 338, 355
Tests; American methods, 51, 221, 222, 225, 230, 243, 256, 259, 319
  B.T.C. Analytical Methods Committee activities, 218, 231
  calibration gas and sampling bag supply, 231
  carburettor cleanliness, 64, 65, 68, 69, 322, 346, 355
  catalysts—furnace, engine testbed, vehicle, and testbed endurance, 232, 337, 348, 358
  combustion chamber configuration, 28, 101, 103, 107, 338, 339, 340, 341, 342, 343, 350, 357, 362
  compression ratio, 21, 28, 102, 107, 269, 338, 339, 340, 341, 342, 343, 350, 357, 362
  coolant circuits, 108
  development and evaluation of testing techniques, 312
  European standards, 220, 226
  fuel composition, 78, 321, 322, 347, 360
  fuel evaporative losses, 286
  fuel properties, 110
  hill climbing, dual fuel operation, 47, 338, 339, 359
  ignition timing and knock, 106
  instrumentation, 220
  knock, 120, 322, 352
  lead emission reduction, 18, 120, 322, 352
  method, Japan, 227, 319
  'moteur compensé', 38, 338, 355
  octane requirement, 122, 322, 352
  recycling, 3, 21, 25, 27, 337, 352
  roadside, 322, 357
  stratified charge engine, emissions, 149, 343, 344, 364
  swirl, 130, 330, 360
  thermal reactor and exhaust recycle, 89
  thermal reactor, Ricardo E6 engine, 348, 363
  Toyota, 269
  valve overlap, 273
  Varicom and Ricardo E6 engines, 342, 357
  water and water–methanol injection, 7, 21, 27, 28, 331, 332, 333, 338, 339, 340, 341, 350, 362, 365
Thermactor package, 264

Thermal efficiency, 43, 338, 339, 359
Thermal reactor and exhaust recycle performance; 89, 92, 93, 99, 334, 335, 348, 358
  effect of ignition timing and mixture strength, 93, 96, 334, 335, 348, 358
Thermal reactors; application to Vauxhall Viva and Saab 99, 253, 254, 257
  ceramic insulation, 335
  corrosion, 179
  design, 2, 94, 178, 334, 335, 336, 337, 348, 358, 363
  durability, 260, 261, 348, 363
  effect on engine performance, 348, 364
  effect on fuel consumption, 91, 99, 334, 335, 348, 364
  Engelhard P.T.X.3, performance, 348, 363
  flow visualization studies, 260
  materials, 178, 308, 336, 375
  RAM, 99
  Ricardo E6 engine, performance, 348, 363
  rotary engine, 308, 341, 359
  spark plugs, timing, 309, 341, 360
  temperature control, 308, 341, 359, 360
Thermodynamics, engine, NO formation, 165, 332
Toyota research, NO production, 265, 269
Transmission, automatic and manual, spark modulation systems, 267
Triumph engine; catalyst tests, 234
  fuel additive tests, 68, 322, 355
Truck, installation of 'moteur compensé', 39, 338, 355
Turbocharging, diesel engine, 42

Uniontown fuel rating technique, 123
United Kingdom, total vehicular emissions, 31

Valve materials, resistance to lead-free fuels, 322, 353
VALVE OVERLAP, INFLUENCE ON OXIDES OF NITROGEN EXHAUST EMISSIONS, 273
Valves; exhaust, effect of variations in closing time, 277
  PCV, effect of fuel additives, 71, 322, 346, 355
Variable dilution sampling, 223
Varicom engine tests, 342, 357
Vauxhall Viva, thermal reactor performance, 254
VEHICLE PARTICULATE EMISSIONS, 14
Venturis, carburettor, 3
Vigom process, 37
Volkswagen cars, lead emission tests, 18

Wankel rotary engine, see Rotary engine
Water and water–methanol injection, diesel engines, 27, 29, 338, 339, 340, 341, 350, 362
WATER INJECTION, INFLUENCE ON NITRIC OXIDE FORMATION IN PETROL ENGINES, 7
Water injection, petrol engines, effect on nitric oxide formation, engine wear, and octane requirements, 7, 328, 331, 332, 333, 365
Wear, engine; effect of unleaded fuels, 322, 353
  with water injection, 332, 365
Weber carburettors, 74, 268, 322, 346, 355
Wedge shape combustion chamber, 101, 341, 342, 343, 357

Zenith 36VN carburetter, performance, 145, 320, 322, 323, 324